GYPSUM CONSTRUCTION HANDBOOK

Published for the Construction Industry
by United States Gypsum Company

90TH
ANNIVERSARY
EDITION

Important Notes to This Edition

This fourth edition of the *Gypsum Construction Handbook* (titled the 90th Anniversary Edition) is a guide to construction procedures for gypsum drywall, veneer plaster, cement board and conventional plaster construction in effect in 1992.

Information, standards, products, product names, properties, application methods, procedures etc., contained herein are subject to change. For the latest available information concerning United States Gypsum Company products, systems or recommended application procedures, contact your local United States Gypsum Company representative.

All information, details, specifications, data, applications, procedures, etc., contained in this *Handbook* are intended as a general guide when using United States Gypsum Company manufactured products. United States Gypsum Company assumes no liability for failure from improper application or installation of its products.

United States Gypsum Company products must not be used in the design or construction of any structure without a complete and detailed evaluation by a qualified architect or other professional to verify the suitability of these products for use in any given structure.

Information from this publication should be used only in conjunction with United States Gypsum Company manufactured products, as physical properties of competitive products may vary.

Results from tests were obtained under controlled laboratory conditions per ASTM procedures in effect in 1992, using only United States Gypsum Company products wherever appropriate. Comparable field performance depends on job conditions, building design and workmanship that may cause variance in job-applied results. Professional design followed by competent supervision of component installation is recommended to achieve desired results.

Editorial Committee
The Editorial Committee for the fourth edition of the *Gypsum Construction Handbook* was headed by John Lieske and included Maurice Marchello, Thomas Sheppard, William Leavitt and Robert Carlson. Al Lukasek was Graphic Designer, Steven Kalter was Technical Artist, and Robert Miller was Production Supervisor.

Published by United States Gypsum Company

Printed in U.S.A.

Nine Decades of Leadership

This 90th Anniversary Edition of the *Gypsum Construction Handbook* is particularly significant because it commemorates nine decades of excellence in construction materials and building systems from United States Gypsum Company.

Since its earliest version in 1905, the *Gypsum Construction Handbook* has been the best reference for gypsum products and systems in the construction industry. Throughout the years it has evolved and changed as the construction industry itself has evolved and changed.

Founded in 1902, United States Gypsum Company's concentration on quality has ensured its success in the development of gypsum plasters and cements for the construction industry. Continued research and innovation enabled the Company to revolutionize the industry in the early 1930s with the introduction of ROCKLATH Gypsum Lath, a paper-bound gypsum board that replaced conventional wood and metal lath as a base material for conventional plaster. Later in the same decade, advancements in production technology and research in gypsum-based compounds resulted in the development of larger gypsum panels. As the Company perfected systems to join the panels together, it introduced the SHEETROCK Drywall Systems that have become the standard of the construction industry throughout the world.

The innovations continue to this day. In this last decade alone, the Company has pioneered major advances in DUROCK Cement Board construction, lightweight SHEETROCK Joint Compounds, high-performance sound-control assemblies, fire-rated systems, high-strength IMPERIAL and DIAMOND Veneer Plasters, and drywall surface preparations.

For 90 years the spirit of innovation has been the guiding force at the Company and will continue to be so in the future. If there's a better way to build, United States Gypsum Company is dedicated to finding it.

Introduction and Contents

The Purpose of the Handbook

The 90th Anniversary Edition of the *Gypsum Construction Handbook* is a guide to good construction procedures for gypsum drywall, veneer plaster, cement board and conventional plaster construction. It contains the newest developments in products and systems including time-saving, lower-cost methods of installation to simplify and speed construction.

The book, which has become a standard handbook in other countries as well as the United States, serves as a valuable reference for those with broad experience and those who wish to learn about gypsum construction.

Architects and Engineers—Technical information on gypsum product construction standards, including system descriptions, fire- and sound-rated construction, limitations and installation procedures.

Contractors, Builders and Dealers—Full data on all aspects of gypsum products and accessories, tools and equipment, and application including information for estimating and planning.

Journeymen—Clear, concise illustrated directions and techniques for applying gypsum products from framing to finish.

Building Inspectors and Code Officials—Fire, sound and physical test data; proper construction procedures for gypsum products to ensure compliance with fire and sound ratings.

How to Use the Handbook

To find the information you want, use the Contents on page 9 or the fully cross-referenced Index on page 502 to find the applicable reference on drywall, veneer or conventional plaster, or cement board construction. The Handbook is organized as follows:

U.S. Gypsum Company Products and Systems

United States Gypsum Company offers a wide variety of quality products and performance-engineered systems. These systems are designed to consider all major factors: cost, sound control, fire resistance, structural capacity, aesthetics and overall utility and function.

Thin, lightweight gypsum board and cement board assemblies are noted for their fast installation and low cost. They are used in the majority of new residential buildings and have gained similar acceptance in commercial buildings.

This Handbook covers framing installation, drywall and veneer plaster construction, joint treatment, veneer plaster finishing, interior and exterior cement board construction, and conventional plaster application, as well as the tools required for each job. It also covers special engineered systems, product application factors, problems and remedies, and various repair and remodeling techniques.

The Manufacture of Gypsum Products

The development of all gypsum products begins with a mined mineral rock, which is gray to white in color and is called gypsum. The basic mineral is composed of calcium sulfate chemically combined with water of crystallization—$CaSO_4 \cdot 2H_2O$. The combined water makes up approximately 20% of the weight of gypsum rock. This is the feature that gives gypsum its fire-resistive qualities and makes it so adaptable for construction purposes.

After gypsum rock is mined or quarried, it is crushed, dried, and ground to flour fineness, then calcined to drive off the greater part of the chemically combined water as steam. This calcined gypsum, commonly called plaster of paris, is then mixed with water and other ingredients and sandwiched between two sheets of specially manufactured paper to form various types of gypsum board or specially formulated and bagged for shipment as gypsum plaster or cement.

Gypsum boards are formed in a highly automated continuous process. After the gypsum core has set, the boards are cut to length, dried, prefinished if required, and packaged for shipment. All processing is in strict accordance with specifications to meet quality standards.

The continued advancement of gypsum construction depends on maintaining quality while reducing construction time and costs. United States Gypsum Company has consistently been at the forefront of this effort. New products for broader uses and new cost-saving systems with improved fire and sound resistance are continually being developed and tested at the USG Corporation Research Center. Once quality is ensured, strategically located operating plants throughout the country produce and/or stock the building materials described herein.

Advantages of Gypsum Product Construction

Life Safety Protection—Fire resistance is inherent in gypsum or cement board construction. Systems provide permanent fire resistance not subject to loss of water pressure or other malfunctions and problems that may occur in sprinkler systems.

Fire Resistance—Neither gypsum nor portland cement panels will support combustion. When attacked by fire, crystalized water in gypsum is released and turns to steam to help retard the spread of flame and heat and protect adjacent constructions. Cement board, too, is an effective fire barrier. Both constructions meet fire resistance and flame spread requirements of all model building codes. Fire resistance ratings up to four hours are available with specific gypsum partition, wall, floor-ceiling, beam and column fireproofing assemblies.

Sound Control—Gypsum and cement board constructions offer excellent resistance to airborne and impact sound transmission without excessive bulk or weight. Resilient attachment of gypsum panels or bases and sound insulating blankets further improve sound ratings, making partitions ideally suited for party walls. Walls and floor-ceiling assemblies are available that meet STC/MTC and IIC requirements of applicable building codes and tenant/owner needs.

Durability—Veneer plaster combines the best features of drywall and conventional plaster. The high-strength and abrasion-resistant features of veneer plaster finishes offer the durability needed in high-traffic areas. Conventional plaster surfaces have high structural integrity and are resistant to impact and abuse. Finished with a U.S. Gypsum Company joint system, gypsum drywall panels form walls and ceilings that are resistant to cracks caused by minor structural movement as well as variations in temperature and humidity. Cement board is an exceptionally durable substrate for both interior and exterior applications that does not deteriorate in water and is not affected by freeze-thaw cycles.

Light Weight—Gypsum and cement board constructions weigh much less than masonry assemblies of the same thickness. They reduce material handling expense and may permit the use of lighter structural members, floors and footings. Veneer plaster construction compares with the weight of gypsum drywall and is considerably lighter than conventional plaster.

Low Installed Cost—Gypsum and cement board systems offer lower installed costs than more massive constructions. The lighter weight systems reduce material-handling costs. The hollow-type constructions provide an ample cavity for thermal and sound insulation, simplifying fixture attachment and mechanical installation. Low material cost and large, quickly erected panels combine to provide a lower cost for gypsum drywall, cement board and veneer plaster systems than for conventional plaster or masonry. Fast veneer plaster finish application plus savings in decorating time make veneer plaster systems competitive to gypsum drywall in many instances.

Fast Installation—Gypsum and cement board construction eliminate costly winter construction delays, permit earlier completion and occupancy of buildings. Cement board, gypsum panels and bases are job-stocked ready for use; easily cut and quickly applied. For high-volume applications, conventional plasters are readily pumped and spray-applied. Veneer plasters, which set in approximately one hour, eliminate drying delays and are usually ready for next-day decorating or painting with breather-type paints.

Easily Decorated—Gypsum construction offers smooth surfaces that readily accept decoration with paint, wallpaper, vinyl coverings or wall tile and permit repeated decoration throughout the life of the building. Plain or aggregated textures are easily applied to gypsum panels or produced during finish coat plastering. The smooth, hard surfaces obtained with veneer plaster finishes and conventional plasters are more sanitary and easier to maintain than exposed concrete block. Cement board can be finished with ceramic tile, thin brick, epoxy matrix or stucco finish.

Versatility—Gypsum and cement board constructions are suitable as divider, corridor and party walls; pipe chase and shaft enclosures; exterior walls and wall furring; and membrane fire-resistant constructions. Adaptable for use in every type of new construction—commercial, institutional, industrial and residential—and in remodeling. They produce attractive joint-free walls and ceilings; easily adapt to most contours, modules and dimensions.

Contents

DRYWALL & VENEER PLASTER CONSTRUCTION

Chapter 1 Products

DRYWALL & VENEER PLASTER CONSTRUCTION (cont.)

DRYWALL & VENEER PLASTER CONSTRUCTION (cont.)

Chapter 2 Framing

Chapter 3 Cladding

DRYWALL & VENEER PLASTER CONSTRUCTION (cont.)

Chapter 4 Finishing

CEMENT BOARD CONSTRUCTION

Chapter 5 Interior & Exterior Systems

CONVENTIONAL LATH & PLASTER CONSTRUCTION

Chapter 6 Products

CONVENTIONAL LATH & PLASTER CONSTRUCTION (cont.)

Chapter 7 Product Application

CONVENTIONAL LATH & PLASTER CONSTRUCTION (cont.)

GENERAL CONSTRUCTION

Chapter 8 System Design Considerations

GENERAL CONSTRUCTION (cont.)

Chapter 9 Planning, Execution & Inspection

Chapter 10 Problems, Remedies & Preventive Measures

GENERAL CONSTRUCTION (cont.)

Appendix

Glossary

Index

Sales Offices

Drywall & Veneer Plaster Construction

CHAPTER 1 Products

Quality Products for Gypsum Construction

Since their introduction over 60 years ago, SHEETROCK brand Gypsum Panels have dominated the drywall revolution and have become the standard for quality interior walls and ceilings. With the addition of veneer plaster bases and finishes, United States Gypsum Company has the nation's largest-selling, broadest line of gypsum products—the highest quality, the best performance.

The gypsum products described in this chapter conform to product standards recommended by United States Gypsum Company and most applicable government and commercial standards. These materials meet the essential requirements of economy, sound isolation, workability, strength, fire resistance and ease of decoration that are characteristic of quality construction.

U.S. Gypsum Company sales and technical representatives are ready to consult with tradespeople, contractors, architects, dealers and code officials on gypsum products and systems, and their application to individual job problems and conditions. For more in-depth information, call the nearest sales office (see inside back cover) or contact United States Gypsum Company headquarters in Chicago, Illinois, at (312) 606-4000.

Gypsum Panel Products

The SHEETROCK brand is still the preferred and most widely used brand of gypsum panels in existence. These panels are available in more specialized forms than any other gypsum panel line. Its high quality standards extend to other U.S. Gypsum Company components, designed to provide high-performance walls and ceilings. Thus, one dependable source offers unit responsibility for the system used.

The SHEETROCK brand Gypsum Panel is a factory-produced panel composed of a noncombustible gypsum core encased in a strong, smooth-finish paper on the face side and a natural-finish paper on the back side. The face paper is folded around the long edges to reinforce and protect the core, and the ends are square-cut and finished smooth. Long edges of panels are usually tapered, allowing joints to be reinforced and concealed with a United States Gypsum Company joint treatment system.

Advantages

Interior walls and ceilings built with SHEETROCK brand Gypsum Panels have a durable surface suitable for most types of decorative treatment and for redecoration during the life of the building.

Dry Construction—Factory-produced panels do not contribute moisture during construction.

Fire Protection—The gypsum core will not support combustion or transmit temperatures greatly in excess of 212°F until completely calcined. Fire-resistance ratings of up to 4 hours for partitions, 3 hours for floor-ceilings and 4 hours for column and shaft fireproofing assemblies are available with specific assemblies. (See Chapter 8 for specific ratings and related assemblies.)

Sound Control—SHEETROCK brand Gypsum Panels are a vital component in sound-resistive partition and floor-ceiling systems. (See Chapter 8 and Appendix for specific rating data.)

Low In-place Cost—The easily cut gypsum panels install quickly. Fixture attachment and installation of electrical and mechanical services are simplified.

Dimensional Stability—Expansion or contraction under normal temperature and humidity changes is small and normally will not result in warping or buckling. With joints properly reinforced, SHEETROCK brand Gypsum Panels are exceptionally resistant to cracking caused by internal or external forces. (See Appendix for thermal and hygrometric coefficients of expansion.)

Availability—More than 20 strategically located United States Gypsum Company manufacturing plants produce gypsum board and related products described herein. Special warehouse facilities, in addition to these plants, increase total distribution and service efficiency to major markets and rural areas from coast to coast. All standard gypsum board products are readily available upon short notice. Certain products are available from USG Corporation subsidiary plants in Mexico and Canada.

Gypsum Panel Limitations

1. Exposure to excessive or continuous moisture and extreme temperatures should be avoided. Not recommended for use in solar or other heating systems when board will be in continuous direct contact with surfaces exceeding 125°F.

2. Must be adequately protected against wetting when used as a base for ceramic or other wall tile (see foil-back panel limitation, page 31). DUROCK Cement Board or SHEETROCK brand Gypsum Panels, Water-Resistant, are the recommended products for partitions in moisture-prone areas.

3. Maximum spacing of framing members: ½" and ⅝" gypsum panels are designed for use on framing centers up to 24"; ¼" and ⅜" panels, on centers up to 16". In both walls and ceilings, when ½" or ⅝" gypsum panels are applied across framing on 24" centers and joints are reinforced, blocking is not required. ⅜" and ¼" SHEETROCK brand Gypsum Panels not recommended for use on steel framing nor as base for water-based texturing materials.

4. Application of SHEETROCK brand Gypsum Panels over ¾" wood furring applied across framing is not recommended since the relative flexibility of the furring under impact of the hammer tends to loosen nails already driven. Furring should be nom. 2x2 minimum (may be nom. 1x3 if panels are to be screw-attached).

5. The application of gypsum panels over an insulating blanket that has first been installed continuously across the face of the framing members is not recommended. Blankets should be recessed and blanket flanges attached to sides of studs or joists.

6. To prevent objectionable sag in new gypsum panel ceilings, the weight of overlaid unsupported insulation should not exceed 1.3 psf for ½" thick panels with frame spacing 24" o.c.; 2.2 psf for ½" panels on 16" o.c. framing (or ½″ SHEETROCK brand Interior Gypsum Ceiling Board on 24″ o.c. framing) and ⅝″ panels 24″ o.c.; ⅜″ thick panels must not be overlaid with unsupported insulation. A vapor retarder should be installed in exterior ceilings, and the plenum or attic space should be properly vented.

During periods of cold or damp weather when a polyethelene vapor retarder is installed on ceilings behind the gypsum board, it is important to install the ceiling insulation before or immediately after installing the ceiling board. Failure to follow this procedure may result in moisture condensation on the back side of the gypsum board, causing the board to sag.

Water-based textures, interior finishing materials and high ambient humidity conditions can produce sag in gypsum ceiling panels *if* adequate vapor and moisture control is not provided. The following precautions must be observed to minimize sagging of ceiling panels:

a) Where vapor retarder is required in cold weather conditions, the temperature of the gypsum ceiling panels and vapor retarder must remain above the interior air dew point temperature during and after the installation of panels and finishing materials.

b) The interior space must be adequately ventilated and air circulation must be provided to remove water vapor from the structure.

Most sag problems are caused by the condensation of water within the gypsum panel. The placement of vapor retarders, insulation levels and ventilation requirements will vary by location and climate and should be reviewed by a qualified engineer if in question.

7. To produce final intended results, certain recommendations regarding surface preparation, painting products and systems must be adhered to for satisfactory performance.

8. Precaution should be taken against creating a double vapor retarder by using gypsum panels as a base for highly water vapor-resistant coverings when the wall already contains a vapor retarder. Moreover, do not create a vapor retarder by such wall coverings on the interior side of exterior walls of air-conditioned buildings in hot-humid climates where conditions dictate a vapor retarder location near the exterior side of the wall. Such conditions require assessment of a qualified mechanical engineer.

Types of tapered edges

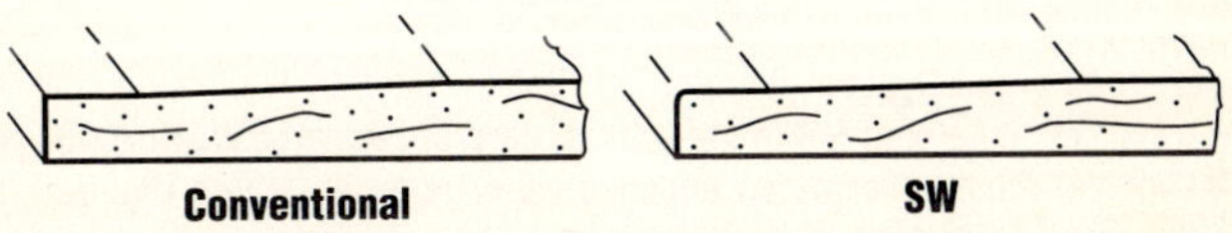

Products Available

SHEETROCK brand Gypsum Panels, Regular—Have long edges tapered on the face side to form a shallow recess (nom. .050″ deep) to accommodate joint reinforcement. Made in four thicknesses for specific purposes:

—**5/8**″, recommended for the finest single-layer drywall construction. The greater thickness provides increased resistance to fire exposure, transmission of sound, and sagging.

—**1/2**″, for single-layer application in typical new construction and remodeling.

—**3/8**″, lightweight, applied principally in repair and remodel work over existing surfaces.

—**1/4**″, a lightweight, low-cost, utility gypsum panel, used as a base layer for improving sound control in multilayer partitions and in covering old wall and ceiling surfaces. Also for forming curved surfaces with short radii.

SHEETROCK brand Gypsum Panels, SW Edge—Have an exclusive tapered rounded edge design to help minimize ridging or beading and other joint imperfections. This edge produces a much stronger joint than a regular tapered edge when finished with joint treatment. Except for the rounded edge, panels are tapered like, and otherwise identical to, regular tapered-edge gypsum panels. Made in ⅝″, ½″ and ⅜″ thicknesses.

SHEETROCK brand Gypsum Panels, FIRECODE Core—⅝″ thick, combine all the advantages of regular panels with additional resistance to fire exposure—the result of a specially formulated core containing special additives that enhance the integrity of the core under fire exposure. Panels comply with ASTM C36 for Type X gypsum board.

SHEETROCK brand Gypsum Panels, FIRECODE C Core—Available in ½″ and ⅝″ thicknesses. Improved formulation exceeds fire requirements of ASTM C36 for Type X gypsum board. Based on tests at Underwriters Laboratories Inc. and other nationally recognized testing agencies, certain partition, floor-ceiling, and column fire-protective assemblies using these special products provide 1-hour to 4-hour fire-resistance ratings. These products are used in 3-hour floor and ceiling and up to 4-hour partition and column fireproofing constructions.

In order to attain fire-resistance ratings, the construction of all such assemblies must be identical to the assembly tested.

SHEETROCK brand Gypsum Panels, ULTRACODE Core—¾″ thick, UL tested to provide a 2-hour fire rating with single-layer construction and a 4-hr. fire rating with double-layer construction in certain specified systems. Because fewer layers are needed to meet fire ratings, ULTRACODE Core panel systems reduce labor material costs.

SHEETROCK brand Gypsum Panels, Foil-Back—Made by laminating special kraft-backed aluminum foil to the back surface of regular, SW, FIRECODE or FIRECODE C Panels. Forms an effective vapor retarder, where required in cold climates, for walls and ceilings when applied with foil surface next to framing on interior side of exterior wall in single-layer application or as the base layer in multi-layer systems. Foil-Back Gypsum Panels provide a water vapor retarder to help prevent interior moisture from entering wall and ceiling spaces. In tests per ASTM C355 (desiccant method), ½" foil-back panels showed a vapor permeability of 0.06 perm. The permeance of the total exterior wall is dependent on the closure of leaks with sealants at periphery and penetrations such as outlet boxes.

These panels are designed for use with furred masonry, wood or steel framing. Thickness: ⅝", ½" and ⅜". Sizes, edges and finish: same as for base panels.

Foil-Back Panel Limitations

1. Not recommended as a base for ceramic or other tile or as base layer for TEXTONE Vinyl-Faced Gypsum Panels in double-layer assemblies.

2. Not to be used in air conditioned buildings in climates having sustained high outside temperature and humidity, such as the Southern Atlantic and Gulf Coast areas. Under these conditions, a qualified mechanical engineer should determine vapor retarder location.

SHEETROCK brand Gypsum Panels, Water-Resistant—A proven water-resistant base for the adhesive application of ceramic and plastic tile and plastic-faced wall panels. Made water-resistant all the way through. The multilayered face and back paper are chemically treated to combat penetration of moisture. The gypsum core is made water-resistant with a special moisture-resistant composition. The panel is easily recognized because of its distinctive green face.

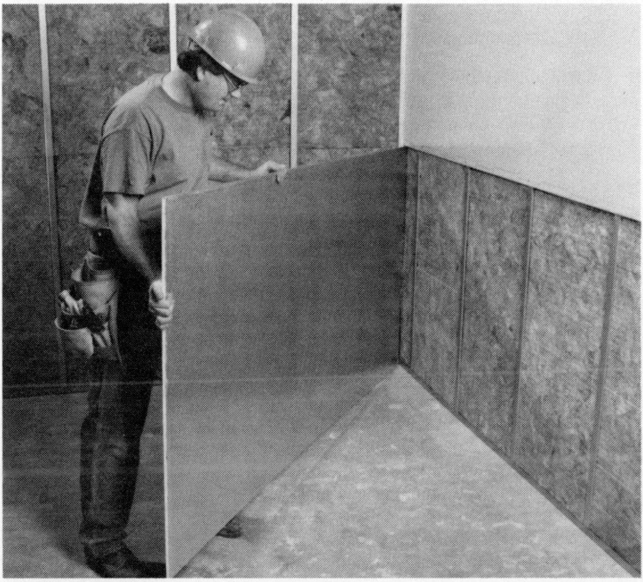

Foil-Back Panels applied to steel framing over the interior of exterior walls provide effective vapor retarder.

These panels are designed for bathrooms, powder rooms, kitchens, utility rooms. In addition, they may be used in modernization work when the existing surfaces are removed and Water-Resistant Panels applied directly to framing. SHEETROCK brand Gypsum Panels, Water-Resistant, FIRECODE and FIRECODE C Core, also are used in fire-rated assemblies that may be exposed to moisture during construction. Panels comply with ASTM C630.

Available in three types with shallow tapered edges:

SHEETROCK brand Gypsum Panels, Water-Resistant, Regular—½″ thickness for single-layer application in residential construction; ⅝″ thickness is also available.

SHEETROCK brand Gypsum Panels, Water-Resistant, FIRECODE Core—in ⅝″ thickness with a Type X core to provide fire resistance for required ratings.

SHEETROCK brand Gypsum Panels, Water-Resistant, FIRECODE C Core— in ½″ thickness with special core to provide fire resistance for required ratings.

Water-Resistant Panel Limitations

1. Adherence to recommendations concerning sealing exposed edges, painting, tile adhesives, framing and installation is necessary for satisfactory performance.

2. Not recommended for ceilings with framing spacing greater than 12″ o.c., for single-layer resilient attachment where tile is to be applied or in remodeling unless applied directly to studs.

3. Panels that would normally receive an impervious finish, such as ceramic tile, should not be installed over a vapor retarder nor on a wall acting as a vapor retarder.

4. Store in an enclosed shelter and protect from exposure to the elements.

5. Panels are not intended for use in areas subject to constant moisture, such as gang showers and commercial food processing—DUROCK Cement Board is recommended for these uses.

SHEETROCK brand Exterior Gypsum Ceiling Board— A weather-resistant board designed for use on the soffit side of eaves, canopies and carports and other commercial and residential exterior applications with indirect exposure to the weather. Noncombustible core is simply scored and snapped for quick application. Panels can be painted and provide good sag resistance.

Installed conventionally in wood and metal-framed soffits; batten strips or mouldings can be used over butt joints or joints can be treated; backing strips required for small vent openings. Natural finish. Available in ½″ thickness with regular core and in ⅝″ thickness with fire-rated core—both with SW tapered edges. Board complies with ASTM C931.

SHEETROCK brand Interior Gypsum Ceiling Board—½″ thick board provides high strength and sag resistance in interior ceiling applications. Ideal for ceilings where heavy, water-based textures will be applied and framing is 24″ o.c. Edges: tapered. Complies with ASTM C36.

Specifications—Gypsum Panel Products

	Thickness in	mm	Length ft[1]	Approx. wt. lb/ft²	kg/m²
SHEETROCK brand Regular Panels[2]	¼	6.4	8 and 10	1.2	5.9
	⅜	9.5	8, 9, 10, 12,14	1.4	6.8
	½	12.7	8, 9, 10, 12,14	1.7	8.3
	⅝	15.9	8, 9, 10, 12,14	2.3	11.2
FIRECODE Core Panels	⅝	15.9	8, 9, 10, 12,14	2.2	10.6
FIRECODE C Core Panels	½	12.7	8, 9, 10, 12,14	1.9	9.4
	⅝	15.9	8, 9, 10, 12,14	2.5	12.2
ULTRACODE Core Panels	¾	19.0	8, 9, 10, 12	2.8	13.7
Water-Resistant Panels	½	12.7	8, 10, 12	1.8	8.8
	⅝	15.9	8, 10, 12	2.2	11.0
Water-Resistant FIRECODE Core Panels	⅝	15.9	8, 10, 12	2.2	10.6
Water-Resistent FIRECODE C Core Panels	½	12.7	10	1.9	9.4
Exterior Ceiling Board Regular Board	½	12.7	8, 12	1.9	9.3
	⅝	15.9	8, 12	2.4	11.7
FIRECODE Core Board	⅝	15.9	8, 12	2.4	11.7
Interior Ceiling Board	½	12.7	8, 12	1.9	9.3

(1) Metric lengths: 8 ft. = 2440 mm; 9 ft. = 2745 mm; 10 ft. = 3050 mm; 12 ft. = 3660 mm; 14 ft. = 4270 mm. (2) Also available in Foil-Back Panels. NOTE: See p. 35 for information on gypsum bases.

Veneer Plaster Gypsum Base Products

Gypsum bases finished with veneer plasters are recommended for interior walls and ceilings in all types of construction. For these interiors, a veneer of specially formulated gypsum plaster is applied in one coat (1/16″ to 3/32″ thick) or two coats (⅛″ to 3/16″ thick) over the base. The resulting smooth or textured monolithic surfaces are preferred for hard-wear locations where durability and resistance to abrasion are required.

IMPERIAL Gypsum Bases are large-size gypsum board panels (4′ width), rigid and fire-resistant. A gypsum core is faced with specially treated, multilayered paper (blue) designed to provide a maximum bond to veneer plaster finishes. The paper's absorbent outer layers quickly and uniformly draw moisture from the veneer plaster finish for proper application and finishing; the moisture-resistant inner layers keep the core dry and rigid to resist sagging. The face paper is folded around the long edges. Ends are square-cut and finished smooth.

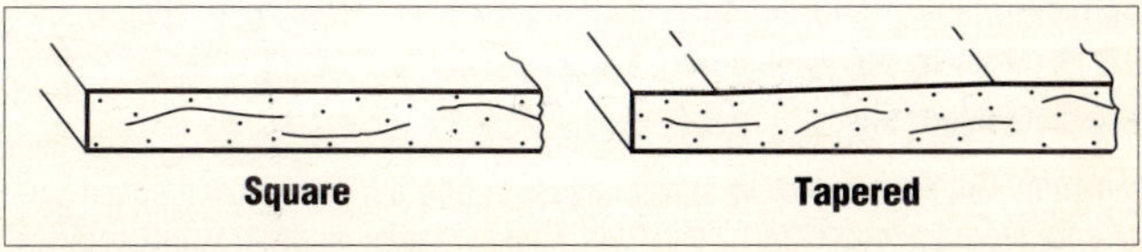

Types of Edges

Advantages

Gypsum bases, in conjunction with selected veneer plaster finishes, provide the lasting beauty of plaster walls and ceilings at a lower cost and with less weight and residual moisture than conventional plastering.

Rapid Installation—Construction schedules are shortened. Walls and ceilings can be completed in 3 to 4 days, from bare framing through decorated interiors.

Fire Resistance—Ratings of up to 4 hours for partitions, 3 hours for floor-ceilings and 4 hours for column fire protection assemblies have been obtained.

Sound Control—Gypsum base partitions faced with veneer plaster finishes on both sides have high resistance to sound transmission. Resilient attachment of base and use of THERMAFIBER SAFB Insulation further improve sound isolation.

Durability—Hard, high-strength surfaces provide excellent abrasion resistance resulting in minimum maintenance, even in high-traffic areas.

Easily Decorated—Smooth-surfaced interiors accept paints, texture, fabric and wallpaper. Veneer plaster finishes also may be textured. IMPERIAL Finish Plaster can be painted with breather-type paints the day following application.

Gypsum Base Limitations

1. Maximum frame and fastener spacing is dependent on thickness and type of base used.

2. Recommended for use with IMPERIAL Basecoat and Finish Plaster, and DIAMOND Basecoat and Interior Finish Plaster. Do not apply gauged lime-putty finishes or portland cement plaster directly to base; bond failure is likely.

3. Not recommended for use in areas exposed to excessive moisture for extended periods or as a base for adhesive application of ceramic tile in wet areas (SHEETROCK brand Gypsum Panels, Water-Resistant, or DUROCK Cement Board are recommended for this use).

4. Gypsum base that has faded from the original light blue color from exposure to sunlight should be treated with either a plaster bonding agent or spray-applied alum solution before DIAMOND Interior Finish Plaster or any veneer plaster finish containing lime is applied. IMPERIAL Basecoat and Finish Plaster, or IMPERIAL and DIAMOND Basecoat Plasters, do not contain lime and are not susceptible to bond failure over faded base.

5. Joints must be treated with SHEETROCK Joint Tape and SHEETROCK Setting-Type Joint Compound (DURABOND) when framing is spaced 24" o.c. in one-layer applications.

Products Available

IMPERIAL Gypsum Base—A special gypsum board that has been specifically engineered for use with IMPERIAL Finish Plaster and DIAMOND Interior Finish Plaster, or IMPERIAL and DIAMOND Basecoat Plasters. It provides the strength and absorption characteristics necessary for top-quality veneer plaster finishing performance. Large sheets minimize the number of joints and speed installation. The high-density, fire-resistant gypsum core has a superior controlled-absorption paper lightly tinted blue on the face side and a strong liner paper on the back side. Available in two thicknesses with square or tapered edges: ½″ for single-layer application in new light construction; ⅝″ recommended for the finest high-strength veneer plaster finish construction. The greater thickness provides increased resistance to fire exposure and sound transmission and allows 24″ o.c. spacing of wood framing. IMPERIAL Gypsum Base may be used with DIAMOND Interior Finish Plaster to embed cables for radiant heat ceilings.

IMPERIAL FIRECODE and FIRECODE C Gypsum Bases—IMPERIAL FIRECODE in ⅝″ thickness and IMPERIAL FIRECODE C in ½″ and ⅝″ thicknesses combine all the advantages of Regular IMPERIAL Gypsum Base with additional resistance to fire exposure—the result of specially formulated mineral cores. Listed under UL Label Service for certain fire-tested partition, floor-ceiling and column constructions.

Foil-Back IMPERIAL Gypsum Bases—Bright aluminum foil laminated to the back side acts as a vapor retarder. Available in Regular, FIRECODE and FIRECODE C Bases.

Foil-Back Base Limitation: do not use as a base for ceramic or other tile or as a face layer in multilayer systems.

Specifications—Gypsum Bases

Product	Thickness in	Thickness mm	Length ft(1)	Approx. wt. lb/ft²	Approx. wt. kg/m²
IMPERIAL Base(2)					
Regular	½	12.7	8, 9, 10, 12, 14	1.8	8.8
Regular	⅝	15.9	8, 9, 10, 12, 14	2.3	11.2
FIRECODE	⅝	15.9	8, 9, 10, 12, 14	2.3	11.2
FIRECODE C	½	12.7	8, 9, 10, 12, 14	2.0	9.8
FIRECODE C	⅝	15.9	8, 9, 10, 12, 14	2.5	12.2

(1) Metric lengths: 8 ft. = 2440 mm; 9 ft. = 2745 mm; 10 ft. = 3050 mm; 12 ft. = 3660 mm; 14 ft. = 4270 mm. (2) Also available in Foil-Back Base.

Gypsum Liner and Sheathing Products

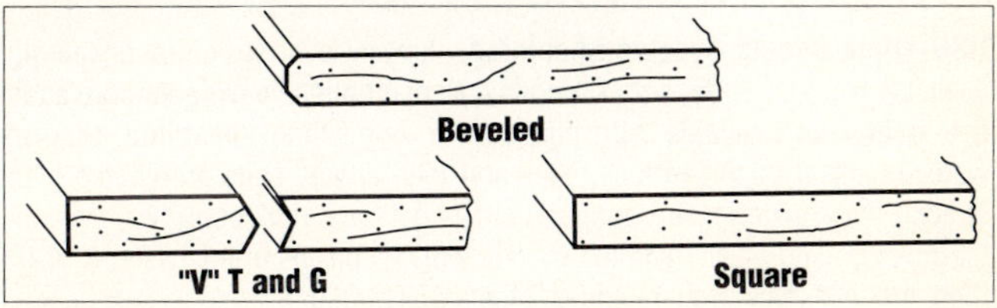

Types of Edges

SHEETROCK brand Gypsum Liner Panel—A 1″ thick, special fire-resistant gypsum core encased in multilayered, moisture-resistant green paper. Panels are used in USG Cavity Shaft Walls, vent shafts, Area Separation

Walls, USG High Sound-Attenuation Double Wall Systems, select floor assemblies and infill panel systems for exterior curtain walls. Panels have beveled edges for easy insertion between the supporting flanges of steel C-H studs, E-studs or H-studs.

SHEETROCK brand Gypsum Coreboard—A 1″ thick gypsum core encased in strong liner paper on both sides. It is used in laminated gypsum partitions with additional layers of gypsum panels applied to the coreboard to complete the wall assembly. Manufactured with V-shaped T&G edges for use in solid partitions or with square edges and prescored 6″ or 8″ o.c. for use in semi-solid partitions. Coreboard strips are then easily snapped and separated from this coreboard panel. Coreboard complies with ASTM C442.

SHEETROCK brand Gypsum Sheathing—A fire-resistant gypsum board with a water-resistant gypsum core encased in specially treated water-repellent paper on both sides and long edges. Its weather resistance, water repellency, fire resistance and low applied cost make it suitable for use in exterior wall construction of garden apartments and light commercial buildings as well as in homes. Also used in steel stud curtain wall construction.

SHEETROCK brand Gypsum Sheathing is suitable for a wide range of exterior finishes such as, but not limited to, masonry veneer, wood, vinyl and aluminum siding, wood shingles, stucco and exterior insulation finish systems (EIFS). All these exterior finish attachments are limited to mechanical fastening through sheathing into the framing.

Made ½″ thick in two economical types: 2′ wide, 8′ long with V-shaped T&G edges for horizontal applications; 4′ wide, 8′ and 9′ long with square edges for vertical application. Also made in ⅝″ thick FIRECODE Type X core, 4′ wide, 8′ and 9′ long with square edges for vertical application. Meets ASTM C79.

GYP-LAP Gypsum Sheathing—A low-cost, weather and fire-resistant board designed to combine excellent performance with excellent economy. Noncombustible gypsum core adds fire safety not available with plywood or wood fiber sheathing. Clad in water-repellent paper on face and back surfaces. Lightweight and easily handled by one worker. Panels are 2′ wide, 8′ long with V-shaped T&G long edges for horizontal application, and 4′ wide, 8′ long with square edges for vertical application. Thickness is ½″ or ⅝″; FIRECODE Type X core available in ⅝″ thickness. Meets ASTM C79.

USG Triple-Sealed Gypsum Sheathing—Lowest cost structural sheathing available from U. S. Gypsum Company. Noncombustible gypsum core adds fire safety not available with plywood or wood-fiber sheathing. Clad in water-repellent paper on face, back and long edges; ends are coated with special waterproofing compound, but panels are not totally waterproof. Lightweight and easily handled by one worker, panels are 4′ wide, 8 or 9′ long and .40″ thick. Square edges. Meets ASTM C79.

Sheathing Limitations

1. Sheathing may be stored outside for up to one month, but must be stored off the ground and must have a protective covering.

2. Maximum stud spacing is 24″ o.c.

3. When applied to a structure, sheathing must not be left exposed to the elements for more than one month unless the procedure as outlined in Limitation 6 is followed.

4. Exterior finish systems must be properly caulked for the life of the job, particularly around all cuts and penetrations.

5. Exterior finish systems applied over gypsum sheathing must be applied with mechanical fasteners through the sheathing into the wall framing. Alternative methods of application are not endorsed and their performance and that of the substrate are solely the responsibility of the specifier. Direct application of paint, texture finishes and coatings over gypsum sheathing is not recommended.

6. For in-place exposure up to six months, all gaps resulting from cuts, corners, joints and machine end cuts of the sheathing should be filled with exterior caulk *at time of erection.*

7. For curtain wall construction, cover the gypsum sheathing with No. 15 asphalt felt within 30 days of sheathing installation. Felt should be applied horizontally with 2″ overlap and immediately anchored with metal lath, masonry ties or corrosion-resistant screws or staples. (See SA-923 Technical Folder for additional curtain wall details.)

8. Sheathing is not recommended for exterior ceilings and soffits, unless covered with metal lath and exterior stucco.

Specifications—Liner and Sheathing Products

	Thickness		Width				Approx.wt.	
Product	in	mm	in	mm	Edges	Length ft	lb/ft^2	kg/m^2
SHEETROCK brand Liner Panels	1	25.4	24	610	Bevel	up to 16	4.1	20.0
SHEETROCK brand Coreboard	1	25.4	24	610	V—T&G	8, 9, 10, 12(1)	4.1	20.0
SHEETROCK brand Sheathing	½	12.7	24	610	V—T&G	8	2.0	9.8
	½	12.7	48	1219	Square	8, 9	2.0	9.8
	⅝	15.9	48	1219	Square	8,9	2.4	11.7
SHEETROCK brand FIRECODE Sheathing	⅝	15.9	48	1219	Square	8, 9	2.4	11.7
GYP-LAP Sheathing	½	12.7	24	610	V—T&G	8	—	—
	½	12.7	48	1219	Square	8	—	—
USG Triple-Sealed	4/10	10.2	48	1219	Square	8, 9	1.6	7.8

(1) Prescored coreboards available in 7′8″ lengths only.

Predecorated Panel Products

TEXTONE Vinyl-Faced Gypsum Panels—Conventional gypsum board with factory-applied vinyl facings in a wide range of coordinated decorator colors. The facings provide a broad choice of color, texture and pattern for mix-and-match versatility. The tough vinyl covering is durable and easily cleaned. Panels have beveled long edges which form a shallow V-groove joint.

TEXTONE Panels, together with TEXTONE Mouldings factory-wrapped in TEXTONE vinyl, fasteners, adhesives and other conventional drywall components, are used for predecorated permanent partitions, demountable partitions and in remodeling work. Not recommended for ceilings because end joints are difficult to conceal.

The rugged, scuff-resistant vinyl is embossed for texture and woodgrain effects.

TEXTONE Panel Limitations

1. For adhesive application of TEXTONE Panels, only water-thinned adhesives are recommended. Other adhesives may not be compatible and could result in delamination and discoloration of vinyl surface.

2. If TEXTONE FIRECODE Panels are used in a fire-rated assembly instead of a non-vinyl-faced product such as SHEETROCK brand Gypsum Panels, FIRECODE Core, the applicable fire test must permit exposed joints or battens.

3. Not recommended for use over foil-back panels or other vapor retarder in exterior walls.

4. Avoid exposure to excessive or continuous moisture and extreme temperatures.

5. Do not apply TEXTONE Vinyl-Faced Gypsum Panels or field laminate non-permeable vinyls over gypsum panels on exterior walls in hot, humid climates without suitable vapor control or dry air circulation behind the panels.

Technical Data

TEXTONE Panels meet ASTM C960, gypsum panels comply with ASTM C36. Light-reflectance values available on request. (Refer to table for vapor permeance values and surface burning characteristics.)

Panels are manufactured ½″ thick, 4′ wide, and 8′, 9′ and 10′ long. They may also be specially ordered in ⅜″ and ⅝″ thicknesses, 2′ widths and custom lengths from 6′ to 14′ TEXTONE FIRECODE Core Panels with special core for fire-rated construction are available in ½″ and ⅝″ thicknesses, 4′ wide. (See current Technical Folder SA-928 for pattern and color selections. Contact sales representative for custom colors and patterns also available.)

Vinyl covering of TEXTONE Panels is directly attached to gypsum panel without sheeting, except Brittany and Textile patterns, which have a cotton sheeting backing.

Panel surface burning characteristics[1] and vapor permeance[2]

TEXTONE pattern	Film thickness or weight	Flame spread	Smoke dev.	Vapor perm.
Pumice	6 mils	20	25	0.8
Suede	6 mils	15	25	0.6
Presidio	6 mils	15	25	0.6
Granite	6 mils	15	25	0.6
Woodgrain	6 mils	20	15	0.6
Linen	8 mils	15	25	0.5
Country Weave	10 mils	20	35	0.8
Textile (Type 1, Fabric-Backed)[3]	10.7 oz./yd^2	25	65	1.0
Brittany (Type 1, Fabric-Backed)[3]	10.0 oz./yd^2	25	55	2.1

(1) Tested in accordance with ASTM E84-80. (2) Tested in accordance with ASTM E96-90.
(3) Comply with Federal Specification CCC-2-408C, Type I.

TEXTONE Vinyl Wallcovering—Provides a commercial match with TEXTONE Panels on adjacent walls and columns. TEXTONE Wallcoverings are supported on cotton sheet backing and weigh 15 oz./yd^2. The cotton sheeting, whose primary purpose is to facilitate field installation, weighs 1.3 oz./yd^2. Available in rolls 54″ wide by 30 lin. yds.

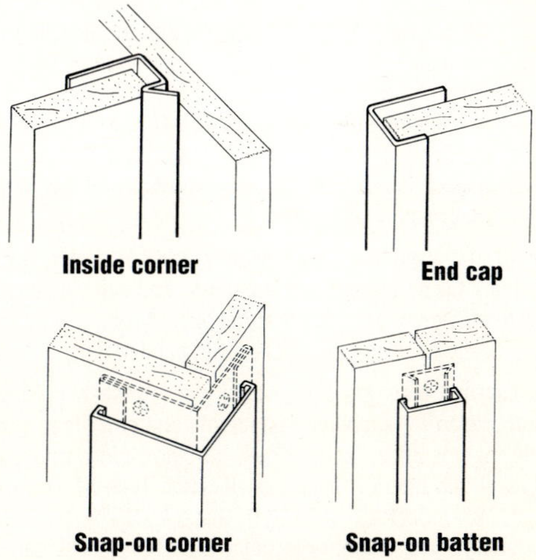

RP series vinyl trim

TEXTONE Mouldings—Cover joints and edges, protect corners. Available to match or contrast with TEXTONE Panels in ½″ and ⅝″ sizes, they are low-cost, precision-extruded vinyl mouldings in five shapes and two finishes. Available plain (RP) in standard Almond or Ash Blue colors or factory-laminated with matching TEXTONE Vinyl (RPV) Series.

Specifications—TEXTONE Mouldings

Product	Size in(1)	Length ft(1)	Approx. wt. lb/1000 ft²	Approx. wt. kg/100m²
RP-2, RPV-2 Inside Corner	½, ⅝	8,9,10	77	11
RP-4, RPV-4 End Cap	½, ⅝	8,9,10	66	10
RP-5, RPV-5 Snap-on Corner	(2)	8,9,10	184	27
RP-7, RPV-7 Snap-on Batten	(2)	8,9,10	95	14
RP-46, RPV-46 Ceiling Drive-in Trim	½	10	155	23

(1) Metric conversions: ½ in. = 12.7 mm; ⅝ in. = 15.9 mm; 8 ft. = 2440 mm; 9 ft. = 2745 mm; 10 ft. = 3050 mm; 12 ft. = 3660 mm. (2) One size fits all panel thicknesses.

Suspended Ceiling Products

Suspended ceilings offer the advantages of variable ceiling height and expanded plenum usage that are not always available with conventional ceiling construction. Several products are available for suspended ceiling construction that provide superior performance in the areas of fire resistance and sound attenuation.

SHEETROCK Lay-In Ceiling Panels

Lay-in panels are designed for use in standard ceiling suspension systems for exceptional economy, ease of installation and accessibility to the plenum. Panels also qualify for UL design fire-rated assemblies to 1½ hours (UL design G222) and 2 hours (UL design G259) when used with fire-rated steel suspension systems such as the DONN DXL, DXLA or ZXLA exposed grid systems. Lay-in panels are made of ½″ FIRECODE C Core gypsum board in both 2′x2′ or 2′x4′ sizes. Both sizes are available with either laminated white vinyl facing or natural paper facing.

Vinyl facing is embossed in a stipple pattern for a soft, lightly textured look. It is 2-mils thick for toughness and durability, and can withstand repeated washings with no sign of abrasion. Natural paper facing can be left plain for utilitarian applications or special ordered with painted finish.

SHEETROCK Lay-In Panels are safe, sanitary and washable. They are FDA- and USDA-approved for kitchens, restaurants and other food service areas and suitable for hospitals, laboratories, nursing homes and other health care facilities. Attain interior finish classification Type III, Form A, Class 3 (Federal Spec. SS-L-30D/ASTM E1264); Class A (NFiPA 101). Panels with white vinyl facing achieve light reflectance LR1. Panels also can be used in exterior applications such as covered entryways and parking garages.

Advantages

Conventional Installation—Panels install quickly and easily in standard exposed grid.

Easy Maintenance—Embossed vinyl facing is washable to keep surface bright and light-reflecting.

Outdoor Applications—Excellent in protected areas when used with compatible suspension system, such as DONN Environmental ZXA grid, which features 25-gauge, hot-dipped galvanized steel with corrosion-resistant aluminum face. 4′ hanger spacing achieves intermediate-duty rating vs. 3′ spacing for aluminum grids.

Sound Attenuation—Panels provide STC range of 45-49.

Performance—Panels qualify for fire-rated assemblies. Surface burning characteristics: Class A rated on all products. Flame spread/smoke developed rating is 15/0 (ASTM E84 test procedure). Thermal performance up to R-0.45. Weight 2.00 lb/ft^2.

		Regular			FIRECODE		
Size	Edge	Item no.	NRC range	STC range	Item no.	NRC range	STC range
Stipple Pattern							
2′x2′x½″	Square	N/A	—	—	3260	N/A	45-49
2′x4′x½″	Square	N/A	—	—	3270	N/A	45-49
Unfinished Paper Facing							
2′x2′x½″	Square	N/A	—	—	3440	N/A	45-49
2′x4′x½″	Square	N/A	—	—	3450	N/A	45-49

DONN brand RIGID X Suspension System

The DONN brand RIGID X Drywall Suspension System provides a fast and economical method of installing a gypsum drywall ceiling while supplying support for lighting and air handling accessories. The system is designed for direct screw attachment of gypsum drywall.

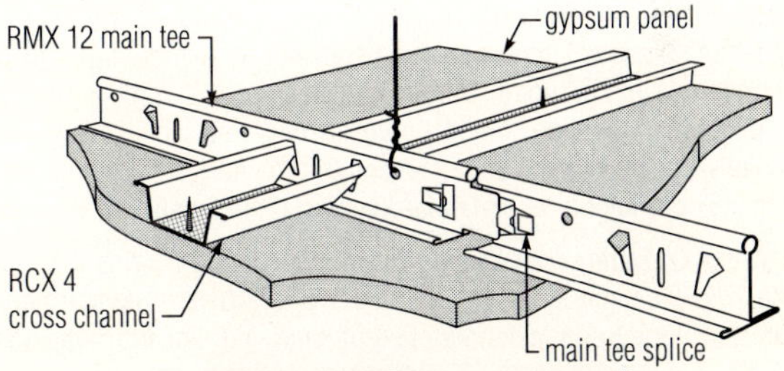

RIGID X with RCX 4 cross channel

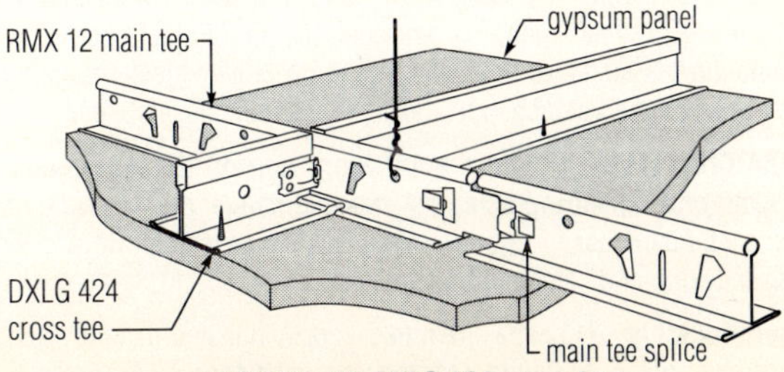

RIGID X with DXLG 424 cross tee

The RIGID X Suspension System is made of commercial quality, cold-rolled steel. Main tees are 1½″ high x 144″ long with a round top bulb and ¹⁵⁄₁₆″ wide flange. The system offers the option of using 1½″ wide-faced furring cross channels or 1½″ wide furring cross tees for gypsum panel attachment. Face of both cross channels and cross tees is knurled to improve fastening of drywall screws. Surface of all components is painted with manufacturer's standard enamel paint. Also available are cross tees with ¹⁵⁄₁₆″ exposed flange and baked enamel finish to be used with light fixtures.

Direct-hung DONN brand RIGID X Suspension System is used in UL designs with fire ratings of 1, 1½, 2 and 3 hours. *1-hour UL designs:* L502, L525, L526, L529, P508, P509, P510. *1½-hour UL designs:* G528, P507, P506, P510, P513. *2-hour UL designs:* G523, G526, G529, J502. *3-hour UL designs:* G523, G529. Consult UL Fire Resistance Directory for further information and construction details.

Advantages

Labor Saving—Factory-controlled module spacing and snap-lock connection of cross channels and cross tees with main tees cuts installation time.

Cost Saving—Components are low in cost compared with conventional construction to achieve the same result.

Strength—Strong metal components are designed with interlocking tabs and splicing mechanisms to resist twisting of assembly.

Accommodates Light Fixtures—Accepts NEMA type G light fixtures.

Added Fire Protection—Uniquely designed integral reversible splice provides for fire expansion relief.

System Components

RMX Main Tee—Conforms to ASTM C635 Heavy-Duty Main Tee Classification. Designed to support gypsum board ceiling with maximum deflection of ¹⁄₃₆₀ of the span. Single-web design, 1½″ high x 144″ long, round top bulb, ¹⁵⁄₁₆″ wide flange, integral reversible end splice. Furring cross channel holes 4″ from ends, spaced 8″ o.c., hanger wire holes 4″ o.c. Prefinished in manufacturer's standard enamel paint finish.

RCX 4 Cross Channel—Hat-shaped formed section, 2⅞″ wide x ⅞″ high x 48″ long. 1½″ wide knurled screw surface, manufacturer's standard enamel paint finish, integral end locks stamped at each end. For fire-rated assemblies.

DXLG 424 Cross Tee—1½″ high x 48″ long, roll formed into double web design with rectangular bulb, 1½″ knurled face and a steel cap prefinished in manufacturer's standard enamel finish, high-tensile-steel double-locking and self-indexing end clenched to web. For fire-rated assemblies.

DXL 424 Cross Tee—1½″ high x 48″ long, roll-formed into double-web design with rectangular top bulb, ¹⁵⁄₁₆″ exposed flange, prefinished in manufacturer's standard baked enamel finish, high-tensile-steel double-locking and self-indexing end clenched to web.

Channel Moulding—U-shape, hemmed edges, finished in manufacturer's standard paint finish, 1″ flange x 1″ opening x ½″ flange x 144″ long.

Hanger Wire—Galvanized carbon steel, soft temper, prestretched, yield stress load at least three times design load, but not less than 12-gauge wire.

Trim Accessories

United States Gypsum Company sells and distributes UNIMAST construction, drywall and plastering steel products. These trim accessory products include corner reinforcements, beads, trims, control joints and decorative mouldings.

Corner Reinforcement

Corner beads permit construction of true, concealed external angles with gypsum base and panels. The exposed nose of the bead helps prevent damage from impact and provides a screed for finishing. Offered in the following styles, all part of the SHEETROCK family of metal products:

DUR-A-BEAD Corner Bead—A specially galvanized steel reinforcement for protecting external corners in drywall construction. It is nailed to framing through the panels and concealed with United States Gypsum Company joint compounds as a smooth, finished corner. Flanges also may be attached with clinch-on tool or staples. Available in two flange widths: No. 103, 1¼" x 1¼"; No. 104, 1⅛" x 1⅛".

SHEETROCK No. 800 Corner Bead—A galvanized steel external corner reinforcement with 1¼" wide fine-mesh expanded flanges, tapered along outer edges to enhance concealment. It is easily nailed or stapled. Provides superior

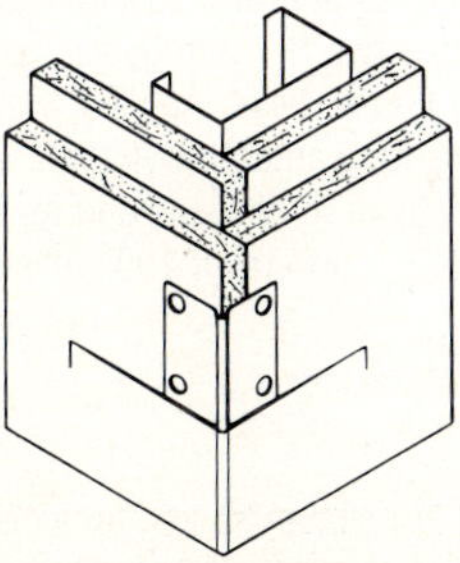

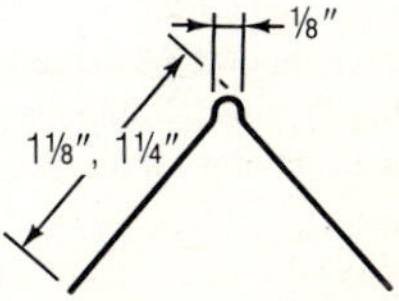

DUR-A-BEAD corner bead

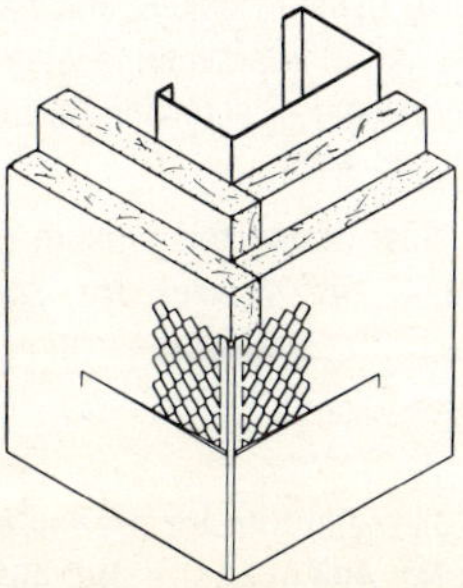

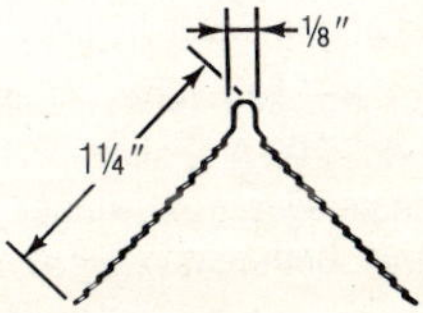

No. 800 & 900 corner bead

reinforcement with joint compound and veneer plaster finishes through approx. 90 keys per lin. ft. It also provides the proper 1⁄16″ grounds for one-coat veneer finishes.

SHEETROCK No. 900 Corner Bead—Used with two-coat veneer plaster systems. It provides 3⁄32″ grounds and its 1¼″ fine-mesh flanges can be either stapled or nailed. Provides reinforcement equivalent to No. 800.

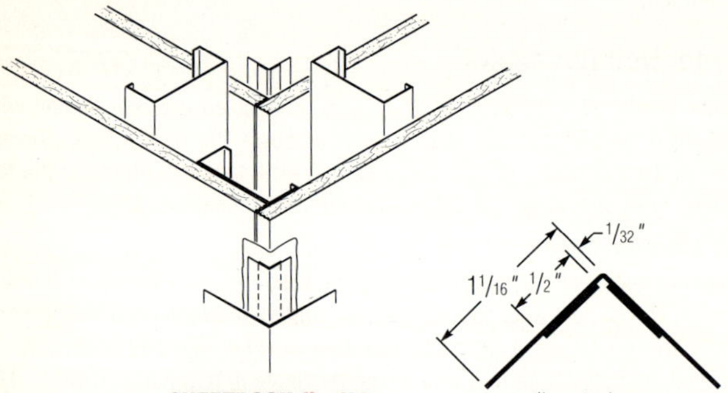

SHEETROCK flexible metal corner tape

SHEETROCK Flexible Metal Corner Tape—A flexible reinforcing that ensures straight, sharp corners on any angle. Provides durable corner protection on cathedral and drop ceilings, arches and around bay windows. Tape is 2⅟16″ wide and has 1⁄16″ gap between two ½″ wide galvanized steel strips. When folded, tape forms a strong corner bead. Applied with standard joint compound feathered at the edges for a smooth wall surface. Also used to join drywall partition to plastered wall in remodeling and for repairing chipped and cracked corners. Available in convenient 100′ rolls in dispenser box.

Metal Trim

Metal trims, part of the SHEETROCK family of metal products, provide maximum protection and neat finished edges to gypsum panels and bases at window and door jambs, at internal angles and at intersections where panels abut other materials. Easily installed by nailing or screwing through the proper leg of trim. Made in following types and sizes:

SHEETROCK No. 200 series—Galvanized steel casing for gypsum panels, includes **No. 200-A** U-shaped channel in ½″ and ⅝″ sizes; **No. 200-B** L-shaped angle edge trim without back flange to simplify application, in ½″ and ⅝″ sizes. Both require finishing with U.S. Gypsum Company joint compounds.

SHEETROCK No. 400 series—Reveal type all-metal trim for drywall panels, requires no finishing compound, includes **No. 400** in ⅜″ size, **No. 401** in ½″ size, **No. 402** in ⅝″ size.

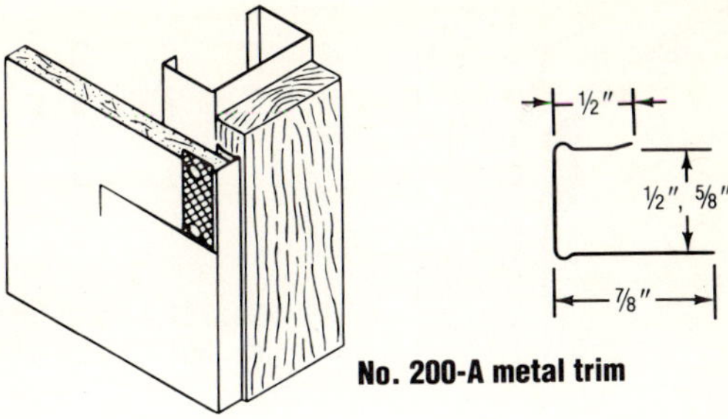

No. 200-A metal trim

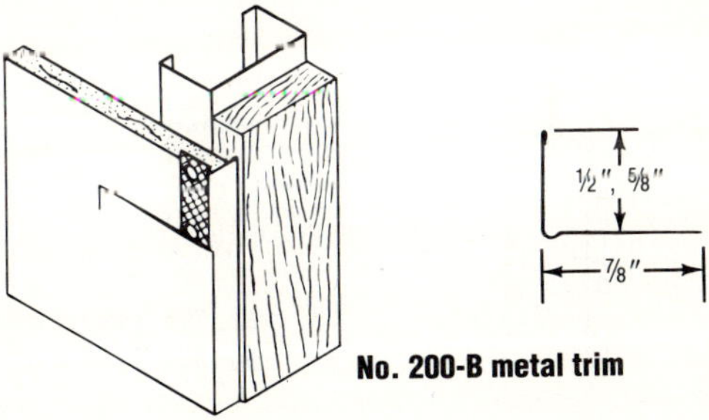

No. 200-B metal trim

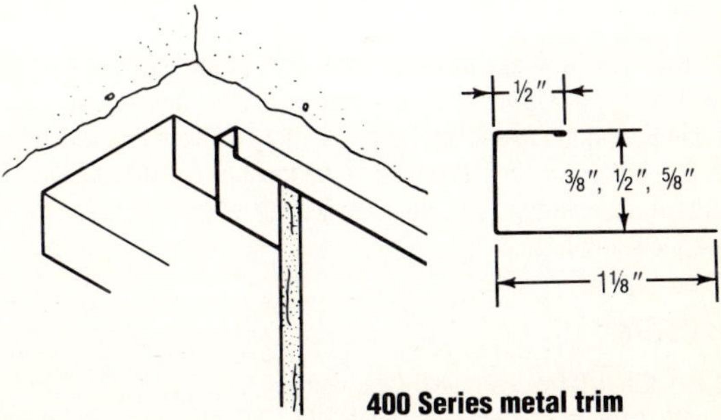

400 Series metal trim

Sheetrock No. 700 series—All-metal trim provides neat edge protection for two-coat veneer plaster finishes at cased openings and ceiling or wall intersections. Fine-mesh expanded flanges strengthen veneer plaster reinforcement and eliminate shadowing. **No. 701-A** channel-type and **No. 701-B** angle edge trim provide $\frac{3}{32}$″ grounds; sizes for ½″ and ⅝″ thick gypsum base.

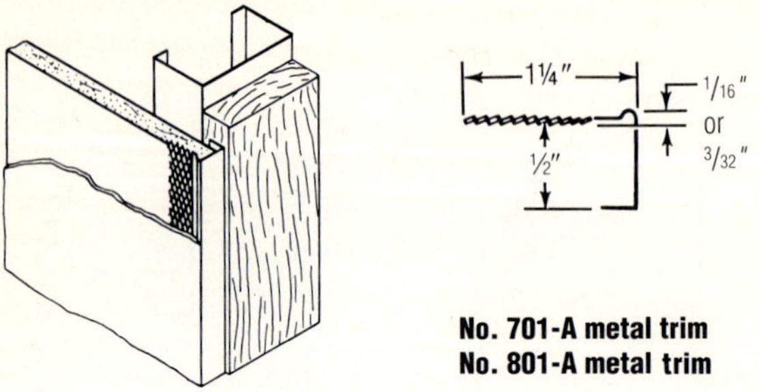

No. 701-A metal trim
No. 801-A metal trim

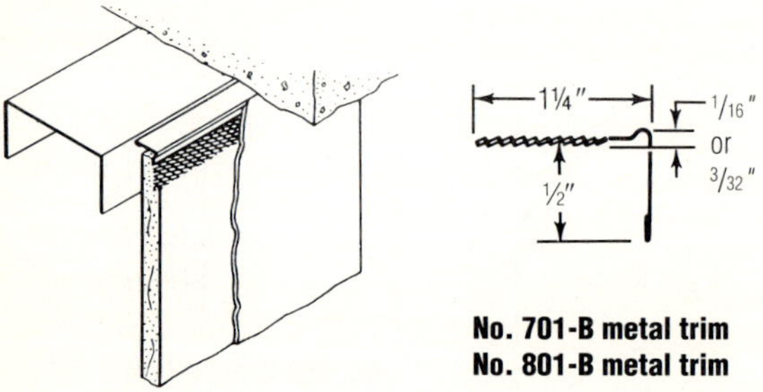

No. 701-B metal trim
No. 801-B metal trim

SHEETROCK No. 800 series—All-metal trim companion to No. 700 series but with 1/16" grounds for one-coat veneer plaster finishes or finishing with joint compound in drywall applications. The 1¼" wide fine-mesh flange provides a superior key and eliminates shadowing, is easily nailed or stapled. **No. 801-A** channel-type and **No. 801-B** angle edge trim; sizes for ½" and ⅝" thick panels and bases.

Vinyl Trim

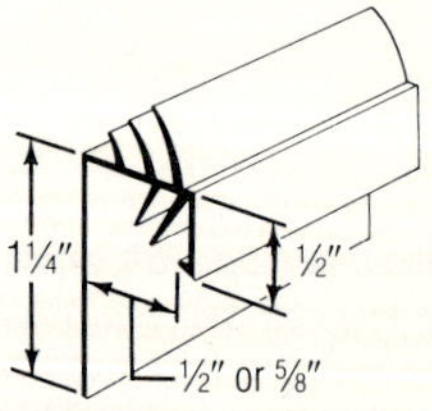

USG P-1 Vinyl Trim—A reveal-type white trim with flanges of rigid vinyl and integral flexible vinyl fins that compress on installation. Fins form permanent, flexible seal to reduce sound transmission, replace caulking and provide structural-stress relief at panel perimeter. Fits tightly over edges of either ½" or ⅝" SHEETROCK brand Gypsum Panels or IMPERIAL Gypsum Base. Helps to stop condensation where board terminates at exterior metal surfaces such as window mullions. Requires no finishing compound; paints easily; includes **P-1A** in ½" size, **P-1B** in ⅝" size.

Product	in	mm	ft-in	lb/1000 ft.	kg/m
DUR-A-BEAD Corner Bead					
No. 103	1¼x1¼	31.8x31.8	6′-10″, 8′, 10′	131	0.19
No. 104	1⅛x1⅛	28.6x28.6	6′-10″, 8′	117	0.17
Expanded Flange Corner Bead					
No. 800	1¼x1¼	31.8x31.8	8′, 10′	83	0.12
No. 900	1¼x1¼	31.8x31.8	8′, 10′	90	0.13
USG Metal Trim					
U-Shaped	½	12.7	8′, 10′	103	0.15
No. 200-A	⅝	15.9	8′, 10′	110	0.16
L-Shaped	½	12.7	8′, 10′	80	0.12
No. 200-B	⅝	15.9	8′, 10′	87	0.13
Reveal Type					
No. 400	⅜	9.5	8′, 10′	118	0.18
No. 401	½	12.7	8′, 10′	125	0.19
No. 402	⅝	15.9	8′, 10′	131	0.19
Channel Type	½	12.7	10′	98	0.15
No. 701-A	⅝	15.9	10′	106	0.16
Angle Type	½	12.7	10′	74	0.11
No. 701-B	⅝	15.9	10′	80	0.12
Channel Type	½	12.7	10′	95	0.14
No. 801-A	⅝	15.9	10′	103	0.15
Angle Type	½	12.7	10′	71	0.11
No. 801-B	⅝	15.9	10′	77	0.11
USG Vinyl Trim					
P-1A	½	12.7	8′, 10′	100	0.15
P-1B	⅝	15.9	8′, 10′	105	0.16
SHEETROCK Control Joint					
No. 093	1¾x7/16	44.4x11.1	10′	115	0.17
No. 75	1¾x¾	44.4x12.9	10′	210	0.31

(1) Metric lengths: 6′10″ = 2080mm; 8′ = 2440 mm; 10′ = 3050 mm.

Control Joints

SHEETROCK Zinc Control Joints are used to relieve stresses induced by expansion and contraction across the control joint in large ceiling and wall expanses in drywall and veneer plaster systems. Used from door header to ceiling or from floor to ceiling in long partitions and wall furring runs; from wall to wall in large ceiling areas. Made from roll-formed zinc to resist corrosion; have a ¼" open slot protected by plastic tape, removed after finishing.

Limitation: Where fire and sound control are prime considerations, a seal must be provided behind the control joint.

SHEETROCK Zinc Control Joint No. 093—For interior applications; provides 3/32" grounds for drywall and veneer plaster finishes. Staple-applied to panel face; requires finishing.

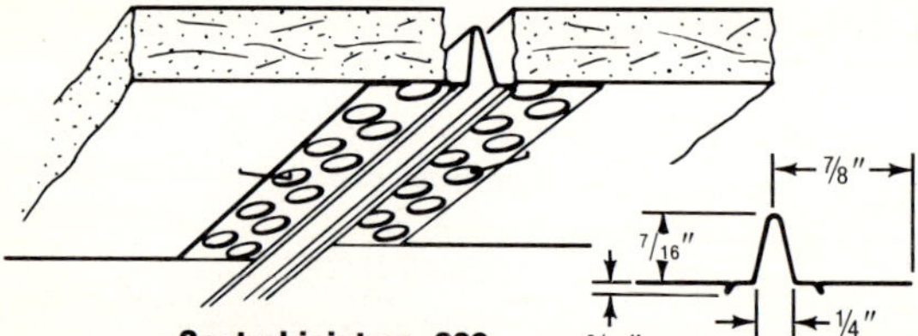

Control joint no. 093

SHEETROCK Control Joint No. 75—Back-mounted for use in veneer plaster radiant heat ceiling systems to minimize surface cracking. Staple or screw-apply as indicated in architectural specifications.

See Chapters 6 and 7 for information on control joints in conventional lath and plaster.

Framing Components

United States Gypsum Company pioneered the development of steel framing components for gypsum construction. They offer the advantages of light weight, low material cost and quick erection, superior strength and versatility in meeting job requirements.

Today, U.S. Gypsum Company sells and distributes steel framing components manufactured by Unimast, Inc. All components are noncombustible, made from corrosion-resistant steel.

It is important that light-gauge steel components such as steel studs and runners, furring channels and resilient channels be adequately protected against rusting in the warehouse and on the job site. In marine areas such as the Caribbean, Florida and the Gulf Coast where salt-air conditions exist with high humidity, components that offer increased protection against corrosion should be used.

UNIMAST Steel Studs and Runners

UNIMAST Studs and Runners are channel-type, roll-formed from corrosion-resistant steel, and designed for quick screw attachment of facing materials. They are strong, non-load bearing components of interior partitions, ceilings and column fireproofing and as framing for exterior curtain wall systems. Heavier thickness members are used in load-bearing construction. Limited chaseways for electrical and plumbing services are provided by punchouts in the stud web. Matching runners for each stud size align and secure studs to floors and ceilings, also function as headers. Made with 1″ and 1¼″ leg as noted. Studs and runners (except ST25 and CR25) are end color-coded at the factory to indicate gauge and help identify products on the job. Available in various styles and widths outlined below:

ST25 Studs and CR25 Runners—Efficient, low-cost 25-ga. members for framing non-load bearing interior assemblies. Studs come in five widths 1⅝″, 2½″, 3⅝″, 4″, 6″—and up to 20′ lengths. Runners come in matching stud widths—10′ lengths.

ST22 Studs and CR22 Runners—Heavier gauge, stronger studs in four widths 2½″, 3⅝″, 4″, 6″—and cut-to-order lengths up to 20′ Runners come in matching stud widths—10′ lengths.

ST20 Studs and CR20 Runners—Heavier 20-ga. members used in framing interior assemblies requiring greater-strength studs and reinforcement for door frames. Also used in curtain wall assemblies. Studs available in 2½″, 3⅝″, 4″, 6″ widths—cut-to-order lengths up to 28′. Runners come in studs widths, (with 1″ unhemmed leg), 10′ lengths.

These items carry a three-part code that identifies the size (212—2½″, 358—3⅝″, etc.) and style (ST—stud and CR—Runner) and steel gauge thickness (25, 22, 20).

SJ Studs and CR Runners—Used for framing load-bearing interior and exterior walls and non-load bearing curtain walls. SJ style with stiffened flanges available in four sizes and CS style in three sizes with unstiffened flanges and less load capacity for greater economy in selecting stud sizes. Products are designated by three-part code that identifies size (40—4″, 362—3⅝″, etc.); style (SJ—stud/joist, CR—C-Runner) and steel gauge thickness.

Thickness—Steel Studs and Runners(1)

Style	Design(2) in	Design(2) mm	Minimum in	Minimum mm	Gauge(3)	End color code
ST, CR25	0.0188	0.48	0.0179	0.45	25	none
ST, CR22	0.0284	0.72	0.0270	0.69	22	blue
ST, CR20	0.0329	0.84	0.0312	0.79	20	white
SJ, CR20	0.0359	0.91	0.0341	0.87	20	white
SJ, CR18	0.0478	1.21	0.0454	1.15	18	yellow
SJ, CR16	0.0598	1.52	0.0568	1.44	16	green
SJ, CR14	0.0747	1.90	0.0710	1.80	14	orange

(1) Uncoated steel thickness; meets ASTM A568. Studs and runners meet ASTM C645. Min. yield strength; all styles 33 ksi, except SJ styles 40 ksi. Coatings are hot-dip galvanized per ASTM A525; aluminized per ASTM A463 (ST, CR25 and 22 only) or aluminum-zinc per ASTM A792 or ASTM A591(weight equivalent of A525). (2) Conforms to AISI Specification for the Design of Cold-Formed Steel Structural Members, 1986 edition. (3) For information only; refer to limiting height tables and structural properties for design data.

There is a serious misconception within the construction industry regarding the substitution of one manufacturer's studs for those of another manufacturer. The assumption is that all studs of a given size and steel thickness are interchangeable. That assumption is completely false. It's possible that the substitution can safely be made, but the decision should not be made until the *structural properties* of the studs involved are compared. Most reliable manufacturers publish structural property tables in their technical literature. U.S. Gypsum Company includes Unimast Inc. data in all architectural technical literature covering steel-framed systems.

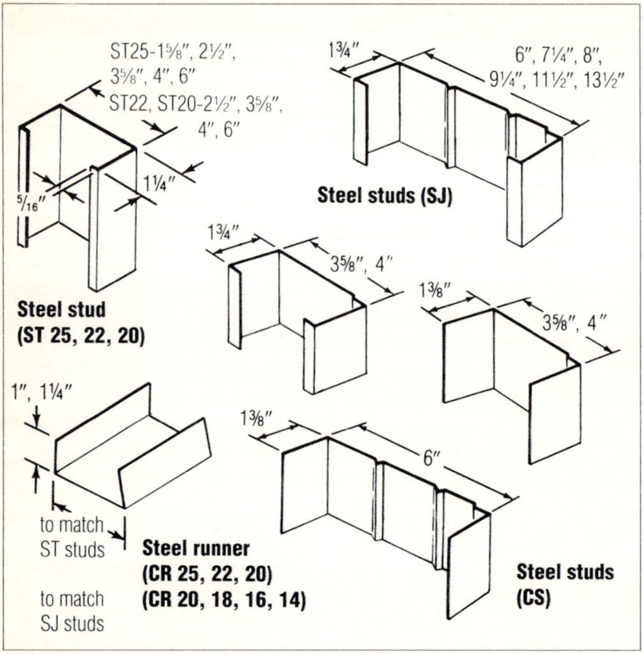

Cavity Shaft Wall, Area Separation Wall, Furring and Double Wall Components

These steel components are lightweight, versatile non-load bearing members of economical, fire and sound-barrier systems used in place of masonry for: (1) Area Separation Walls between units in multifamily wood-frame buildings; (2) Shaft Walls around elevator and mechanical shafts, return air ducts, stairwells and smoke shafts in multistory buildings; (3) Furring and Double Walls for high-performance sound control partition systems used as party, chase and furring walls. Components are formed from corrosion-resistant steel. C-H Stud base metal meets structural performance standards in ASTM A446, Grade A. Corrosion-resistant coatings supplied are: G60 hot-dip galvanized meeting ASTM A525, or aluminum-zinc meeting ASTM A792. Items are end color-coated at the factory to indicate thickness as follows: 25 ga. black, 24 ga. red, 22 ga. blue, 20 ga. white.

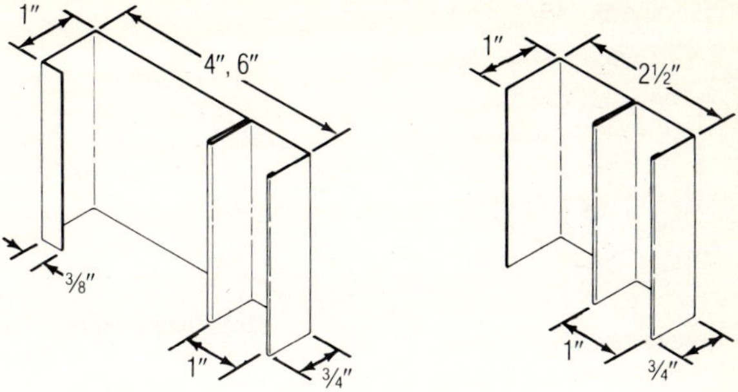

Steel E-studs

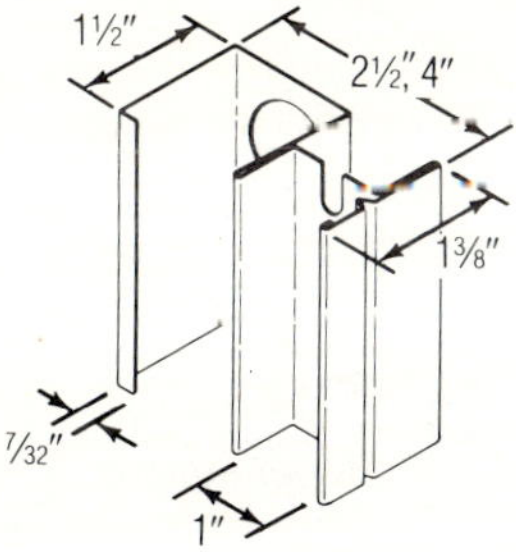

Steel C-H stud

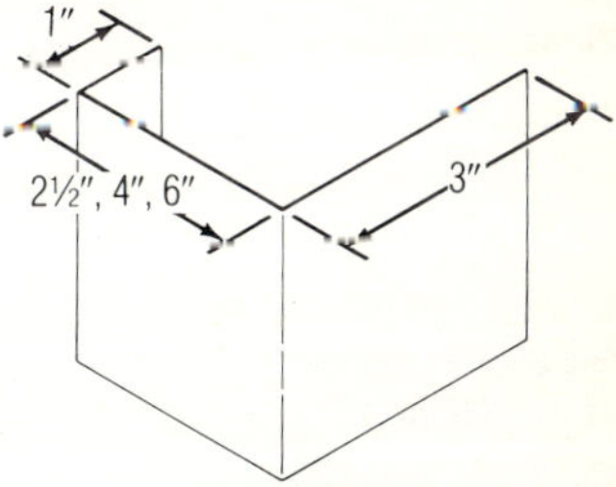

Steel jamb strut (20 ga.)

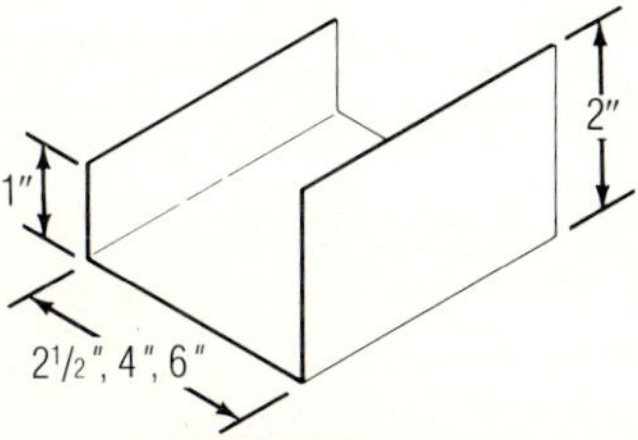

Steel J-runner

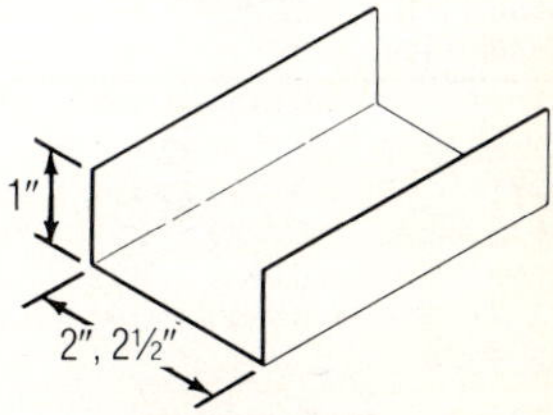

Steel C-runner

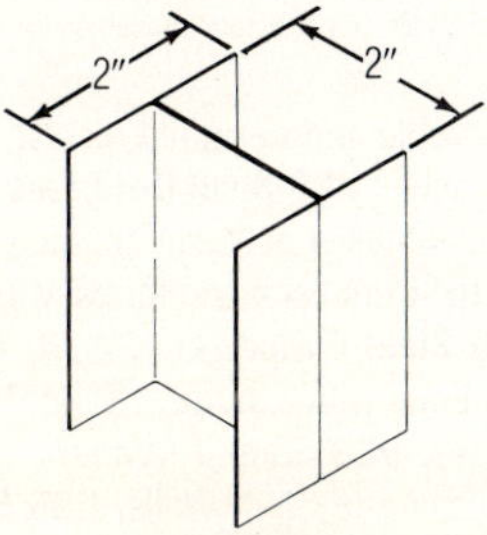

Steel H-stud (two piece)

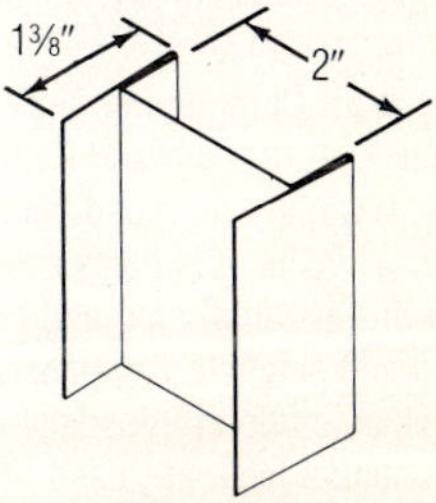

Steel H-stud (one piece)

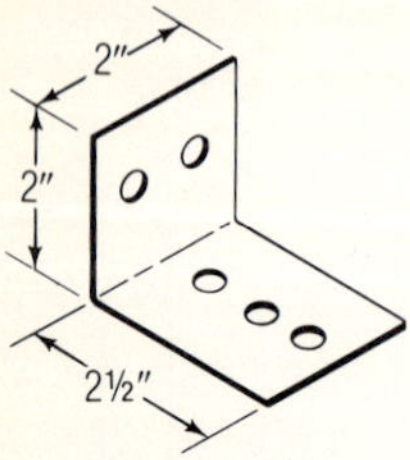

Breakaway clip

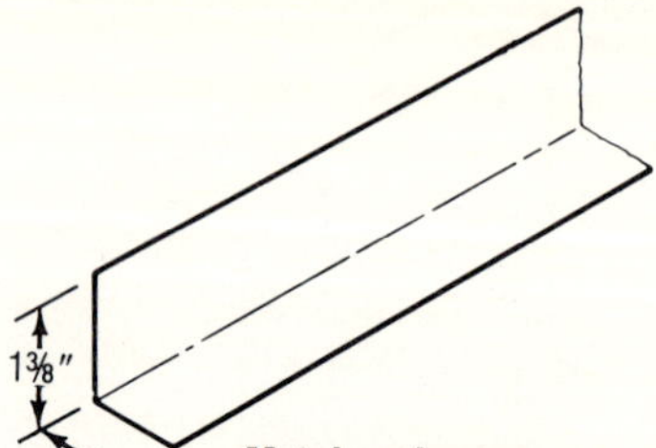

Metal angle runner

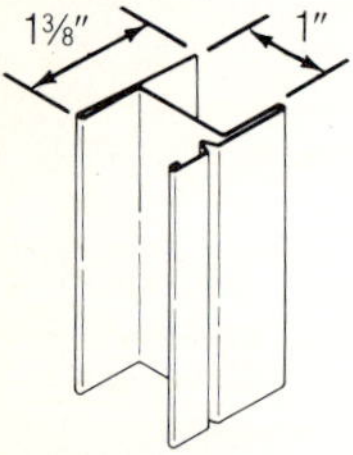

Steel H-spline (20 ga.)

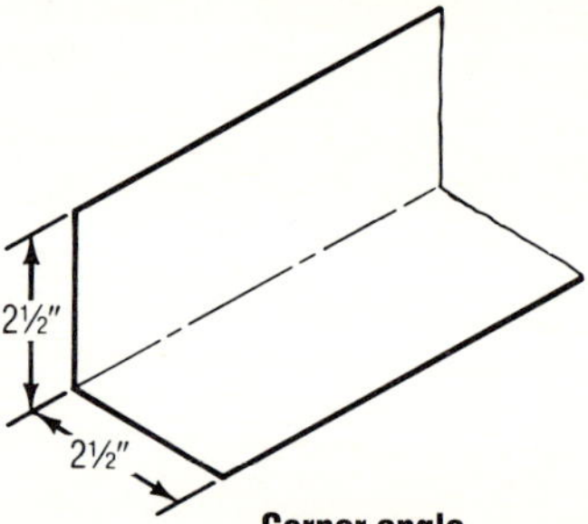

Corner angle

Thickness—Area Separation, Shaft Wall, H-Spline, Furring and Double Wall Components[1]

Component Designation	Design[2] in	Design[2] mm	Minimum in	Minimum mm	Gauge[3]	Gauge code (end color)
CR, HS, CH, ES25	0.0188	0.48	0.0179	0.45	25	none
JR24	0.0239	0.61	0.0227	0.58	24	red
Metal Angles	0.0239	0.61	0.0227	0.58	24	none
CH22	0.0310	0.79	0.0294	0.75	22	blue
ES, JR, JS, CH, HS20	0.0359	0.91	0.0341	0.87	20	white

(1) Uncoated steel thickness; meets ASTM A568. Studs and runners meet ASTM C645. Base metal meets ASTM A446 standards for structural performance. Min. yield strength 33 ksi. Coatings are hot-dip galvanized per ASTM A525; aluminized per ASTM A463, or aluminum-zinc per ASTM A792. (2) Conforms to AISI Specification for the Design of Cold-Formed Steel Structural Members, 1986 edition. (3) For information only; refer to limiting height tables and structural properties for design data.

Cavity Wall Components—2½″, 4″ and 6″ wide and designed for use with 1″ thick SHEETROCK brand Gypsum Liner Panels. **USG Steel C-H Studs,** 2½″ and 4″ wide, are non-load bearing sections installed between abutting liner panels. They have 1″ holes spaced 12″ to 16″ from each end for easy installation of horizontal pipe and conduit. **USG Steel E-Studs** are 2½″, 4″ or 6″ wide, used singly to cap panels at intersections with exterior walls or back-to-back as studs in unusually high partitions. **USG Steel J-Runners**, made with unequal legs, are used at floor and ceiling in Shaft Walls. **USG Steel C-Runners** are used singly at terminals, top and bottom of wall and back-to-back between vertical liner panels at intermediate floors, in Area Separation Walls. **USG Steel J-Struts**, 2½″, 4″ and 6″ wide, are used in jamb framing for fire-rated elevator doors.

Solid Wall Components—2″ wide and used with two layers of 1″ SHEETROCK brand Gypsum Liner Panels. **USG Steel H-Studs** fit over and engage edges of adjacent liner panels. **USG Steel C-Runners** are used in Area Separation Walls as floor and top runners and back-to-back between liner panels at intermediate floors. Also used singly to cap Area Separation Walls.

USG Aluminum Breakaway Clip—A 2″ wide, angle clip made of .063 thick aluminum. Used to attach Area Separation Walls to intermediate floor and roof framing. Clips are designed to melt and break away when exposed to fire. Size 2½″x 2″; approx. wt. 60 lb./1,000 pcs.

Double Wall Components—1″ wide for use with 1″ thick SHEETROCK brand Gypsum Liner Panels. **USG H-Spline**, roll-formed from 20-ga. steel, slides over and engages edges of adjacent 1″ gypsum liner panels. **USG Metal Angle Runners**, 1⅜″ x ⅞″ x 24-ga., are used as top and bottom runners. **USG Corner Angles**, 2½″ x 2½″ x 24-ga. steel angle sections brace and secure 1″ gypsum liner panels at corner intersections.

Specifications—Area Separation Wall & Shaft Wall Components

Component Designation(1)	Section Depth in	mm	Length ft	mm	Approx. Weight lb/1000 ft	kg/100 m
C-H Studs						
212CH25	2½	63.5	8 to 24	2440 to 7315	519	77.2
212CH22	2½	63.5	8 to 24	2440 to 7315	850	126.5
212CH20	2½	63.5	8 to 24	2440 to 7315	1000	148.8
400CH25	4	101.6	8 to 24	2440 to 7315	612	91.1
400CH20	4	101.6	8 to 24	2440 to 7315	1245	185.3
E-Studs						
212ES25	2½	63.5	8 to 28	2440 to 8530	358	53.3
212ES20	2½	63.5	8 to 28	2440 to 8530	729	108.5
400ES25	4	101.6	8 to 28	2440 to 8530	472	70.2
400ES20	4	101.6	8 to 28	2440 to 8530	970	144.3
600ES25	6	152.4	8 to 28	2440 to 8530	689	102.5
600ES20	6	152.4	8 to 28	2440 to 8530	1285	191.2
J-Runners						
212JR24	2½	63.5	10	3050	535	79.6
212JR20	2½	63.5	10	3050	736	109.5
400JR24	4	101.6	10	3050	680	101.2
400JR20	4	101.6	10	3050	937	139.4
600JR24	6	152.4	10	3050	860	128.0
600JR20	6	152.4	10	3050	1191	177.2
H-Studs						
100HS20	1	25.4	8 to 12	2440 to 3660	514	76.5
200HS25	2	50.8	8 to 16	2440 to 4880	405	60.3
C-Runners						
200CR25	2	50.8	10	3050	270	40.1
Metal Angles						
2½″x2½″	2½	63.5	10	3050	425	63.2
1⅜″x ⅞″	1⅜	34.9	10	3050	190	28.3
Jamb Strut						
212JS20	2½	63.5	8 to 12	2440 to 3660	826	122.9
400JS20	4	101.6	8 to 12	2440 to 3660	1026	152.7
600JS20	6	152.4	8 to 12	2440 to 3660	1256	186.9

(1) All components shipped unbundled, additional charge for bundling.

Framing & Furring Accessories

UNIMAST Metal Angles—Made of 24-ga. galvanized steel in two standard sizes. The 1⅜"x ⅞" size is used to secure 1" coreboard at floor and ceiling in laminated gypsum drywall partitions. The 2½"x2½" size is used to secure 1" SHEETROCK brand Gypsum Liner Panels in Double Wall Systems. Length: 10′; approx. wt./1000 ft: 190 lb. (1⅜"x⅞"), 425 lb. (2½"x2½"). Angles in other sizes and gauges available on request.

UNIMAST Cold-Rolled Channels—Made of 16-ga. steel. Used in furred walls and suspended ceilings. Available either galvanized or black asphaltum painted. Sizes ¾" with ½" flange, 1½" and 2" with 17⁄32" flange; lengths 16′ and 20′; approx. wt. ¾"—300 lb./1,000 ft., 1½"—500 lb./1,000 ft., 2"—590 lb./1000 ft.

RC-1 Resilient Channels—Made of 25-ga. corrosion resistant steel. One of the most effective, lowest-cost methods of improving sound transmission loss through wood and steel-frame partitions and ceilings. Used for resilient attachment of SHEETROCK brand Gypsum Panels and IMPERIAL Gypsum Bases. Prepunched holes 4" o.c. in the flange facilitate screw attachment to framing; facing materials are screw-attached to channels. Size ½"x 2½"; length 12′; approx. wt. 200 lb./1,000 ft.

Limitation: not for use beneath highly flexible floor joists; should be attached to ceilings with 1¼" Type W or S screws only—nails must not be used. See framing requirements in Chapter 2.

SHEETROCK Z-Furring Channels—Made of min. 24-ga. corrosion-resistant steel used to mechanically attach THERMAFIBER FS-15 Insulating Blankets, polystyrene insulation (or other rigid insulation) and gypsum panels or base to interior side of monolithic concrete and masonry walls. Sizes 1", 1½", 2", 3"; length 8'6", approx. wt. (lb/1000 ft): 224 (1"), 269 (1½"), 313 (2"), 400 (3").

UNIMAST Metal Furring Channels—Roll-formed, hat-shaped sections made of 20- and 25-ga. corrosion-resistant steel. They are designed for screw attachment of gypsum panels and gypsum base in wall and ceiling furring. Size ⅞"x 2⁹⁄₁₆"; length 12′; approx. wt. (lb./1,000 ft.): 276(DWC-25), 515(DWC-20).

UNIMAST Furring Channel Clips—Made of galvanized wire and used in attaching UNIMAST Metal Furring Channels to 1½" cold-rolled channel ceiling grillwork. For use with single-layer gypsum panels or base. Clips are installed on alternate sides of 1½" channels; where clips cannot be alternated, wire-tying is recommended. Size 1½"x 2¾"; approx. wt. 38 lb./1,000 pcs.

UNIMAST Adjustable Wall Furring Brackets—Used for attaching ¾" cold-rolled channels and UNIMAST Metal Furring Channels to interior side of exterior masonry walls. Made of 20-ga. galvanized steel with corrugated edges, brackets spaced not more than 32" o.c. horizontally and 48" o.c. vertically are attached to masonry and wire-tied to horizontal channel stiffeners in braced furring systems. Permits adjustment from ¼" to 2¼" plus depth of channel. Approx. wt. 56 lb./1,000 pcs.

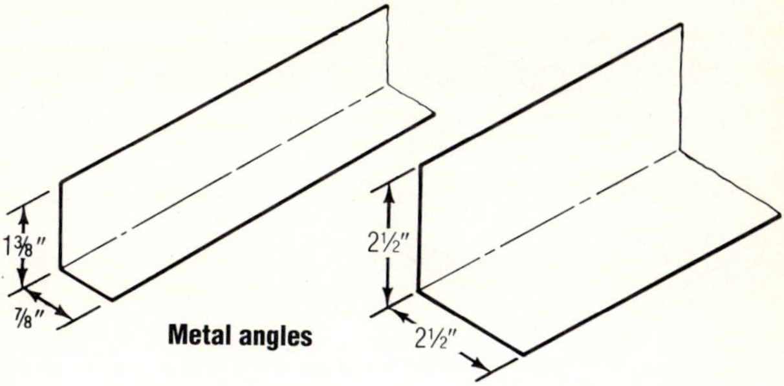

Metal angles

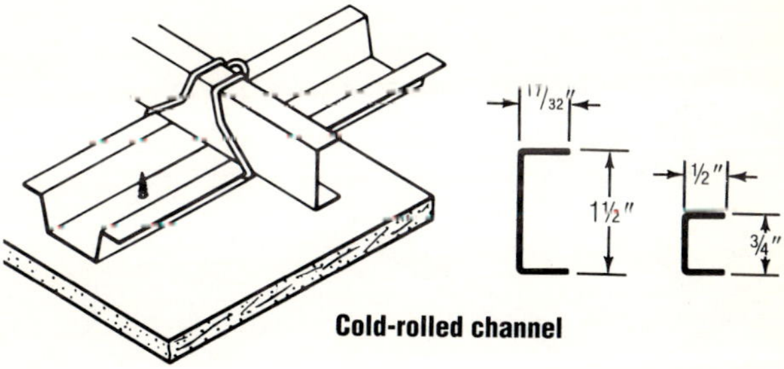

Cold-rolled channel

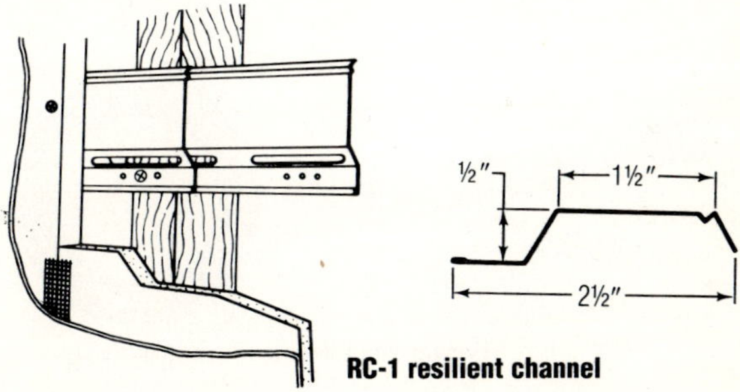

RC-1 resilient channel

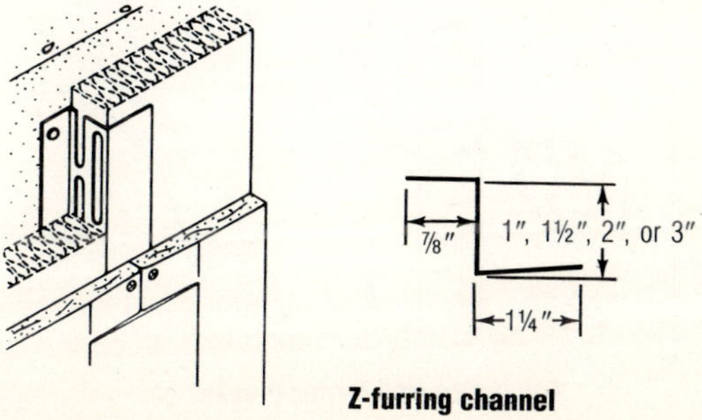

Z-furring channel

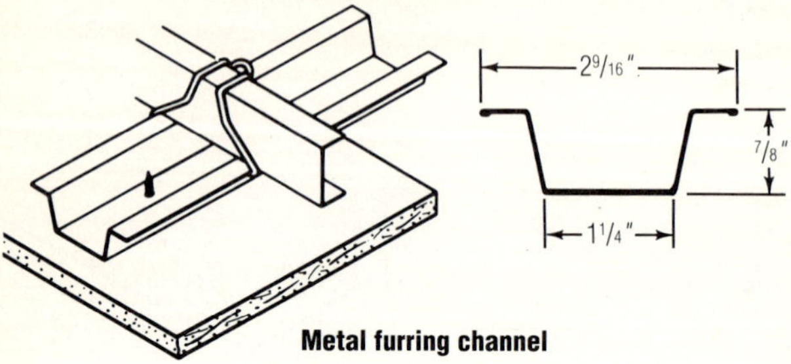

Metal furring channel

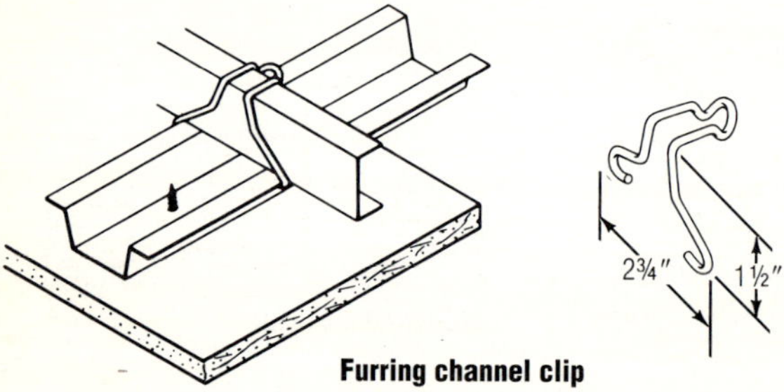

Furring channel clip

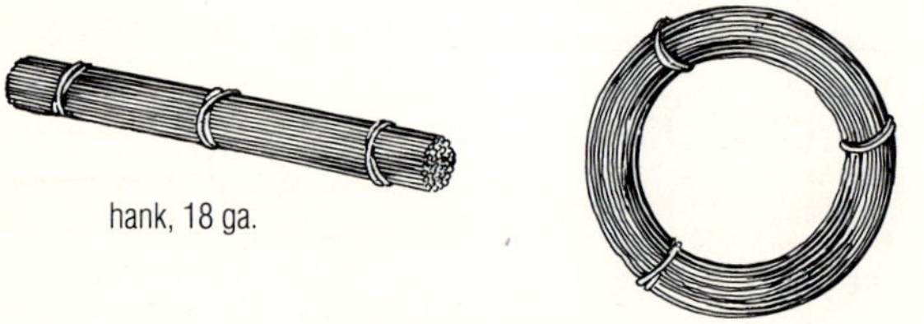

Hanger and tie wire

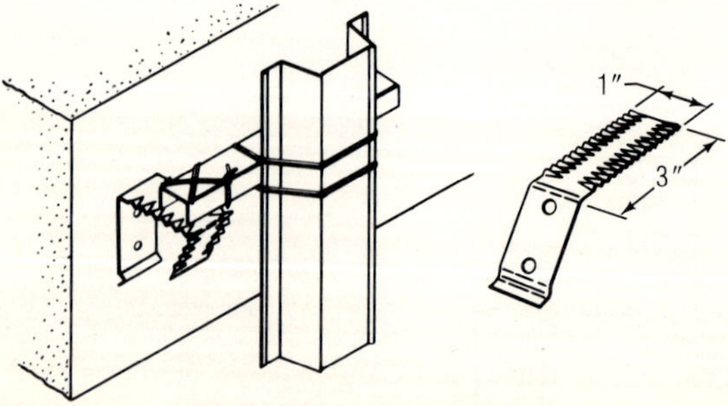

Adjustable wall furring bracket

UNIMAST Hanger and Tie Wire—Galvanized soft annealed wire available in two sizes: 8-ga. wire, used for hangers in suspended ceiling grill work, available in 50-lb. coils (approx. 730′); 18-ga. wire, used for wire-tying channels in wall furring and ceiling construction, available in 50-lb. coils (approx. 8,310′) and 25-lb. hanks (48″ straight lengths—4,148′ total).

Sound Control and Insulation Products

Adequate sound control and energy conservation are among the most important requirements in today's buildings. The public has become sufficiently aware of these factors to demand effective measures to control unwanted sound and heat transfer in both commercial and residential construction. With its advanced research, U.S. Gypsum has been a leader in developing new systems and products for efficient, low-cost sound control products and thermal insulation for new construction and remodeling.

THERMAFIBER Mineral Fiber Insulation products meet every important insulation need—thermal, acoustical and fire protection. They provide superior resistance to heat and sound transmission, resilience that ensures full installed thickness and outstanding durability.

THERMAFIBER Insulation products consist of spun mineral fibers formed into mats of varying dimensions and densities, or into nodules for blowing into framing spaces.

The use of THERMAFIBER Insulation products increases fire ratings of certain partition assemblies—provides greater fire resistance than low-melt point, glass-fiber insulation. Products without facings are rated noncombustible as defined by NFiPA 220 when tested per ASTM E136-73.

THERMAFIBER Insulation Blankets offer excellent sound-absorbing properties, in addition to providing thermal values. When used in partition cavities, THERMAFIBER Insulation improved STC ratings up to nine points. All THERMAFIBER Insulation products are asbestos-free. They resist decay, corrosion, moisture and will not support vermin.

Products Available

SHEETROCK Acoustical Sealant—A highly elastic, water-base caulking compound for sealing sound leaks around partition perimeters, cutouts and electrical boxes. Easily applied in beads or may be worked with a knife over flat surfaces such as the outside of electrical boxes. Provides excellent adherence to most surfaces. Highly resilient, permanently flexible, shrink and stain-resistant, long life expectancy. Accepted for use in 1 to 3-hour fire-rated assemblies with no adverse effect on assembly fire performance. Complies with ASTM C919. (Refer to Appendix for surface burning characteristics.)

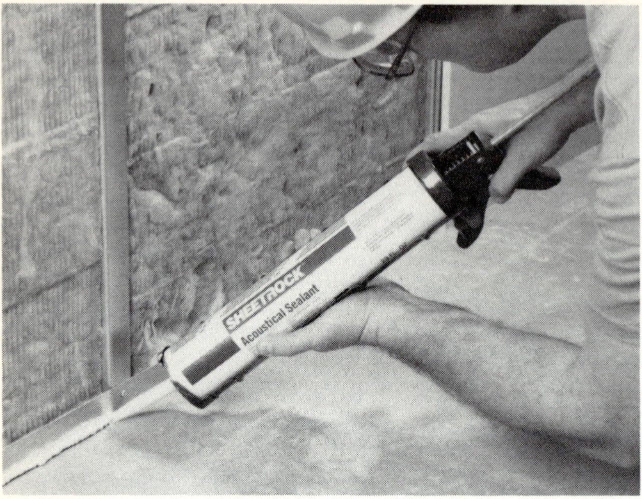

SHEETROCK Acoustical Sealant at partition perimeters plugs leaks to help deliver tested sound attenuation.

Coverage—SHEETROCK Acoustical Sealant

Product	Bead size in.	Bead size mm	Approx. coverage
SHEETROCK Acoustical Sealant	¼	6.4	392 ft./gal
	⅜	9.5	174 ft./gal.
	½	12.7	98 ft./gal.

THERMAFIBER Sound Attenuation Fire Blankets (SAFB)—Paperless, semi-rigid spun mineral fiber mat that substantially improves STC ratings when used in stud cavities of U.S. Gypsum Company partition assemblies. Each blanket has a dense, highly complex labyrinthine structure composed of fibers which produce millions of sound-retarding air pockets. Easily handled and cut; simple to install. Meet ASTM C665, Type I.

THERMAFIBER Fire Safety FS-15 Insulating Blankets—Eliminate need for staple fastening because they are paperless. Made slightly wider than normal to give snug friction fit between wood studs. FS-15 Insulating Blankets require a separate vapor retarder, such as Foil-back SHEETROCK brand Gypsum Panels or Foil-back ROCKLATH or IMPERIAL Gypsum Base, or a 6-mil polyethylene film. Specially designed for use with extruded polystyrene insulation. With this product, 3½″ THERMAFIBER FS-15 Blankets are used in load-bearing, wood-framed exterior walls having a 1-hr. fire rating. Also designed for insulating exterior furring and curtain wall assemblies that utilize steel studs and in load-bearing steel-frame exterior walls offering a 1-hr. fire rating. Used with Z-furring channels as an effective fire-resistant, semi-rigid insulating material where Class A construction is required. Use for sidewalls only. Meet ASTM C665, Type I.

THERMAFIBER Fire Safety FS-25 Blankets—Open-faced, foil-covered on vapor retarder side. The foil-kraft laminate is applied with a special flame-resistant adhesive. Blankets require a minimum air space—same as Foil-Faced Blankets—for improved installed thermal resistance. Used for ceilings, floors, walls for flame-resistant insulation and vapor control or where insulation will be exposed. (See Appendix for surface burning characteristics.) Meet ASTM C665, Type III.

THERMAFIBER Sound Attenuation Fire Blanket

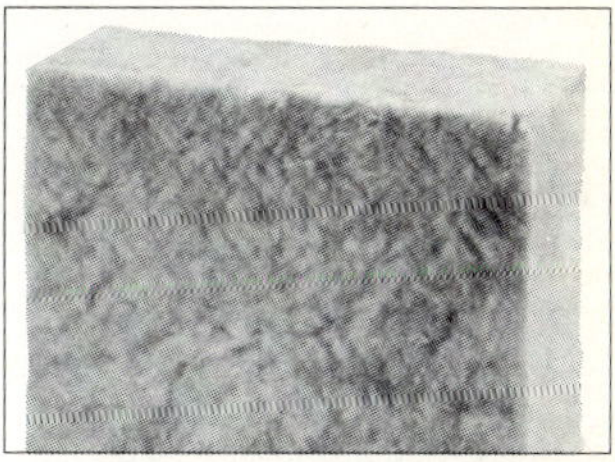

THERMAFIBER Fire Safety FS-15 Blanket

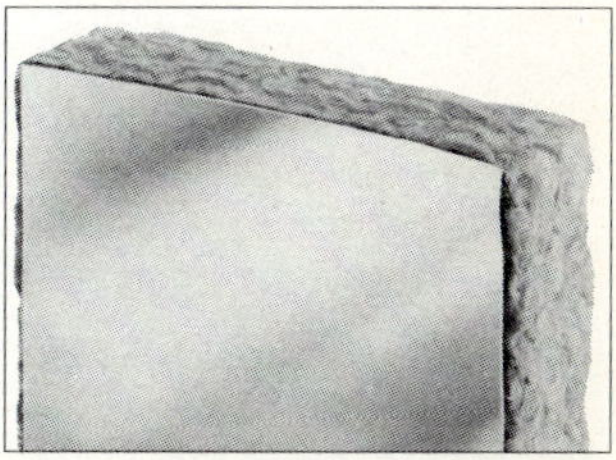

THERMAFIBER Fire Safety FS-25 Blanket

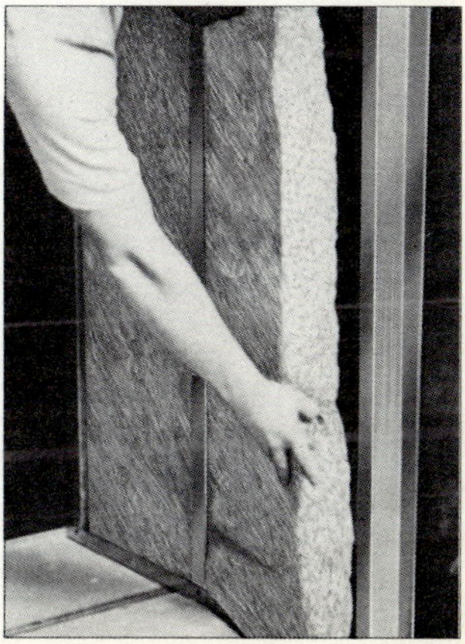

THERMAFIBER Sound Attenuation Fire Blankets fit snugly between steel studs.

Specifications—THERMAFIBER Blankets(1)

	Thickness		Width		Length		Nom. Density		Thermal resistance(2) R(3)	
Product	in	mm	in	mm	ft	m	lb/ft^3	kg/m^3	$F\text{-}h\text{-}ft^{2(4)}$ Btu	$K\text{-}m^{2(4)}$ W
Sound	1	25	16, 24	406, 610	4	1.22	4.0	0.25	4	0.7
Attenuation	1½	38	16, 24	406, 610	4	1.22	2.5	0.16	5.6	1.1
Fire	2	51	16, 24	406, 610	4	1.22	2.5	0.16	7.4	1.4
(SAFB)	3	76	16, 24	406, 610	4	1.22	2.5	0.16	11.1	2.1
Fire Safety	1	25	15, 23	381, 584	4	1.22	4.0	0.25	4	0.7
FS-15	1½	38	15, 23	381, 584	4	1.22	2.5	0.16	5.6	1.1
	2	51	15, 23	381, 584	4	1.22	2.5	0.16	7.4	1.4
	3	76	15, 23	381, 584	4	1.22	2.5	0.16	11.1	2.1
	3½	89	16, 24	406, 610	4	1.22	2.5	0.16	13	2.3
	5¼	133	16, 24	406, 610	4	1.22	2.5	0.16	19.4	3.3
	6	152	16, 24	406, 610	4	1.22	2.5	0.16	22.2	3.9
Fire Safety	3	76	16, 24	406, 610	4	1.22	2.5	0.12	11.1	1.9
FS-25	3½	89	16, 24	406, 610	4	1.22	2.5	0.12	13	2.3
	5¼	133	16, 24	406, 610	4	1.22	2.5	0.12	19.4	3.3
	6	152	16, 24	406, 610	4	1.22	2.5	0.12	22.2	3.9

(1) Check local availability of package sizes.

(2) C factor = $\frac{1}{R}$; K factor = $\frac{1}{R/thickness}$

(3) "R" value at 75°F (24°C), without facing.

(4) Symbols: °F = degrees Fahrenheit; h = hour; Btu = British thermal units; K = Kelvins; W = watts.

THERMAFIBER Regular Curtain Wall Insulation

THERMAFIBER Dark Curtain Wall Insulation

THERMAFIBER FSP Curtain Wall Insulation

THERMAFIBER **Curtain Wall Insulation**—Developed for fire-containing curtain wall systems. Positive fire-stopping material, has a significantly higher melt point than glass fiber insulation. Backs and protects spandrel components—typically made of aluminum, porcelainized steel, structural glass, concrete, marble or granite—from fire exposure. Also provides thermal control. Designed for quick mechanical attachment. Function equally well in exterior column covers and in window and track fillers. Blankets are available in three different densities: 4 lb./cu. ft.—CW 40; 6 lb./cu.ft.—CW 70; 8 lb./cu. ft.—CW 90. Contact a U.S. Gypsum Company representative for thicknesses and availability from various plant locations.

Regular Curtain Wall Insulation comes as semi-rigid blankets of unfaced felt for backing spandrel panels of opaque material, where no vapor retarder is needed.

FSP Curtain Wall Insulation is the same as regular curtain wall insulation, but with a tough scrim-reinforced foil facing that serves as a vapor retarder. The facing also adds durability for field installation.

Dark Curtain Wall Insulation is similar to regular curtain wall insulation, except it has a darker color for backing dark-colored glass spandrel panels. Using dark insulation instead of light insulation improves the look of the assembly.

Safing and Fire Stop Products

THERMAFIBER **Safing Insulation**—Fills the void between slab edges and curtain wall insulation to contain fire. Foil-faced insulation also impedes the passage of smoke and noxious gasses. THERMAFIBER Insulation also is the

THERMAFIBER Safing is cut wider than the opening to ensure compression fit. Installs easily with wire support brackets or safing clips. Available with foil backing to impede smoke and noxious gases.

principal fire-resistant material used to fill poke-through openings. Blankets come in batts 4″ thick by 24″ wide and are designed for field cutting and installation using special impaling clips or wire support brackets. Strips of insulation must be cut min. ½″ wider than the opening to ensure a compression fit. See Chapter 8 for information on fire stop systems.

THERMAFIBER SMOKE SEAL Compound—A specially designed fire and smoke resistant compound. Applied with a caulking gun to seal foil backing of curtain wall insulation to foil backing of safing insulation, and safing insulation to perimeter floor slab. Also can be trowel applied to seal insulation-filled penetration holes with 2″ layer of SMOKE SEAL Compound. Effectively blocks particulate, smoke and air movement. Smoke Stop system carries UL Classification #165 for through-penetration fire stops, 2-hr. and 3-hr. ratings. Comes in 30-oz. cartridges and 3½ or 5-gal. pails.

FIRECODE Compound—Non-toxic compound developed for use with THERMAFIBER Safing Insulation to provide wall and floor through-penetration fire stop systems that combine exceptional economy and performance. Rated non-combustible as defined by NFiPA Standard 220 when tested in accordance with ASTM E136. Surface burning characteristics: flame spread 0, smoke developed 0, when tested in accordance with ASTM E84. Effectively seals openings around pipe and cable poke-through openings. Comes in 15-lb. bags to mix easily with water at jobsite. More economical to use than tube and pail products, especially in large scale jobs. See page 359 on USG Fire Stop Systems for Floor and Wall Penetrations (U.L. Systems Nos. 449, 450, 510). Complies with ASTM E814 and UL 1479.

Coverage—FIRECODE Compound*

Dry Powder (lbs.)	Approx. Water Additions (pts.)	Wet Mixed Compound (lbs.)	Approx. Applied Fire Stop (in²/1″ deep)
1	0.5	1.5	33.6
5	2.5	7.7	172.5
7.5	3.8	11.5	257.6
10	5.0	15.4	344.9
15	7.5	23.1	517.4

*Based on approximately 7.5 pints water per 15 lb. bag for wall penetrations. For floor penetrations, approximately 8.3 pints water per 15 lb. bag is recommended and yields approximately 537 cu. in. of applied fire stop.

THERMAFIBER Mineral Fiber Fireproofing—Designed to protect structural elements in high-rise buildings. Readily attach by impaling on steel wire studs welded to columns, clip-on studs or strap-attached barbed battens. UL Classified for use in assemblies fire-rated from 1½ to 4 hr. See Technical Folder SA-707 for design numbers and descriptions. Batts range from 1″ to 4″ in thickness. Contact representative for availability.

THERMAFIBER Mineral Fiber Kitchen Duct Fireproofing—Semi-rigid batts come 2½″ thick, 8-lb./cu. ft. density and are faced with aluminum foil. Is applied around 16-ga. steel duct, impaled on mechanically attached pins and secured with sheet metal shields. Tested in accordance with Section 64.67(6) of the State of Wisconsin Code (CEG Report 9-12-77), assembly provides a 2-hr. rating.

Fasteners

Gypsum Board Screws

BUILDEX Screws—A complete line of special self-drilling, self-tapping steel screws, including types with a double-lead thread design which produces up to 30% faster penetration, less screw stripping, and greater holding power and pull-through resistance than conventional fasteners.

Screws are corrosion-resistant and all (except Hex Washer Head type) have a Phillips-head recess for rapid installation with a special bit and power-driven screwdriver. The bugle head spins the face paper into the cavity under the screwhead for greater holding power and helps prevent damage to the gypsum core and face paper. Defects associated with improper nail dimpling are eliminated. Other head types are designed specifically for attaching metal to metal and installing wood and metal trim. Screws meet ASTM C1002 (Type S and W) and ASTM C954 (Type S-12).

Type S screws have specially designed drill point and threads that minimize stripping, provide maximum holding power and pull-through resistance in steel studs and runners. Type S screws are designed for use with steel up to .04″ thick; Type S-12 screws for steel from .04″ to .07″ thick (see table, below). The special threads on Type G and Type W screws offer superior holding power in attachment to gypsum boards and wood framing, respectively. TAPCON Anchors provide fast, safe attachment of steel components to poured concrete and concrete block surfaces. Special 1$\frac{15}{16}$″ Type S-12 bugle head pilot point screws are designed for attachment of plywood to steel joists and studs.

Steel Thickness/Screw Type

Steel thickness /in.(1)	Gauge	Screw type	Steel thickness /in.(1)	Gauge	Screw type
.02	25	S	.05	18	S-12
.03	22	S	.06	16	S-12
.04	20	S or S-12	.07	14	S-12
			.10	12	S-4

(1) Nom. uncoated steel.

The superior pull-through resistance of Type W screws has virtually eliminated loose panel attachment and nail pops in wood frame construction. Tests have shown the Type W screw to have 350% greater pullout strength than GWB-54 nails. Fewer screws than nails are generally required, and the speed of installation using electric screwguns compares favorably with nailing.

SUPER-TITE Screws—High quality, economical screws for interior framing applications. These self-drilling, self-tapping steel screws have specially designed drill point and threads to ensure fast penetration into steel and wood framing. Sizes available: 1″, 1⅛″, 1¼″, 1⅝″, 2¼″, 3″ bugle head for attaching gypsum panels to 20 and 25-ga. steel framing; 1¼″ Type W bugle head for attaching panels to wood framing; $\frac{7}{16}$″ pan head for securing studs to runners.

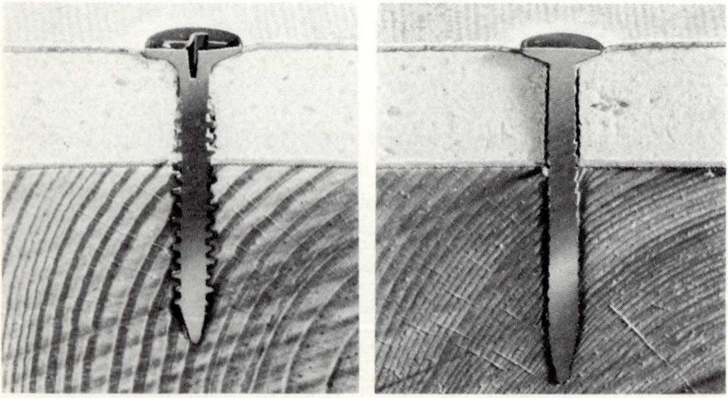

Secret to superiority of screw attachment is shown by comparative diagrams. Bugle-head screw (left) depresses face paper of gypsum panel without tearing; threads cut into and deform wood to hold tightly. Longer drywall nail (right) grips with friction, loosens hold as wood shrinks, which may pop nailhead above surface to create callback situation.

Selector Guide for Screws

Fastening application	Fastener used	Fig[2]
Gypsum panels to steel framing[1]		
½" single-layer panels to steel studs, runners, channels	1" Type S bugle head	1
⅝" single-layer panels to steel studs, runners, channels	1" Type S bugle head 1⅛" Type S bugle head	1 1
¾" single-layer panels to steel studs, runners, channels	1¼" Type S bugle head	1
1" coreboard to metal angle runners in solid partitions	1⅝" Type S bugle head	1
½" double-layer panels to steel studs, runners, channels	1⅝" Type S bugle head	2
⅝" double-layer panels to steel studs, runners, channels	1⅝" Type S bugle head	2
¾" double-layer panels to steel studs, runners, channels	2¼" Type S bugle head	2
½" panels through coreboard to metal angle runners in solid partitions	1⅞" Type S bugle head	2
⅝" panels through coreboard to metal angle runners in solid partitions	2¼" Type S bugle head 3" Type S bugle head	2 2
1" double-layer coreboard to steel studs, runners	2⅝" Type S bugle head	2
Wood to steel framing		
Wood trim over single-layer panels to steel studs, runners	1" Type S trim head 1⅝" Type S trim head	5 5
Wood trim over double-layer panels to steel studs, runners	2¼" Type S trim head	5
Steel cabinets, brackets through single-layer panels to steel studs	1¼" Type S oval head	6
Wood cabinets through single-layer panels to steel studs	1⅝ " Type S oval head	6
Wood cabinets through double-layer panels to steel studs	2¼", 2⅞", 3¾", Type S oval head	6
Steel studs to door frames, runners		
Steel studs to runners ST-25 & 22	⅜" Type S pan head	9
Steel studs to runners		
Steel studs to door frame jamb anchors ST-20	⅜" Type S-12 pan head ⅝" Type S-12 low-profile head	10 11
Other metal-to-metal attachment (12-ga. max.)		
Steel studs to door frame jamb anchors (heavier shank ensures entry in anchors of hard steel)	½" Type S-12 pan head ⅝" Type S-12 low-profile head	10 11

continued on next page

Selector Guide for Screws *continued*

Fastening application	Fastener used	Fig[2]
Steel studs to door frames, runners		
Strut studs to door frame anchors, rails, other attachments in ULTRAWALL movable partitions	½″ Type S-16 pan head with anti-corrosive coating	10
Metal-to-metal connections up to double thickness of 12-ga. steel	¾″ S-4 hex washer head with anti-corrosive coating	12
Gypsum panels to 12-ga. (max.) steel framing		
½″ and ⅝″ panels and gypsum sheathing to steel studs and runners; specify screws with anti-corrosive coating for exterior curtain wall applications	1″ Type S-12 bugle head	3
Self-furring metal lath and brick wall ties through gypsum sheathing to steel studs and runners; specify screws with anti-corrosive coating for exterior curtain wall applications	1¼″ Type S-12 bugle head 1¼″ Type S-12 pancake head	4 13
½″ and ⅝″ double-layer gypsum panels to steel studs and runners	1⅝″ Type S-12 bugle head	4
Multilayer gypsum panels and other materials to steel studs and runners	1⅞″, 2″, 2⅜″, 2⅝″, 3″ Type S-12 bugle head	4
Cement board to steel framing		
DUROCK Cement Board or Exterior Cement Board direct to steel studs, runners	1¼″, 1⅝″ DUROCK Steel Screws	17
Rigid foam insulation to steel framing		
Rigid foam insulation panels to steel studs and runners; Type R for 20-, 25-ga. steel	1½″, 2″, 2½″, 3″ Type S-12 or R wafer head	15
Aluminum trim to steel framing		
Trim and door hinges to steel studs and runners (screw matches hardware and trim)	⅞″ Type S-18 Oval Head with anti-corrosive coating	7
Batten strips to steel studs in demountable partitions	1⅛″ Type S bugle head	1
Aluminum trim to steel framing in demountable and ULTRAWALL partitions	1¼″ Type S bugle head with anti-corrosive coating	1
Gypsum panels to wood framing		
⅜″, ½″ and ⅝″ single-layer panels to wood studs, joists	1¼″ Type W bugle head	8
Cement board to wood framing		
DUROCK Cement Board or Exterior Cement Board to wood framing	1¼″, 1⅝″, 2¼″ DUROCK Wood Screws	18
RC-1 Resilient channels to wood framing		
Screw attachment required for both ceilings and partitions	1¼″ Type W bugle head 1¼″ Type S bugle head	8 1
For fire-rated construction	1¼″ Type S bugle head	1

continued

Selector Guide for Screws *continued*

Fastening application	Fastener used	Fig[2]
Gypsum panels to gypsum panels		
Multilayer adhesively laminated gypsum-to-gypsum partitions (not recommended for double-layer ⅜″ panels)	1½″ Type G bugle head	8
Plywood to steel joists		
⅝″ to ¾″ plywood to steel joists (penetrates double thickness 14-ga.)	1¹⁵⁄₁₆″ Type S-12 bugle head, pilot point	16
Steel to poured concrete or block		
Attachment of steel framing components to poured concrete and concrete block surfaces	³⁄₁₆″ x 1¾″ acorn slotted HWH TAPCON Anchor	14

Notes: (1) Includes UNIMAST Steel Studs and Runners, ST25, ST22 and ST20; Metal Angles; Metal Furring Channels; RC-1 Resilient Channels. If channel resiliency makes screw penetration difficult, use screws ⅛″ longer than shown to attach panels to RC-1 channels. For other gauges of studs and runners, always use Type S-12 screws. For steel applications not shown, select a screw length which is at least ⅜″ longer than total thickness of materials to be fastened. Use screws with anti-corrosive coating for exterior applications. (2) Figures refer to following screw illustrations.

Bit Tips

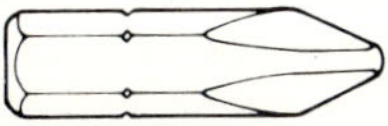

No. 1 bit for trim and pancake heads

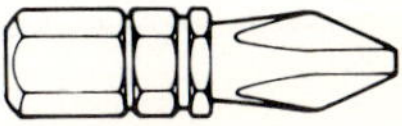

No. 2 bit for bugle, pan, water, low profile & oval heads

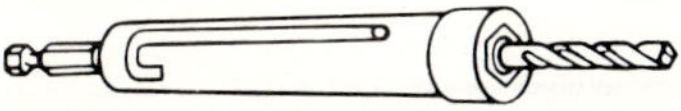

Tool/bit for HWH TAPCON Anchors

Note: Hex-head bit not illustrated

Basic Types of Screws

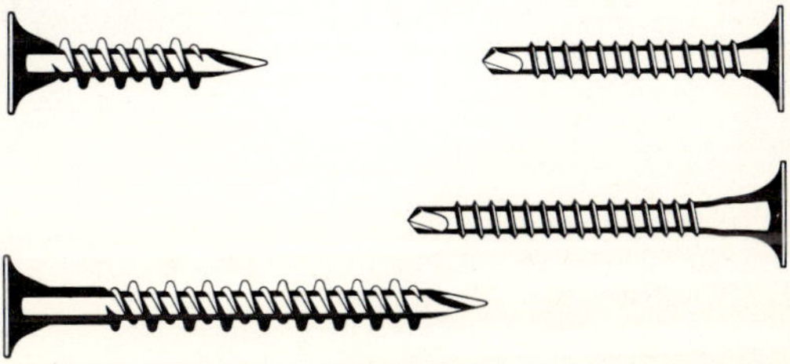

#1 #3 #4 #2

continued on next page

Basic Types of Screws *continued*

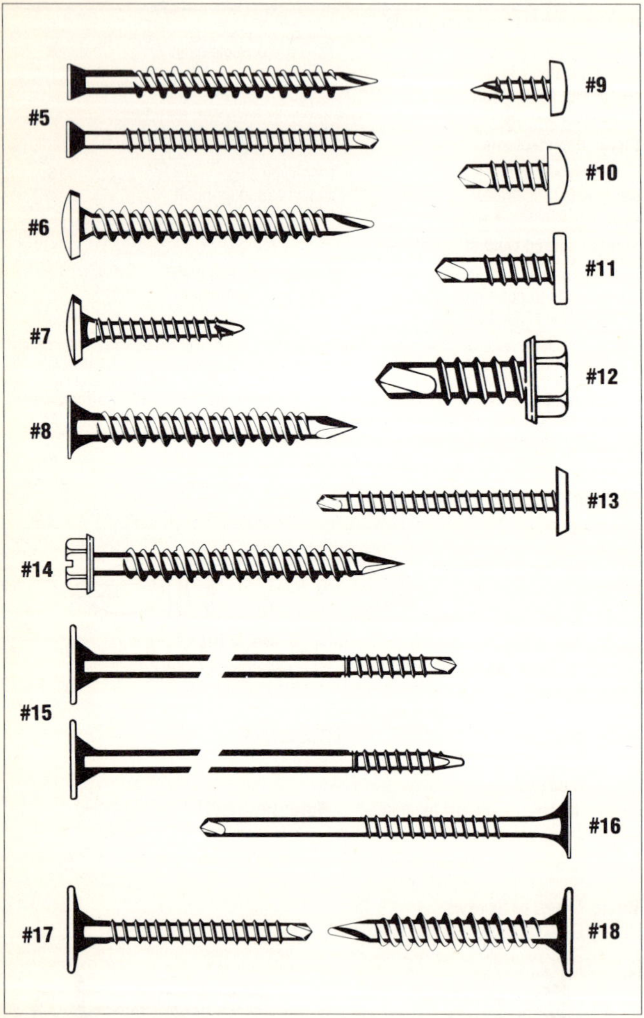

Specifications—Screws

Description				Bulk packaging			Master carton packing			
Length				Qty.	Approx. wt.		Qty.		Approx. wt.	
in	mm	Type	Head	1,000 pc	lb	kg	1,000 pc	Pkgs	lb/ctn	kg/ctn
Base screws										
1	25.4	S	bugle	10	36	16.3	16	8	52	23.5
1⅛	28.6	S	bugle	8	32	14.5	16	8	57	25.8
1¼	31.8	S	bugle	8	34	15.4	16	8	62	28.1
1⅝	41.3	S	bugle	5	28	12.7	8	8	43	19.5
1⅞	47.6	S	bugle	4	24	10.9	8	8	47	21.3
2¼	57.2	S	bugle	2.5	19	8.6	4	8	29	13.1
2⅝	66.7	S	bugle	2.5	24	10.9	4	8	36	16.3
3	76.2	S	bugle	2	23	10.4	4	8	42	19.0
Specialty screws										
⅜	9.5	S	pan	16	47	21.3	20	4	57	25.8
⅜	9.5	S-12	pan	16	47	21.3	20	4	60	27.2
½	12.7	S-12	pan	14	48	21.8	12	4	39	17.6
½	12.7	S-12	pancake	12	49	22.2	12	4	48	21.7
½	12.7	S-16	pan[1]	14	48	21.8	12	4	45	20.4
⅝	15.9	S-12	low-profile	10	36	16.3	10	4	40	18.1
¾	19.1	S-4	hex washer[1]	3	40	18.1	—	—	—	—
⅞	22.2	S-18	oval[1]	10	50	22.6	10	4	54	24.4
1	25.4	S	trim	12	33	14.9	16	4	44	19.9
1	25.4	S-12	trim	12	33	14.9	16	4	44	19.9
1	25.4	S-12	bugle	10	35	15.8	16	8	60	27.2
1¼	31.8	S-12	bugle	8	35	15.8	16	8	68	30.8
1¼	31.8	S	bugle[1]	8	34	15.4	16	8	62	28.1
1¼	31.8	W	bugle	8	33	14.9	16	8	60	27.2
1¼	31.8	S-12	pancake	5	34	15.4	8	8	56	25.4
1¼	31.8	S	oval	—	—	—	8	4	29	13.1
1½	38.1	G	bugle	5	33	14.9	8	8	50	22.6
1½	38.1	R	wafer	4	30	13.6	—	—	—	—
1½	38.1	S-12	wafer	4	30	13.6	—	—	—	—
1⅝	41.3	S	oval	—	—	—	8	8	44	19.9
1⅝	41.3	S	trim	6	28	12.7	8	8	36	16.3
1⅝	41.3	S-12	bugle	5	27	12.2	8	8	43	19.5
1⅝	41.3	S-12	trim	6	29	13.1	8	8	37	16.7
1⅞	47.6	S-12	bugle	4	25	11.3	8	8	49	22.2
1 15/16	49.2	S-12	bugle, pilot pt.	3	24	16.3	—	—	—	—
2	50.8	S-12	bugle	4	26	11.7	8	8	49	22.2
2	50.8	R	wafer	3	32	14.5	—	—	—	—
2	50.8	S-12	wafer	3	32	14.5	—	—	—	—
2¼	57.2	S	trim	4	27	12.2	8	8	51	23.1
2¼	57.2	S	oval	—	—	—	4	4	30	13.6
2¼	57.2	S-12	trim	4	27	12.2	8	8	53	24.0
2⅜	60.3	S-12	bugle	3	24	10.9	4	8	51	23.1
2½	63.5	R	wafer	2	27	12.2	—	—	—	—
2½	63.5	S-12	wafer	2	27	12.2	—	—	—	—
2⅝	66.7	S-12	bugle	2.5	24	10.9	4	8	51	23.1
2⅞	73.0	S	oval	—	—	—	4	4	40	18.1
3	76.2	S-12	bugle	2	23	10.4	4	8	44	19.9
3	76.2	R	wafer	1.5	23	10.4	—	—	—	—
3	76.2	S-12	wafer	1.5	23	10.4	—	—	—	—
3¾	95.3	S	oval	—	—	—	1	1	12	5.4
TAPCON screw										
1¾	44.5	conc.	hex	1	10	4.5	—	—	—	—

(1) Corrosion-resistant coating.

SUPER-TITE Screws—Provide excellent results for most interior framing requirements at economical cost. A deep, sharp Phillips-head recess mates with the SUPER-TITE Bit Tip for positive, nonslip control, eliminates spinouts. Fast penetration afforded by specially designed threads and self-drilling point ensure lower in-place cost. Bugle head screws in six sizes attach gypsum panels to 20- or 25-ga. steel framing, and a 1¼″ Type W bugle head screw also is available for attaching gypsum panels to wood framing. The 7/16″ pan head screw for attaching steel studs to runners can penetrate double-thickness 25-ga. steel. SUPER-TITE screws meet ASTM C1002.

				Packaging		
				Bulk		
	Application	Part no.	Screw size and length (no. x inches)	Quantity (1000 pcs.)	Approx. wt. lb	(kg)
	**1″ (25.4 mm) Bugle Head** Attaches ½″ or ⅝″ single-layer gypsum panels and bases to steel framing	1ES	6x1	10	31	(14.1)
	**1⅛″ (28.6 mm) Bugle Head** Attaches ⅝″ gypsum panels and bases to RC-1 Channels or other steel framing, also batten strips for demountable partitions	2ES	6x1⅛	10	35	(15.9)
	**1¼″ (31.8 mm) Bugle Head** Attaches 1″ coreboard to steel runners Attaches ½″ and ⅝″ gypsum panels and bases to wood studs	3ES	6x1¼	8	30	(13.6)
	**1⅝″ (41.3 mm) Bugle Head** Attaches double-layer gypsum panels to steel framing	4ES	6x1⅝	5	24	(10.8)

continued

SUPER-TITE Screws *continued*	**Application**	**Part no.**	**Screw size and length (no. x inches)**	**Packaging: Bulk Quantity (1000 pcs.)**	**Approx. wt. lb**	**(kg)**
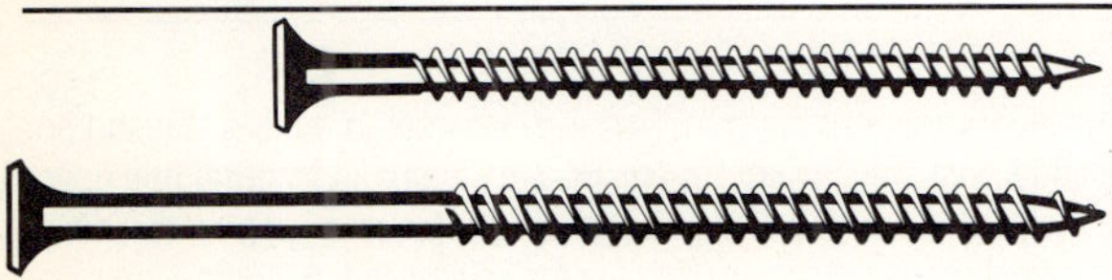	**2″ (50.8 mm) Bugle Head**	5RS	6x2	3.5	22	(10.0)
	2¼″ (57.2 mm) Bugle Head	6RS	6x2¼	3	23	(10.4)
	2½″ (63.5 mm) Bugle Head	7RS	7x2½	2.5	24	(10.8)
	3″ (76.2 mm) Bugle Head	8RS	8x3	2	24	(10.8)
	Attaches multiple layers of gypsum panels and other compatible materials to steel framing					
	**1¼″ (31.8 mm) Type W Bugle Head**	3CS	6x1¼	8	30	(13.6)
	Attaches ½″ or ⅝″ single-layer gypsum panels, bases, or RC-1 Channels to wood framing					
* 	**7⁄16″ (11.1 mm) Pan Head**	19SS	6x7⁄16	10	27	(12.2)
	*Special Framer	20S	7x7⁄16	10	22	(10.0)
	Attaches 25-ga. steel studs to runners					
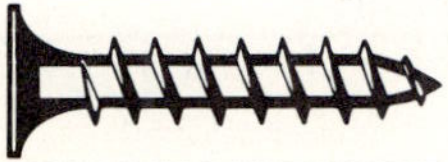	**1½″ (38.1 mm) Bugle Head-Laminating**	8S	10x1½	5	40	(18.1)
	Temporary attachment of gypsum to gypsum					
	**1⅝″ (41.3 mm) Trim Head**	9S	6x1⅝	5	24	(10.8)
	2¼″ (57.2 mm) Trim Head	10S	6x2¼	3	20	(10.0)
	Attaches wood trim to 20 to 25-ga. steel framing					

continued on next page

SUPER-TITE II Screws—High-and-low thread screws for attaching gypsum board to 20- and 25-ga. interior steel framing. SUPER-TITE II Screws have a specially designed drill point and threads to provide rapid penetration and positive attachment to reduce in-place cost. SUPER-TITE II Screws meet ASTM C1002.

Screw type/application	Part no.	Screw size and length (no.x Inches)	Packaging – Master carton – Carton qty. pieces (1000's)	Approx. wt./ctn. lb	(kg)
Bugle Head	201ES	6x1	10	31	(14.1)
Attaches gypsum board	202ES	6x1⅛	10	35	(15.9)
to 20- to 25-ga.	203ES	6x1¼	8	30	(13.6)
steel framing	204ES	6x1⅝	5	24	(10.8)
	205RS	6x2	3.5	22	(10.0)
	206RS	6x2¼	3	23	(10.4)
	207RS	7x2½	2.5	26	(11.8)
	208RS	8x3	2	24	(10.8)

SUPER-TITE DRILLERS—A quality line of self-drilling screws for numerous steel framing applications. SUPER-TITE DRILLERS penetrate up to 14-ga. steel (.0747 inches). Sharp drilling flutes provide immediate drilling action and removal of chips. Low-driving torque ensures easy application and positive seating of fastener. SUPER-TITE DRILLERS Screws meet ASTM C954.

Screw type/application	Part no.	Screw size and length (no.x Inches)	Packaging – Master carton – Carton qty. pieces (1000's)	Approx. wt./ctn. lb	(kg)
Bugle Head	11S	6x1	10	35	(15.9)
Attaches single-layer gypsum	11AS	6x1⅛	10	36	(16.3)
board steel framing up	12S	6x1¼	8	35	(15.9)
to 14-ga.					

continued

SUPER-TITE DRILLERS *continued*	Screw type/application	Part no.	Screw size and length (no.x Inches)	Packaging / Master carton: Carton qty. pieces (1000's)	Approx. wt./ctn. lb	(kg)
	**Bugle-Head** Attaches multilayer gypsum board to steel framing up to 14-ga.	13S	6x1⅝	5	28	(12.7)
		14S	6x1⅞	4	25	(11.3)
		16S	8x2⅛	2.5	22	(10.0)
		17S	8x2⅝	1.6	19	(8.6)
		18S	8x3	1.4	19	(8.6)
	**Pan Head** Attaches stud to runner up to 14-ga.	23S	7x7/16	10	24	(10.8)
		24S	8x⅛	10	32	(14.5)
 	**Hex Washer Head** Attaches steel to steel up to 14-ga.	25S	8x½	10	37	(16.8)
		21S	8x⅝	10	39	(17.7)
		26S	8x¾	10	40	(18.1)
		27S	8x1	8	44	(20.0)
	**Modified Truss Head** Attaches metal lath to steel framing up to 14-ga.	30S	8x½	10	35	(15.9)
		31S	8x¾	10	47	(21.3)
		32S	8x1	8	44	(20.0)
		33S	8x1¼	6	36	(16.3)

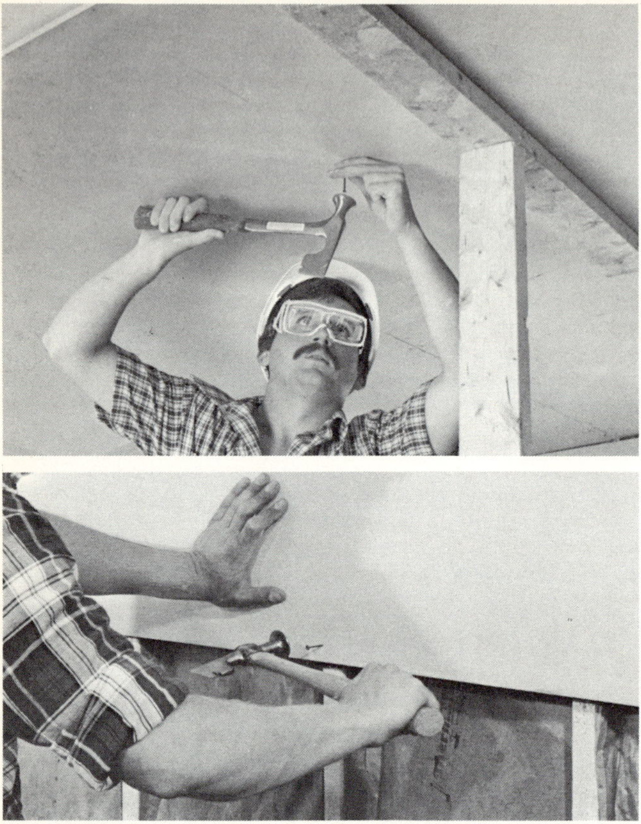

Hand pressure is applied to panel as nail is driven.

Gypsum Board Nails

Gypsum board nails have been vastly improved since the relationship of wood shrinkage to nail popping was discovered. Nails have been developed to concentrate maximum holding power over the shortest possible length—notably the annular ring type nail which has about 20% greater holding power than a smooth-shank nail of the same length and shank diameter. However, under lengthy, extreme drying conditions, such as a cold dry winter or in arid climates, resultant wood shrinkage may cause fastener pops even with the shorter annular ring nail.

As with screws, specification of the proper nail for each application is extremely important, particularly for fire-rated construction where nails of the specified length and diameter only will provide proper performance. When wood-frame gypsum panel systems are subjected to fire, nails on surface attain temperatures that tend to char the wood, thereby reducing their holding power. Nails used in gypsum construction should comply with performance standards of ASTM C514. Nails are not available from U.S. Gypsum Company.

Selector Guide for Gypsum Board Nails[1]

Fastener description[2]	Fastener length		Total thickness of surfacing materials[3]										Approx. usage	
			in	1/4	3/8	1/2	5/8	3/4	7/8	1	1-1/4	1-3/8		
	in	mm	mm	6.4	9.5	12.7	15.9	19.1	22.2	25.4	31.8	34.9	lb/1,000 ft²	kg/100m²
Annular Ring Drywall Nail 12½ ga. (2.50 mm) ¼" (6.35 mm) diam. head, med. diamond point 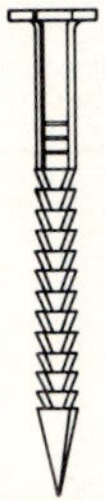	1¼	31.8		X	X	X							4.50	2.20
	1⅜	34.9					X						5.00	2.44
	1½	38.1						X					5.25	2.56
	1⅝	41.3							X				5.75	2.81
Same as above except 19/64" (7.54 mm) diam. head	1¼	31.8		X	X	X							4.50	2.20
	1⅜	34.9					X						5.00	2.44
	1½	38.1						X					5.25	2.56
	1⅝	41.3							X				5.75	2.81
	1¾	44.5								X			6.00	2.93
	2	50.8									X		7.00	3.42
12½ ga. (2.50 mm) 19/64" (7.54 mm) diam. head 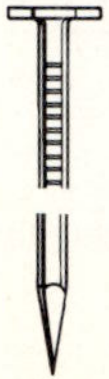	1¼	31.8		X	X	X							4.50	2.20
	1⅜	34.9					X						5.00	2.44
	1½	38.1						X					5.25	2.56
	1⅝	41.3							X				5.75	2.81

Selector Guide for Gypsum Board Nails(1)

			Total thickness of surfacing materials(3)											
	Fastener length		in	1/4	3/8	1/2	5/8	3/4	7/8	1	1-1/4	1-3/8	Approx. usage	
Fastener description(2)	in	mm	mm	6.4	9.5	12.7	15.9	19.1	22.2	25.4	31.8	34.9	lb/1,000 ft²	kg/100m²
Same as preceding except ¼″ (6.35 mm) diam. head														
14 ga. (2.03 mm)	1⅜ (4d)	34.9					X						3.50	1.71
13½ ga. (2.18 mm)	1⅝ (5d)	41.3						X					4.50	2.20
13 ga. (2.32 mm)	1⅞ (6d)	47.6								X			5.75	2.81
13½ ga. (2.18 mm)	2⅛ (7d)	54.0										X	7.50	3.66

(1) For wood framing 16″ o.c., nails 8″ o.c. for walls, 7″ o.c. for ceilings
(2) All nails treated to prevent rust with joint compounds or veneer plaster finishes. Fire-rated assemblies generally require greater nail penetration; therefore, for fire-rated assemblies, use exact nail length and diameter specified for rated assembly (see Fire Test Report).
(3) In laminated double-layer construction, base layer is attached in same manner as single layer.

Adhesives

Drywall adhesives make an important contribution to gypsum panel attachment where the finest room interiors are desired. Their use greatly reduces the nail or screw fastening otherwise required, thus saves labor on spotting and sanding also minimizes nail pops and other fastener imperfections.

U.S. Gypsum Company offers reliable, field-tested adhesives. Each is formulated to produce superior attachment, freedom from fastener imperfections and high-quality results. Recommended for laminating gypsum panels in multilayer fire-rated or non-rated partitions and ceilings are **SHEETROCK Setting-Type (DURABOND) or Lightweight Setting-Type (EASY SAND) Joint Compounds**—dry powder products, applied by spreader, requiring mixing and temporary fastening in application or **SHEETROCK Ready-Mixed Joint Compound—All Purpose or Taping**. All provide tight bond when dry yet permit adjustment of panels after contact.

SHEETROCK Setting-Type (DURABOND) or Lightweight Setting-Type (EASY SAND) Joint Compounds—Dry, powder products to be mixed with water, used for laminating gypsum panels in multilayer fire-rated or non-rated partitions and ceilings. Spreader-applied, require temporary fastening in application. Provide tight bond when dry, yet permit panel adjustment after contact. Meet ASTM C475.

SHEETROCK Ready-Mixed Joint Compound—Taping or All Purpose—Compounds formulated to a creamy smooth consistency for fast spreader application. Used for laminating gypsum panels in multilayer fire-rated or non-rated partitions and ceilings. Offer ready-to-use convenience, eliminate extensive mixing and waste. Provide good bond and strength when dry. Use above grade; keep from freezing. Meets ASTM C475.

Commercial Adhesives—Available in drywall stud and construction types meeting ASTM C557. Used in non-fire rated gypsum construction. Bridge minor irregularities in the base or framing, make it easier to form true joints and level surfaces. The use of adhesive adds strength to an assembly, reduces fasteners required, helps eliminate loose panels and nail pops.

Coverage—Laminating Adhesives

Product(1)	Type of laminating	Approx. coverage(2) Lam. blade notch spacing 2″ o.c.	1½″ o.c.
SHEETROCK Ready-Mixed Joint Compounds—Taping or All Purpose	sheet strip	340 170	465 230
SHEETROCK Lightweight All Purpose Ready-Mixed Joint Compound (PLUS 3)	sheet strip	23.0 11.5	31.7 15.5
SHEETROCK Setting-Type Joint Compounds	sheet strip	184 93	246 123
SHEETROCK Lightweight Setting Type Joint Compounds (EASY SAND)	sheet strip	134 68	179 90

(1) See page 83 for standard package sizes. (2) Coverage in lb./1000 ft.² of packaged product not including water necessary to achieve working consistency. Exception: PLUS 3 is gal./1000 ft.².

Joint Compounds

Today's complete U.S. Gypsum Company joint compound line includes both ready-mixed and powder products in drying and setting (hardening) types. All are formulated without asbestos and therefore meet all OSHA and Consumer Product Safety Standards pertaining to asbestos. In addition to conventional joint finishing and fastener spotting, certain of these products are designed for repairing cracks, patching, spackling, back-blocking, texturing and for laminating gypsum panels in double-layer systems. Products comply with ASTM C475.

Advantages

Low Cost—High-quality products reduce preparation time, save application labor, prevent expensive callbacks.

Versatility—Job-tested compounds available in specialized types to meet finishing requirements.

Non-Asbestos Formulations—Safe to handle and use; meet OSHA and Consumer Product Safety Standards.

Use of U.S. Gypsum Company joint compounds brings the important added advantage of dealing with one manufacturer who is responsible for *all* components of the finished walls and ceilings—formulated in U.S. Gypsum Company laboratories, manufactured in Company plants for maximum system performance.

General Limitations

1. U.S Gypsum Company joint compounds are not compatible with and should not be intermixed with any other compounds.

2. For interior use only except SHEETROCK Setting-Type (DURABOND) and SHEETROCK Lightweight Setting-Type (EASY SAND) Joint Compounds.

3. Not recommended for laminating except SHEETROCK Setting-Type (DURABOND) and SHEETROCK Lightweight Setting-Type (EASY SAND) Compounds and SHEETROCK Ready-Mixed Compounds—All Purpose and Taping.

4. Protect bagged and cartoned products against wetting; protect ready-mixed products from freezing and extreme heat.

5. Each compound coat must be dry before the next is applied (except SHEETROCK Setting-Type (DURABOND) and SHEETROCK Lightweight Setting-Type (EASY SAND) Compounds), and completed joint treatment must be thoroughly dry before decorating.

6. Use only SHEETROCK Setting-Type (DURABOND 90) or SHEETROCK Lightweight Setting-Type (EASY SAND 90) with 85-130 min. hardening time, or SHEETROCK Setting-Type (DURABOND 45) or SHEETROCK Lightweight Setting-Type (EASY SAND 45) with 30-80 min. hardening time for treating joints of Water-Resistant Gypsum Panels to be covered with ceramic or plastic tile.

7. For SHEETROCK Lightweight All Purpose Joint Compound (PLUS 3), SHEETROCK Topping Joint Compound Ready-Mixed and SHEETROCK Light-

weight All Purpose Joint Compound (AP LITE): If smoothing by dry sanding, use nothing coarser than 150 grit sandpaper or 220 grit abrasive mesh cloth.

8. For painting and decorating, follow manufacturer's directions for materials used. All surfaces must be thoroughly dry, dust-free and not glossy before decorating. SHEETROCK First Coat or good quality, high-solids, interior latex flat wall paint—used undiluted—should be applied and allowed to dry before decorating.

9. Gypsum panel surface should be skim coated with joint compound to equalize suction before painting in areas where gypsum panel walls and ceilings will be subjected to severe artificial or natural side lighting and be decorated with a gloss paint (egg shell, semi-gloss or gloss).

10. If dry sanding is used to smooth the joint compound, avoid roughening the gypsum panel face paper.

11. Do not use topping compound for taping or as first coat over bead.

12. Children can fall into bucket and drown. Keep children away from bucket with even a small amount of liquid. Do not reuse bucket.

SHEETROCK Ready-Mixed Drying-Type Joint Compounds

SHEETROCK Ready-Mixed Joint Compounds are drying-type products that are vastly superior to ordinary ready-mixed compounds and are preferred for consistently high-quality work. These vinyl-based formulations are specially premixed to a creamy, smooth consistency essentially free of crater-causing air bubbles. They offer excellent slip and bond, easy workability. Available for hand or machine-tool applications.

Limitation: must protect wet joints and container from freezing.

Four specialized products:

SHEETROCK Taping Joint Compound Ready-Mixed—A high-performance product for embedding tape and for laminating.

SHEETROCK Topping Joint Compound Ready-Mixed—A low-shrinkage, easily applied and sanded product recommended for second and third

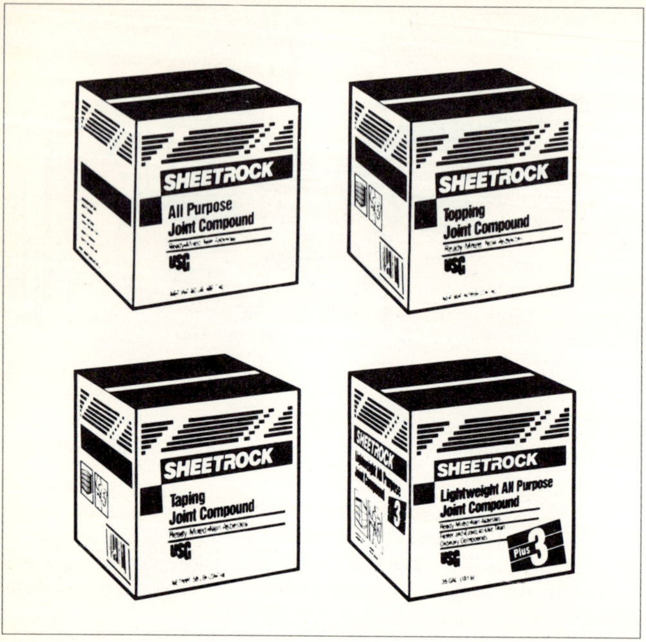

coats over SHEETROCK Taping and All Purpose Joint Compounds. Also used for simple texturing or skim coating. Not suitable for embedding tape or as first coat over metal corners, trim and fasteners.

SHEETROCK All Purpose Joint Compound Ready Mixed—Used for embedding, finishing, simple texturing and for laminating. Combines single-package convenience with good taping and topping characteristics. Recommended for finishing SHEETROCK brand Gypsum Panels, SW Edge joints over prefill of SHEETROCK Setting-Type (DURABOND) or Lightweight Setting-Type (EASY SAND) Compound; also for repairing cracks in interior plaster and masonry not subject to moisture.

SHEETROCK Lightweight All Purpose Joint Compound Ready-Mixed (PLUS 3)—Offers all the benefits of an all purpose compound, plus three exclusive advantages: up to 35% less weight, less shrinkage, and exceptional ease of sanding. Usually needs only two coats over metal. Eliminates need for separate taping and topping compounds—sands with ease of topping compound, bonds like taping compound. Used for laminating gypsum panels in fire-rated construction, for coating interior concrete ceilings and columns above grade, and for patching plaster cracks. Also good for texturing.

SHEETROCK Powder Drying-Type Joint Compounds

SHEETROCK Powder Joint Compounds are top-quality, non-asbestos, drying-type products providing easy mixing, smooth application and ample working time. Designed for embedding tape, for fill coats and finishing over drywall joints, corner bead, trim and fasteners. Also used for simple texture finishes. Included in line:

SHEETROCK Taping Joint Compound—Designed for embedding tape and for first fill coat on metal corner beads, drywall trims and fasteners; also used for patching plaster cracks. Offers excellent bond and resistance to cracking.

SHEETROCK Topping Joint Compound—A smooth-sanding material for second and third coats over taping or all-purpose compound. Produces excellent feathering and superior finishing results.

SHEETROCK All Purpose Joint Compound—Incorporates good taping and topping characteristics in a single product, for use where finest results of the specialized compounds are not necessary. Also has good texturing properties.

SHEETROCK Lightweight All Purpose Joint Compound (AP LITE)—All purpose compound weighs 20% less than conventional compounds; offers lower shrinkage, better crack resistance, easier mixing, application and sanding.

SHEETROCK Powder Setting-Type Joint Compounds

These setting-type powder products were developed to provide faster finishing of drywall interiors, even under slow drying conditions. Rapid chemical hardening and low shrinkage permit same-day finishing and usually next-day decoration. Features exceptional bond; virtually unaffected by humidity extremes. Ideal for laminating double-layer systems, particularly fire-rated assemblies, and for adhering gypsum panels to above-grade concrete surfaces. May be used for skim coating and surface texturing and for filling, smoothing and finishing interior above-grade concrete. Also used to treat joints in exterior gypsum ceiling board; as prefill material for SHEETROCK brand Gypsum Panels, SW Edge; for treating joints of SHEETROCK brand Gypsum Panels, Water-Resistant; treating fastener heads in areas to receive ceramic or plastic tile; and (except for SHEETROCK Lightweight Setting-Type Joint Compound) to embed tape and fill beads in veneer plaster finish systems when rapid drying conditions exist.

SHEETROCK Lightweight Setting-Type Joint Compound (EASY SAND)—Weighs 25% less than conventional setting-type compounds for easier handling, faster application and improved productivity on the job. Provides sanding ease similar to a ready-mixed, all purpose joint compound. Offers varied setting times of 20 to 30 min. (EASY SAND 20); 30 to 80 min. (EASY SAND 45); 85 to 130 min. (EASY SAND 90); 180 to 240 min. (EASY SAND 210); and 240 to 360 min. (EASY SAND 300).

SHEETROCK Setting-Type Joint Compound (DURABOND)—Available in a number of setting times to meet varying job requirements: 20-30 min. (DURABOND 20); 30 to 80 min. (DURABOND 45); 85 to 130 min. (DURABOND 90); 180 to 240 min. (DURABOND 210); and 240 to 360 min. (DURABOND 300).

SHEETROCK Setting-Type Joint Compound (DURABOND LC)—A low-consistency compound that permits same-day joint finishing and same-day decorating of drywall interiors. When applied over IMPERIAL Type P Tape, it provides as strong a joint in just two coats as regular SHEETROCK Setting-Type Compound and SHEETROCK Joint Tape provide in three coats. Also ideal for finishing joints in exterior gypsum ceiling boards and for coating interior concrete surfaces.

Setting-Type Joint Compound Limitations

1. Not to be applied over moist surfaces or surfaces likely to become moist, on below-grade surfaces, or on other surfaces subject to moisture exposure, pitting or popping.

2. SHEETROCK Setting-Type Compounds (DURABOND) are difficult to sand after drying and must be smoothed before complete setting.

Joint Compound Selection

Choosing the right joint compound for a specific job requires an understanding of a number of factors: job conditions, shop practices, applicators' preferences, types of available joint systems, characteristics of products considered, and recommended product combinations.

Joint compound products are usually named according to function, such as taping, topping and all-purpose. Taping typically performs as the highest shrinking, strongest bonding, hardest sanding of the three compounds, and is used for embedding tape. Topping usually is the lowest shrinking, easiest applying and sanding of the compounds for use in second and third coats; may occasionally be designed for texturing. Taping and topping are usually designed as companion products to give the highest quality workmanship. All purpose is generally a compromise of taping and topping and may be used as a texturing material. Lightweight All Purpose Joint Compound is also an all-purpose compound, but is lighter, shrinks less, sands easier.

Types of Joint Compounds

Two-Compound Systems—Formulated for superior performance in each joint finishing step. Separate taping compounds develop the greatest bond strength and crack resistance. Separate topping compounds have the best sanding characteristics, lower shrinkage and smoothest finishing.

All Purpose Compounds—Good performance in all joint finishing steps—do not have the outstanding bond strength, workability and sandability of separate taping and topping compounds. However, all-purpose compounds minimize inventories—avoid jobsite mix-ups—especially good for scattered jobs.

Ready-Mixed Compounds—Open-and-use convenience—save time and mistakes in mixing—minimum waste. Require minimal water supply at the job. Ready-mixed compounds have the best working qualities of all compounds—excellent performance plus factory-controlled batch consistency. Superior storage life compared to mixed powders.

These compounds do require heated storage. Should they freeze they can be slowply thawed at room temperature, mixed to an even viscosity and used without damaging effect. However, repeated freeze/thaw cycles cause remixing to become more difficult.

Powder Compounds—Have the special advantage of being storable (dry) at any temperature. If they are stored in a cold warehouse, however, they should be moved to a warm mixing room the day before they are to be mixed. Best results require strict adherence to proportioning of powder and water.

Specifications—SHEETROCK Joint Compounds

Product	Container size	Approx. coverage
SHEETROCK Ready-Mixed Joint Compound—Taping, Topping, All Purpose	12-lb. (5.4 kg), 42-lb. (19 kg) or 61.7-lb. (28 kg) pail; 48-lb. (21.8 kg), 50-lb. (22.7 kg) or 61.7-lb. (28 kg) carton	138 lb./1,000 ft.^{2} (67.4 kg/100 m^2)
SHEETROCK Lightweight All Purpose Joint Compound, Ready-Mixed (PLUS 3)	1 gal. (3.8L) or 4.5 gal. (17L) pail; 4.5 gal. (17L) or 3.5-gal. (13L) carton	9.4 gal./1,000 ft.^{2} (38.3L/100 m^2)
SHEETROCK Powder Joint Compound-Taping, Topping, All Purpose	25-lb. (11.3 kg) bag	83 lb./1,000 ft^2 (40.5 kg/100 m^2)
SHEETROCK Lightweight All Purpose Joint Compound, (AP LITE)	20-lb. (9 kg) bag	67 lb./1,000 ft.^{2} (32.7 kg/100 m^2)
SHEETROCK Setting-Type Joint Compound (DURABOND) 20, 45, 90, 210, 300, LC	25-lb. (11.3 kg) bag	72 lb./1,000 ft^2 (35.2 kg/100 m^2)
SHEETROCK Setting-Type Joint Compound (EASY SAND) 20, 45, 90, 210, 300	18-lb. (8.1 kg) bag	52 lb./1,000 ft^2 (25.3 kg/100 m^2)

Concrete Coatings

COVER COAT Compound—A vinyl-base product, formulated for filling and smoothing monolithic concrete ceilings, walls and columns located above grade—no extra bonding agent needed. Supplied in ready-mixed form

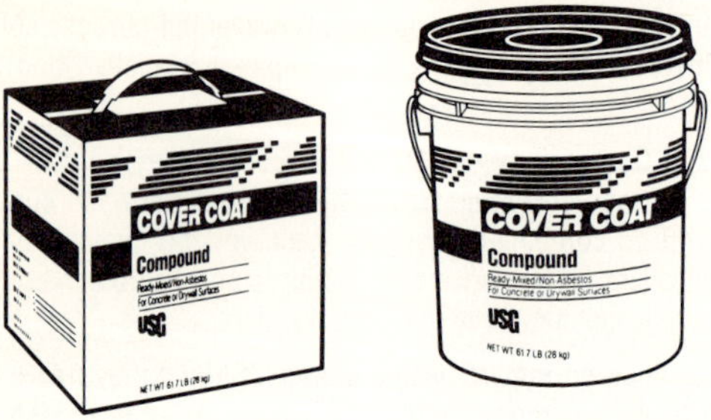

(sand can be added), easily applied with drywall tools in two or more coats. Dries to a fine white surface usually making ceiling painting unnecessary. Also can be used for embedding tape, for first coat over metal bead and trim, and for skim coating over gypsum panels.

Limitation: Not to be applied over moist surfaces or surfaces likely to become moist (from condensation or other source), on ceiling areas below grade, on surfaces that project outside the building, or any area that might be subject to moisture, freezing, efflorescence, pitting or popping.

SHEETROCK Setting-Type (DURABOND) and Lightweight Setting-Type (EASY SAND) Joint Compounds—These setting-type compounds are ideally suited to fill offsets and voids left in concrete; produce a hard finish in various shades of white. Overpainting may be required.

Where deep fills are required, SHEETROCK Setting-Type (DURABOND) and SHEETROCK Lightweight Setting-Type (EASY SAND) Compounds are especially recommended for the first coat, then followed by COVER COAT Compound. This practice minimizes check cracking.

Limitation: same as for COVER COAT Compound.

Reinforcing Tapes

From the originator of modern joint finishing, U.S. Gypsum Company reinforcing tapes add strength and crack resistance for smooth concealment of flat joints and inside corners. Two products—both quickly and easily applied—are available for specialized uses: paper tape for treatment with joint compounds; glass-fiber tape for veneer plaster finishes.

SHEETROCK Joint Tape—A special high-strength fiber tape for use with U.S. Gypsum Company joint compounds in reinforcing joints and corners in gypsum drywall and veneer plaster finish interiors. Exceptional wet and dry strength; resists stretching, wrinkling and other distortions; lies flat and resists tearing under tools. The wafer-thin tape is lightly sanded for increased bond and lies flat for easy concealment on next coat. Precision-processed with positive center creasing, which simplifies application in corners; uniform winding provides accurate, trouble-free attachment to angles and to flat joints.

Preferred for its consistent high performance in gypsum drywall finishing, SHEETROCK Joint Tape with SHEETROCK Setting-Type (DURABOND) Joint Compounds is also used with veneer plaster finish systems. Available nom. 1 31/32″ (50 mm) and 2 1/16″ (52 mm) wide in 75′, 250′, 500′ rolls. Approx. coverage: 370 lin. ft. tape per 1,000 sq. ft. panels.

A drywall joint treatment system (reinforcing tape and joint compound) must provide joints as strong as the gypsum board itself. Otherwise, normal structural movement in a wall or ceiling assembly can result in the development of cracks over the finished joint.

Repeated joint strength tests conducted at the USG Corporation Research Center have shown that joints taped and finished with conventional fiberglass leno-weave mesh tape and conventional joint compounds (ready-mixed, powder, and chemically setting) will crack when subjected to only one half the load and deflection as joints finished with paper tape and conventional joint compounds. This is because conventional fiberglass leno-weave mesh tapes tend to stretch under a load, even after being covered with these compounds.

Permanent repair of these cracks is difficult. Usually the fiberglass mesh tape must be removed and joints retaped and finished with paper tape. Accordingly, U.S. Gypsum Company does not recommend using conventional fiberglass leno-weave mesh tape with conventional ready-mixed, powder or chemically setting compounds for general drywall joint finishing.

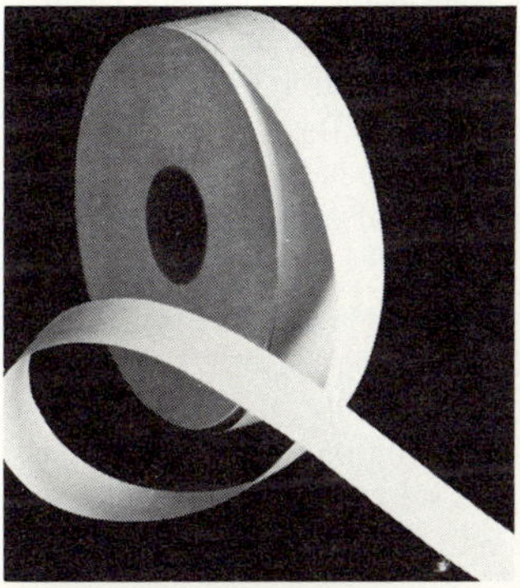

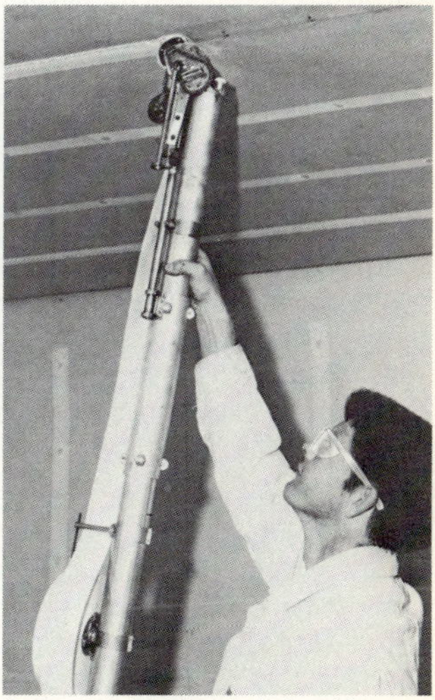

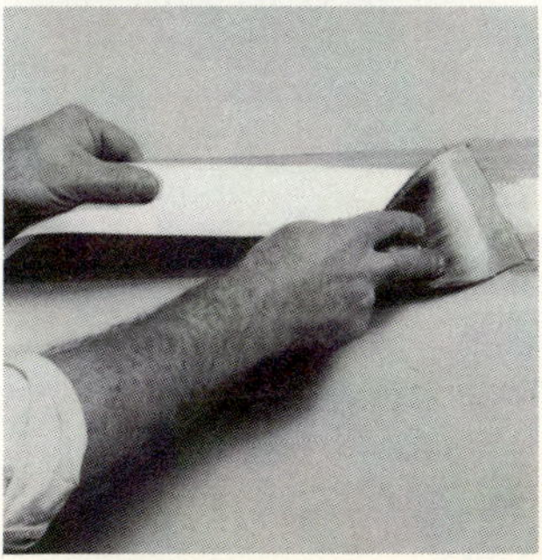

SHEETROCK Joint Tape is designed for both embedding by hand (left) and application with mechanical taping tool (above).

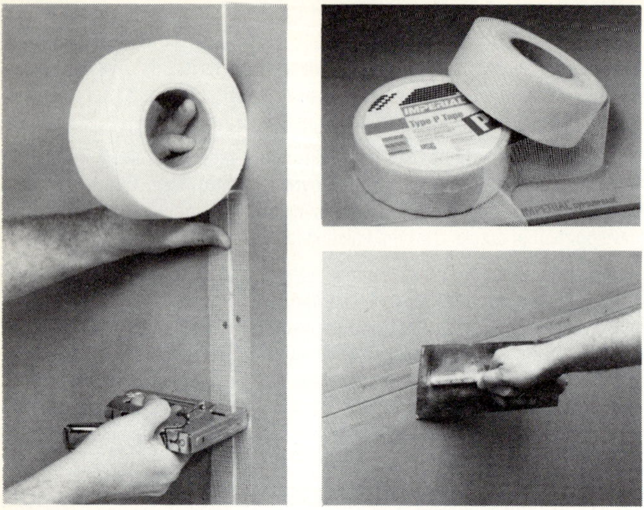

Both types of glass-fiber IMPERIAL Tape are quickly applied—Type S with ⅜″ staples at staggered 24″ intervals (left), self-stick Type P by light hand pressure and bonding with finishing knife or trowel (right). Use of Type P Tape cuts taping time up to 50%, simplifies embedding and saves cost of staples.

Where use of conventional fiberglass leno-weave mesh tape is desired, U.S. Gypsum Company has developed a system which consists of IMPERIAL Type P fiberglass tape and DURABOND LC, a special high-strength, low-consistency, chemically-setting powder compound. This system provides a satisfactory joint using only two coats, and equals the strength of joints finished with regular SHEETROCK Setting-Type (DURABOND) Compound and SHEETROCK Joint Tape in three coats.

IMPERIAL Tape—A strong, glass-fiber tape used in wood frame construction to conceal and reinforce joints and interior angles of IMPERIAL Gypsum Base prior to veneer plaster finishing with IMPERIAL Basecoat, IMPERIAL Finish Plaster, DIAMOND Veneer Basecoat and DIAMOND Interior Finish Plaster. High-tensile strength glass fibers are woven into an open mesh, coated with binder and slit to roll width.

The open weave of IMPERIAL Tape (100 meshes per sq. in.) provides excellent reinforcing and keying of plaster to resist cracking. The glass fibers lay flat and minimize stretching for wrinkle-free attachment without springback or distortion. Spirally woven (leno) long strands and the binder coating reduce edge raveling and fraying, keep loose threads from defacing finished surfaces. Tape flexes readily to permit fast application to flat joints and corners. Available in two types:

Type P with pressure-sensitive adhesive backing. Selected for quick, self-stick hand application; saves installation time and fastener cost.

Type S with plain back, fastened with staples. Lower in cost than Type P.

Available 2″ and 2½″ wide in 300′ rolls, 12 rolls per ctn. Approx. coverage: 370 lin. ft. tape per 1,000 sq. ft. gypsum base.

Veneer Plaster Finishes

Veneer plaster finishes offer the opportunity to trim days from interior finishing schedules and provide strong, highly abrasion-resistant surfaces. These products are designed for one or two-coat work over gypsum bases or directly to concrete block or properly prepared monolithic concrete. Formulated for hand or machine application (except for IMPERIAL Finish Plaster and DIAMOND Interior Finish Plaster hand only), they provide a thin, lightweight veneer plaster that sets rapidly.

Conventional plaster is the best system to attain a uniform, monolithic, blemish-free, smooth surface with excellent wear resistance. By contrast, one- and two-coat veneer plaster systems take advantage of the large size gypsum panels to improve speed of installation, while providing more monolithic appearing, harder, abuse-resistant surfaces than are achievable with drywall. Plaster thickness is reduced from the standard ½″ associated with conventional plaster to a mere ¹⁄₁₆″ to ⅛″ using high-strength gypsum in the product formulations. While Keenes cement-lime-sand provides the most universal texture finish in two-coat application, IMPERIAL Finish Plaster and DIAMOND Interior Finish Plaster provide better surface hardness, abrasion-resistance and wearability. Final finish is in as little as 48 hours and is easily textured. (See “Comparing Plaster Systems” on page 471.)

Advantages

Rugged, Abuse-resistant Surfaces—High-strength IMPERIAL Finishes (3,000 psi compressive strength) provide hard, durable interiors that require minimum maintenance.

Quicker Completion/Faster Occupancy—Veneer plaster finishes apply rapidly, set fast, dry quickly to save days in finishing interior walls and ceilings. DIAMOND Interior Finish Plaster can be decorated in 24 hrs. with breather-type paint or left undecorated if desired.

Competitive Costs—Veneer plaster finishes are easily applied and cover more area per ton than conventional plasters. Joints and interior angles are pre-set with the same veneer plaster finish that goes on the walls and ceilings.

Easily Decorated—Veneer plasters are readily finished in smooth-trowel, float or texture surfaces. The hard, smooth surface is decorated easily and economically with paint, fabric, wallpaper or texture.

Versatile—A wide choice of assemblies is available to meet design requirements. Fire and sound-rated systems for wood or steel framing, hard and abuse-resistant surfaces for high-traffic areas, electrically heated ceilings.

Products Available

IMPERIAL Basecoat—For use as a basecoat in two-coat veneer plaster application finished with proper lime or gypsum finishes. Can be applied to either IMPERIAL Gypsum Base, directly to concrete block, or over a bonding agent on monolithic concrete. Formulated as the basecoat for high-strength

IMPERIAL Finish Plaster, gauged lime putty, DIAMOND Interior Finish Plaster, STRUCTO-GAUGE-lime-smooth trowel, or Keenes-lime-sand float finishes.

Available in hand and machine-application formulations. Complies with ASTM C587. Available in 80-lb. bags.

IMPERIAL Finish Plaster—For single-coat application composed of scratch coat and immediate doubling back directly over special IMPERIAL Gypsum Base, glass-fiber tape or SHEETROCK Joint Tape and SHEETROCK Setting-Type Joint Compound (DURABOND). Also used over IMPERIAL Basecoat Plaster in a two-coat system. Available for hand application—provides a smooth-trowel or float or spray-texture finish ready for decoration. Complies with ASTM C587. Available in 80-lb. bags.

IMPERIAL Basecoat and Finishes—Coverage

	ft²/ton		m²/ton (metric)(1)	
Product	**gypsum base**	**masonry**	**gypsum base**	**masonry**
IMPERIAL Basecoat	3250-4250	2700-3600	335-435	275-370
IMPERIAL (1-coat) Finish	3500-4000	not recommended	360-410	not recommended
IMPERIAL (2-coat) Finish	3200-3600	not applicable	330-370	not applicable

(1) Coverage rounded to nearest 5m² per metric ton.

DIAMOND Veneer Basecoat Plaster—Provides quality walls and ceilings for residential and commercial construction where superior strength of IMPERIAL Basecoat Plaster is not essential. Offers superior workability—ease and speed of application. Formulated to receive a variety of finishes. Apply to IMPERIAL Gypsum Base, concrete block or monolithic concrete. Complies with ASTM C58. Available in 79.4-lb. (36 kg) bags.

DIAMOND Basecoat—Coverage

	ft²/ton		m²/ton (metric)(1)	
Product	**gypsum base**	**masonry**	**gypsum base**	**masonry**
DIAMOND Basecoat	4000-5000	3500-4500	410-510	360-460

DIAMOND Interior Finish Plaster—A white finish formulated for hand application directly to IMPERIAL Gypsum Base or over a bonding agent on monolithic concrete. Also suitable in a two-coat system over IMPERIAL or DIAMOND Basecoat or a sanded gypsum basecoat. Applied to a nom. 1/16" thickness, this finish is unaggregated for a smooth or skip-trowel finish; may be job-aggregated with up to an equal part by weight of clean silica sand for Spanish, swirl, float or other textures. Not recommended for use over portland cement basecoat or masonry surfaces. Complies with ASTM C587. Available in 50-lb. bags.

DIAMOND Interior Finish Plaster should be applied only to IMPERIAL Gypsum Base having blue face paper. Faded base must be treated with alum solution or a bonding agent before finish is applied to prevent possible bond failure. See page 212 for specific instructions.

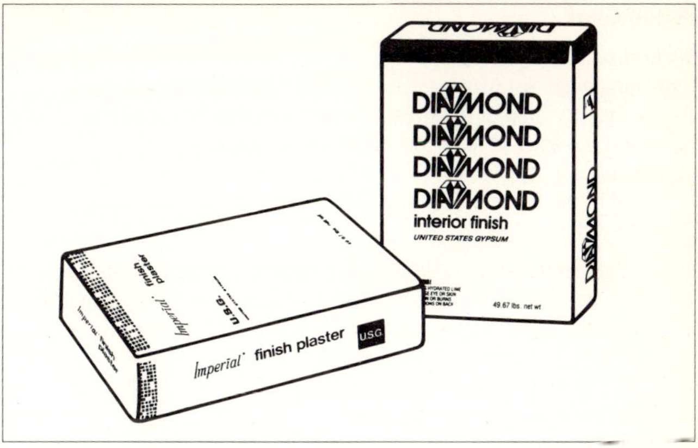

DIAMOND Interior Finish Plaster is also suitable for use with electric cable ceilings. Allows higher operating temperatures than with other products, provides more heat transmission and greater resistance to heat deterioration. Finish is job-sanded and hand-applied 3/16″ thick to cover cable. A finish coat of the same material is applied 1/16″ to 3/32″ thick to bring the total plaster thickness to 1/4″. Applied over IMPERIAL Base attached to wood joists, to metal furring channel or suspended metal grillage; or over a bonding agent directly to monolithic concrete ceilings (5/16″ fill coat plus finish coat for 3/8″ total thickness).

DIAMOND Interior Finish Plaster Coverage

Conventional walls and ceilings

Surface applied to	neat		sand float finish sanded 1:2 (sand: DIF)(1)		heavy texture finish sanded 1:1 (sand: DIF)(1)	
	ft²/ton	m²/ton(2)	ft²/ton	m²/ton(2)	ft²/ton	m²/ton
IMPERIAL Gypsum Base	6000	610	4660	475	3500	355
IMPERIAL or DIAMOND Basecoat	5500	560	4330	440	3250	330
Sanded RED TOP Basecoat	5000	510	4000	410	3000	305
Monolithic concrete	5500	560	4330	440	3250	330

Electric cable heat ceilings

Surface applied to	fill coat(3) sanded 1:1 (sand: DIF)(1)		1/16″ finish coat sanded 1:4 (sand: DIF)(1)		1/16″ finish coat sanded 1:1 (sand: DIF)(1)	
	ft²/ton	m²/ton(2)	ft²/ton	m²/ton(2)	ft²/ton	m²/ton(2)
IMPERIAL Gypsum Base	2300	235	5000	510	3250	330
Monolithic concrete	900	84	5500	560	4500	418

(1) Coverage based on one ton of aggregated mixture (combined weight of sand and DIAMOND Interior Finish Plaster). (2) Coverage rounded to nearest 5m² per metric ton. (3) Fill coat over gypsum base is 3/16″ thick, over monolithic concrete is 5/16″.

Prime Coat

SHEETROCK First Coat—Decorating problems such as "joint banding" or "photographing" are usually caused by differences between the porosities and surface textures of the gypsum board face paper or concrete on one hand, and the finished joint compound on the other. SHEETROCK First Coat for priming is a flat latex basecoat paint-type product especially formulated to provide a superior first (prime) coat over interior gypsum board and concrete surfaces.

In contrast to a sealer, SHEETROCK First Coat does not form a film that seals the substrate surface. Instead, it minimizes porosity differences by providing a base that equalizes the absorption rates of the drywall face paper and the finished joint compound when painted. SHEETROCK First Coat also provides the proper type and amount of pigments and fillers, lacking from conventional primers and sealers, that minimize surface texture variations between the gypsum board face paper and the finished joint compound.

SHEETROCK First Coat is designed for fast, low-cost application. Applies with brush, roller or airless or conventional sprayer. Dries to the touch in less than 30 minutes under 72° F/50% R.H. conditions. White finish is ready for decoration in an hour. Non-asbestos; safe to handle and use. Not intended as a final coating—it should be overpainted when dry. The product comes ready-mixed in 5-gal. and 1-gal. pails or packaged dry in 25-lb. bags.

Interior Texture Finishes

Texture finishes from U.S. Gypsum Company offer a wide variety of possible texture patterns to provide distinctive interior styling. Fast, easy application; quick drying. Hide minor surface blemishes to reduce surface preparation needed. Save labor time to preserve job profits. All products are non-asbestos.

Powder Texture Products

IMPERIAL QT Spray Texture Finish (F-Fine) (P-Medium) (PC-Coarse) (PS-Super Coarse)—A powder product with polystyrene aggregate, available in four finishes. Produces a handsome simulated acoustical ceiling finish but with no acoustical correction. Requires only addition of water and short soaking period at job site. Produces excellent bonding qualities for application to gypsum panels, concrete, plaster or wood. High wet and dry-hide masks minor surface defects. Dries to exceptional whiteness, usually left unpainted but may be overpainted if desired. Not washable but can be painted when redecoration is needed.

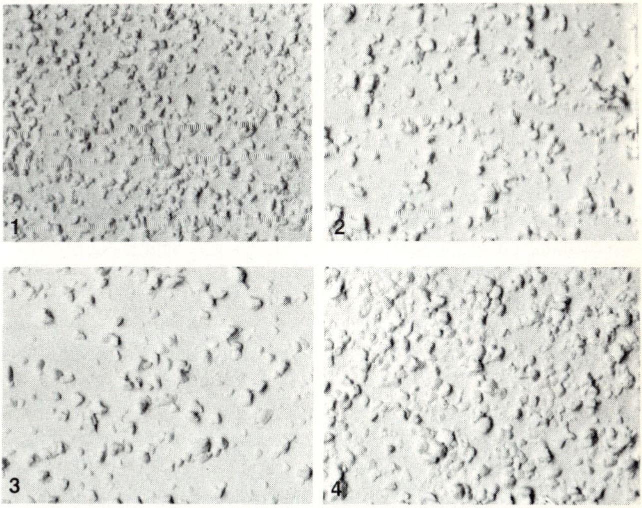

Examples of high-style textures produced by IMPERIAL QT Spray Texture Finish: (1) Fine Finish for light effect; (2) Medium Finish for striking texture; (3) Coarse Finish for unusual decorating effect; (4) Super Coarse Finish for dramatic treatment.

USG Spray Texture Finish—A product available in aggregated and unaggregated forms for texture variety on most interior wall surfaces. Produces light spatter, fog-and-spatter and light orange peel texture with spray application. Superior wet and dry bond ensures good holdout of finish film over

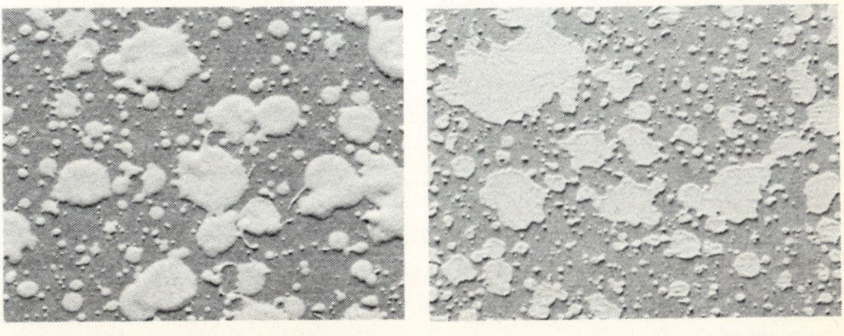

Surface designs available with USG Spray Texture Finish include Spatter Finish (left) and Spatter/Knockdown Finish (right).

dry joints. Often applied in two coats with second coat producing texture such as fog-and-spatter finish. Dries to a soft-tone white surface with good concealment. Should be overpainted when dry on walls. Can be left unpainted on ceilings if adequate amount of material is applied to provide sufficient hiding properties. Not washable unpainted.

SHEETROCK Wall and Ceiling Spray Texture (TUF TEX)—An unaggregated texture coating. Produces a variety of texture patterns from bold spatter/knockdown to light orange peel. Dries to a hard, white finish. Helps conceal minor substrate defects. Not intended as a final coating—should be overpainted when dry. Not washable unpainted.

Distinctive medium stipple texture is achieved with TUF TEX Spray Texture.

USG Multi-Purpose Texture Finish—An economical, unaggregated, powder product, to be mixed with water for desired texturing consistency. Excellent for producing a variety of light to medium-light textures on drywall or other interior surfaces. Textured effect obtained by brush, roller or spray application. Helps conceal minor surface defects; dries to a soft-tone white finish; should be overpainted on walls; can be left unpainted on ceilings when adequate material is applied. Not washable unpainted.

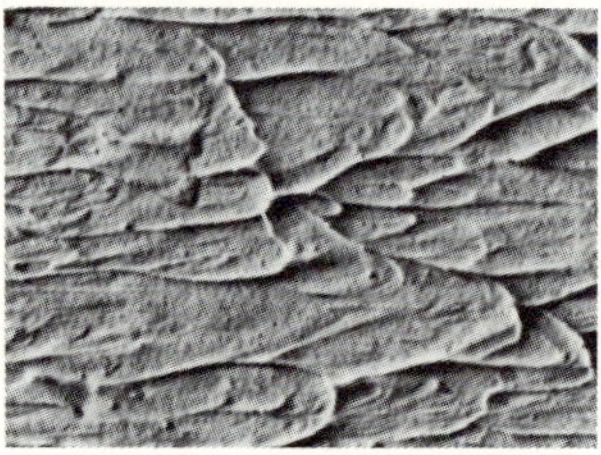

Variety of effects obtained with USG Multi-Purpose Texture Finish include (clockwise from above) bold shadowing with roller application, medium-light finish applied by spray, lightly stippled surface applied with small brush or roller-stippler.

USG Texture XII Drywall Surfacer—A white coating for spray application. Available in aggregated form to achieve a sand-finish effect. Conceals minor surface defects—provides a uniform texture with good hide. Ideal base for wall paints. Can be left unpainted on ceilings. Not washable unpainted.

Close-up view shows typical sand-effect finish obtained with aggregated USG Texture XII Drywall Surfacer. In application, fan technique is used on walls, cross-spray on ceilings.

SHEETROCK Powder Joint Compound (All Purpose)—Easy-mixing, smooth-working products that can be used to produce attractive light to medium textures. Color is white but may vary in degree of whiteness. Surfaces should be painted. Applied with brush, roller or trowel. Not washable unpainted.

Simple roller-applied texture is obtained with vinyl-base SHEETROCK Powder Joint Compounds. Same products can be used for joint finishing and texturing on job.

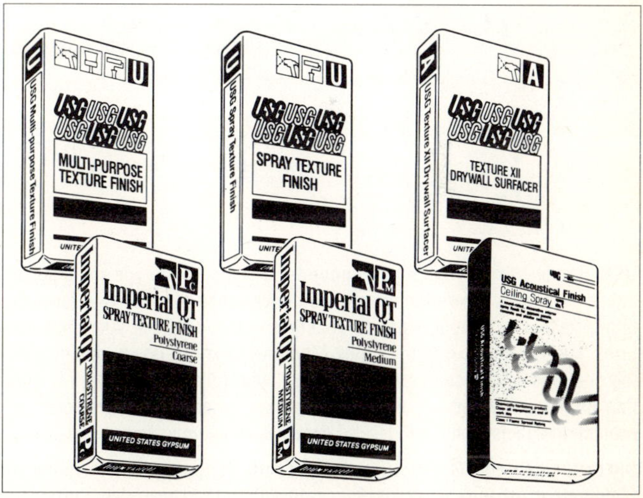

Sound-Rated Texture Finish

USG Acoustical Finish—An exclusive interior spray-on texture that gives a sound-absorbing, sound-rated decorative finish to gypsum panels, concrete and nonveneer-type plaster ceilings and other noncontact surfaces. Produces a handsome, natural-white, evenly textured finish. Requires no application of a bonding agent. Reduces surface preparation time and costs in new construction or remodeling. Use on noncontact surfaces only. USG Acoustical Finish is a chemically setting-type product that is totally different from conventional drying-type simulated acoustic texture. Sound rated: NRC .50 for gypsum panels, .55 for concrete, .55 for conventional plaster at ½″ finish thickness. Designated Class A for surface flame spread per ASTM E1042-85. Flame spread 10, smoke developed 25. Not washable but can be painted when redecoration is needed.

USG Acoustical Finish absorbs sound and gives dramatic appeal to ceilings and other noncontact surfaces.

Ready-Mixed Texture Products

USG QUIK & EASY Ready-to-Use Wall and Ceiling Texture—Offers unique super thickness with just one coat, plus the fast start of ready-mixed material. Massive thickness in just one pass eliminates doubling back. This white, latex-type finish develops a tough, durable surface with stubborn resistance to fissure cracks. Trowel, roller, brush or spray application. Bonds well with excellent hide over many surfaces—gypsum panels, concrete, primed plaster, interior masonry and non-staining wood surfaces. Can be aggregated on the job. Painting not required on non-contact surfaces. Overpaint if desired. Not washable unpainted.

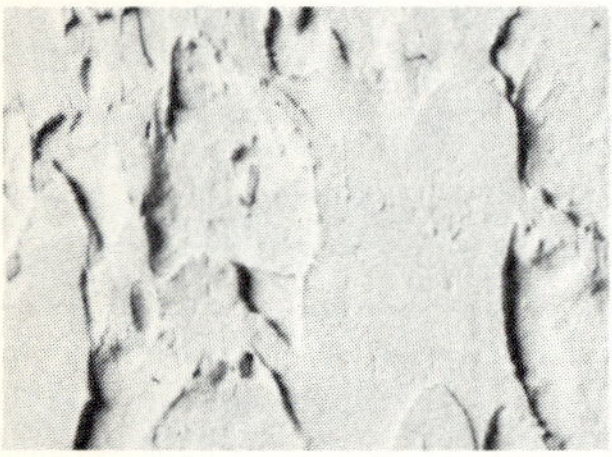

Extra-thick finish applied in one coat provides eye appeal and decorative charm.

USG Ready-Mixed Texture Compound—A white, latex-type material for interior surfaces, offers the speed and convenience of ready-mix formulation. Provides extra tough and durable finish up to $\frac{1}{16}$″ thick. Attractive textures, such as stomp or crow's foot, Monterey or knock-down, orange peel, fog coat and others can be readily created with roller, brush or spray gun. Can be left unpainted on remote noncontact surfaces, or overpainted for protection against soiling, if desired. Not washable unpainted.

SHEETROCK Ready-Mixed Joint Compounds (Topping, All Purpose or Lightweight All Purpose-PLUS 3)—Virtually ready to use, these products will produce textures ranging from light to medium depending upon method of application. Color is white, but may vary. Surfaces should be painted. Applied with brush, roller or trowel. Not washable unpainted. Check local sales office for suitability of joint compound formulation for texturing in your area.

Tables on the following pages give complete information on texture uses and the finishes produced.

all & Veneer
ter Construction

TER 2 Framing

Texture/ Compound Selector

PRODUCT	SURFACES: prime coat required	SURFACES: ceilings	SURFACES: walls	PROPERTIES: type of aggregate	PROPERTIES: aggregate size	PROPERTIES: ability to hide substrate imperfections	PROPERTIES: water dilution gal/lb	PROPERTIES: solution time	APPLICATION: machine	APPLICATION: hand	SPRAY EQUIPMENT: pole gun	SPRAY EQUIPMENT: 7E2 type texture gun	SPRAY EQUIPMENT: 18D type texture gun	SPRAY EQUIPMENT: hopper gun	SPRAY EQUIPMENT: aggregate fallout (bounce)	SPRAY EQUIPMENT: abrasiveness on equipment	FEATURES: drying time	FEATURES: bond of dry aggregate	FEATURES: dried whiteness	FEATURES: crack resistance	FEATURES: coverage ft²/lb-spray(1)	FEATURES: coverage ft²/lb-hand(1)
IMPERIAL QT Spray Texture Finishes																						
Polystyrene	yes	yes	no	poly-styrene	fine	good	var-ies**	very good	yes	no	yes	yes	no	yes	min. to mod.	min.	slow-med.	mod.	excel-lent	good	up to 8	N/A
Polystyrene	yes	yes	no	poly-styrene	med.	excel-lent	var-ies**	very good	yes	no	yes	yes	no	yes	min. to mod.	min.	slow-med.	mod.	excel-lent	good	up to 8	N/A
Polystyrene	yes	yes	no	poly-styrene	coarse	excel-lent	var-ies**	very good	yes	no	yes	yes	no	yes	min. to mod.	min.	slow-med.	mod.	excel-lent	good	up to 8	N/A
Polystyrene	yes	yes	no	poly-styrene	super coarse	excel-lent	var-ies**	very good	yes	no	yes	yes	no	yes	min. to mod.	min.	slow-med.	mod.	excel-lent	good	up to 8	N/A

** Varies—see Chapter 4. Also see footnotes, page 97.

continued on next page

Texture/ Compound Selector
continued

PRODUCT	SURFACES: prime coat required	ceilings	walls	PROPERTIES: type of aggregate	aggregate size	ability to hide substrate imperfections	water dilution gal/lb[4]	APPLICATION: solution time	machine	hand	SPRAY EQUIPMENT: pole gun	7E2 type texture gun	18D type texture gun	hopper gun	aggregate fallout (bounce)	abrasiveness on equipment	FEATURES: drying time	bond of dry aggregate	dried whiteness	crack resistance	coverage ft²/lb-spray[1]	coverage ft²/lb-hand[1]
USG Texture Finishes																						
USG Spray Texture (agg.)	yes	yes	yes	per-lite	fine-med.	very good	4-5/50 3-4/40	good	yes	yes	no	yes	yes	yes	min.	mod.	very fast	good	good	good	up to 40	N/A
USG Spray Texture (unagg.)	yes	yes	yes	N/A	N/A	good	4-5/50 3-4/40	good	yes	yes	no	yes	yes	yes	N/A	min.	fast	N/A	good	good	†	N/A
SHEETROCK Wall & Ceiling Spray (TUF TEX)	yes	yes	yes	N/A	N/A	good	4-4.8/40	good	yes	yes	no	yes	yes	yes	N/A	min.	fast	N/A	good	good	up to 40	10-20
USG Multi-Purpose	yes	yes	yes	N/A	N/A	good	2-3/25 3-4/40	good	yes	yes	no	yes	yes	yes	N/A	min.	fast	N/A	good	good	up to 20	10-15
USG Drywall Surfacer Texture XII	yes	yes	yes	per-lite	fine	good	2½-3¼ /25	good	yes	no	no	no	yes	yes	min.	mod.	fast	excel-lent	very good	good	20-35	N/A
USG Acoustical Finish	yes	yes	no	poly-styrene	fine-med.	excel-lent	3.5 gal. 30 lb.	good	yes	no	yes	yes	no	no	min.-mod.	min.	slow	very good	good	excel-lent	1½-3	N/A
USG QUIK & EASY Ready-to-Use Wall & Ceiling Texture	yes*	yes	yes	N/A	N/A	excel-lent	up to 3 pts./ 3.5 gal.	N/A	yes	yes	yes	yes	yes	yes	N/A	min.	slow-med.	N/A	very good	good	70 sq. ft./ gal.	25-70 sq. ft./ gal.

Texture/ Compound Selector
continued

PRODUCT	SURFACES: prime coat required	ceilings	walls	PROPERTIES: type of aggregate	aggregate size	ability to hide substrate imperfections	water dilution gal/lb[4]	APPLICATION: solution time	machine	hand	SPRAY EQUIPMENT: pole gun	7E2 type texture gun	18D type texture gun	hopper gun	aggregate fallout (bounce)	abrasiveness on equipment	FEATURES: drying time	bond of dry aggregate	dried whiteness	crack resistance	coverage ft²/lb-spray[1]	coverage ft²/lb-hand[1]
USG Texture Finishes (continued)																						
USG Ready-Mixed Texture Compound	yes*	yes	yes	N/A	N/A	very good	½-2 /50	N/A N/A	yes	yes	no	yes	yes	yes	N/A	min.	slow-med.	N/A	fair	good	7-8	4-6
Joint Compounds[2]																						
SHEETROCK Powder All Purpose	yes	yes	yes	N/A	N/A	very good	2¼-2¾ /25	good	N/A	yes	N/A	N/A	N/A	N/A	N/A	min.	slow-	N/A				
SHEETROCK Ready-Mixed Topping[3] or All Purpose	yes*	yes	yes	N/A	N/A	very good	1-1½ /62	N/A	N/A	yes	N/A	N/A	N/A									
SHEETROCK Lightweight All Purpose Joint Compound (PLUS 3)	yes*	yes	yes	N/A	N/A																	

Dryw
Plas
CHAP

General Requirements

The choice and installation of framing depends on a number of factors. In the case of wood framing these include the species, size and grade of lumber used. In the case of steel framing, the configuration of the frame member, size, and the thickness and grade of steel must be considered. Equally important are frame spacing and the maximum span of the surfacing material. Selection of steel stud size is usually derived from limiting height tables given in U.S. Gypsum Company technical literature based on capacity of steel and allowable deflection of finish surfaces.

Loads—Framing members and their installation must be selected according to their ability to withstand the loads to which they will be subjected. These include live loads (contributed by the occupancy and elements such as wind, snow and earthquake) and dead loads (weight of the structure itself). Minimum load for interior partition is 5 psf; for exterior walls 15 psf to 45 psf or greater depending on building height and geographic location.

Deflection—Even though an assembly is structurally capable of withstanding a given load, its use may be restricted if the amount of deflection that would occur when the load is applied exceeds that which the surfacing materials can sustain without damage. Obviously, this deflection factor influences the selection of surfacing materials.

For drywall assemblies it is desirable to limit deflection to L/240 and to never exceed L/120 (L/180 in some codes). The preferred limit for veneer plaster assemblies is L/360 and should not exceed L/240. Using L/240 as an example, and where the length of a span (distance between framing members) is 10′, deflection is figured as follows:

$$D = \text{Deflection Limit} = \frac{L}{240}$$

$$D = \frac{120}{240}$$

$$D = 0.5 \text{ in.}$$

Bending Stress—Framing members also must withstand any unit force exerted that will break or distort the stud based on the capacity of the studs acting alone.

End Reaction Shear—This factor is determined by the amount of force applied to the stud that will bend or shear the runner, or cripple the stud.

Frame Spacing—A factor in load-carrying capability and deflection, it also is a limiting factor for the finishing materials. Every finishing or surfacing material is subject to a span limitation—the maximum distance between frame members that a material can span without undue sagging. For that reason, "maximum frame spacing" tables for the various board products are included in this Chapter. However, where frame spacing exceeds maximum limits, furring members can be installed to provide necessary support for the surfacing material (covered in this Chapter under wall and ceiling furring).

Insulation and Services—Chase walls provide vertical shafts where greater core widths are needed for pipe chase enclosures and other service installations. They consist of a double row of studs with gypsum panel cross braces between rows. Plumbing, electrical and other fixtures, and mechanicals within the framing cavities must be flush with or inside the plane of the framing. Fasteners used to assemble the framing must be driven reasonably flush with the surfaces.

In wood frame construction, the flanges of batt-type insulation must be attached to the sides of frame members and not to their faces. Any obstruction on the face of frame members that will prevent firm contact between the gypsum board and framing can result in loose or damaged board and fastener imperfections.

Wood Framing

Wood framing meeting the following minimum requirements is necessary for proper performance of all gypsum drywall and veneer plaster base assemblies:

1. Framework should meet the minimum requirements of applicable building codes.

2. Framing members should be straight, true and of uniform dimension. Studs and joists must be in true alignment; bridging, fire stops, soil pipes, etc., must not protrude beyond framing.

3. All framing lumber should be the correct grade for the intended use, and 2x4 nominal size or larger and should bear the grade mark of a recognized inspection agency.

Nailing of "scab" to straighten warped wood stud.

4. All framing lumber should have a moisture content not in excess of 19% at time of gypsum board application.

Failure to observe these minimum framing requirements, which are applicable to screw, nail and adhesive attachment, will materially increase the possibility of fastener failure and surface distortion due to warping or dimensional changes. This is particularly true if the framing lumber has greater than normal tendencies to warp or shrink after erection.

The moisture content of wood framing should be allowed to adjust as closely as possible to the level it will reach in service before gypsum panel or veneer plaster base application begins. After the building is enclosed, delay board application as long as possible (consistent with schedule requirements) to allow this moisture content adjustment to take place.

Framing should be designed to accommodate shrinkage in wide dimensional lumber such as is used for floor joists or headers. Gypsum wallboard and veneer plaster surfaces can buckle or crack if firmly anchored across the flat grain of these wide wood members as shrinkage occurs. With high uninterrupted walls such as are a part of cathedral ceiling designs or in two-story stairwells, regular or modified balloon framing can minimize the problem.

Framing Corrections—If joists are out of alignment, 2″ x 6″ leveling plates attached perpendicular to and across top of ceiling joists may be used. Toe-nailing into joists pulls framing into true horizontal alignment and ensures a smooth, level ceiling surface. Bowed or warped studs in non-load bearing partitions may be straightened by sawing the hollow sides at the middle of the bow, pulling stud back into line and inserting a wedge into the saw kerf to fill the void. Reinforcement of the stud is accomplished by securely nailing 1″ x 4″ wood strips or "scabs" on each side of the cut.

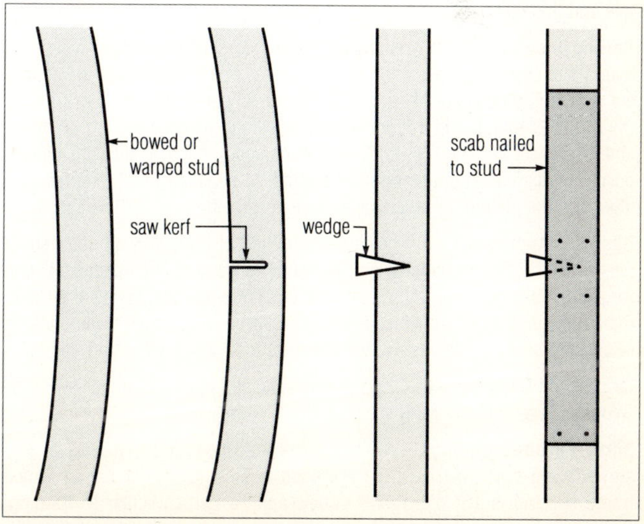

Maximum Frame Spacing—Drywall Construction

Direct Application

Panel thickness(1)	Location	Application method(2)	Max. frame spacing o.c.	
SINGLE-LAYER APPLICATION			in	mm
⅜″ (9.5 mm)	ceilings(3)	perpendicular(4)	16	406
	sidewalls	parallel or perpendicular	16	406
½″ (12.7 mm)	ceilings	parallel(4)	16	406
		perpendicular	24(5)(6)	610
	sidewalls	parallel or perpendicular	24	610
⅝″ (15.9 mm)	ceilings(6)	parallel(4)	16	406
		perpendicular	24	610
	sidewalls	parallel or perpendicular	24	610
DOUBLE-LAYER APPLICATION				
⅜″ (9.5 mm)	ceilings(7)	perpendicular	16	406
	sidewalls	perpendicular or parallel	24(8)	610
½″ & ⅝″ (12.7 & 15.9 mm)	ceilings	perpendicular	24(8)	610
	sidewalls	perpendicular or parallel	24(8)	610

(1) A ⅝″ thickness is recommended for the finest single-layer construction, providing increased resistance to fire and transmission of sound; ½″ for single-layer application in new residential construction and remodeling; and ⅜″ for repair and remodeling over existing surfaces. (2) Long edge position relative to framing. (3) Not recommended below unheated spaces. (4) Not recommended if water-based texturing material is to be applied or any surface treatment that will dry slowly. (5) Max. spacing 16″ if water-based texturing material to be applied. (6) If ½″ SHEETROCK brand Interior Gypsum Ceiling Board may be used in place of ⅝″ gypsum panels, max. spacing is 24″ o.c. for perpendicular application. (7) Adhesive must be used to laminate ⅜″ board for double-layer ceilings. (8) Max spacing 16″ o.c. if fire rating required.

Installation

For wood framing installed in the conventional manner with lumber meeting requirements outlined above, maximum frame spacing is as shown on this and the next page.

Ceiling Insulation—To prevent objectionable sag in ceilings, weight of overlaid unsupported insulation should not exceed 1.3 psf for ½″ thick panels with frame spacing 24″ o.c.; 2.2 psf for ½″ panels on 16″ o.c. framing (or ½″ SHEETROCK brand Interior Gypsum Ceiling Board on 24″ o.c. framing) and ⅝″ panels 24″ o.c.; ⅜″ thick panels must not be overlaid with unsupported insulation. A vapor retarder should be installed in all exterior ceilings, and the plenum or attic space properly vented.

Resilient Application—On ceiling assemblies of both drywall and veneer plaster, install RC-1 Channels perpendicular to framing and spaced 24″ o.c. for joists 16″ o.c.; 16″ o.c. for joists 24″ o.c. For sidewalls, install at 24″ o.c. max. See single-layer sections in tables and preceding pages for limitations of specific board thickness. Fasten channels to framing with screws only.

Cable Heat Ceilings—Maximum frame spacing is 16″ o.c. for ½″ IMPERIAL Gypsum Base; 24″ o.c. for ⅝″ base.

Spray-Textured Ceilings—Where water-based texturing materials or any slow-drying surface treatment are used over single-layer panels, max. frame spacing is 16″ o.c. for ½″ panels applied perpendicular to framing.

Maximum Frame Spacing—Veneer Plaster Construction

Direct Application

Gypsum base thickness	Construction	Location	Application method[1]	Max. frame spacing o.c. in	mm
	one layer 1-coat finish	ceilings	perpendicular	16	406
		sidewalls	perpendicular or parallel	16	406
½″ (12.7 mm)	one layer, 2-coat finish	ceilings	perpendicular	16 or 24[2]	406 or 610[2]
		sidewalls	perpendicular or parallel	16 or 24[2]	406 or 610[2]
	two layer, 1 & 2-coat finish	ceilings	perpendicular	24	610
		sidewalls	perpendicular or parallel	24	610
	one layer, 1-coat finish	ceilings	perpendicular	16 or 24([2]	406 or 610[2]
		sidewalls	perpendicular or parallel	16 or 24([2]	406 or 610[2]
⅝″ (15.9 mm)	one layer, 2-coat finish	ceilings	perpendicular	24[2]	610[2]
		sidewalls	perpendicular or parallel	24[2]	610[2]
	two layer, 1 & 2-coat finish	ceilings	perpendicular	24	610
		sidewalls	perpendicular or parallel	24	610

(1) Perpendicular preferred on all applications for maximum strength. Where fire rating is involved, application must be identical to that in assembly tested. Parallel application not recommended for ceilings. (2) 24″ o.c. frame spacing with either one- or two-coat veneer application requires SHEETROCK Joint Tape and SHEETROCK Setting-Type (DURABOND) Joint Compound.

Parallel application is not recommended, nor is use of ⅜″ thick panels. For best results use SHEETROCK brand Interior Gypsum Ceiling Board with max. spacing 24″ o.c. *Note:* Airless spraying of latex paint in one heavy application (10 to 14 mil.) will also sag ceilings.

See "Ceiling Sag Precautions" on page 362 for more information on the application of water-based textures and interior finishing materials.

Partition Layout

Properly position partitions according to layout. Snap chalk lines at ceiling and floor. Be certain that partitions will be plumb. Where partitions occur parallel to and between joists, ladder blocking must be installed between ceiling joists.

Steel Framing

Steel stud framing for non-load bearing interior partitions is secured to floors and ceilings with runners fastened to the supporting structure.

Runner Installation

Securely attach runners:

1. To concrete and masonry—use stub nails, power-driven fasteners, or the TAPCON Concrete Fastening System.

2. To foam-backed metal (max. 14-ga.) concrete inserts—use ⅜″ Type S-12 pan head screws.

3. To suspended ceilings—use expandable hollow wall anchors or toggle bolts.

4. To wood framing—use 1¼″ Type S oval head screws or 12d nails.

To all substrates, secure runners with fasteners located 2″ from each end and spaced max. 24″ o.c. Attach runner ends at door frames with two anchors when 3-piece frames are used. (One-piece frames should be supplied with welded-in-place floor anchor plates, pre-punched for two anchors into substrate.)

At partition corners, extend one runner to end of corner and butt or overlap other runner to it, allowing necessary clearance for gypsum panel thickness. Runners should not be mitered.

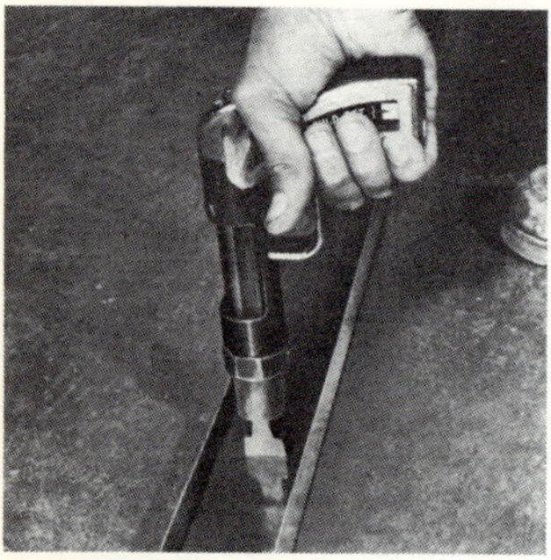

Fastening channel runners

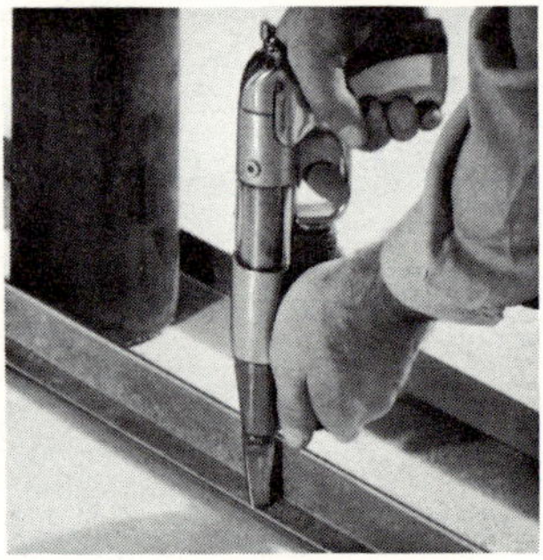

Fastening angles

Stud Installation

Insert floor-to-ceiling UNIMAST Steel Studs between runners, twisting them into position. Position studs vertically, with open side facing in same direction and web punch-outs aligned properly, engaging floor and ceiling runners and spaced 16″ or 24″ o.c. max. as required. Proper alignment will provide for proper bracing, utility runs and prevention of stepped or uneven joint surfaces. Anchor all studs adjacent to door and borrowed light frames, partition intersections and corners to floor and ceiling runners. Intermediate partition studs should not be anchored to runners.

Place studs in direct contact with all door frame jambs, abutting partitions, partition corners and existing construction elements. Grouting of door frames is always recommended and is required where heavy or oversize doors are used.

Where a stud directly abuts an exterior wall and there is a possibility of condensation or water penetration through the wall, place a No. 15 asphalt felt strip between stud and wall surface.

Over metal doors and borrowed light frames, place a section of runner horizontally with a web-flange bent at each end. Secure runner to strut-studs with two screws in each bent web. At the location of vertical joints over the

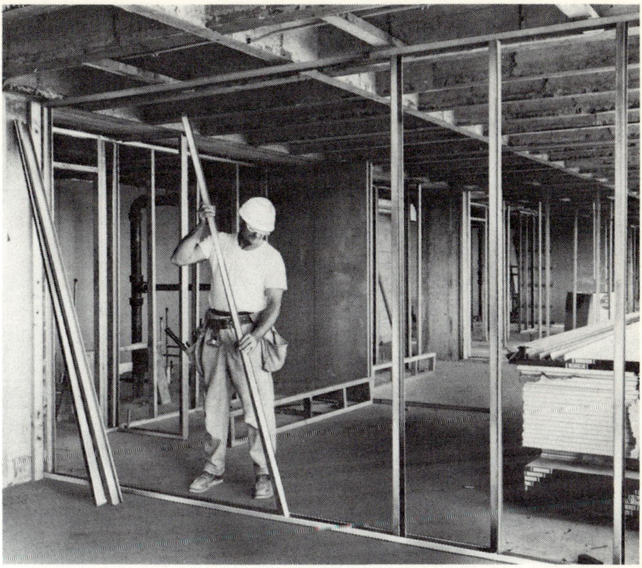

Steel studs are positioned in floor and ceiling runners.

door frame header, position a cut-to-length stud extending to the ceiling runner. (See page 119, "Door and Window Openings.")

Fasten together with two ⅜″ Type S pan head screws in each flange. Locate each screw no more than 1″ from ends of splice.

UNIMAST Steel Studs may be conveniently spliced together when required. To splice two studs, nest one into the other forming a box section, to a depth of at least 8″.

Resilient Channel Framing

Stud System Installation—Attach steel runners at floor and ceiling to structural elements with suitable fasteners located 2″ from each end and spaced 24″ o.c. Position studs vertically, with open side facing in same direction, engaging floor and ceiling runners, and spaced 24″ o.c. For non-fire-rated resilient channel system, anchor studs to floor and ceiling runners on the resilient side of partition. Fasten runner to stud flange with ⅜″ Type S pan head screw.

Resilient Channel Installation—Position RC-1 Resilient Channel at right angles to steel studs, space 24″ o.c. and attach to stud flanges with ⅜″ Type S pan head screws driven through holes in channel mounting flange. Install channels with mounting flange down, except at floor to accommodate attachment. Locate channels 2″ from floor and within 6″ of ceiling. Splice channel by nesting directly over stud, screw-attach through both flanges. Reinforce with screws located at both ends of splice.

Chase Wall Framing

Align two parallel rows of floor and ceiling runners according to partition layout. Spacing between outside flanges of each pair of runners must not exceed 24″. Follow instructions above for attaching runners.

Position steel studs vertically in runners, with flanges in the same direction, and with studs on opposite sides of chase directly across from each other. Except in fire-rated walls, anchor all studs to floor and ceiling runner flanges with ⅜″ or ½″ Type S pan head screws.

Cut cross-bracing to be placed between rows of studs from gypsum board 12″ high by chase wall width. Space braces 48″ o.c. vertically and attach to stud web with screws spaced 8″ o.c. max. per brace.

Bracing of 2½″ min. steel studs may be used in place of gypsum board. Anchor web at each end of metal brace to stud web with two ⅜″ pan head screws. When chase wall studs are not opposite, install steel stud cross-braces 24″ o.c. horizontally, and securely anchor each end to a continuous horizontal 2½″ runner screw-attached to chase wall studs within the cavity.

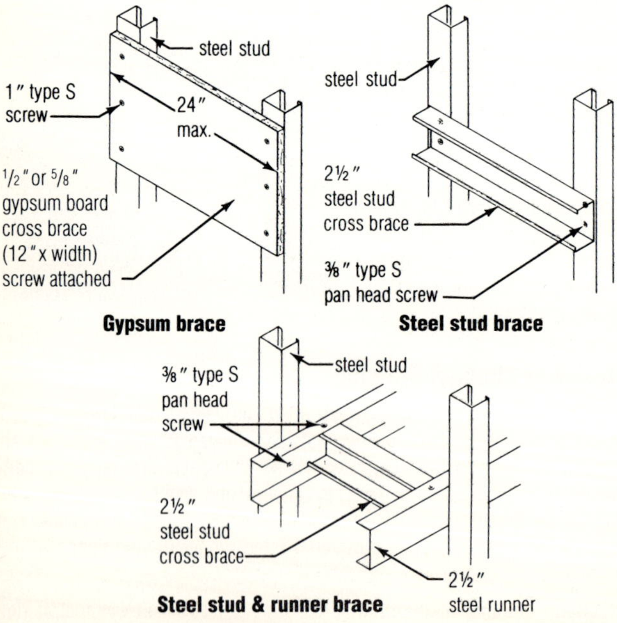

Furred Ceiling Framing

Space UNIMAST Metal Furring Channels 24″ o.c. at right angles to bar joists or other structural members. As an alternative, 1⅝″ steel studs may be used as furring. Saddle-tie furring channels to bar joists with double-strand 18-ga. tie wire at each intersection. Provide 1″ clearance between furring ends and abutting walls and partitions. At splices, nest furring channels with at least 8″ overlap and securely wire-tie each end with double-strand 18-ga. tie wire. Frame around openings such as light troffers with additional furring channels and wire-tie to bar joists.

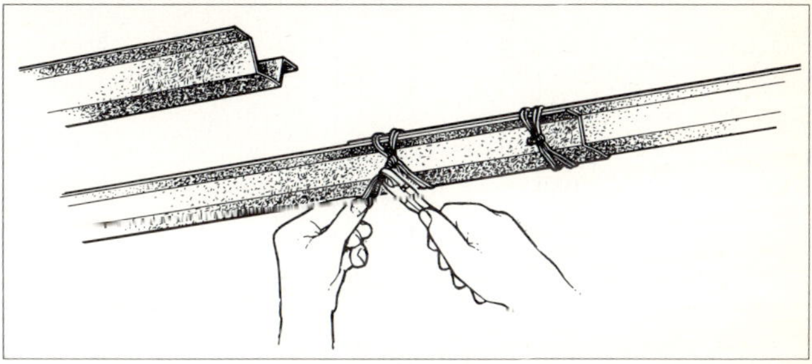

Max. allowable spacing for UNIMAST Metal Furring Channel is 24″ o.c. for ½″ and ⅝″ thick gypsum panels or plaster base. See table below for limiting spans.

For bar joist spacing up to 60″, steel studs may be used as furring channels. Wire-tie studs to supporting framing. Position 1⅝″ studs with open side up; position larger studs with opening to side. See table on next page for stud spacings and limiting spans.

Limiting Span(1)—Metal Furring Members(2)

Type furring member	Member spacing (in o.c.)	Single-layer panels (2.5 psf max.)		Double-layer panels (5.0 psf max.)	
		1-span	3-span	1-span	3-span
DWC-25	16	5′9″	7′1″	4′7″	5′8″
(hemmed)	24	5′0″	6′2″	4′0″	4′11″
DWC-20	16	6′11″	8′6″	5′5″	6′9″
(unhemmed)	24	6′0″	7′5″	4′9″	5′11″
158ST25 stud	16	7′2″	8′10″	5′8″	7′0″
	24	6′3″	7′9″	5′0″	6′2″

(1) For beams, joists, purlins, sub-purlins; not including 1½″ cold rolled channel support spaced 4′0″ max. (2) Limiting spans for ½″ and ⅝″ thick panels, max. L/240 deflection and uniform load shown. Investigate concentrated loads such as light fixtures and exhaust fans separately.

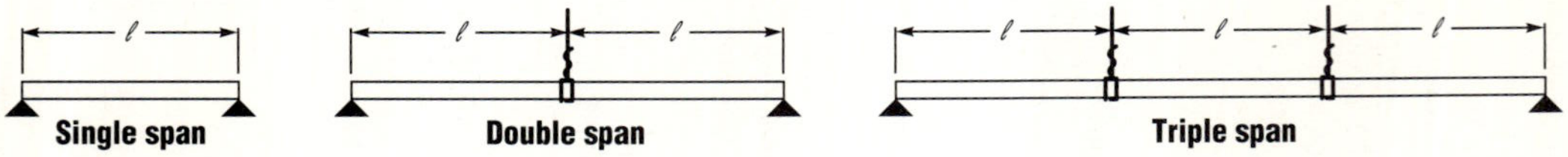

Limiting Span(ℓ)—UNIMAST Steel Stud Ceiling System(1)

Stud style		212ST25			358ST25(2)			400ST25(2)			212ST20			358ST20			400ST20			600ST20		
Stud spacing-in.		12	16	24	12	16	24	12	16	24	12	16	24	12	16	24	12	16	24	12	16	24
Single span																						
Uniform load —psf	5	10′11″	9′11″	8′8″	14′7″	13′3″	11′7″	15′9″	14′4″	12′6″	13′2″	11′11″	10′5″	17′6″	15′11″	13′11″	19′0″	17′3″	15′0″	26′3″	23′10	20′10″
	10	8′8″	7′11″	6′9″	11′7″	10′6″	7′3″	12′6″	11′0″	9′0″	10′5″	9′6″	8′3″	13′11″	12′8″	11′0″	15′0″	13′8″	11′11″	20′10″	18′11″	16′6″
	15	7′7″	6′10″	4′9″	9′8″	7′3″	4′9″	10′4″	9′0″	6′8″	9′1″	8′3″	7′3″	12′2″	11′0″	9′8″	13′2″	11′11″	10′4″	18′2″	16′6″	13′11″
	20	6′9″	5′4″	—	7′3″	5′5″	—	9′0″	7′6″	5′0″	8′3″	7′6″	6′4″	11′0″	10′0″	8′4″	11′11″	10′10″	9′0″	16′6″	14′9″	12′0″
Double and triple span (distance between supports)																						
Uniform load —psf	5	13′6″	12′4″	10′2″	17′5″	14′8″	11′2″	17′6″	14′7″	11′0″	16′4″	14′10″	12′11″	21′9″	19′9″	17′8″	23′6″	21′4″	18′8″	32′6″	29′6″	25′9″
	10	10′2″	8′8″	6′11″	11′2″	9′2″	6′8″	11′0″	8′9″	6′3″	12′11″	11′9″	10′1″	17′3″	15′8″	13′3″	18′8″	16′11″	14′3″	25′9″	21′10″	16′10″
	15	8′2″	6′11″	5′9″	8′4″	6′8″	4′9″	8′0″	6′3″	4′4″	11′4″	10′0″	8′2″	15′0″	13′3″	10′10″	16′3″	14′3″	11′7″	20′3″	16′10″	13′10″
	20	6′11″	5′9″	4′4″	6′8″	5′3″	—	6′3″	4′10″	—	10′0″	8′9″	7′1″	13′3″	11′6″	9′4″	14′3″	12′4″	9′9″	16′10″	13′10″	10′2″

(1) Based on L/240 allowable deflection. Bracing of top flanges is required and must not exceed 48″ o.c. (2) Stud end stiffening required. Additional hangers are necessary when span area exceeds 16 ft.^{2}.

Steel stud framing system

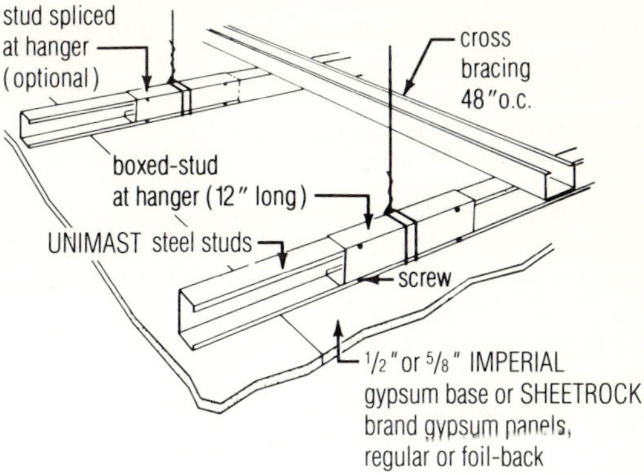

Direct suspension system

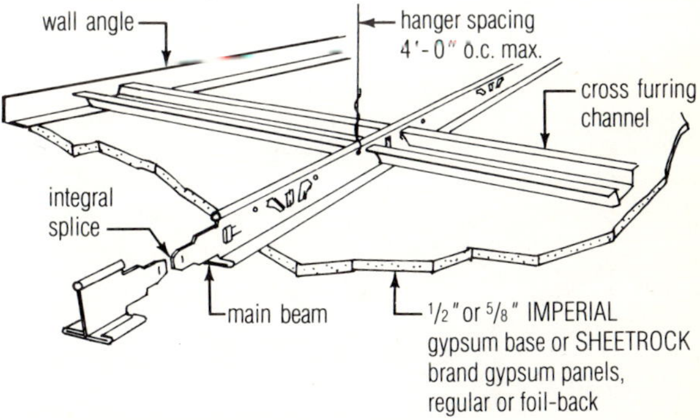

Direct Suspension System

For direct suspension system, attach gypsum panels to wall angles at perimeter. Locate cross furring channels within 6″ of walls without wall angles and within 8″ of panel end joints.

Suspended Ceiling Grillage Erection

Space 8-ga. hanger wires 48″ o.c. along carrying channels and within 6″ of ends of carrying-channel runs. In concrete, anchor hangers by attachment to reinforcing steel, by loops embedded at least 2″ or by approved inserts. For steel construction, wrap hanger around or through beams or joists.

Install 1½″ carrying channels 48″ o.c. (spaced as tested for fire-rated construction) and within 6″ of walls. Position channels for proper ceiling height, level and secure with hanger wire saddle tied along channels (see illustration). Provide 1″ clearance between runners and abutting walls and partitions. At channel splices, interlock flanges, overlap ends 12″ and secure each end with double-strand 18-ga. tie wire.

Erect metal furring channels at right angles to 1½″ carrying channels. Space furring within 6″ of walls. Provide 1″ clearance between furring ends and abutting walls and partitions. Attach furring channels to 1½″ channels with wire ties or UNIMAST Furring Channel Clips installed on alternate sides of carrying channel. Saddle tie furring to channels with double-strand 18-ga. tie wire when clips cannot be alternated. At splices, nest furring channels with at least 8″ overlap and securely wire tie each end with double-strand 18-ga. tie wire.

Where required in fire-rated assemblies, install double furring channels to support gypsum panel ends and back block with gypsum board strip. When staggered end joints are not required, control joints may be used.

UNIMAST metal furring channel

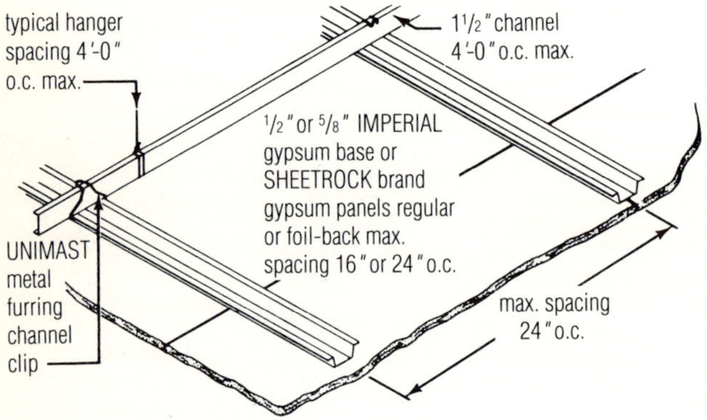

UNIMAST steel stud furring

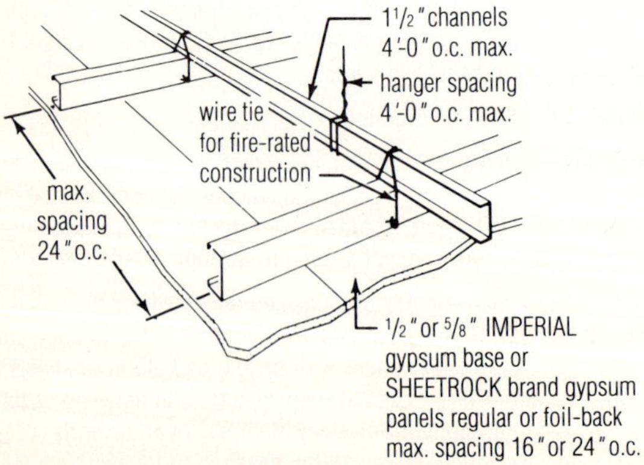

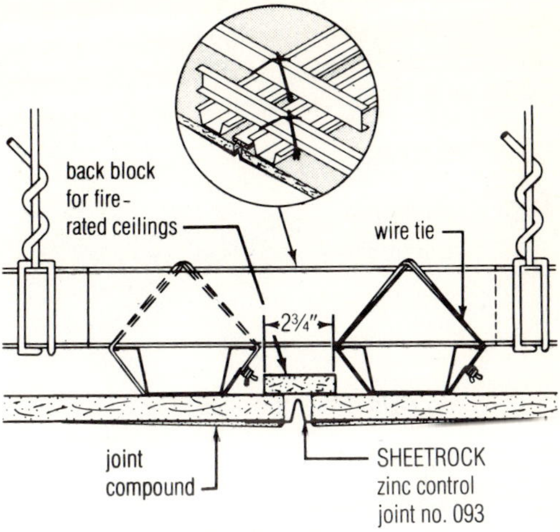

Control joint

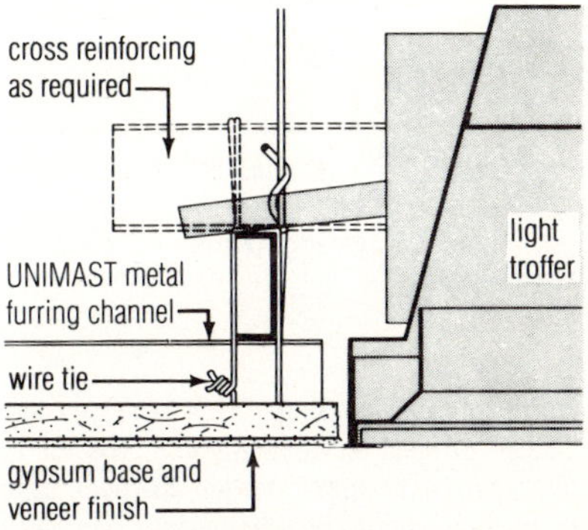

Lighting fixture

At light troffers or any openings that interrupt the carrying or furring channels, install additional cross-reinforcing to restore the lateral stability of grillage.

Light Fixture Protection—Use over recessed lighting fixtures installed in direct suspension grid when required in fire-rated construction. Cut pieces of ½″ or ⅝″ SHEETROCK brand Gypsum Panels or IMPERIAL Gypsum Base with FIRECODE C Core to form a five-sided enclosure, trapezoidal in cross-section (see detail). Fabricate box larger than the fixture to provide at least ½″ clearance between the box and the fixture, and in accordance with fire test report.

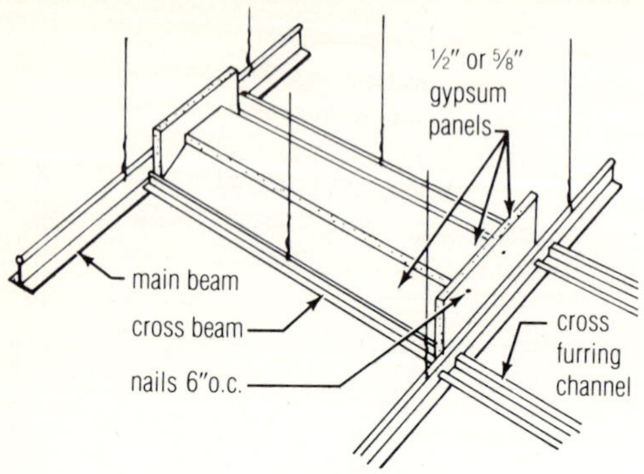

Light fixture fire protection

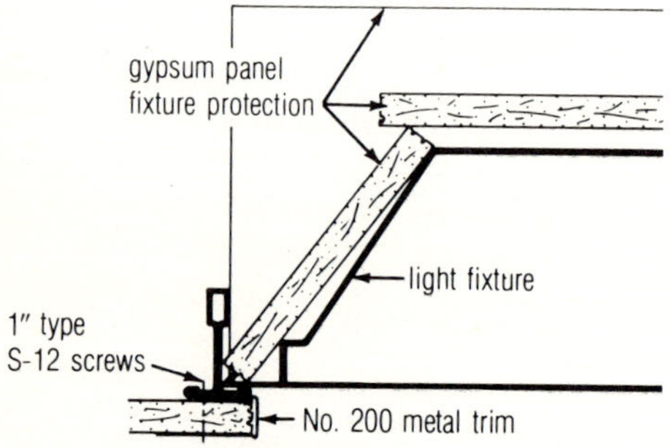

Lighting fixture

Wall Furring

Exterior walls are readily furred using ½″ foil-back gypsum panels/plaster bases screw-attached to steel or wood framing. The foil-back board provides an effective, low-cost vapor retarder. In these systems, different framing methods may be used to provide a vapor retarder, thermal insulation, and chase space for pipes, conduits and ducts. Vinyl wall coverings are not recommended in furred walls containing foil-back gypsum drywall or plaster base. The need for and location of a vapor retarder should be determined by a qualified mechanical engineer.

UNIMAST Metal Furring Channels are fastened directly to interiors of exterior walls or monolithic concrete and virtually any type of masonry—brick, concrete block, tile. With foil-back gypsum panel or plaster base screw-attached to channels, and appropriate sealants at periphery and penetrations, this economical system provides an excellent vapor retarder and a durable, easily decorated interior surface.

SHEETROCK Z-Furring Channels are used to mechanically attach THERMAFIBER Fire Safety FS-15 Blankets or rigid plastic foam insulation to interiors of exterior walls. The insulation panels are applied progressively as Z-furring channels are attached to the wall. Gypsum panels/plaster bases are screw-attached to channel flanges to provide an interior surface isolated to a great degree from the masonry wall. In new construction and remodeling, this system provides a highly insulative self-furring solid back-up for gypsum boards.

UNIMAST Steel Studs erected vertically between floor and ceiling runners serve as free-standing furring for foil-back gypsum panels/plaster bases screw-attached to one side of studs. This free-standing system with 1⅝″ studs provides maximum clear chase space and minimizes possibilities for photographing or shadowing to occur. When heights greater than 12′0″ are required, the stud framing is secured to the exterior wall with adjustable wall furring brackets at mid-height. Other furring providing greater height may be constructed with wider and heavier steel studs.

Temperature differentials on the interior surface of exterior walls may result in collection of dust on the colder areas of the surface. Consequently, shadowing (accumulated dust) may occur at locations of fasteners or furring channels where surface temperatures usually are lowest. U.S. Gypsum Company cannot be held responsible for surface discoloration of this nature. Where temperature, humidity and soiling conditions are expected to cause objectionable blemishes, use free-standing furring with insulation against the exterior wall.

Wall elevation–furring

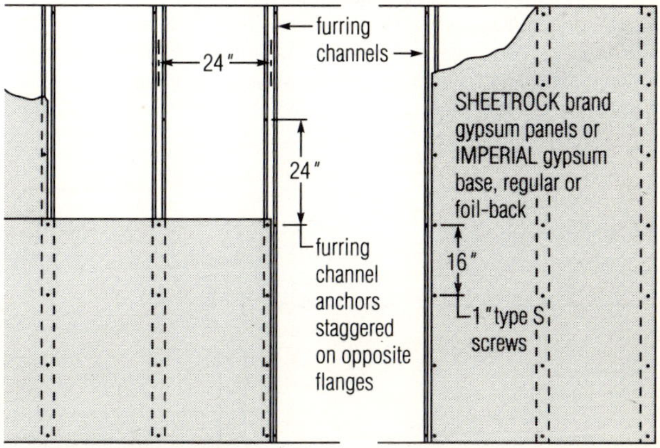

Furring Channel Erection—Direct Attachment

Attach metal furring channels to masonry or concrete surfaces, either vertically (preferred) or horizontally (for spacing, see frame spacing tables). For channels positioned horizontally, attach a furring channel not more than 4″ from both the floor line and the ceiling line. Secure channels with fasteners placed on alternate channel flanges and spaced 24″ o.c. Use a 2″ cut nail in mortar joints of brick, clay tile or concrete block or in the field of lightweight aggregate block; ⅝″ concrete stub nail, TAPCON Anchors or other power-driven fasteners in monolithic concrete.

Channels may be furred using UNIMAST Adjustable Wall Furring Brackets and ¾″ cold-rolled channels to provide additional space for pipes, conduits or ducts.

At window locations, attach furring channels horizontally over masonry returns to support gypsum board at corners (see detail).

Free-standing Furring

Free-standing furring consists of 1⅝″ UNIMAST Steel Studs in 1⅝″ UNIMAST Steel Runners. To erect, plumb and align runners at the desired distance away from the exterior wall. Fasten runners to floor and ceiling with suitable anchors. Snap studs into place in runners (see framing spacing tables for required stud spacing).

If greater height is required than can be attained with 1⅝″ studs, wider or heavier gauge studs can be used. However, if space is critical, heights greater than 12′0″ can be attained with 1⅝″ studs by bracing them to the exterior wall at midheight or more frequently. For bracing, install UNIMAST Adjustable Furring Brackets to the exterior wall and attach to the stud webs with ⅜″ Type S pan head screws.

Z-furring Channel Erection

Erect insulation vertically and hold in place with Z-furring channels spaced 24″ o.c. Except at exterior corners, attach narrow flanges of furring channels to wall with concrete stub nails or power-driven fasteners spaced 24 o.c. At exterior corners, attach wide flange of furring channel to wall with short flange extending beyond corner. On adjacent wall surface, screw-attach short flange of furring channel to web of attached channel. Start

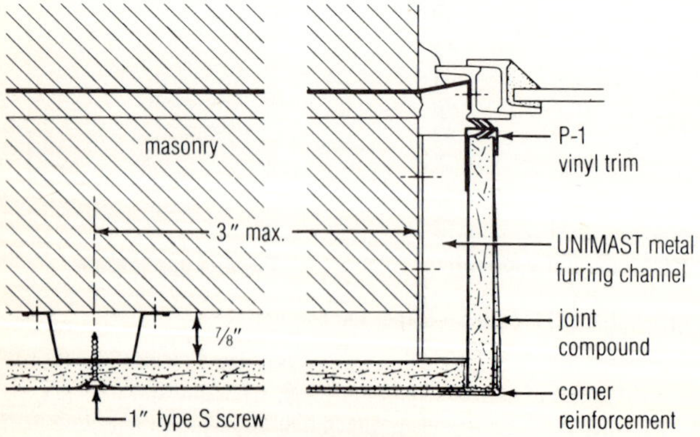

Metal window–jamb

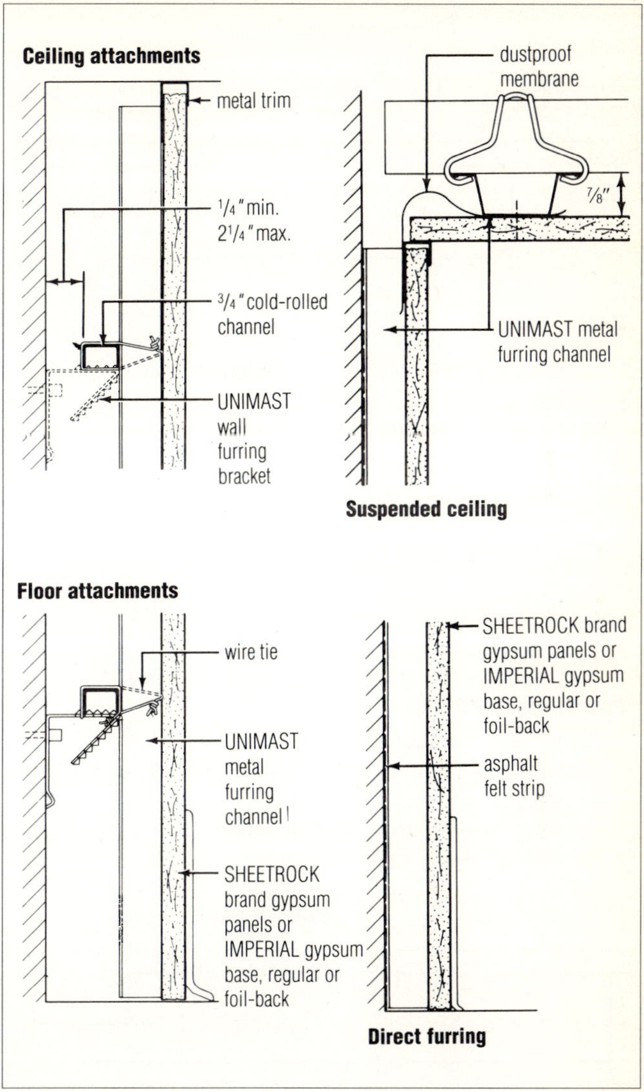

from this furring channel with a standard width insulation panel and continue in regular manner. At interior corners, space second channel no more than 12″ from corner and cut insulation to fit. Hold mineral-fiber insulation in place until gypsum panels/plaster bases are installed with 10″ long staple field-fabricated from 18-ga. tie wire and inserted through slot in channel. Apply wood blocking around window and door openings and as required for attachment and support of fixtures and furnishings.

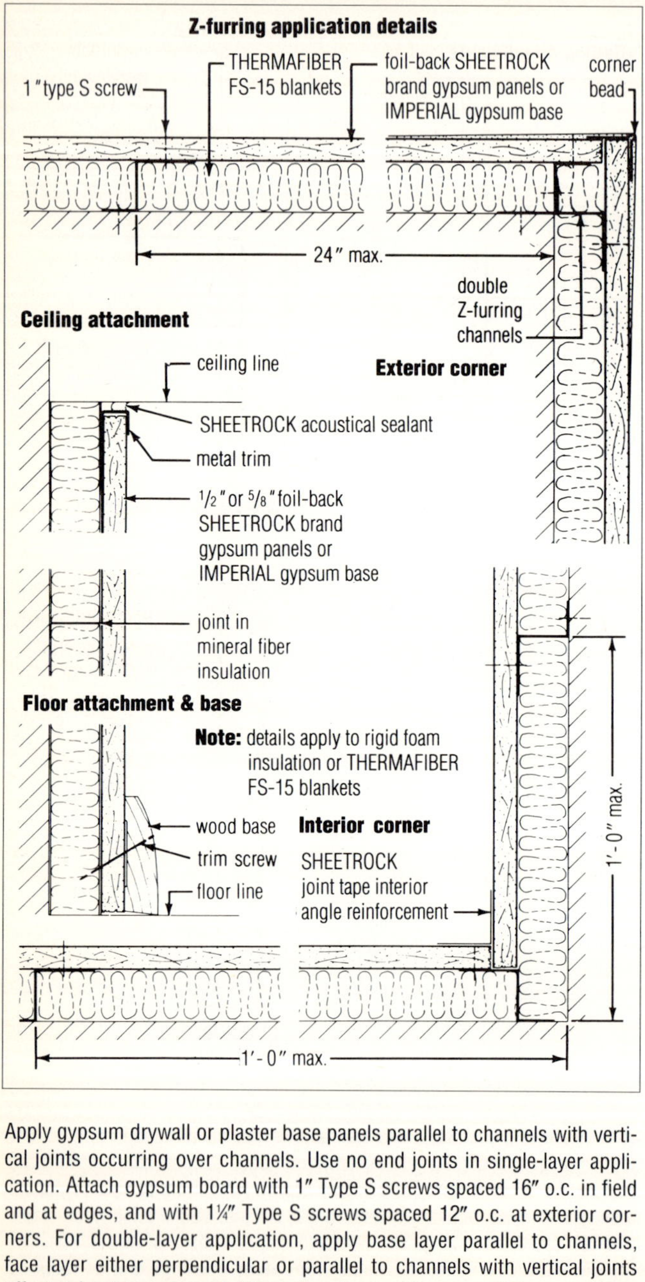

Apply gypsum drywall or plaster base panels parallel to channels with vertical joints occurring over channels. Use no end joints in single-layer application. Attach gypsum board with 1″ Type S screws spaced 16″ o.c. in field and at edges, and with 1¼″ Type S screws spaced 12″ o.c. at exterior corners. For double-layer application, apply base layer parallel to channels, face layer either perpendicular or parallel to channels with vertical joints offset at least one channel. Attach base layer with screws 24″ o.c. and face layer with 1⅝″ screws 16″ o.c.

Z-furring application details

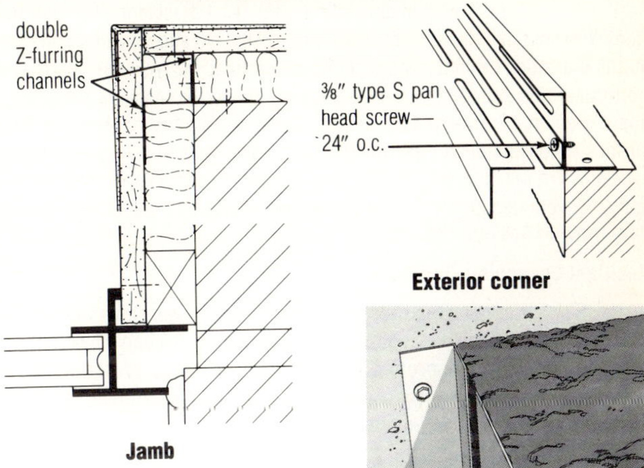

Design of SHEETROCK Z-furring Channel helps prevent wicking of moisture to inside surfaces, avoids faults of metal-over-insulation systems where "fishhooking" of pins can occur. Channel is available in four depths, 1" to 3".

Wood Furring Erection

Wood furring strips over wood framing must be 2x2 (nom.) min. size for nail-on application. Strips may be 1x3 (nom.) if gypsum board is to be screw-attached.

When panels are to be applied parallel to furring strips securely attached to masonry walls, use strips 2x3 or 1x3 (nom.) min. size; where long edges of board are to be applied across the furring, use strips 2x2 or 1x2 (nom.) min. size. Space furring strips as specified by frame spacing tables. For board application select a screw length that will not penetrate through furring.

Where there is a possibility of water penetration through the walls, install a layer of asphalt felt between furring strips and wall surface.

Note: Nail application of gypsum board over 1" (nom.) thickness wood furring applied across framing members is not recommended since the relative flexibility of undersize furring prevents proper fastening and tends to loosen nails already driven.

Resilient Framing—Wood Frame

Resilient attachment of gypsum board with RC-1 Resilient Channels provides low-cost, highly efficient, sound-rated drywall and veneer plaster partitions and floor-ceilings. The steel channels float the gypsum board away from the studs and joists; provide a spring action that isolates the gypsum board from the framing. This spring action also tends to level the panel surface when installed over uneven framing. Additional features include excellent fire resistance (from the total assembly) and simple, fast installation for overall economy. For fire and sound-resistant assemblies, refer to Construction Selector, SA-100.

Resilient Channels—Partitions

Attach RC-1 Resilient Channels with attachment flange down and at right angles (perpendicular) to wood studs. Position bottom channel with attachment flange up for ease of attachment. Use 1¼″ Type W screws driven through the flanges for attachment. Nails are not recommended. Fasten channels to studs at each intersection with the slotted hole directly over a framing member.

Locate channels 2″ max. up from floor, within 6″ of the ceiling and at no more than 24″ intervals. (For some veneer plaster assemblies max. channel spacing is 16″ o.c. Refer to frame spacing tables earlier in this chapter.) Extend channels into all corners and attach to corner framing. Splice channels directly over studs by nesting (not butting) the channels and driving fastener through both flanges into the support.

Where cabinets are to be installed, attach RC-1 Channels to studs directly behind cabinet hanger brackets. When distance between hangers exceeds 24″ o.c., install additional channel at midpoint between hangers.

For cabinet installation with resilient framing, refer to page 180 in Chapter 3.

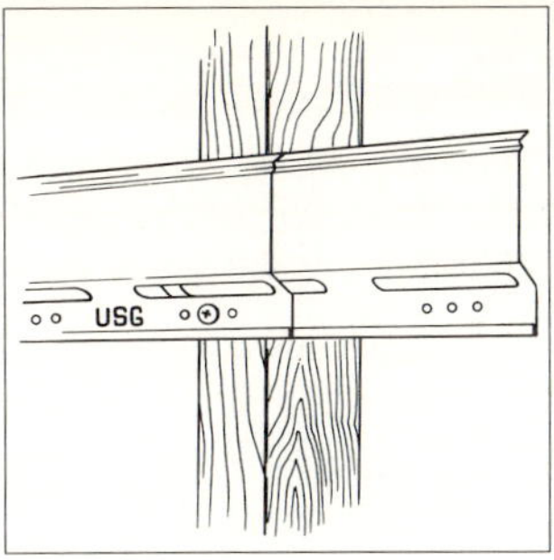

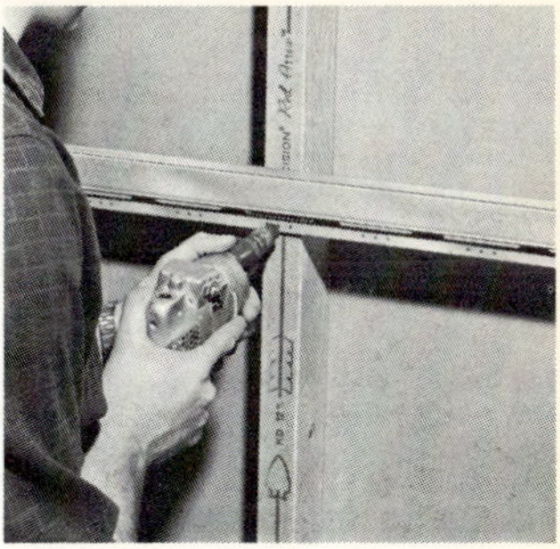

Resilient Channel—Ceilings

Attach RC-1 Resilient Channels at right angles to wood joists. Use 1¼″ Type W or 1¼″ Type S screws driven through channel attachment flange for single-layer construction. Fasten channels to joists at each intersection. Do not use nails to attach channels to joists in either single or double-layer assemblies. THERMAFIBER SAFB is required when sound control is needed.

A 2-hr. floor/ceiling system with STC/MTC ratings as high as 60/54 is achievable with a ceiling of double-layer ⅝″ SHEETROCK brand Gypsum Panels FIRECODE C core attached to RC-1 Channel mounted across joists and 3″ THERMAFIBER SAFB in the cavity (see UL Design L541 on page 337).

For fire-rated, double-layer assembly, apply RC-1 Channels over base layer and attach with 1⅞″ Type S screws driven through channel flange and base layer into joist (see UL Design L511—not recommended when sound control is a major consideration).

Door and Window Openings

Rough framing for most door and window openings is the same for gypsum panels and gypsum base veneer plaster systems.

Wood Framing

Install additional cripple studs above header and ½″ from bearing studs where control joints are required. Do not anchor cripple stud to bearing stud, header or plate.

In long runs, treat window openings in same manner as shown for doors.

Steel Framing

Door and borrowed light openings should be rough framed with steel studs and runners. Position floor-to-ceiling-height strut-studs vertically, adjacent to frames, and anchor securely to top and bottom runners with screws. Where heavy or oversize doors are used, install additional strut-stud at jambs. Fabricate sill and header sections from UNIMAST Steel Runners and install over less-than-ceiling-height door frames and above and below borrowed light frames. Fabricate from a section of runner cut-to-length approx. 6″ longer than rough opening. Slit flanges and bend web to allow flanges to overlap adjacent vertical strut-studs. Securely attach to strut-

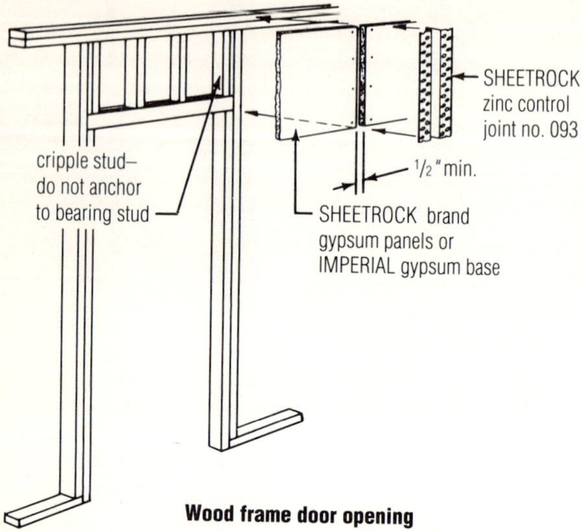

Wood frame door opening

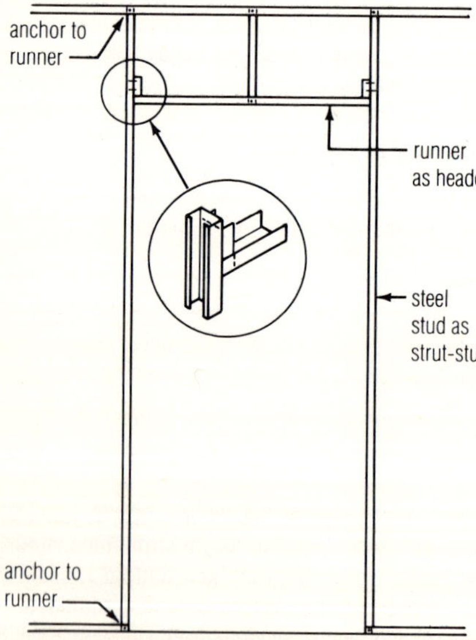

Door frame as steel runner as header

studs with screws. For frames with jamb anchor clips, fasten clips to strut-studs with two ⅜″ Type S-12 pan head screws. Install cripple studs in the center above the door opening and above and below borrowed light openings spaced 24″ o.c. max.

Where control joints in header boards are required, install cripple studs away from strut-studs but do not attach cripple to runners or strut-studs.

Note: 3-piece frames are recommended for drywall and veneer plaster construction since these frames are installed after drywall or plaster base is in place. One-piece frames, which must be installed before the gypsum board, are more difficult to use because the panels must be inserted under the frame returns as it is installed.

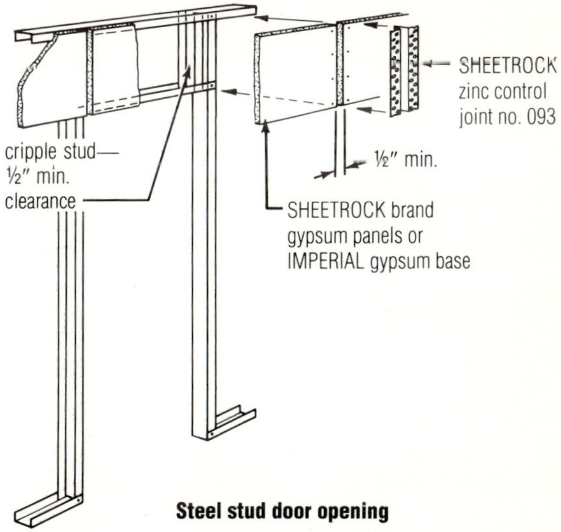

Steel stud door opening

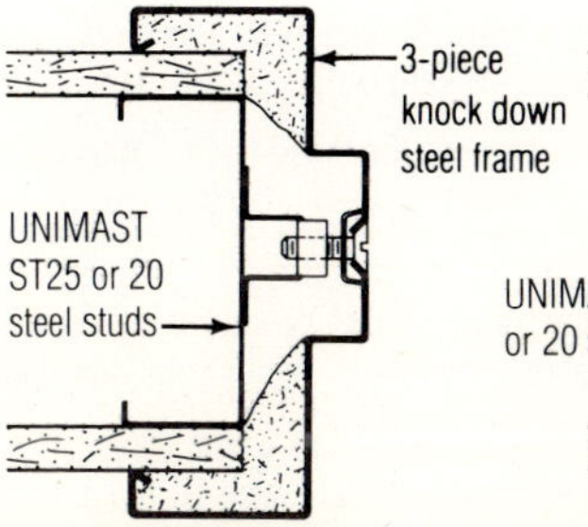

Jamb standard door

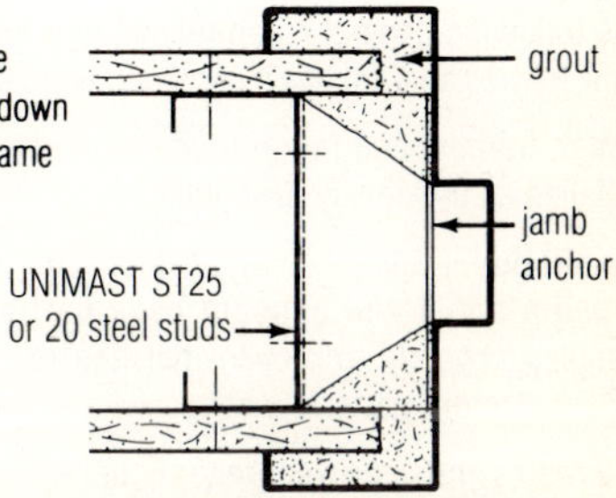

Jamb standard door

Framing for Heavy and Oversize Doors

The steel framing method described above is suitable for standard doors up to 2′8″ wide, weighing not more than 100 lb. max. ST25 steel studs and runners may be used for framing the opening. For wider or heavier doors, the framing must be reinforced.

For solid-core doors and hollow-core doors 2′8″ to 4′0″ wide (200 lb. max.), rough framing should be ST20 steel studs and runners. For heavy doors up to 4′0″ wide (300 lb. max.), two ST20 studs should be used. For doors over 4′0″ wide, double doors and extra-heavy doors (over 300 lb.), framing should be specially designed to meet load conditions. Rough framing for all doors in fire-rated partitions should be ST20 studs and runners.

For added door frame restraint, spot grouting of the frame is recommended and is required for solid-core doors and doors over 2′8″ wide. Apply SHEETROCK Setting-Type (DURABOND) Joint Compound or RED TOP Gypsum Plaster mixed with sand just before inserting board into frame. Do not terminate gypsum board against trim return.

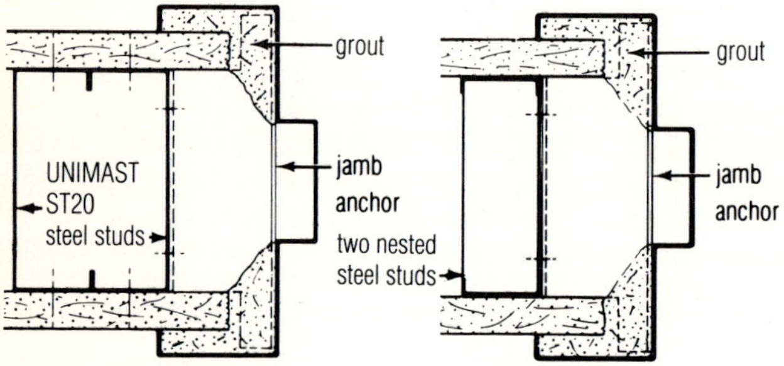

Jamb heavy doors 2′8″ to 4′0″

Alternate jamb

Door Frame Installation

The following general recommendations apply to one-piece and three-piece door frames and are basic considerations for satisfactory performance.

Rough framing and rough frame reinforcement for these frames should be installed as previously described.

Installation—One-piece metal door (and borrowed light) frames used with gypsum panel and gypsum base partitions must be constructed and installed properly to prevent twisting or movement. Basic considerations for satisfactory performance are:

1. Frames must be securely anchored. If frames are free to twist upon impact, or trim returns are free to vibrate, movement of the frame will tend to pinch gypsum board face paper and crush core, resulting in unsightly cracks in the finish and loose frames.

2. Partition must fit securely in frame so that wall and frame work as a unit. Impact stresses on frame will then be dissipated over entire partition surface and local damage minimized.

3. The frame must have a throat opening between trim returns that accurately fits the overall thickness of the partition. The face-layer panels should be enclosed by the trim and not butted against the trim return. This throat opening measurement is critical, as too large a tolerance between panels and trim return will cause door frame to twist and vibrate against the panels. Too small a tolerance will prevent the panels from fully entering frame opening; as a result, the door frame will not be held securely by the partition.

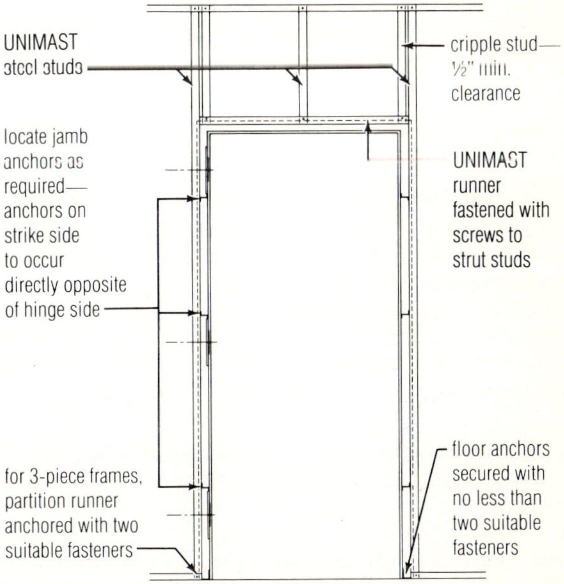

Frame for standard door

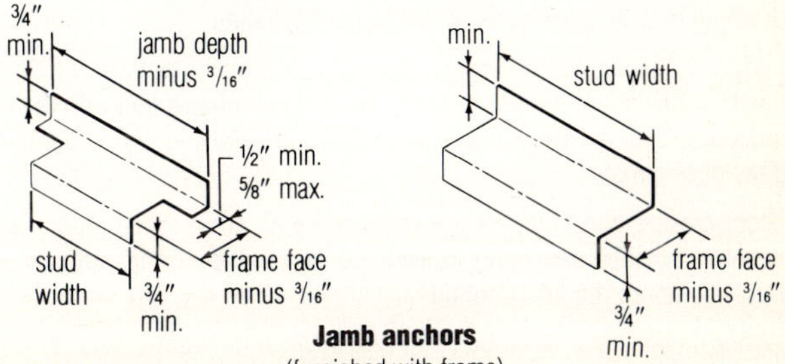

Jamb anchors
(furnished with frame)

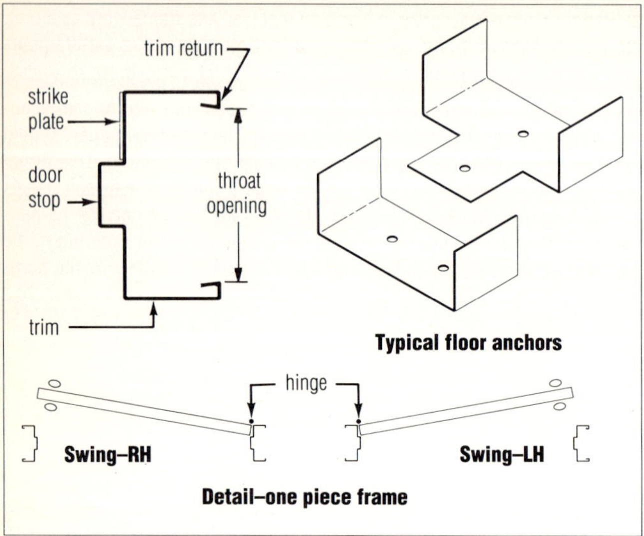

4. One-piece metal door (and borrowed light) frames should be formed from 18-ga. steel min., shop-primed. Floor anchor plates for door frames should be 16-ga. steel min., designed with two anchor holes to prevent rotation, and shop-welded to frame rabbets to dampen door impact vibrations. Floor anchorage should be by two power-driven anchors or equivalent per plate. Jamb anchors should be formed of 18-ga. steel min., fit tightly in jambs, and screw-attached to the stud. A min. three anchor clips per jamb is recommended with locations at approximate hinge points.

5. Spot grouting of one-piece door jambs will increase the rigidity of the frame and improve resistance to frame rotation caused by the weight of the door. To spot grout, apply SHEETROCK Setting-Type Joint Compound (DURABOND) or RED TOP Gypsum Plaster Job-Aggregated mixed in accordance with bag instructions to each jamb anchor filling the inside face of the jamb at each point. Immediately insert the gypsum panels into the jamb and attach to framing. Do not terminate the gypsum board against the trim.

Full grouting of the jambs flush with the jamb anchors prior to installation of framing may be used as an option to spot grouting.

To improve the sound seal around door frames, apply a bead of SHEETROCK Acoustical Sealant to the return of the jamb at the intersection with the gypsum board. Tool the bead of sealant smooth and allow to dry before finishing the door jamb.

6. Door closers and bumpers are required on all doors where door weight (including attached hardware) exceeds 50 lb., or where door width exceeds 36″. These doors require grouting.

7. When installing a three-piece knock-down door frame, secure runner ends with two floor anchors and allow space in the rough framing for the adjustment shoes in the frame.

8. When ordering metal door frames, the factors to be considered include: gauge of frame; width and height of door; swing direction of door; type and thickness of door; stud size, and overall thickness of partition.

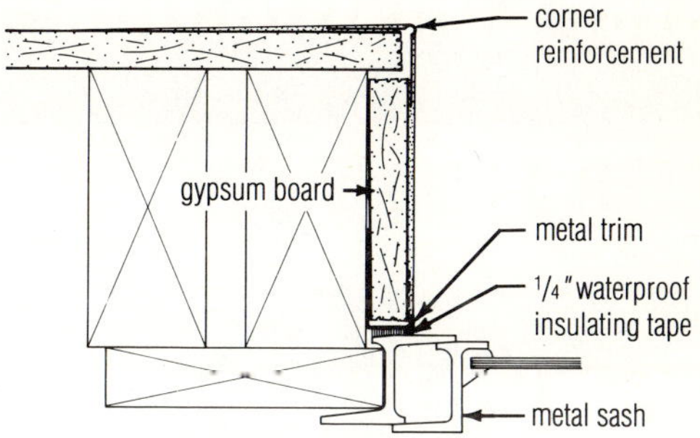

Detail–window trim

Metal Window Framing

In climates where extremes in summer or winter temperatures may result in condensation on metal frames, gypsum panels/plaster bases should be isolated from direct contact with the frame.

By placing metal trim (SHEETROCK No. 200-B, No. 801-A and B and No. 400 for drywall assemblies; No. 701-A and B or 801-A and B for veneer plaster assemblies) between the gypsum board and window frame, protection against moisture damage is provided.

Waterproof insulating tape, ¼" thick and ½" wide, or a waterproof acrylic caulk is required to separate metal sash and metal trim and will provide some measure of insulation between the two different metals. Direct contact of an aluminum frame and steel trim in the presence of condensation moisture may cause electrolytic deterioration of aluminum frame.

Drywall & Veneer Plaster Construction

CHAPTER 3 Cladding

General Planning Procedures

In most instances, job planning requirements and the application techniques used for the installation of large-size gypsum board apply equally to gypsum panels and gypsum bases. For that reason the term "gypsum board" is used throughout this Chapter wherever the recommendations apply to both types of products. Where the requirements differ, the products are treated separately.

Planning the Job

Advance planning by the wall and ceiling contractor can mean savings in time and material cost and result in a better-appearing job. Proper planning achieves the most effective use of materials, elimination of unnecessary joints, and the placement of necessary joints in the least conspicuous locations. One gypsum board should span the entire length or width of the wall or ceiling, if possible. By using the longest practical board lengths obtainable, end joints are kept to a minimum. Where they do occur, end joints should be staggered.

In double-layer construction, end joints in the face layer must be offset at least 10″ from parallel joints in the base layer. Layout of the base layer must be planned to accommodate this offset and still provide optimum joint-finishing conditions and efficient use of materials in the face layer.

Estimating Materials

Gypsum Board—From practical experience, professional estimators have developed methods for determining footage required to complete various types of jobs. Basically, these methods stem from the simple principle of "scaling a plan," and determining the length and width and ceiling height of each room on the plan. Frequently, door and window openings are "figured solid" with no openings considered. Exceptions may be large picture windows and large door openings. From these dimensions the estimator determines the square footage of each room. The footage of each room is added to determine total footage required. From these figures the number of gypsum boards needed may be determined. (Refer to Chapter 1 for available lengths of each panel.)

Screws—For single-layer wall application to 16″ o.c., approx. 1,000 Type W screws for wood or Type S for steel are required per 1,000 ft.^{2} of gypsum board; approx. 850 for 24″ o.c. framing. See page 464 in Appendix for complete information on estimating screws.

Fastener usage for other assemblies varies with the construction and spacing.

Nails—Usage for nails is shown in the Selector Guide for Gypsum Board Nails in on page 75.

Acoustical Sealant—The approx. lin. ft. of bead realized per gal. of SHEETROCK Acoustical Sealant is: 392 for ¼″ bead, 174 for ⅜″ bead, 98 for ½″ bead.

Adhesive—The table on the next page shows the amount of adhesive needed per 1000 ft.^{2} of laminated board surface.

Coverage Adhesives for Lamination

		Approx. quantity			
		lb/1000 ft.^{2}		kg/100 m^2	
		Lam. blade ¼" notch spacing			
Product	**Application**	**2"**	**1½"**	**50 mm**	**38 mm**
SHEETROCK Ready-Mixed Joint Compound	Strip lam.	170	230	83	112
	Sheet lam.	340	465	166	227
SHEETROCK Setting Type (DURABOND) Joint Compound	Strip lam	93	123	45	60
	Sheet lam.	184	246	90	120
SHEETROCK Lightweight Setting Type (EASY SAND)	Strip lam.	68	90	33	44
	Sheet lam.	134	179	66	87
		gal./1000 ft.^{2}		**l/100 m^2**	
		2"	**1⅛"**	**50 mm**	**34 mm**
SHEETROCK Lightweight All Purpose Ready-Mixed (PLUS 3)	Strip lam.	11.5	15.5	45.6	63
	Sheet lam.	23.0	31.7	93.5	129

Joint Treatment—Gypsum Panels—Approximate quantities required for finishing 1,000 sq. ft. of gypsum panels: 370′ of SHEETROCK Joint Tape; 83 lb. of conventional drying-type powders, 67 lb. of lightweight drying-type powder (AP LITE); 72 lb. of conventional setting-type powders, 52 lb. of lightweight setting-type powder (EASY SAND); and 9.4 gallons of lightweight ready-mixed compound (PLUS 3).

Joint Treatment—Gypsum Base—For regular application over wood framing, approx. 370′ of either Type P or Type S IMPERIAL Tape is required per 1,000 sq. ft. of base surface.

For application over metal framing, approx. 370′ SHEETROCK Joint Tape and 72 lb. of SHEETROCK Setting Type (DURABOND) Joint Compound are required per 1,000 ft.^{2} of surface. This application is also required for certain spacing requirements and when building temperature-humidity conditions fall in the "rapid drying" area of graph on page 205.

Handling and Storage

When drywall and veneer plaster construction moved into high-rise buildings, it brought with it the new challenge of moving large gypsum boards from the ground to the point of use, stories above. Inefficient materials handling at the job site can add cost and reduce profit. Time and money savings can be substantial when correct handling procedures are used.

Your U.S. Gypsum Company representative can help in determining job-site handling costs and methods suited to particular job conditions.

Gypsum board products should be ordered for delivery several days in advance of installation. Materials stored on the job for a long period of time are subject to damage. Gypsum boards, like millwork, must be handled with care to avoid damage. Since joint compounds and veneer plaster finishes are subject to aging, they must not be stored for extended periods.

Board should be placed inside under cover and stacked flat on a clean floor in the centers of the largest rooms. It is often desirable to place the necessary number of boards in the location where they will be used. All materials used on the job should remain in their packaging until ready for actual use.

Gypsum boards intended for use on ceilings should be placed on top of pile for removal first. Avoid stacking long lengths atop short lengths.

All successful veneer plaster finish jobs require adequate equipment: power mixers, mortar boards, scaffolding and tools. Ample scaffolding should be provided. Rather than ship all veneer plaster finish to the job at one time, fresh material should be sent to the job every few days as needed. Plaster stored for long periods is subject to damage, variable moisture conditions and aging that probably will produce variations in setting time and create performance problems.

Store veneer plaster products inside, in a dry location and away from heavy traffic areas. Stack bags on planks or platforms away from damp floors and walls. Protect metal corner beads, casing beads and trim from being bent or damaged. All materials used on the job should remain in their packaging until used.

Environmental Conditions

In cold weather (outdoor temperature less than 55°F or 13°C), controlled heat in the range of 55° to 70°F (13° to 21°C) must be provided. This heat must be maintained both day and night, 24 hours before, during and after entire gypsum board joint finishing and until the permanent heating system is in operation or the building is occupied. Minimum temperature of 50°F (10°C) should be maintained during gypsum board application.

Methods for Applying Drywall and Plaster Bases

Gypsum panels and gypsum bases may be applied in one or two layers directly to wood framing members, to steel studs or channels, or to interior masonry walls with adhesive. Use of stilts will provide convenience in application.

Single Layer vs. Double Layer

Single-Layer Application—This basic construction is used to surface interior walls and ceilings where economy, fast erection and fire resistance are required. It is equally suitable for remodeling, altering and resurfacing cracked and defaced areas.

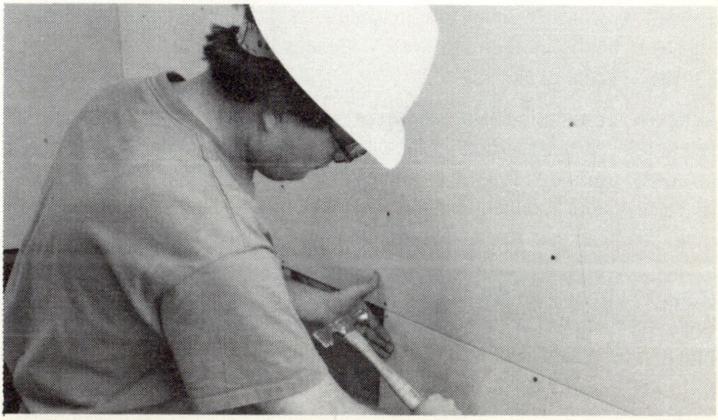

Nailing technique for single-layer application

Double-Layer Application—Consists of a face layer of gypsum board applied over a base layer of gypsum board that is directly attached to framing members. This construction can offer greater strength and higher resistance to fire and to sound transmission than single-layer applications. Double-layer construction when adhesively laminated is especially resistant to cracking and provides the finest, strongest wall available. Also, these adhesively laminated constructions are highly resistant to sag and joint deformation. In double-layer application, always apply all base-layer board in each room before beginning face-layer application.

Attachment Methods

Gypsum boards are attached to framing by several methods depending on the type of framing and the results desired.

Single Nailing—Conventional attachment for wood framing.

Double Nailing—Minimizes defects due to loose board.

Screw Attachment—Screws are excellent insurance against fastener pops caused by loosely attached board. Screws are recommended for wood frame attachment, and required for attachment to steel framing and resilient channels. When mounting to resilient channels, take care not to locate screws directly over studs, thereby "shorting out" or negating the resiliency.

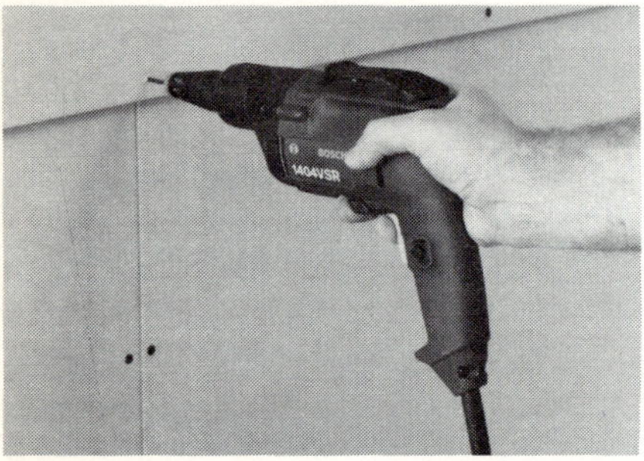

Screw attachment along vertical edges of face-layer board in double-layer application

Adhesive Attachment—A continuous bead of drywall stud adhesive applied to wood framing plus supplemental nail or screw attachment improves bond strength and greatly reduces the number of face nails or screws needed.

Adhesive Lamination (Double Layer)—Produces the finest interior surfaces. Adhesive attachment of face layer to base layer in double-layer construction and of single-layer board to interior masonry walls usually requires only supplemental mechanical fastening until adhesive attains full bond. Reduces nails or screws required, saves finishing labor and minimizes fastener pops and joint ridging. A SHEETROCK Setting-Type (DURABOND) or Lightweight Setting-Type (EASY SAND) Joint Compound or SHEETROCK Ready-Mixed Joint Compound—Taping or All Purpose is required for adhesive lamination with fire-rated assemblies.

Perpendicular vs. Parallel Application

Gypsum board may be applied perpendicular (long edges of board at right angles to the framing members) or parallel (long edges parallel to framing). Fire-rated partitions may require parallel application (see Chapter 8 for information on fire-rated systems).

Perpendicular application generally is preferred because it offers the following advantages:

1. Reduces the lineal footage of joints to be treated up to 25%.

2. Strongest dimension of board runs across framing members.

3. Bridges irregularities in alignment and spacing of frame members.

4. Better bracing strength—each board ties more frame members together than does parallel application.

5. Horizontal joints on wall are at a convenient height for finishing.

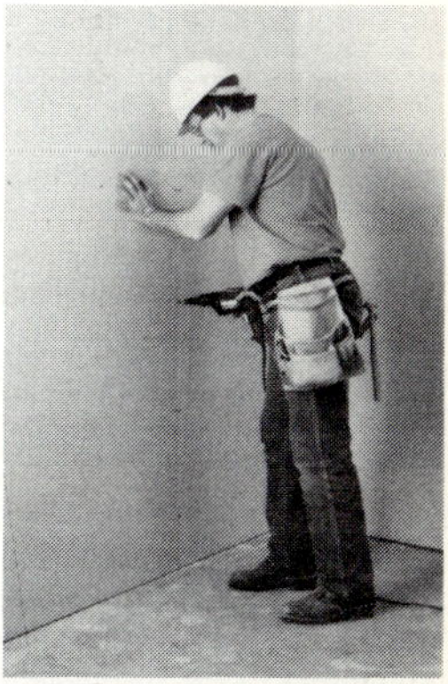

Starting at ceiling line, horizontal board is screw-attached (left). Parallel application (right) is used in special situations.

For wall application, if ceiling height is 8′1″ or less, perpendicular application results in fewer joints, easier handling and less cutting. If ceiling height is greater than 8′1″ or wall is 4′ wide or less, parallel application is more practical. For ceiling application, use whichever method results in fewer joints, or is required by frame spacing limitations.

For double-layer ceiling application, apply base-layer boards perpendicular to frame members; apply face layer parallel to framing with joints offset. On wall, apply base layer parallel with long edges centered on framing; apply face layer perpendicular. Exception: when using TEXTONE Panels for face layer, apply base-layer boards at right angles to studs.

Gypsum Drywall and Plaster Base Application

General recommendations for gypsum panels applied to wood and steel framing:

1. Apply ceiling boards first.

2. Cut boards so that they slip easily into place.

3. Butt all joints loosely. Never force panels into position.

4. Whenever possible place tapered or wrapped edges next to one another.

5. Whenever possible apply boards perpendicular to framing and in lengths that will span ceilings and walls without creating end (butt) joints. If butt joints do occur, stagger and locate them as far from the center of walls and ceilings as possible.

6. Support all ends and edges of gypsum board on framing, except long edges at right angles to framing and where end joints are to be floated between frame members and back-blocked (back-blocking is covered on page 144).

7. When fastening, apply hand pressure on panel next to fastener being driven to ensure panel is in tight contact with framing member.

8. If metal trim is to be installed around edges, doors, or windows, determine if trim is to be installed before panel application. Refer to Chapter 1 for description of products.

9. Do not anchor panel surfaces across the flat grain of wide dimensional lumber such as floor joists and headers. Float panels over these members or provide a control joint to compensate for wood shrinkage.

10. To ensure level surfaces at joints, arrange board application so that the leading edge of each board is attached to the open or unsupported edge of a steel stud flange. To do this, all studs must be placed so that their flanges point in the same direction. Board application is then planned to advance the direction opposite to flange direction. When this simple procedure is followed, attachment of each board holds the stud flange at the joint in a rigid position for attachment of the following board.

If the leading edge of gypsum board is attached to the web edge of a flange, the open edge of the flange can deflect under the pressure of attachment of the following gypsum board. Friction between the tightly abutted board edges can then cause them to bind, preventing return of the second board to the surface plane of the first. A stepped or uneven joint surface results.

This recommended application procedure is absolutely essential for good results in steel-framed veneer plaster and drywall assemblies. (See drawings on next page for correct methods.)

Measurements—All measurements must be accurate. Make two measurements as a check. This procedure will usually warn of partitions or door openings that are out of plumb or out of square. Then, framing corrections can be made before the board is hung. A 12′ to 25′ steel power tape is recommended. Tools for measuring and cutting are shown in Chapter 11.

Cutting—Make straight-line cuts across full width or length of board by scoring the face paper, snapping the board core and then cutting the back paper. The common tool used to score and cut gypsum board is a utility knife with replaceable blade. Regardless of the type knife used, its blade should be kept sharp so that score will be made through paper without tearing or rolling it up, and into the gypsum core.

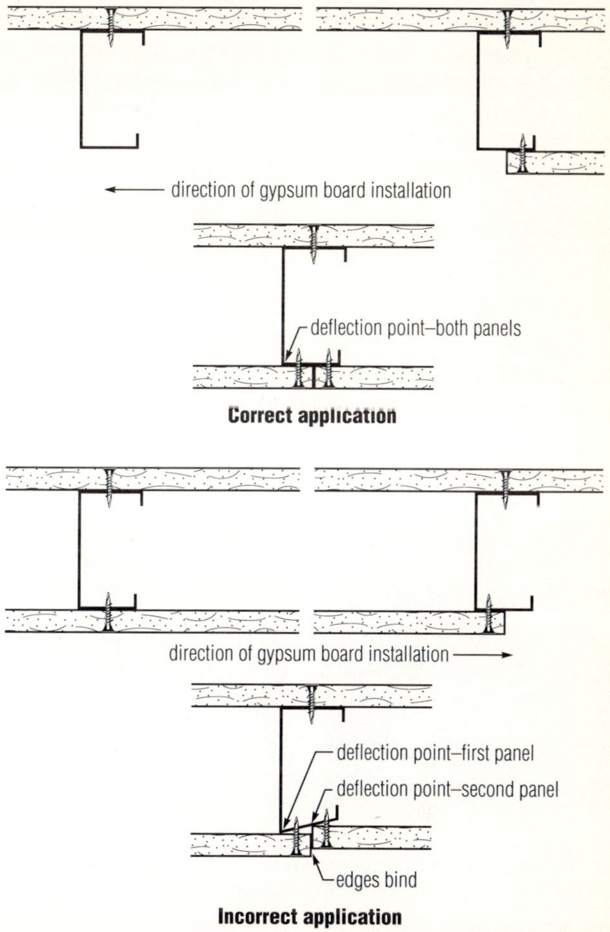

For cuts across the board width, a straightedge is recommended. An aluminum 4′ drywall T-square, ruled on both edges, facilitates clean, straight cuts. For cuts along the long length of the board, use a steel tape with an adjustable edge guide and a tip that accepts the utility knife blade. With this tape the edge guide is set for the desired width and placed against the board edge. The knife blade is then inserted into the slotted tape tip, and by moving both hands together the tool is drawn down the full length of the board to make a smooth and accurate cut.

Cut and fit board neatly for pipes, electrical outlet boxes, medicine cabinets, etc. Holes for electrical outlet boxes can be made with a special outlet box cutting tool. For circular holes, an adjustable circular cutting tool is available. Keyhole saws can be used for any type of cutout. After cutting hole, remove any loose face paper at cut. Refer to Chapter 11 for examples of panel stripper, cutting tools, etc.

Left to right, gypsum board is cut by scoring with utility knife against drywall T-square, then snapping toward back (top), cutting back paper with same knife and separating sections (bottom)—quick method to obtain clean edges and precise fit.

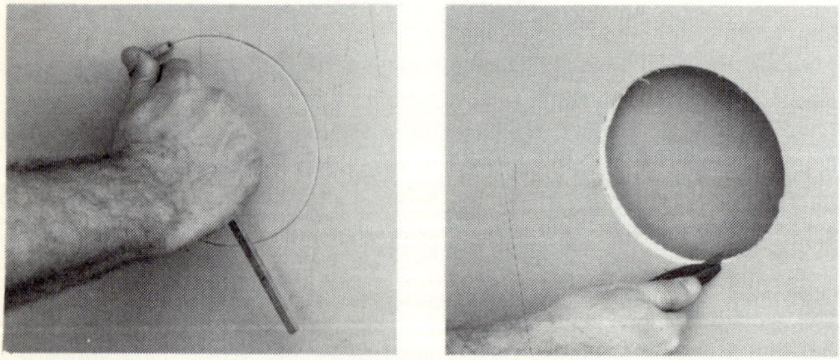

Adjustable cutting tool makes quick work of circular holes, as cutter wheel on calibrated shaft rotates from center pin (left). Edges are trimmed with utility knife (right). Stiff drywall saw and other tools are used to make odd-shaped cuts.

Cut edges of board are smoothed with a rasp, coarse sandpaper or piece of metal lath stapled around wood block (right). Measurements for cutouts are carefully made with flexible rule (below).

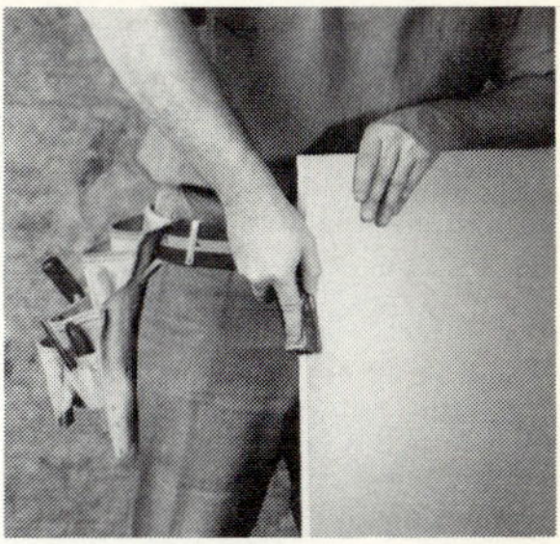

Screw Application

Screws are applied with a positive-clutch electric power tool, commonly called an electric screwgun, equipped with adjustable screw-depth control head and a Phillips bit. The use of screws provides a positive mechanical attachment of gypsum board to either wood or steel framing.

Adjust Screwgun—Set adjustment for proper screw depth. For gypsum panels (drywall), screwhead must be driven slightly below face of panel (max. 1⁄32″), but not deep enough to break the paper. For gypsum bases (veneer plaster), screwhead is set flush with the base surface. To adjust depth, rotate control head to provide proper screw depth. When proper adjustment has been made, secure control head to maintain adjustment.

Place Screw—Phillips head tip holds drywall screw for driving. Bit tip does not rotate until pressure is applied to gypsum board during application.

Start Screw Straight—Firm hand grip on electric screwgun is important for straight line of entry. To avoid stress on wrist, hold gun as shown on next page, not by the pistol grip. Screw must enter perpendicular to board face for proper performance. Drive screws at least 3⁄8″ from ends or edges of board.

Operate electric screwgun constantly during usage. When screwhead is driven solidly against board, screwgun head will automatically stop turning as the positive clutch disengages.

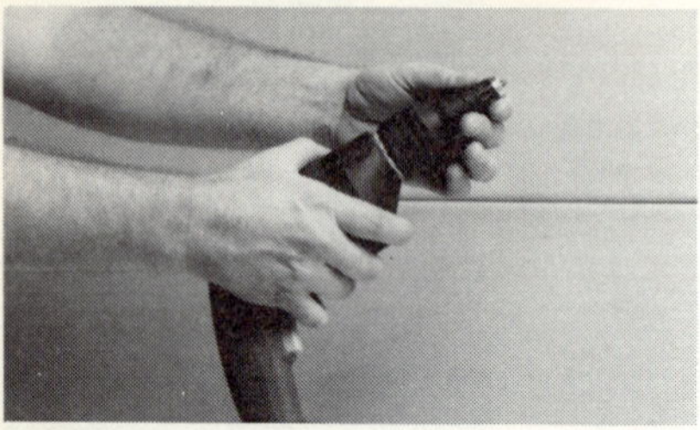

Secure control head to maintain adjustment.

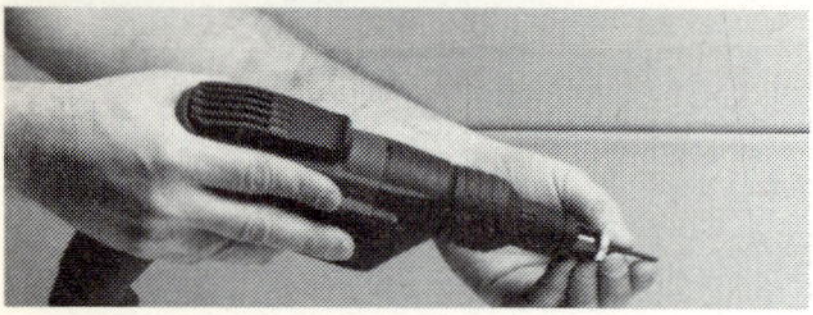

Phillips head tip holds drywall screw for driving.

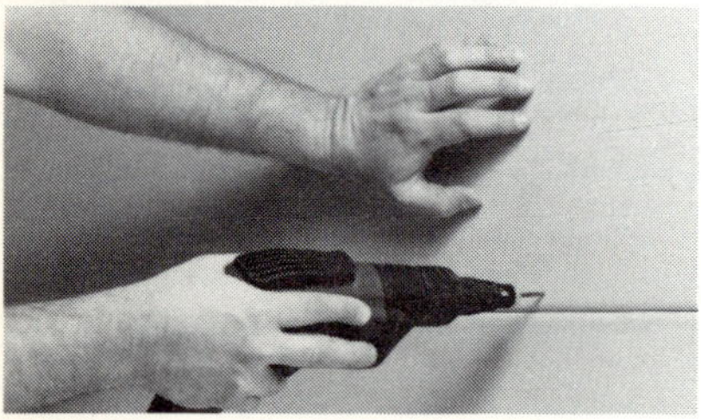

Hold screwgun as shown, not by pistol grip, to avoid stress on wrist.

The electric screwgun technique is relatively simple and a proficiency with the tool can be developed after a few hours of use. For description of screws, see Chapter 1; for screw spacing, see the fastener spacing table on next page.

Staple Application

Staples are recommended only for attaching base layer boards to wood framing in double layer assemblies. Staples should be 16-ga. flattened galvanized wire with 7⁄16″ wide crown divergent points and leg lengths to provide min. 5⁄8″ penetration into supports. Drive staples with crown perpendicular to gypsum board edges except where edges fall on supports. Drive staples so crown bears tightly against board but does not cut paper.

Nail Application

Single-Nailing Application

1. Begin nailing from abutting edge of board and proceed toward opposite ends or edges. Do not nail perimeter before nailing field of board. Ceiling application may cause board to deflect or sag in center and prevent firm fastening.

2. Position nails on adjacent ends or edges opposite each other.

3. Drive nails at least 3⁄8″ from ends or edges of gypsum board.

4. Apply hand pressure on board adjacent to nail being driven to ensure that board is in tight contact with framing member.

5. Drive nails with shank perpendicular to face of board.

6. Use a drywall hammer with crowned head for gypsum panels.

7a. For gypsum panels (drywall), seat nail so head is in a shallow, uniform dimple formed by last blow of hammer. Do not break paper or crush core at nailhead or around circumference of dimple by over-driving. Never use a nail set. Depth of dimple should not exceed 1/32″ for gypsum panels.

b. For gypsum bases (veneer plaster) nail heads should be driven flush with the board surface without dimpling.

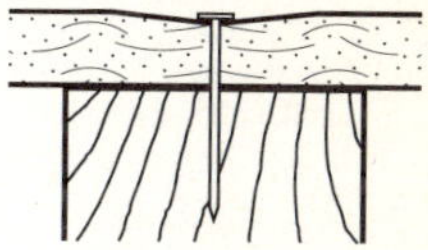

Drywall panel

Maximum Fastener Spacing—Gypsum Panel/Base Constructions[1]

Framing	Type const.	Fastener type	Location	Max. fastener spacing: Gypsum panels in	Gypsum panels mm	Gypsum base in	Gypsum base mm
wood	single layer[2] —mechanically attached	nails	ceilings	7	178	7	178
			sidewalls	8	203	8	203
		screws	ceilings	12	305	12	305
			sidewalls	16[3]	406	12	305
		screws—with RC-1 channels	ceilings	12	305	12	305
			sidewalls	12	305	12	305
	single layer—adhesively attached	nails/ screws	ceilings (perpendicular)	16″ or 406 mm o.c. at ends, edges—1 field fastener per frame member at mid-width of board		same as for gypsum panels	
			ceilings (parallel)	16″ or 406 mm o.c. along each edge and 24″ or 610 mm o.c. along intermediate framing		same as for gypsum panels	
			walls (perpendicular)	16″ or 406 mm o.c. at ends, edges—1 field fastener per frame member at mid-width of board		same as for gypsum panels	
			walls (parallel)	16″ or 406 mm o.c. along each edge		same as for gypsum panels	
	base layer of double layer—both layers mechanically attached	nails	ceilings	24	610	24	610
			sidewalls	24	610	24	610
		screws	ceilings	24	610	24	610
			sidewalls	24	610	24	610
		staples	ceilings	16	406	16	406
			sidewalls	16	406	16	406

continued on next page

Maximum Fastener Spacing—Gypsum Panel/Base Constructions[1]
continued

Framing	Type const.	Fastener type	Location	Max. fastener spacing: Gypsum panels in	Gypsum panels mm	Gypsum base in	Gypsum base mm
wood	face layer of double layer—both layers mechanically	nails	ceilings	7	178	7	178
			sidewalls	8	203	8	203
		screws	ceilings	12	305	12	305
			sidewalls	16	406	12	305
	base layer of double layer—face layer adhesively attached	nails/ screws	ceilings	7	178	7	178
			sidewalls	8	203	8	203
		screws	ceilings	12	305	12	305
			sidewalls	16	406	12	305
		staples	ceilings	7	178	7	178
			sidewalls	7	178	7	178
	face layer of double layer—face layer adhesively attached	nails/ screws	ceilings	16″ or 406 mm o.c. at ends and edges—1 field fastener per frame member at mid-width of board		same as for gypsum panels	
			sidewalls	fasten top and[4] bottom as required		same as[4] for gypsum panels	
steel	single layer	screws	ceilings	12	305	12	305
			sidewalls	16[3]	406	12	305
	base layer of double layer—both layers mechanically attached	screws	ceilings	16	406	16	406
			sidewalls	24	610	24	610
	face layer of double layer—both layers mechanically attached	screws	ceilings	12	305	12	305
			sidewalls	16	406	12	305
	base layer of double layer—face layer adhesively attached	screws	ceilings	12[4]	305[5]	12[5]	305[5]
			sidewalls	16[4]	406[5]	12[5]	305[5]
	face layer of double layer—face layer adhesively attached	screws	ceilings	16″ or 406 mm o.c. at ends and edges—1 field fastener per frame member at mid-width of board		same as for gypsum panels	
			sidewalls	fasten top and[4] bottom as required		same as[4] for gypsum panels	

(1) Fastener spacings based on wood framing 16″ o.c., steel framing 24″ o.c. Spacings are not for fire-rated assemblies; see test reports for fastener spacing for specific fire-rated assemblies. (2) See page 142 for fastener spacing using adhesive. (3) Water-resistant board spacing is 12″ o.c. (4) When board has been prebowed. For flat boards, use temporary nails or Type G screws called for in sheet or strip lamination. (5) Spacing is 8″ o.c. at joint edges.

Double-Nailing Application

In the double-nailing method for attaching gypsum board to wood framing, space the first nails 12″ o.c. along the supports in the field of the board and around the perimeter spaced 7″ o.c. for ceilings and 8″ o.c. for walls. Drive second nails about 2″ from first in field of board and make sure first nails are properly seated.

This application method helps prevent loose boards and resultant nail pops that may occur when boards are not applied correctly and drawn tightly to framing. This method will not reduce the incidence or severity of nail pops due to wood shrinkage.

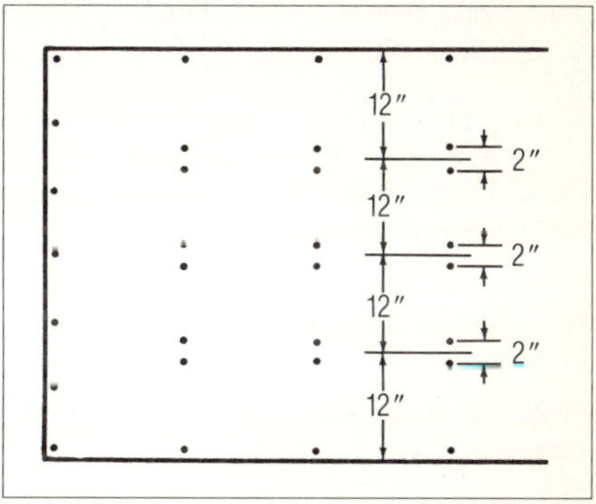

Double-nailing in field of board

Adhesive Application

In the adhesive method, a continuous bead of drywall stud or construction adhesive is applied to the face of wood framing. Adhesives should meet ASTM C557 standards. Gypsum boards are applied and attached with only a minimum number of supplementary fasteners compared to conventional fastening methods (see preceding table for fastener spacing required).

Spacing of framing members is the same as that used for conventional attachment.

Advantages of attachment with adhesives are:

1. Reduces up to 75% of the number of fasteners used and consequent problems.

2. Stronger than conventional nail application—up to 100% more tensile strength, up to 50% more shear strength.

3. Unaffected by moisture, high or low temperature; vermin-resistant.

4. Fewer loose panels caused by improper fastening.

5. Bridges minor framing irregularities.

6. Will not stain or bleed through most finishes.

Adhesives are readily available in 29-oz. cartridges and applied with hand or powered guns.

General Directions

The following recommendations will help explain the proper use of adhesives and the conditions which may affect the quality of the finished job.

1. Select the proper adhesive for specific job requirements. Read container directions carefully.

2. Make sure that all substrates are clean, sound and free from oil, dirt or contamination.

3. Exercise care regarding open flames when using flammable solvent adhesives in poorly ventilated areas.

4. Prevent freezing of adhesives.

5. Apply adhesives at temperatures between 50°F (10°C) and 100°F (38°C) except as directed by the manufacturer. Extremely high temperatures may cause solvent-base products to evaporate rapidly, shortening open time and damaging bond characteristics.

6. Close containers whenever adhesive is not in use. Evaporation (or escape) of vehicle can affect adhesive's wetting, bonding and application properties.

7. Do not exceed open time specified by manufacturer. Disregarding of directions may cause poor bonding.

8. Follow manufacturer's recommendation on proper amounts of adhesive to be applied. Too small or too large a bead will lead to performance problems or waste.

9. Apply adhesive with proper tools and as recommended by the manufacturer.

for walls **for ceilings**

Nozzle cuts

Cartridge Preparation

Cut the cartridge tip in two different ways: for walls, make a chevron or "V" cut in order to produce a round, uniform bead. The cut edge of the nozzle then rides along the stud easily.

For ceilings, use a single, angled slash across the nozzle. This gives a wipe-on effect on the ceiling joist to minimize dripping.

With a 3/8″ bead, approx. 3 to 5 gal. of adhesive will prepare framing for 1,000 sq. ft. of gypsum board. See adhesive manufacturer for specific product coverage.

Proper nozzle opening and gun position are required to obtain the right size and shape of bead for satisfactory results. Initial height of bead over framing should be 3/8″ and of sufficient volume to provide 1/16″ thickness of adhesive over the entire support when compressed.

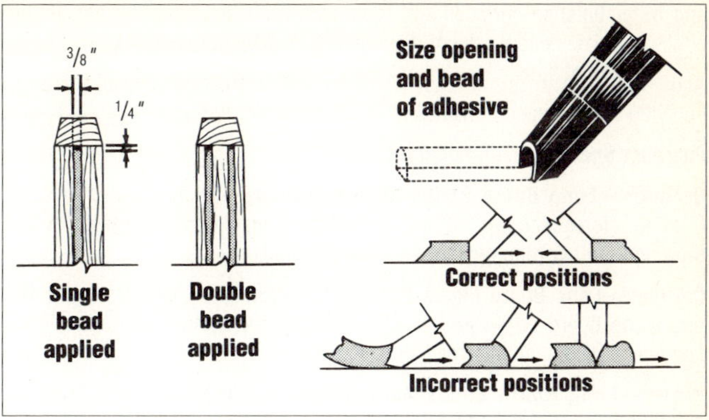

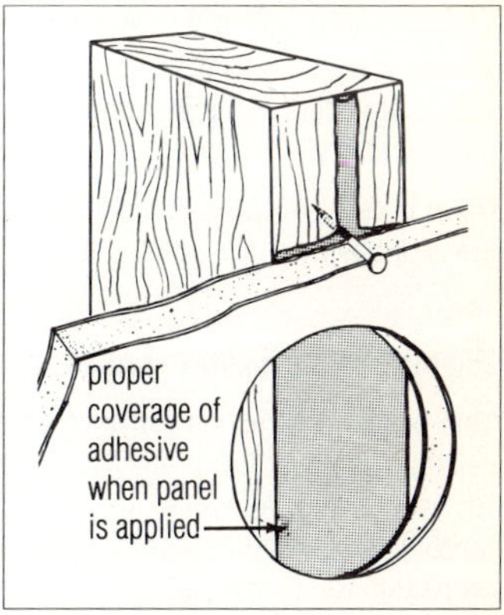

Apply adhesive in a continuous ⅜" bead in center of attachment face (at right) and to within 6" of ends of all framing members. Where two gypsum boards meet on a framing member, apply two continuous ⅜" beads to framing members at extreme edges of face, to ensure adequate contact with paper on back of board. Do not apply adhesive to members such as bridging, diagonal bracing, etc. into which no supplemental fasteners will be driven. Adhesive is not required at inside corners, top and bottom plates, bracing or fire stops and is not ordinarily used in closets.

Place gypsum boards shortly after adhesive bead is applied and fasten immediately, using proper screws or nails. After board has been fastened, impact by hand along each stud or joist to insure good contact at all points.

Where fasteners at vertical joints are objectionable (such as with predecorated panels), boards may be prebowed and adhesively attached with fasteners at top and bottom only.

Prebow boards by stacking face up with ends resting on nom. 2x4 lumber or other blocks and with center of boards resting on floor. Allow to remain overnight or until boards have a 2" permanent bow. (Under very humid conditions, board may be too flexible to assume stiff bow needed to provide adequate pressure against framing.)

To ensure good bond, no more adhesive should be applied than can be covered in 15 minutes. If adhesive is left exposed to the air for longer periods, the volatile materials will evaporate, causing surface hardness or skimming that prevents a full bond. Remove excess adhesive from board and other finished surfaces and tools with a solvent-base cleaner before adhesive dries. Follow solvent manufacturer's safety procedures.

Allow adhesive to dry at least 48 hours before treating drywall joints or applying veneer plaster finishes.

Fastener Spacing Using Adhesive

Ceilings—Long Board Edges Across Framing—Fasten board at each framing intersection and 16″ o.c. at each end. Install one field fastener per framing member required at mid-width of board.

Ceilings—Long Board Edges Parallel to Framing—Space fasteners 16″ o.c. along board edges and at each framing intersection on ends. Space fasteners 24″ o.c. on intermediate supports.

Walls—Long Board Edges Across Framing Application—Same as "Ceilings" above, except that no field fasteners are required.

Walls—Long Board Edges Parallel to Framing—Same as "Ceilings" above, except that no fasteners are required on intermediate supports. Where fasteners at the vertical joints are objectionable, prebow the gypsum board and apply fasteners 16″ o.c. only at the top and bottom of the board.

Note: If using vinyl foam tape as a temporary supplementary fastener, follow manufacturer's directions for additional fasteners required.

Wood Frame Single-Layer Application

This basic construction provides economical, quickly completed walls and ceilings wherever fire protection is desired with wood framing; also usable for wall furring. All types of gypsum boards, including predecorated vinyl-faced panels, may be used in the assembly. For measuring and cutting, perpendicular or parallel application, framing requirements and fastening, refer to sections found earlier in this Chapter. For complete information on fire and sound-resistant assemblies, refer to Construction Selector, SA-100.

Installation

Wood Studs and Joists—Apply gypsum boards so that ends and edges occur over framing members, except when joints are at right angles to the framing members as in perpendicular application or when the end joints are to be back-blocked.

To minimize end joints, use boards of maximum practical lengths. When end joints occur, they should be staggered. Arrange joints on opposite sides of a partition so they occur on different studs.

Apply gypsum boards first to the ceiling and then to the walls. If foil-back gypsum boards are used, apply foil side against framing. Fit ends and edges closely but do not force boards into place. Cut boards accurately to fit around pipes and fixtures.

Usually two men are required to install long-length board on ceilings. Fasten boards with screws or nails starting from abutting edges and working toward the opposite ends and edges. While fasteners are being driven,

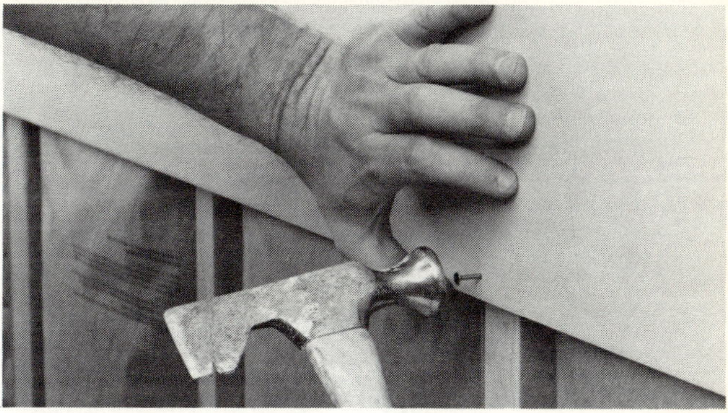

Fasteners are placed at least ⅜" from edges and ends.

the boards must be held in firm contact with the framing or joists. When single fasteners are used, attach boards to framing with screws or nails spaced as shown in the fastener spacing table on page 137. Drive fasteners at least ⅜" from edges and ends of board. On ceiling application, do not fasten perimeter before fastening field to avoid bowing of board and associated problems.

Apply gypsum boards to the sidewalls after ceilings are erected. Where long panel edges are across studs (perpendicular application), apply top wall board first, butted against ceiling. When long edges are parallel to studs (parallel application), span sidewall from ceiling to floor with a single length of board. Use parallel application where ceiling height is over 8′1″ or where this method reduces waste and joint treatment.

On sidewalls, space screws 16″ o.c. max. for gypsum panels, 12″ o.c. max. for gypsum base. Space nails 8″ o.c.

Wherever possible, use board of sufficient length to span wall areas. If joints occur near an opening, apply boards so vertical joints are centered, if possible, over opening. Keep vertical joints at least 8″ away from external corners of windows, doors, or similar openings except at interior or exterior angles within the room or when control joints are used.

After installation, exert hand pressure against wall and ceiling surfaces to detect loose fasteners. If loose fasteners are found, drive them tight. Whenever nails or screws have punctured paper, hold board tight against framing and install another fastener properly, about 1½″ from screw or nailhead that punctured paper. Remove the faulty fasteners. When nailing boards to second side of a partition, check opposite side for nails loosened by pounding and drive them tight again.

With platform framing and sidewall expanses exceeding one floor in height, fur the gypsum boards over floor joists using RC-1 Resilient Channels. As an alternative, install a horizontal control joint between gypsum boards at the junction of the bottom of top plates and the first-floor studs. Do not fasten gypsum boards to the side face of joists or headers. See details on next page.

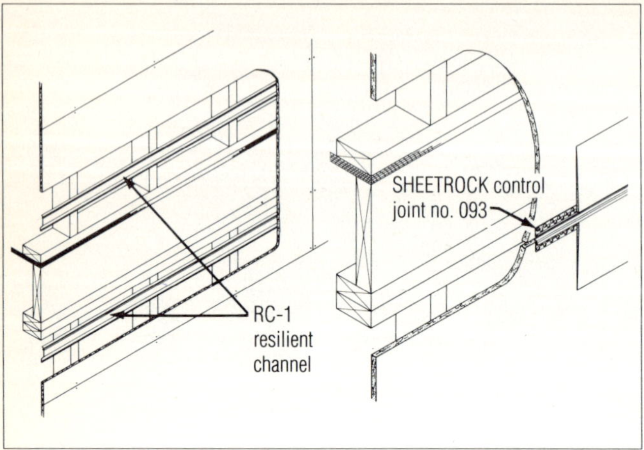

Acoustical Sealant Application

To prevent flanking and loss of the sound-control characteristics of rated partitions, SHEETROCK Acoustical Sealant must be used at all wood and steel floor runners to seal bottom edge of gypsum board and at wall angles where dissimilar materials meet. Caulking at possible leaks in all U.S. Gypsum Company systems is required to obtain comparable sound reduction to that obtained in the laboratory.

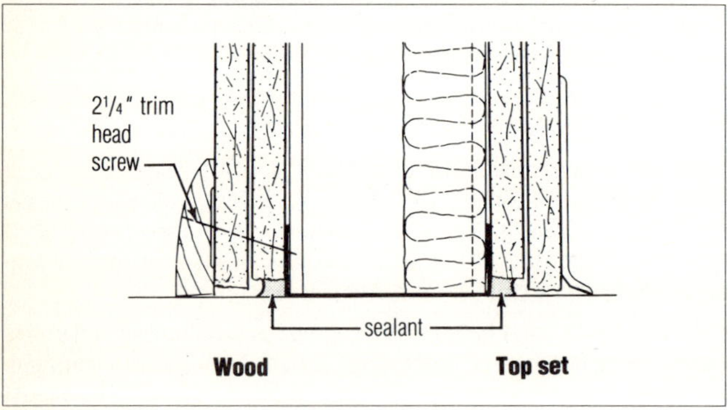

Back-Blocking Application

Back-blocking is a system designed to minimize an inherent joint deformation ("ridging") in single-layer gypsum board construction, which sometimes occurs under a combination of adverse job and weather conditions. The Back-Blocking System, developed by U.S. Gypsum Company, enables floating of end joints between studs or joists and makes it easier to form a good surface over a twisted stud or joist. The system has been widely used for years and produces outstanding results.

Back-blocking consists of laminating cut-to-size pieces of gypsum board to the back surface of boards directly behind joints to provide resistance to ridging. To install the system, follow these steps:

1. Cut backing blocks 8″ wide and long enough to fit loosely between framing.

2. Install separate gypsum strips along sides of studs, set back enough to accommodate block thickness and keep face of blocks flush with or slightly behind stud faces.

3. Spread the surface of the blocks with SHEETROCK Setting-Type (DURABOND) or Lightweight Setting-Type (EASY SAND) Joint Compound or SHEETROCK Ready-Mixed Joint Compound—Taping or All Purpose. Apply the compound in beads ½″ high, ⅜″ wide at the base, spaced 1½″ o.c.

4. Apply gypsum boards horizontally with long edges at right angles to joists. Place backing blocks along full length of edge and ends of board.

5. Immediately after all blocks are in place, erect the next board, butting ends loosely.

6. Upon fastening the abutting board, install a block and bracing as shown in the cross-section illustration. This method forms a taper that remains after bracing strips are removed.

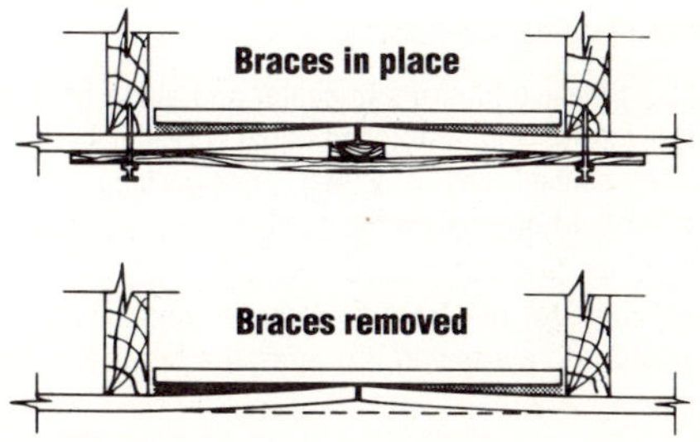

Double-Layer Adhesive Lamination

In adhesive application, face-layer gypsum boards or predecorated TEXTONE Vinyl-Faced Gypsum Panels are job-laminated to a base layer of gypsum board or interior masonry partitions.

In multilayer adhesive systems, the base layer must be attached with the same fastener, fastener spacing, and framing spacing as for a single-layer assembly of the same thickness as the base layer.

In fire-rated assemblies, permanent fasteners and the type of board used must be the same as in the particular tested assembly (see fire test report for complete description).

Application of the base layer may have long edges either parallel or perpendicular to the framing. Plan the layout of the face layer so that all joints are offset a minimum of 10″ from parallel base-layer joints. It is preferable to apply the face layer perpendicular to the base layer. At inside vertical angles, only the overlapping base layer should be attached to the framing to provide a floating corner. Omit all face-layer fasteners within 8″ of vertical angles.

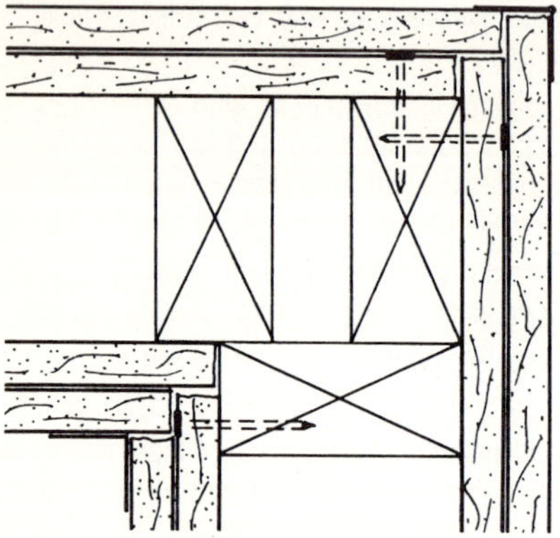

Corner detail

Application—Laminating Adhesive

Apply laminating adhesive in strips to center and along both edges of face layer board. Apply strips with a notched metal spreader having four ¼″ x ¼″ minimum notches spaced max. of 2″ o.c. Position face layer against base layer; fasten at top and bottom (vertical application) as required. For laminated ceilings, space fasteners 16″ o.c. along edges and ends, with one permanent field fastener per framing member installed at mid-width of board. Press board into place with firm pressure to ensure bond; reimpact within 24 hr. if necessary.

Application—Joint Compound Used as Laminating Adhesive

All SHEETROCK Setting-Type (DURABOND) and Lightweight Setting-Type (EASY SAND) Joint Compounds and SHEETROCK Ready-Mixed Joint Compounds—All Purpose and Taping may be used for two methods of lamination: sheet lamination and strip lamination.

When using SHEETROCK Setting-Type (DURABOND) or Lightweight Setting-Type (EASY SAND) Compounds, supplemental or temporary fasteners or supports are required until compound has hardened (minimum three hours depending on which type of compound is used). Because the compound is of heavy consistency, it provides a leveling action not obtainable with thinner-bodied adhesives.

When using SHEETROCK Ready-Mixed Joint Compounds for laminating, temporary nailing or permanent Type G screws are needed until the compound is dry (usually overnight). In cold weather, provide heat to keep compound from freezing until adhesive is dry.

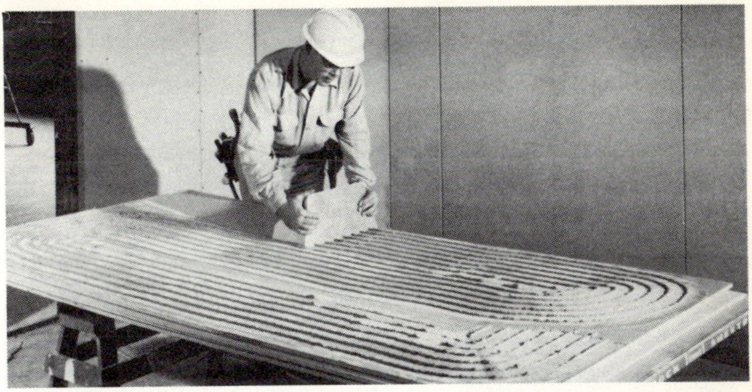

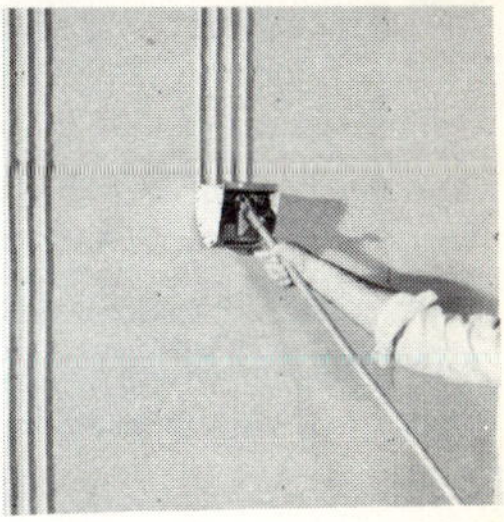

In sheet lamination (above), notched spreader is used to spread compound over entire back surface of face-layer board. In strip lamination of vertical sidewall boards (right), adhesive can be applied to either base surface or face panel. Mechanical tool used is Ames Laminating Spreader.

Mixing—SHEETROCK Setting-Type Joint Compounds

1. Mix in a clean plastic container.

2. Use only clean, drinkable water.

3. Mix according to bag directions, making sure compound is uniformly damp.

4. Do not contaminate with previously mixed SHEETROCK Setting-Type (DURABOND) or Lightweight Setting-Type (EASY SAND) Joint Compound or other compounds or dirty water as it will affect the setting time.

5. Mix only as much compound as can be used within the time period indicated on the bag—usually one hour for SHEETROCK Setting-Type (DURABOND) or Lightweight Setting-Type (EASY SAND) 90 and two hours for 210, for example.

6. The addition of extra water (retempering) will not prevent set or increase working time with SHEETROCK Setting-Type Joint Compounds.

Mixing—SHEETROCK Ready-Mixed Compounds

Use the compound at package consistency for best leveling action. If a thinner adhesive is desired, add cool water in half-pint increments to avoid overthinning. Remix lightly with a potato masher type mixer and test apply after each water addition. If compound becomes too thin, add thicker compound from the container and remix.

Application—Sheet Lamination

Cut and fit face boards prior to compound application. Spread compound over entire back surface of face layer. Use a metal spreader-blade having ⅜″ wide by ½″ high min. size notches spaced 1½″ to 2″ o.c. An Ames laminating spreader may also be used. Caution must be taken with the Ames spreader to prevent hardening of compound on tool. Clean immediately after use.

Immediately after compound is applied, position face layer and drive fasteners or install temporary bracing.

On all laminated ceilings, face layer must be permanently attached with fasteners spaced 16″ o.c. max. at ends and edges, plus one field fastener in each frame member at mid-width of board. Nails must penetrate wood framing a minimum of ¾″. Screws must penetrate steel framing a minimum of ⅜″.

On walls, permanently attach top and bottom of the face layer with fasteners driven 24″ o.c. max. (except prebowed boards). Provide temporary support fasteners, or Type G screws 24″ o.c. max. in the field of the board.

1. Temporary Nailing—Use double-headed scaffold nails driven through wood or gypsum board scraps so that nail penetrates framing a minimum of ¾″.

2. Type G Screws—Permanently attach face layer with screws driven into base layer to avoid framing. Apply compound just prior to face board erection to prevent wetting of base layer that would reduce holding power of screws. Press face layer firmly against base layer when driving screw. Compound should be thin enough to spread as screw is driven. Type G screws should not be used with base-layer boards less than ½″ thick.

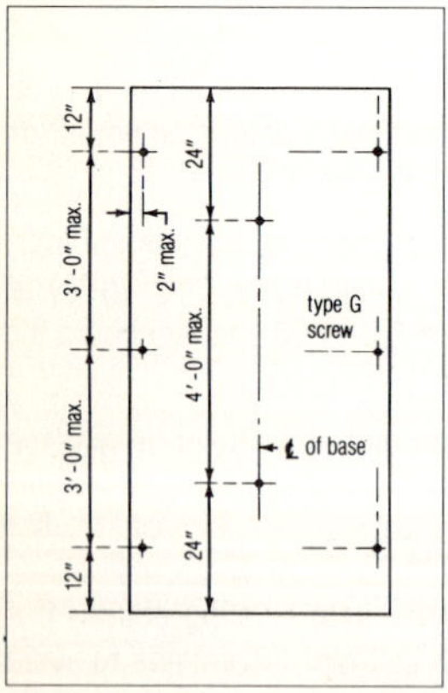

Strip Lamination (vertical face layer, sidewalls only)—This method is often preferred because it requires less compound and improves sound attenuation. Apply strips (four beads, each ⅜″ wide by ½″ high spaced 1½″ o.c.) 24″ o.c. max. Place permanent fasteners 24″ o.c. max. at each end of face layer. Drive Type G screws as shown in diagram.

Application—Liquid Contact Adhesive

Apply liquid contact adhesive according to manufacturer's directions. Use a short nap paint roller to cover both contact surfaces. Let adhesive air dry to the touch. Apply boards as soon as possible after drying occurs. On walls, fasten 16″ o.c. at top and bottom (vertical application) as required. In ceiling lamination, apply permanent supplementary fasteners at each corner of board, and along edges spaced max. 48″ o.c. Press board into place with firm pressure to ensure bond.

Resilient Board Application

Gypsum Board—Sidewalls

Apply gypsum boards perpendicular to framing with long dimension of boards parallel to resilient channels. To avoid compromising sound insulation, lift panels off floor and ensure ⅛″ relief around perimeter to be filled later with SHEETROCK Acoustical Sealant. Attach boards with 1″ Type S screws spaced 12″ o.c. along channels. Center horizontal abutting edges of boards over screw flange of channel, and screw-attach. Where channel resiliency makes screw placement difficult, use the next longer screw. Take particular care that these longer screws do not penetrate the resilient channels and enter studs since this "grounding" will nullify the resilient properties of the channels. Certain fire-rated constructions require floating vertical butt joints between studs and backing this joint with a length of RC-1 Channel. Where fire rating is not required, board may be applied with long dimension vertical.

For two-layer application of gypsum board, apply base layer vertically and attach to resilient channels with 1″ Type S screws spaced 24″ o.c. Apply face layer with long dimension at right angles to long edges of base layer and fasten with Type S screws spaced 16″ o.c., and of sufficient length to penetrate channels ⅜″ min.

Gypsum Board—Ceilings

Single Layer—Apply boards of maximum practical length with long dimension at right angles to resilient channels and end joints staggered. To avoid compromising sound insulation, ensure ⅛″ relief around perimeter to be filled later with SHEETROCK Acoustical Sealant. End joints may occur over resilient channels or midway between channels with joint floated and back-blocked with sections of RC-1 Channels. Fit ends and edges closely, but not forced together. Fasten boards to channels with 1″ Type S screws spaced 12″ o.c. in field of boards and along abutting ends. Cut boards neatly and provide support around cutouts and openings.

Two Plane—Base—For fire-rated assemblies UL Design L510 and L511, apply base layer of gypsum boards with long edges across joists and end joints staggered. Fasten boards to framing with 8d cement-coated nails spaced 7″ o.c. Attach resilient channel through base layer perpendicular to framing with 1⅞″ Type S screws. **Face**—Apply face boards in same manner as for single layer but at right angles to base layer. Fasten boards to resilient channels with 1″ Type S screws spaced 12″ o.c.

Double Layer—For fire-rated assemblies UL Design L527 and L541, apply RC-1 Channel 16″ o.c. perpendicular to joists. Attach base layer of ⅝″ gypsum boards to RC-1 Channel with 1″ Type S screws 24″ o.c. Attach face layer at right angles to base layer using 1⅝″ Type S screws 12″ o.c. in field and 1⅝″ Type G screws 8″ o.c. at butt joints. For added sound control and fire protection, install 3″ THERMAFIBER SAFB in cavity.

Steel Frame Single-Layer Application

This noncombustible assembly has won wide acceptance because of its sound attenuation, low cost, speed of erection and light weight—only 4 to 6 lb./ft.^{2}. Partitions are ideal for space division within units. Ceilings, both suspended and furred, conceal overhead structural and mechanical elements and provide a surface ready for either final decoration or adhesively applied acoustical tile.

With long edges of panels applied parallel to steel framing, worker places screws at 16″ o.c. for drywall, 12″ o.c. for gypsum base.

Gypsum Board Erection

Apply gypsum boards with long dimension parallel or perpendicular to framing. (See frame spacing tables on page 137 for limitations.) Use maximum practical lengths to minimize end joints. Position boards so all abutting ends and edges (except edges with perpendicular application) will be located in center of stud flanges. Plan direction of board installation so that lead edge or end of board is attached to open end of stud flange first. Be certain that joints are neatly fitted and staggered on opposite sides of the partition so they occur on different studs. Cut boards to fit neatly around all outlets and switch boxes.

For single-layer application, fasten boards to supports with 1″ Type S screws spaced according to fastener spacing table. Stagger screws on abutting edges or ends.

For fire-rated construction, apply gypsum boards and fasten as specified in the fire-tested assembly (see test report).

Steel Frame Double-Layer Application

Double-layer construction using steel studs offers some of the best performances in both fire and sound resistance—up to 2-hr. fire ratings and 55 STC sound rating. These economical, lightweight partitions are adaptable as party walls or corridor walls in virtually every type of new construction.

In these assemblies a face layer of gypsum board is job-laminated to the base layer or screw-attached through the base-layer gypsum board to steel studs. The installation of UNIMAST Steel Studs and Runners is the same as for single-layer application.

Base-layer Erection

Apply gypsum board with long dimension parallel to studs. Position board so abutting edges will be located in center of stud flanges. Be certain joints are neatly fitted and staggered on opposite sides of partition so they occur on different studs. For double-layer screw attachment (both layers screw-attached), fasten panels to studs with Type S screws spaced 24″ o.c. Use 1″ screws for ½″ and ⅝″ thick board. For double-layer adhesively laminated construction, fasten board with 1″ screws spaced 8″ o.c. at joint edges and 16″ o.c. in field for panels, 12″ o.c. for gypsum base. For fire-rated construction, fasten board as specified in the fire-tested design being erected (see test report).

Face-layer Erection

Apply gypsum board with long dimension parallel to studs. Position board so abutting edges will be located in center of stud flanges. Stagger joints from those in base layer and on opposite sides of partition. For double-layer screw attachment (both layers screw-attached), fasten face layer to studs with Type S screws spaced 16″ o.c. for gypsum panels, 12″ o.c. for gypsum base. Use 1⅝″ screws for ½″ and ⅝″ thick board. (As a rule of thumb, screws should be a min. ⅜″ longer than the total thickness of material to be attached to steel studs.) For double-layer laminated construction, attach face layer using adhesive lamination described earlier in this chapter. For fire-rated construction, fasten gypsum boards with screws as specified in the fire-tested design (see test report).

Steel Frame Multilayer Application

Multilayer construction, using steel studs, 1½″ or greater THERMAFIBER SAFB Insulation and ½″ or ⅝″ SHEETROCK brand Gypsum Panels, FIRECODE C Core or ½″ IMPERIAL Gypsum Base, FIRECODE C Core, offers 3- and 4-hour fire ratings and up to 65 STC sound rating. These superior assemblies are low cost, much lighter weight and thinner than concrete block partitions offering equivalent performance.

3-Layer Application

Apply gypsum panels vertically with long dimension parallel to studs (except face layer may be applied horizontally across studs). Position base so abutting edges are located in center of stud flanges. Stagger joints from those in adjacent layers and on opposite sides of the partition.

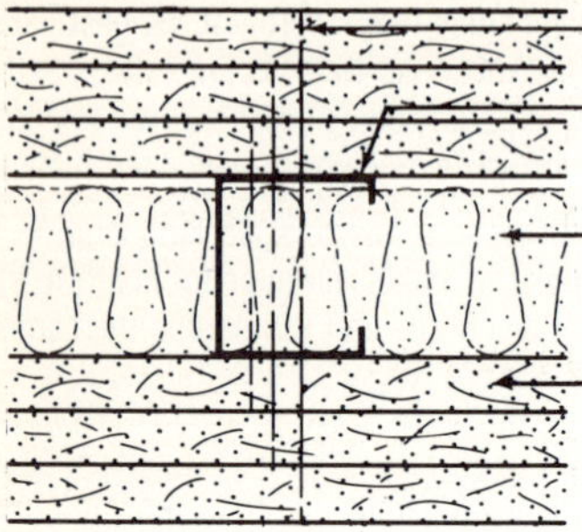

Fasten first layer to studs with 1″ Type S screws spaced 48″ o.c. Fasten second layer to studs with 1⅝″ Type S screws spaced 48″ o.c. Fasten face layer to studs with 2¼″ Type S screws spaced 12″ o.c. Horizontally applied face layer requires 1″ Type G screws in base between studs and 1½″ from horizontal joints.

4-Layer Application

Apply gypsum boards vertically with long dimension parallel to studs (except face layer may be applied horizontally across studs). Position base so abutting edges are located in center of stud flanges. Stagger joints from those in adjacent layers and on opposite sides of the partition.

Fasten first layer to studs with 1″ Type S screws spaced 48″ o.c. Fasten second layer to studs with 1⅝″ Type S screws spaced 48″ o.c. Fasten third layer to studs with 2¼″ Type S screws spaced 12″ o.c. Horizontally applied face layer requires 1½″ Type G screws in base between studs and 1″ from horizontal joints.

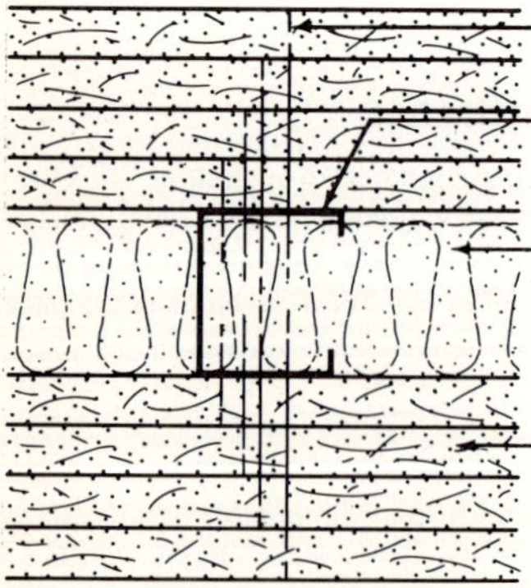

Furred Framing Board Application

Apply gypsum board of maximum practical length with long dimension at right angles to furring channel. Center end joints over channel web; fit joints neatly and accurately; stagger end joints from those in adjacent rows of board. Fasten boards to furring channels with Type S screws spaced according to fastener spacing table on page 137. Use 1″ screw length for ½″ or ⅝″ thick boards.

Masonry Single-Layer Direct Application

Gypsum boards adhesively applied directly to interior, above-grade monolithic concrete or unit masonry are laminated using a SHEETROCK Setting-Type (DURABOND) or Lightweight Setting-Type (EASY SAND) Joint Compound or SHEETROCK Ready-Mixed Joint Compound—All Purpose or Taping, or a certified subfloor plywood construction adhesive. Either regular or predecorated TEXTONE Vinyl-Faced Gypsum Panels may be applied. Use metal furring channel or Z-furring channel system for gypsum board application to interior of exterior and below-grade wall surfaces. The inside of exterior cavity walls having a continuous (1″ min.) clear air space, with the outside surfaces of the interior masonry well dampproofed, may be considered here as an interior wall surface.

Preparation

Mortar joints on surface of unit masonry to which gypsum boards are to be bonded should be cut flush with the masonry to provide a level surface. The wall surface should be plumb and true. Grind off rough or protruding areas before lamination is started. Fill pockets or holes greater than 4″ in diameter and ⅛″ deep with grout, mortar, or SHEETROCK Setting-Type (DURABOND) or Lightweight Setting-Type (EASY SAND) Joint Compound. Allow to dry before laminating.

The masonry surface must have all form oils, grease and other release agents removed. It must be dry and free of dust, loose particles and efflorescence. If masonry has been coated or painted, test by attaching a small section of board to surface. Pull from surface after allowing sufficient time for adhesive to bond. If attachment fails at bond line to masonry, the surface coating must be removed or a furring system used.

If wood base is used, attach a wood nailer to the wall with mechanical fasteners before laminating gypsum boards. Nailer should be equal to the board thickness and at least 1½″ high (or ¾″ less than wood base height).

Board Installation With Adhesive

Cut face boards to allow continuous clearance (⅛″ to ¼″) at floor. Apply SHEETROCK Setting-Type (DURABOND) or Lightweight Setting-Type (EASY SAND) Joint Compound, SHEETROCK Ready-Mixed Joint Compound—All Purpose or Taping, at center and near each board edge in strips consisting of 4 beads, ⅜″ wide by ½″ high and spaced 1½″ to 2″ o.c. Position boards vertically over wall surface, press into place and provide temporary support until adhesive is hardened.

Trim Accessory Application

Trim accessories simplify and enhance the finishing of gypsum board assemblies. The accessories are low in cost, easily applied and designed to work together for long-lasting, trouble-free construction. All are suitable for steel-frame and wood-frame construction.

Corner Bead Application

SHEETROCK Corner Reinforcements provide strong, durable protection for outside angle corners, uncased openings, pilasters, beams and soffits. The exposed nose of the bead resists impact and forms a screed for finishing. Expanded flange for veneer plaster construction provides exceptional keying and reinforcement. Corner bead should be installed in one piece unless length of corner exceeds stock bead lengths. Install as noted for each product following.

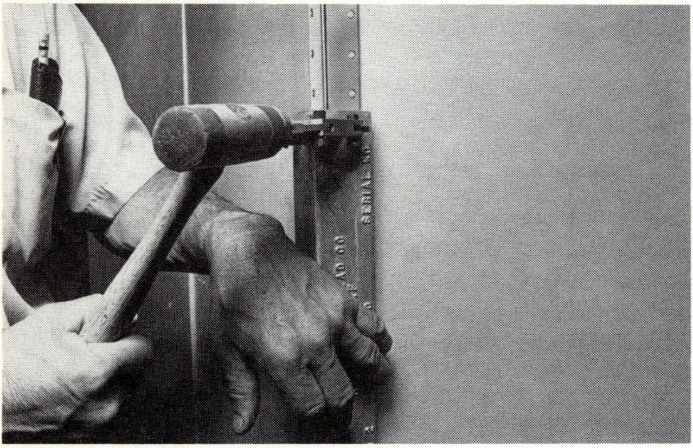

Clinch-on tool crimps solid-flange beads into place.

Nos. 800 and 900 Corner Bead and DUR-A-BEAD Corner Reinforcement

Drywall Construction—Apply DUR-A-BEAD or SHEETROCK No. 800 Corner Reinforcement with drywall nails or staples spaced 9″ o.c. in both flanges and placed opposite. Fasten to framing through board. Solid-flange beads also may be attached with a "clinch-on" tool. For No.104 (1⅛″ x 1⅛″) and No.103 (1¼″ x 1¼″) DUR-A-BEAD Corner Reinforcement, use correct size tool for flange width. Finish corner with three coats of joint compound; only two coats required with SHEETROCK Lightweight All Purpose Joint Compound Ready-Mixed (PLUS 3), SHEETROCK Setting-Type (DURABOND) or Lightweight Setting-Type (EASY SAND) Joint Compound, and SHEETROCK Lightweight All Purpose Powder Joint Compound (AP LITE).

Veneer Plaster Construction—Apply SHEETROCK No. 800 or No. 900 Corner Bead with nails or 9⁄16″ galvanized staples spaced 12″ o.c. through both flanges. Use No. 800 for one-coat applications, No. 900 for two-coat applications. On masonry corners, hold bead firmly against corner and grout both flanges with veneer plaster. On monolithic concrete apply a high-grade bonding agent over corner before placing bead and grouting. Preset all beads with a veneer finish.

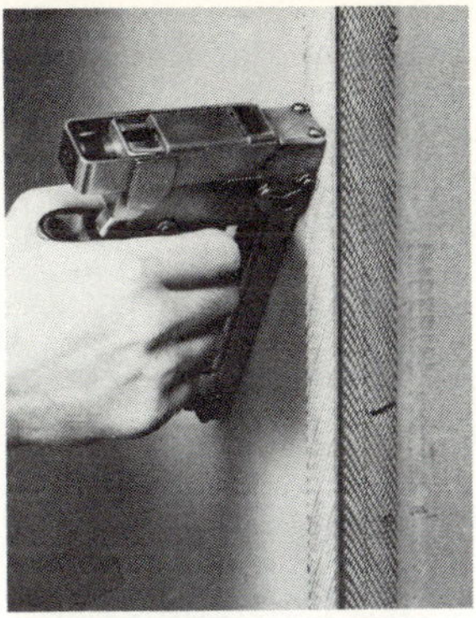

Stapling is standard way to attach SHEETROCK No. 800 Corner Bead.

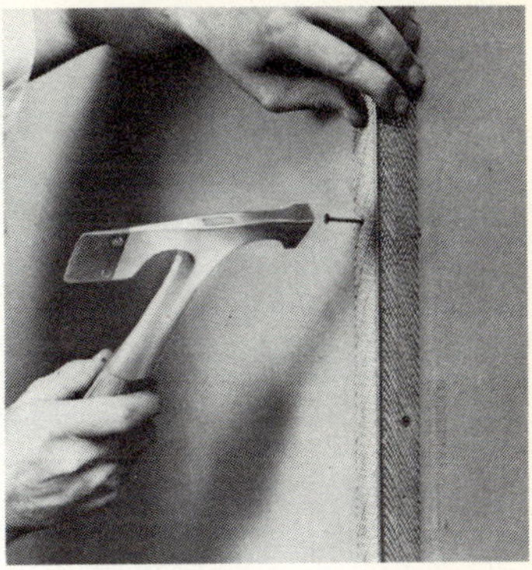

For wood studs, nails in both bead flanges are also satisfactory.

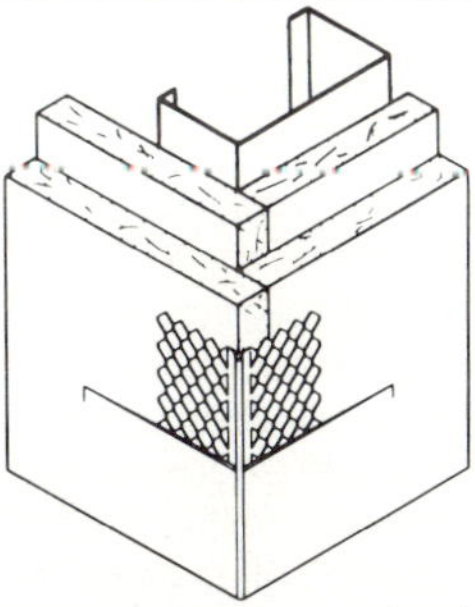

No. 800 & 900 corner bead

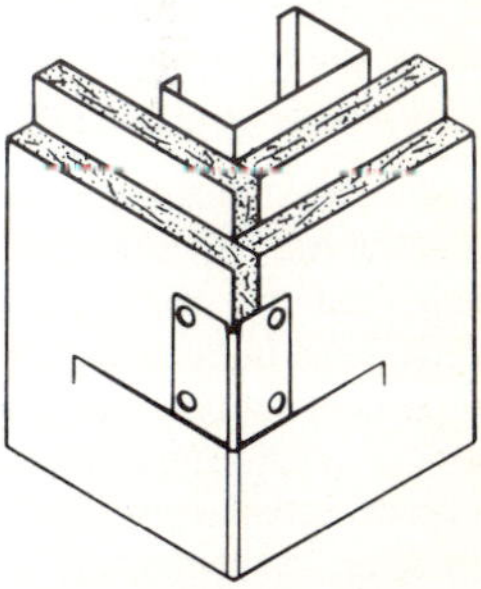

DUR-A-BEAD corner bead

SHEETROCK Flexible Metal Corner Tape

Cut tape to length desired with snips or score with knife and bend. Notch or angle cut for arches and window returns. Do not overlap at intersection or corners. Apply joint compound to both sides of corner angle, fold tape at center to form bead and press metal strip side into joint compound. Follow immediately with a thin coat of compound over tape and let dry. Finish corner in conventional manner with additional coats of joint compound.

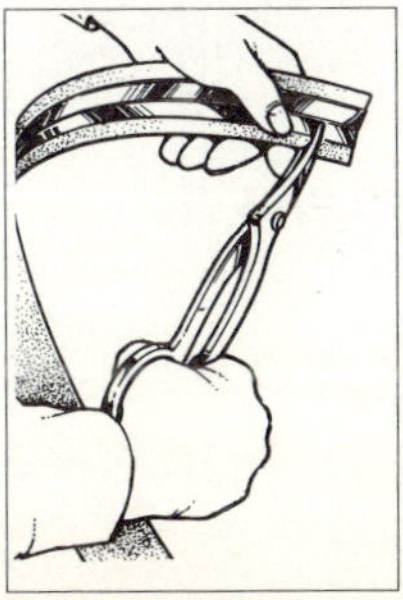

Cut tape with snips.

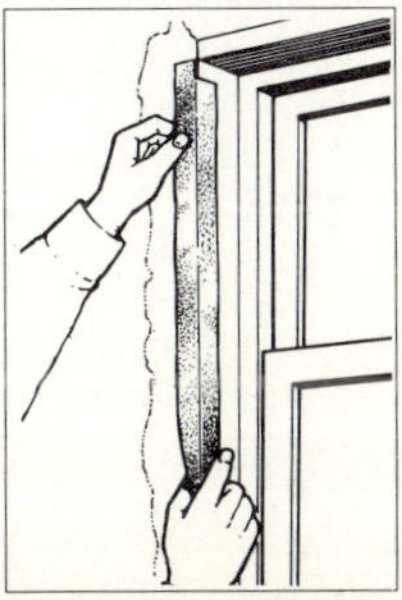

Embed in joint compound.

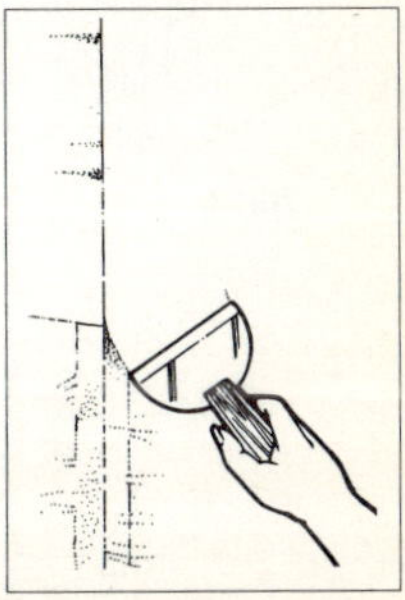

Finish strong to corner.

Metal Trim Application

SHEETROCK Metal Trim serves to protect and finish gypsum boards at window framing and door jambs; also used at ceiling-wall intersections and partition perimeters to form recess for acoustical sealant. Also serves as a relief joint at the intersection of dissimilar constructions such as gypsum board to concrete.

Drywall Construction

No. 200-A SHEETROCK Metal Trim (½″ and ⅝″ size)—Apply gypsum panels, omitting fasteners at framing member where trim is to be installed. Leave a space ⅜″ to ½″ wide between edge of panel and face of jamb. This provides space for installation of hardware. Slip trim over edge of panel with wide knurled flange on room side and fasten trim and panel to framing. Use same type fasteners used to attach panels; space fasteners 9″ o.c. max. Finish with three coats of conventional joint compound; only two coats required with SHEETROCK Lightweight All Purpose (PLUS 3) Ready-Mixed Joint Compound, SHEETROCK Setting-Type (DURABOND) or Lightweight Setting-Type (EASY SAND) Joint Compound, and SHEETROCK Lightweight All Purpose Powder Joint Compound (AP LITE).

No. 200-B SHEETROCK Metal Trim (½″ and ⅝″ size)—Apply gypsum panels same as for No. 200-A Trim, omitting fasteners and leaving ⅜″ to ½″ space at jamb. Place trim over edge of panel with knurled flange exposed. Attach trim and panel to framing with fasteners spaced 9″ o.c. max. Finish with three coats of conventional joint compound; only two coats required with SHEETROCK Lightweight All Purpose (PLUS 3) Ready-Mixed Joint Compound, SHEETROCK Setting-Type (DURABOND) or Lightweight Setting-Type (EASY SAND) Joint Compound, and SHEETROCK Lightweight All Purpose Powder Joint Compound (AP LITE).

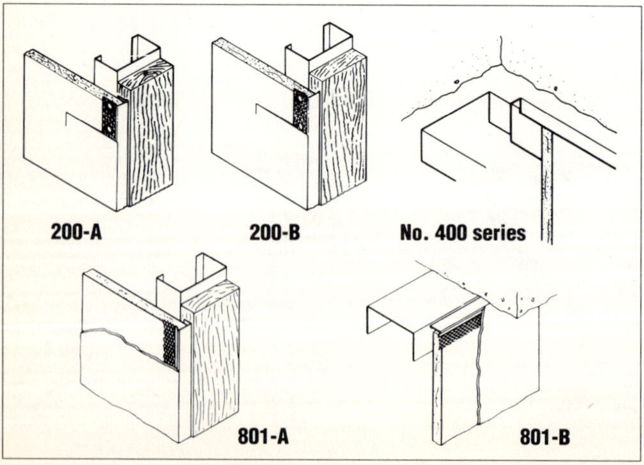

Nos. 400, 401 and 402 SHEETROCK Metal Trim (⅜″, ½″ and ⅝″ size)—Apply trim to wall before gypsum panels go up, by nailing through trim flange into framing; board is held firmly in place by short leg of trim. No additional edge fastening is necessary. Space fasteners 9″ o.c.

Veneer Plaster Construction

No. 701-A and 701-B SHEETROCK Metal Trim (½″ and ⅝″ size)—Slip channel-type 701-A Trim over edge of base or position L-shaped 701-B Trim on edge of base with expanded flange on room side. Fasten with staples or nails 12″ o.c. max. Both trims are designed for two-coat veneer plaster finish application.

Drywall or Veneer Plaster Construction

No. 801-A and 801-B SHEETROCK Metal Trim (½″ and ⅝″ size)—Slip channel-type 801-A Trim over edge of board, or position L-shaped 801-B Trim on edge of board with expanded flange on room side. Fasten with staples or nails 9″ o.c. max. for drywall applications, 12″ o.c. max. for veneer plaster assemblies. For drywall, finish with three coats of conventional joint compound; only two coats required with SHEETROCK Lightweight All Purpose Joint Compound Ready-Mixed (PLUS 3), SHEETROCK Setting-Type (DURABOND) or Lightweight Setting-Type (EASY SAND) Joint Compound, and SHEETROCK Lightweight All Purpose Joint Compound (AP LITE). For veneer plaster, finish with one-coat veneer plaster application.

Vinyl and Plastic Trim Application

For structural relief and effective sound control, USG Vinyl Trim finishes edges of gypsum panels and acts as a seal where the panel adjoins dissimilar structural surfaces. Install as follows:

USG P-1 Vinyl Trim (P-1A, ½″ size; P-1B, ⅝″ size)—Apply trim to edges of gypsum panels that will abut ceilings or walls. Slip trim over edges of panels for friction fit. Position panels, press trim edges against abutting surfaces for a snug contact and attach in conventional manner. Use same procedure for P-1A and P-1B.

TEXTONE Plastic Moulding Application

Solid-Color TEXTONE Plastic Mouldings (RP Series)—Available in solid colors (almond and ash blue), these mouldings are used to cover panel joints and edges, protect corners and as trim around openings. (For installation, see page 163—procedure is same as for TEXTONE Mouldings with predecorated panels.)

Control Joint Application

Proper installation of control joints in wall and ceiling membranes should include breaking the gypsum boards behind the control joint. In ceiling construction, the framing should also be broken, and in partitions, separate studs should be used on each side of control joints. Control joints should be positioned to intersect light fixtures, air diffusers, door openings and other areas of stress concentration.

Gypsum construction should be isolated with control joints where (a) partitions or ceilings of dissimilar construction meet and remain in the same plane; (b) wings of "L," "U" and "T" shaped ceiling areas are joined; and (c) expansion or control joints occur in the base wall construction and/or building structure. Just as important, control joints should be used in the face of gypsum partitions and ceilings when the size of the surface exceeds the following control-joint spacings; **Partitions**, 30′ maximum in either direction; **Interior Ceilings** (with perimeter relief), 50′ maximum in either direction; **Interior Ceilings** (without perimeter relief), 30′ maximum in either direction; and **Exterior Ceilings**, 30′ maximum in either direction.

Control joints will not accommodate transverse shear displacement on opposing sides of joints. A joint detail comprising casing beads on each side of joint opening is typically used to accommodate expansion, contraction and shear. Such joints require special detailing by designer to control sound and fire ratings where applicable as well as dust and air movement. In exterior walls, particular attention is required to resist wind, driving rain, etc., by adequate flashing, backer rods, sealants and gaskets as required.

Ceiling-height door frames may be used as vertical control joints for partitions; however, door frames of lesser height may only be used as control joints if standard control joints extend to the ceiling from both corners of the top of the door frame. When planning locations for control joints in the ceiling, it is recommended that they be located to intersect column penetrations, since movement of columns can impose stresses on the ceiling membrane.

Control joints, when properly insulated and backed by gypsum panels, have been fire-endurance tested and are certified for use in one- and two-hour rated walls.

Installation

At control joint locations:

1. Leave a ½″ continuous opening between gypsum boards for insertion of surface-mounted joint.

2. Interrupt wood floor and ceiling framing with a ½″ gap, wherever there is a control joint in the structure.

3. Provide separate supports for each control joint flange.

4. Provide an adequate seal or safing insulation behind control joint where sound and/or fire ratings are prime considerations.

Fire-rated control joints (WHI-495-PSV-0824, 0824A)

Two-hour rated steel stud partitions

47STC SA-860302

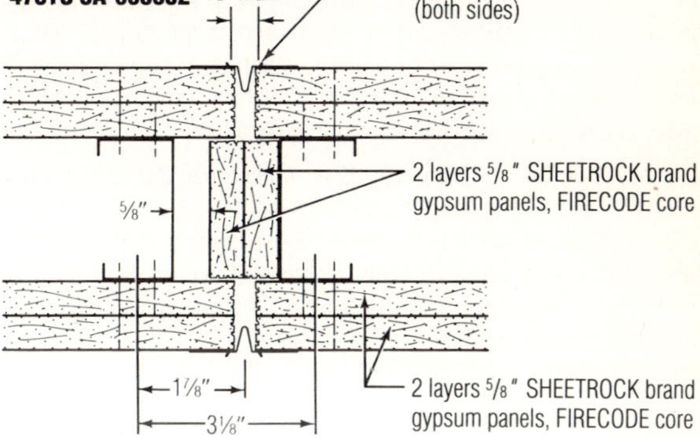

47ST SA-860217

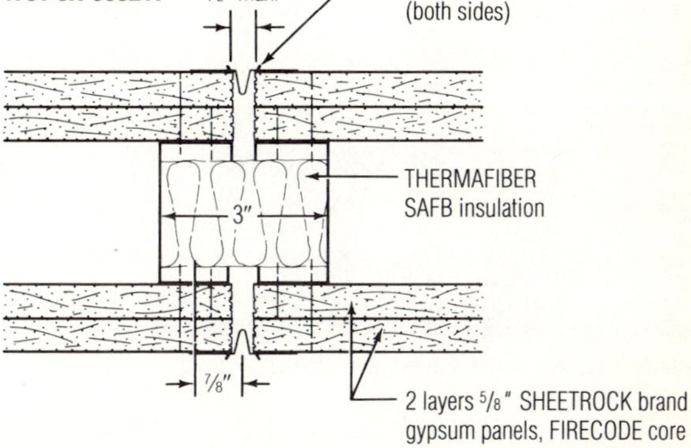

One-hour rated steel stud partitions

47STC SA-860302

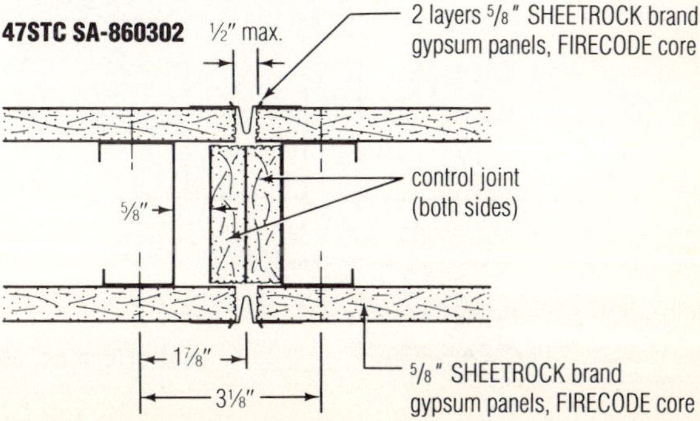

SHEETROCK Zinc Control Joint No. 75—Apply to bottom of double row of wood joists in radiant-heated ceilings before IMPERIAL Gypsum Base is applied. Attach control joint to joists with Bostitch $9/16$″ Type G staples or nails spaced 6″ o.c. max. along each flange. Splice end joints with two pieces 16-ga. galvanized tie wire inserted in the sections. Apply gypsum base over control joint attachment flange and fasten to joist with proper fastener (see fastener selector guide in Chapter 1). Space nails 7″ o.c., screws 12″ o.c. Use control joint as a screed for applying veneer plaster. Remove plastic tape after veneer plaster application.

SHEETROCK Zinc Control Joint No. 093—Apply over face of gypsum board where specified. Cut to length with a fine-toothed hacksaw (32 teeth per in.). Cut end joints square, butt together and align to provide neat fit. Attach control joint to gypsum board with Bostitch $9/16$″ Type G staples, or equivalent, spaced 6″ o.c. max. along each flange. Remove plastic tape after finishing with joint compound or veneer plaster.

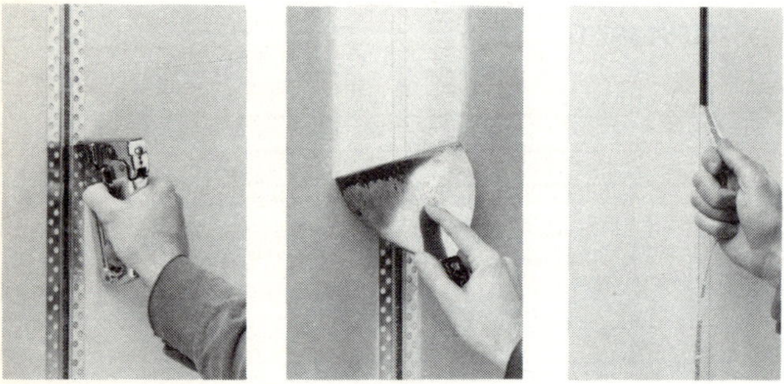

Control Joint No. 093 stapled, finished, tape removed.

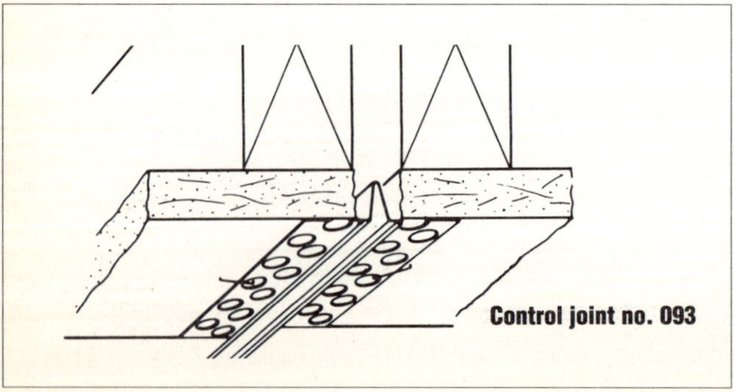

Max. Spacing-Control Joints

Construction & location	Max. single dimension ft	m	Max. single area ft²	m²
Partition-interior	30	9	—	—
Ceiling-interior				
with perimeter relief	50	15	2500	230
without perimeter relief	30	9	900	85
Ceiling-exterior gypsum	30	9	900	85

Predecorated Panel Application

The use of predecorated gypsum panels takes full advantage of the real economy of fire-resistant gypsum panels in providing highly serviceable, quickly installed decorative walls. With TEXTONE Vinyl-Faced Gypsum Panels, walls resist stains and scuffs, are readily washable and colorfast.

TEXTONE Vinyl-Faced Gypsum Panels have good resistance to dimensional change. (See Appendix for hygrometric and thermal coefficients.)

TEXTONE Vinyl-Faced Gypsum Panels are applied vertically to the walls so that ends occur at floor and ceiling lines. The beveled edges form an attractive joint not requiring joint treatment. Panels are not practical as a ceiling finish, as end joints are difficult to conceal. They can be used with wood or steel studs, in single or double-layer application, in new construction or over plaster or gypsum panel surfaces in remodeling; may also be applied to furring attached to masonry. Not recommended for use over foil-back panels in exterior walls. For additional information, fire-rated construction and technical data, see Technical Folders WB-1330 and SA-928.

Panel Installation

When installing patterns other than one-color patterns, such as Pumice, place panels against wall, inverting alternate panels, and rearrange to obtain the best match in pattern and tone; there will be a slight variation from panel to panel. Number backs of panels for proper installation sequence. Panels used in the same area should be of the same lot number for best color match (lot numbers are imprinted on panel backs).

Apply panels vertically. Position less-than-full-width panels with cut edge at corner where the raw edge can be overlapped by the abutting panel or covered with a corner mould. Use color-matched nails for nail-on application. Drive nails with plastic-headed hammer or rawhide mallet. Space 1⅜″ nails at least ⅜″ from ends and edges, 8″ o.c.

Cut panels with a sharp knife. Cut through vinyl film into core; then snap board and cut back paper.

TEXTONE Mouldings in matching colors and patterns are available to finish edges and conceal fasteners in TEXTONE Vinyl-Faced Gypsum Panel installations.

Prebowing—Where fasteners at the vertical joints are objectionable, panels may be prebowed, adhesively applied and fastened at top and bottom only.

Prebow by stacking panels face up with ends resting on nom. 2x4 lumber on edge or gypsum panel slutters and center of panel resting on floor. Allow to remain overnight or until panels show at least a 2″ permanent bow. During high humidity, it may be necessary to elevate ends as much as 8″ to achieve desired permanent bow.

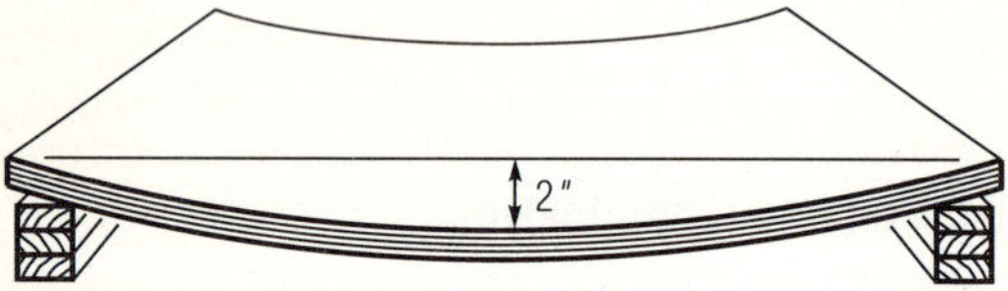

Adhesives—Generally, most water-based adhesives and some solvent-based adhesives may be used to install TEXTONE Vinyl-Faced Gypsum Panels. However, many solvent-based adhesives may not be compatible and could result in delamination and/or discoloration of the vinyl surface. It is recommended that 24 hours before installation, a small piece of paneling be test-laminated to the actual framing or backing with the actual adhesive. If results are acceptable after 24 hours, the full scale job can begin. Also, check the adhesive manufacturer's recommendations before use with vinyl-covered panels.

The following commercially available adhesives may be used for applying TEXTONE Vinyl-Faced Gypsum Panels in non-fire rated assemblies: drywall stud adhesive (meeting ASTM C557) for application to wood or steel studs; laminating adhesive for bonding panels to monolithic concrete, concrete block, wood and mineral-fiber sound deadening board, polystyrene and urethane rigid-foam insulation and most other wall surfaces; contact adhesive for laminating TEXTONE Vinyl-Faced Gypsum Panels to gypsum base-layer panels. Vinyl foam tape may be used with adhesive for supplemental attachment (in lieu of prebowing or temporary shoring) until permanent adhesive attains ultimate strength.

A SHEETROCK Setting-Type (DURABOND) or Lightweight Setting-Type (EASY SAND) Joint Compound or SHEETROCK Ready-Mixed Joint Compound-All Purpose or Taping and mechanical fasteners are required for fire-rated construction (see page 145, “Double-Layer Adhesive Application”).

Adhesive Application to Wood or Steel Studs—Apply 8″ strip of vinyl foam tape to face of each stud, positioned at midpoint of studs up to 8′ long, at third-points on studs up to 12′ long and quarter-points on studs over 12′. Where no mechanical fasteners are to be used at top or bottom of stud, apply an 8″ strip of tape. Apply a continuous ⅜″ bead of drywall stud adhesive to the entire face of studs between vinyl foam tape. Immediately apply TEXTONE Vinyl-Faced Gypsum Panels vertically and apply sufficient pressure to ensure complete contact with both tape and adhesive.

Adhesive Application to Base Layer of Gypsum Panels—Apply liquid contact adhesive to back of TEXTONE Vinyl-Faced Gypsum Panels and face of base layer according to manufacturer's directions. Allow adhesive to air dry, then bring panels in contact. Impact entire surface to ensure complete contact.

Adhesive Application to Base Layer of Masonry, Gypsum Board, Wood or Mineral Fiber Board—For interior masonry walls and gypsum board, apply continuous strips of vinyl foam tape to entire width of panel back at midpoint and ⅜″ from each end. Spread laminating adhesive over entire area of panels between tape using notched metal spreader with ¼″ x ¼″ notches spaced 2″ o.c. Position panel and immediately apply sufficient pressure to ensure complete contact over entire surface. (Mechanical fasteners may be substituted for tape at ends of panels.)

For application of TEXTONE Vinyl-Faced Gypsum Panels to wood or mineral board, prebow panels and apply laminating adhesive over entire back surface. Use mechanical fasteners at top and bottom of panel.

TEXTONE Mouldings Application

Rigid vinyl trim and mouldings are available in solid and matching colors to finish TEXTONE Vinyl-Faced Gypsum Panel installations. Refer to page 39 for product description and colors available.

Installation

General—Store mouldings at room temperature for 24 hr. before installation. Start installation from corner or door. Be sure that starting points are plumb and level. Fasten mouldings with flat-head wire nails, staples or drywall screws 12″ o.c. Fasten snap-on mouldings with nails or screws driven through holes in retainer. Use a fine-toothed hacksaw to cut mouldings. For mitering, use the same procedures as with wood moulding. Cut mouldings ¹⁄₁₆″ short for a loose fit to allow for thermal expansion; never force mouldings into place.

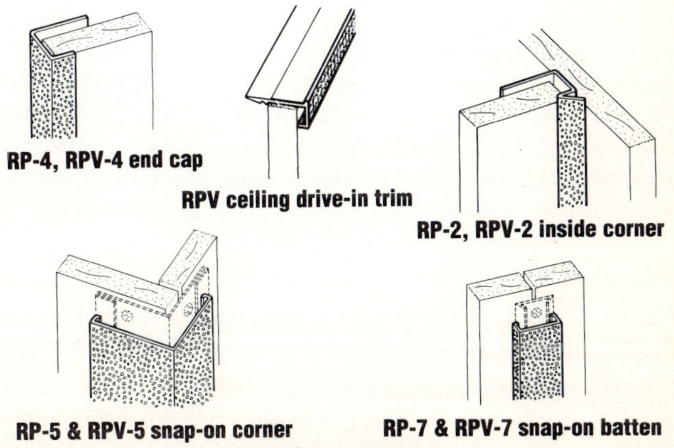

TEXTONE mouldings

RP-2 and RPV-2 Inside Corner—Install first panel so that vertical edge aligns with framing. Apply moulding over first panel, fastening exposed flange to framing. Insert opposite panel into moulding.

RP-4 and RPV-4 End Cap—Align and fasten end cap to framing. Insert panel into moulding and apply panel to wall.

RP-5 and RPV-5 Snap-on Corner—Apply panels, then place retainer strip over joint and fasten with nails or screws through holes provided. Snap corner face over retainer strip.

RP-7 and RPV-7 Snap-on Batten—Apply panels, then place retainer strip over joint and fasten with nails or screws through holes provided. Snap batten face over retainer strip.

RP and RPV-46 Ceiling Drive-in Trim—Use only with steel stud partitions. Install after panels are applied. Insert grooved flange between runner and ceiling; tap trim into place. (Not recommended where perimeter must be acoustically sealed.)

Painting—If mouldings other than TEXTONE Mouldings are used, they should be decorated prior to application over panels. Avoid applying masking tape to mouldings or predecorated panels when decorating.

RP-series mouldings should be used when painting is required. A good quality alkyd enamel or acrylic latex paint is recommended. Apply according to manufacturer's directions.

Water-Resistant Gypsum Panel Application

SHEETROCK brand Gypsum Panels, Water-Resistant, and Water-Resistant FIRECODE or Water-Resistant FIRECODE C Core, are superior bases for the adhesive application of ceramic tile. For use in new construction in wet areas such as bathrooms, powder rooms, kitchens and utility rooms. They install quickly and easily to wood or steel framing or furring using standard attachment methods. Do not apply water-resistant board to ceilings.

Exposed edges and joints in areas to be tiled are treated with a coat of thinned down ceramic tile mastic or an approved waterproof flexible sealant. Joints are treated with SHEETROCK Setting-Type (DURABOND 45 or 90) or SHEETROCK Lightweight Setting-Type (EASY SAND 45 or 90) Joint Compound and SHEETROCK Joint Tape.

Where water-resistant panels are used in remodeling, old wall surfaces must be removed and water-resistant panels applied to exposed studs as in new construction. Refer to page 32 for other limitations.

Installation

Framing—Check alignment of framing. If necessary, fur out studs around shower receptor so that inside face of lip of fixture will be flush with gypsum panel face.

Install appropriate blocking, headers, or supports for tub and other plumbing fixtures, and to receive soap dishes, grab bars, towel racks or similar items. SHEETROCK brand Gypsum Panels, Water-Resistant, are designed for framing 16″ o.c. but not more than 24″ o.c. When framing is spaced more than 16″ o.c., or when ceramic tile more than 5⁄16″ thick will be used, install suitable blocking between studs. Place blocking approx. 1″ above top of tub or receptor and at midpoint between base and ceiling. Blocking is not required on studs spaced 16″ o.c. or less. Vapor retarders must not be installed between water-resistant panels and framing.

Receptors—Install receptors before panels are erected. Shower pans, or receptors, should have an upstanding lip or flange at least 1″ higher than the water dam or threshold at the entry to the shower.

Gypsum Panels—After tub, shower pan or receptor is installed, place temporary ¼″ spacer strips around lip of fixture. Cut panels to required sizes and make necessary cut-outs. Before installing panels, apply thinned ceramic tile mastic to all cut or exposed panel edges at utility holes, joints and intersections.

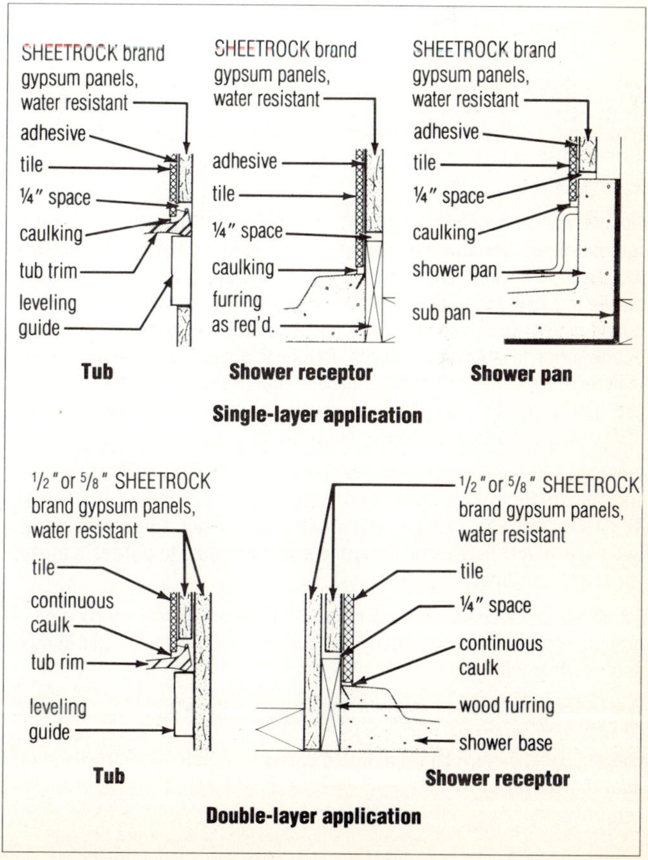

Install panels perpendicular to studs with paper-bound edge abutting top of spacer strip. Fasten panels with nails 8″ o.c. max., or screws 12″ o.c. max. Where ceramic tile more than 5⁄16″ thick will be used, space nails 4″ o.c. max. and screws 8″ o.c. max.

For tile 5⁄16″ thick or less, panels may be installed with stud adhesive (meeting ASTM C557) to wood or steel framing. Apply 3⁄8″ bead to stud faces—two beads on studs where panels join. Do not apply adhesive to blocking where no fasteners will be used. Position panel and drive nails or screws at 16″ intervals around perimeter, 3⁄8″ from edges.

For double-layer applications, both face and base layer must consist of SHEETROCK brand Gypsum Panels, Water-Resistant.

In areas to be tiled, treat all fastener heads with SHEETROCK Setting-Type (DURABOND 45 or 90) or Lightweight Setting-Type (EASY SAND 45 or 90) Joint Compound. Fill tapered edges in gypsum panel completely with compound, embed SHEETROCK Joint Tape firmly, and wipe off excess compound. When hardened, apply a second or skim coat over the taping coat, being careful not to crown the joint or to leave excess compound on panel (DURABOND compound is difficult to sand and remove when dry). For butt joints and interior angles, embed SHEETROCK Joint Tape with SHEETROCK Setting-Type (DURABOND 45 or 90) or SHEETROCK Lightweight Setting-Type (EASY SAND 45 or 90) Joint Compound without crowning the joints. A fill coat is not necessary. Spot fastener heads at least once with SHEETROCK Setting-Type (DURABOND) or SHEETROCK Lightweight Setting-Type (EASY SAND) Compound.

Fill and seal all openings around pipes, fittings and fixtures with a coat of thinned-down ceramic tile mastic or an approved waterproof flexible sealant. To thin water-based mastic, add one-half pint of water per quart of mastic to make a paint-like viscosity. With a brush, apply the thinned compound onto the raw gypsum panel core at cut-outs. Allow areas to dry thoroughly prior to application of tile. Before compound dries, wipe excess material from surface of gypsum panels. Remove spacer strips but do not seal gap at bottom edge of panels. Install tile down to top edge of shower floor or tub and overlapping lip or return of tub or receptor.

For areas not to be tiled, embed tape with SHEETROCK Setting-Type (DURABOND 45 or 90) or SHEETROCK Lightweight Setting-Type (EASY SAND 45 or 90) Joint Compound in the conventional manner. Finish with at least two coats of a U.S. Gypsum Company joint compound to provide a treated surface for painting and wallpapering.

Fill all tile joints continuously with grout. Apply nonsetting caulking compound, such as tub caulk, between wall surfacing material and shower floor, curb or tub rim.

Where SHEETROCK brand Gypsum Panels, Water-Resistant, are to be painted with a gloss enamel and subject to critical lighting, it is recommended that the panel surface be skim coated with a conventional joint compound prior to painting.

Gypsum Sheathing Application

Gypsum sheathing is not intended for use where the exposed surfacing material is to be adhesively applied, with no mechanical fasteners. Refer to page 37 for other limitations.

Installation

SHEETROCK brand Gypsum Sheathing—Apply ½″x24″x8′ SHEETROCK brand Gypsum Sheathing horizontally with face side out (paper on back is lapped at edges). With tongue and groove edge, *place tongue edge up to prevent water penetration at joints.* Use diagonal bracing where necessary. Space 1¾″ 11-ga. galvanized nails 8″ o.c. at each stud.

Apply ½″ or ⅝″x48″x8′ or 9′ SHEETROCK brand Gypsum Sheathing vertically with face side out. Space nails 8″ o.c. on framing members.

For staple or screw application, use same fastener spacing as for nails. Drive staples parallel to long dimension of framing, heads flush with sheathing surface but not breaking face paper.

GYP-LAP Gypsum Sheathing—Apply 2′ wide GYP-LAP Gypsum Sheathing horizontally with tongue edge up, or 4′ wide sheathing vertically, to braced framing. Use 11-ga. galvanized roofing nails 1¾″ long, spaced 8″ o.c.

USG Triple-Sealed Gypsum Sheathing—Apply 4′ wide USG Triple-Sealed Gypsum Sheathing vertically to braced framing. Use 11-ga. galvanized roofing nails 1¾″ long, spaced 8″ o.c.

When required for any of the sheathing products above, install 1″x4″ diagonal corner braces, or equal, at all external corners, let into face of studs, corner posts, sill and plates, or as required by applicable code.

Refer to U.S. Gypsum Company folder GS-116 for complete data on gypsum sheathing. Refer to separate Technical Folder SA-923 for sheathing application to exterior steel framing systems.

Interior Gypsum Ceiling Board Application

½″ SHEETROCK brand Interior Gypsum Ceiling Board is specially formulated to support water-based spray texture paints and overlaid insulation with the same sag resistance as regular ⅝″ gypsum board but without the weight. Can be substituted for regular ½″ board in other applications, reducing waste and lowering in-place cost. Ideal for new construction or renovation over wood or steel framing.

Handling—Store and handle ½″ SHEETROCK brand Interior Gypsum Ceiling Board in the same manner as other gypsum board. Stack flat and store under cover.

Installation—Apply ½″ SHEETROCK brand Interior Gypsum Ceiling Board to ceilings before applying gypsum boards to walls. Joists must be spaced 24″ o.c. or less. Board may be cut by scoring and snapping in the same manner as other gypsum board.

Begin application at one side of room in a corner. Place boards with long dimension perpendicular to framing and fasten with 1¼″ drywall nails

(ASTM C514) spaced 7″ o.c. or 1¼″ Type W screws (ASTM C1002) spaced 12″ o.c. Hold the board firmly in place against framing and drive fasteners beginning at butt end of board and work to opposite end. Adhesive nail-on fastening improves bond strength and reduces face nailing. Drive fasteners so that heads are slightly below the board surface, taking care not to break the face paper or crush the core. Finish with a U.S. Gypsum Company joint treatment system.

In new construction or renovation applications using steel furring channels (RC-1 Resilient Channels, UNIMAST Metal Furring Channels or SHEETROCK Z-Furring Channels), space steel channels a maximum of 24″ o.c. fastened to bottom of joists. Maximum joist spacing for RC-1 Resilient Channel and SHEETROCK Z-Furring Channel is 24″. Refer to page 107 for span capacities of both DWC-20 and DWC-25 metal furring channels.

Surface Preparation—Before texturing, apply a high-quality, undiluted latex or alkyd primer/sealer. Follow manufacturer's directions for application.

Sagging—To minimize sag in new gypsum panel ceilings, the weight of overlaid unsupported insulation should not exceed 1.3 psf with frame spacing 24″ o.c. and 2.2 psf with frame spacing 16″ o.c. (24″ o.c. for SHEETROCK brand Interior Gypsum Ceiling Board). A separate vapor retarder should be installed where required in roofed ceilings, and the plenum or attic space vented with a min. ½-sq. in. net-free vent area per sq. ft. of horizontal space.

See "Ceiling Sag Precautions" on page 362 for more information on the application of water-based textures and interior finishing materials.

Exterior Gypsum Ceiling Board Application

SHEETROCK brand Exterior Gypsum Ceiling Board embodies a specially treated gypsum core encased in chemically treated fiber paper. The result is an ideal surface material for sheltered exterior ceiling areas such as covered walkways and malls, large canopies, open porches, breezeways, carports and exterior soffits.

Weather- and fire-resistant SHEETROCK brand Exterior Gypsum Ceiling Board may be applied directly to wood framing or to cross-furring of wood or UNIMAST Metal Furring Channels attached to main supports.

Special Conditions

Where frame spacing exceeds 16″ o.c. for ½″ board or 24″ o.c. for ⅝″ board, furring is required to provide support for gypsum board.

Wood Framing Requirements—1″ x 3″ wood furring may be used for screw application where support member spacing is 24″ o.c. max. Furring of 2″ nom. thickness should be used for nail application of board or where framing spacing is from 24″ to 48″ max. o.c.

Steel Framing Requirements—Installation of grillage should be the same as for "Steel Frame Single-Layer Application" previously described in this Chapter on page 150.

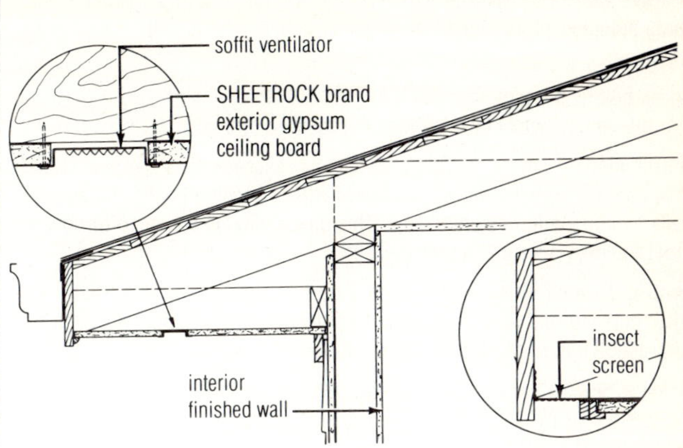

Application details

Ventilation—Where the area above exterior ceiling board opens to an attic space above habitable rooms, the space should be vented to the outside in accordance with accepted recommendations of one sq. ft. free-vent area per 150 sq. ft. of attic area. Where exterior ceiling board is applied directly to rafters or to roof-ceiling joists (as in flat-roof construction) that extend beyond habitable rooms, vents are required at each end of each rafter or joist space. The vents should be screened and be a minimum 2″ wide by full length between rafters (or joists). Vents should be attached through board to min. 1″ x 2″ backing strips installed prior to board application. Vent openings should be framed and located within 6″ of outer edge of eave.

Weather Protection—At the perimeter and at vertical penetrations, the exposed core of panels must be covered with No. 401 Metal Trim or securely fastened mouldings.

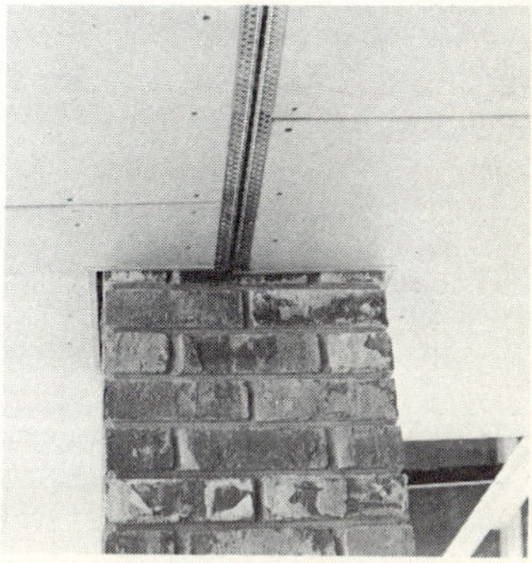

Exterior ceiling board application showing wall intersection and control joint.

In areas subject to freezing temperatures and other severe weather conditions, shingled roofs should be installed in accordance with good roofing practices.

Fascia boards should extend at least ¼″ below the ceiling board or adjacent trim mouldings, whichever is lower to provide a drip edge.

Intersections—Where ceiling board expanse exceeds 4′, a space of at least ¼″ should be provided between edge of exterior ceiling board and adjacent walls, beams, columns and fascia. This space may be screened or covered with moulding but must not be caulked.

Control Joints—SHEETROCK brand Exterior Gypsum Ceiling Board, like other building materials, is subject to structural movement, expansion and contraction due to changes in temperature and humidity.

Install a SHEETROCK Control Joint No. 093 or a control joint consisting of two pieces of No. 401 Metal Trim back-to-back in ceiling board where expansion or control joints occur in the exterior wall or roof. Where aluminum H-mouldings are used, they will serve as control joints provided board is not tightly inserted.

Long narrow areas should have control joints spaced no more than 30′ apart. Wings of "L," "U" and "T"-shaped areas should be separated with

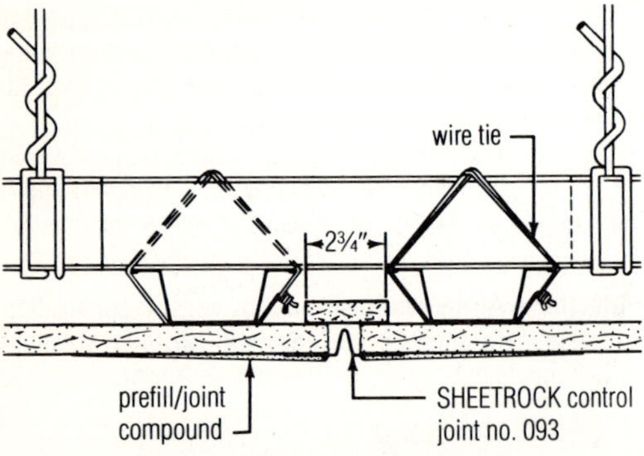

Steel frame canopy (commercial)

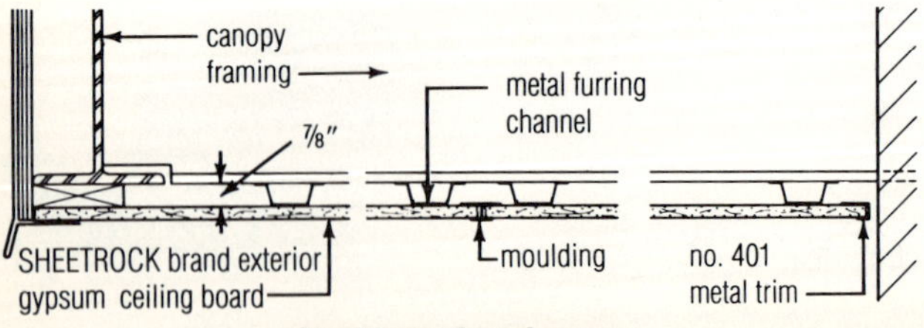

Steel frame furred canopy

control joints. These joints usually are placed to intersect light fixtures, vents, etc. to relieve stress concentrations. Canopy must be designed to resist uplift.

Fixtures—Provide backing or blocking for electrical boxes, vents and heavy fixtures. Cut board neatly and accurately to fit within ¼″ of fixtures and vents. Cover openings with trim.

Installation

Apply exterior ceiling board with long dimension across supports. For ½″ board, max. support spacing is 16″ o.c.; for ⅝″ board, 24″ o.c. max. Position end joints over supports. Use maximum practical lengths to minimize end joints. Allow 1/16″ to ⅛″ space between butted ends of board. Fasten board to supports with screws spaced 12″ o.c. or nails spaced 8″ o.c.

For steel framing, use 1″ Type S corrosion-resistant screws (Type S-12 for 20-ga. and thicker steel.) For wood framing, use 1¼″ Type W screws or 1½″ galvanized box nails or 1½″ aluminum nails. Treat fasteners and joints using a SHEETROCK Setting-Type (DURABOND) or SHEETROCK Lightweight Setting-Type (EASY SAND) Joint Compound. If desired, panel joints may be concealed with batten strips or by installing panels with ends inserted into aluminum H-mouldings. After joint compound has dried, apply one coat oil-based primer-sealer and one coat exterior oil or latex paint or other balanced finishing system recommended by paint manufacturer to all exposed surfaces.

Hotel entrance canopy faced with exterior ceiling board.

Gypsum Board Suspended Ceiling Application

Gypsum board applications for suspended ceilings provide excellent fire protection and appearance with exceptional economy. Applications include SHEETROCK Lay-In Ceiling Panels in standard DONN brand Suspension Grid or surface mounted SHEETROCK brand Gypsum Panels on DONN brand RIGID X Drywall Suspension System.

Lay-In Panels—SHEETROCK Lay-In Ceiling Panels have a FIRECODE C Core and square-cut edges. They are available in 2' x 2' or 2' x 4' sizes and either natural paper facing or a laminated white vinyl facing with stipple pattern. Panels may be installed in DONN brand DX, DXL or DXLA suspension systems for most interior applications or ZXA, ZXLA or AX suspension systems for exterior applications or high humidity areas (see Technical Folder SA-905 for complete information).

Install panels beginning at one corner of the room and work one row at a time. Tilt panel up through opening and lower it to rest squarely on all four tees. Where partial panels are required, use a straight edge and cut face of panels with utility knife, snap at score and cut through backing. Trim rough edges as necessary to fit.

Surface Mounted Panels—SHEETROCK brand Gypsum Panels provide a monolithic ceiling when mounted to DONN brand RIGID X Suspension System. This system offers 1, 1½, 2 and 3-hr. fire ratings when constructed with SHEETROCK brand Gypsum Panels, FIRECODE C Core (consult UL Fire Resistance Directory).

Beginning at one corner of room, mount panels parallel to main tees with butt ends meeting in center of cross channels. Hold panels firmly in place against channels and secure with 1¼″ Type S screws. Complete assembly in the same manner as conventional gypsum board ceiling construction. Finish with a U.S. Gypsum Company joint treatment system and caulk perimeter with SHEETROCK Acoustical Sealant.

Mineral Fiber Blanket Application

Many U.S. Gypsum Company drywall and veneer plaster partitions have been developed to meet the demand for increased privacy between units in residential and commercial construction. Designed for wood stud, steel stud or laminated gypsum board construction, these assemblies offer highly efficient sound-control properties, yet are more economical than other partitions offering equal sound isolation. These improved sound-isolation properties and ratings are obtained by using THERMAFIBER Sound Attenuation Fire Blankets and decoupling the partition faces. Decoupling is achieved with resilient application or with double rows of studs on separate plates. General application procedures for these products follow. See Chapter 1 for product descriptions and SA-100 Construction Selector for sound ratings.

Installation

Install blankets to completely fill height of stud cavity and with the vapor retarder facing inside or outside of wall according to job specifications. If necessary to tightly fill height, cut stock-length blankets with a serrated knife for insertion in the void. Tightly butt ends and sides of blankets within a cavity. Cut small pieces of THERMAFIBER Blankets for narrow stud spaces next to door openings or at partition intersections. Fit blankets carefully behind electrical outlets, bracing, fixture attachments, medicine cabinets, etc.

In ceilings, insulation should be carefully fitted around recessed lighting fixtures. Covering fixtures with insulation causes heat to build up which could possibly result in fire.

Creased THERMAFIBER Sound Insulation Systems

Creased THERMAFIBER assemblies are steel-framed, 1-hr. fire-rated systems that offer high sound ratings (50-55 STC) plus the lower in-place cost of lightweight single-layer gypsum board. The systems consists of ⅝″ SHEETROCK brand Gypsum Panels, FIRECODE C Core; 3⅝″ steel studs spaced 24″ o.c. and set in runners; and THERMAFIBER Sound Attenuation Fire Blankets, 25″ wide.

Since the blanket is 1″ wider than the cavity, it is installed with a slit field-cut down the center and partially through the blanket. This allows the blanket to flex or bow in the center, easing the pressure against the studs and transferring it to the face panel, thereby dampening sound vibrations more effectively. Panels screw-attach directly or resiliently to the steel framing.

Perimeter Isolation

Perimeter relief should be provided for gypsum construction surfaces where (a) partition or furring abuts a structural element (except floor) or dissimilar wall or ceiling; (b) ceiling abuts a structural element, dissimilar partition or other vertical penetration; and (c) ceiling dimensions exceed 30′ in either direction.

Isolation is important to reduce potential cracking in partitions, ceilings, wall, column and beam furring, and reduces the likelihood of sound flanking in rated construction. Generally, methods for isolating surfaces are detailed and specified according to the job. The typical intersection application described on next page may be adapted as required.

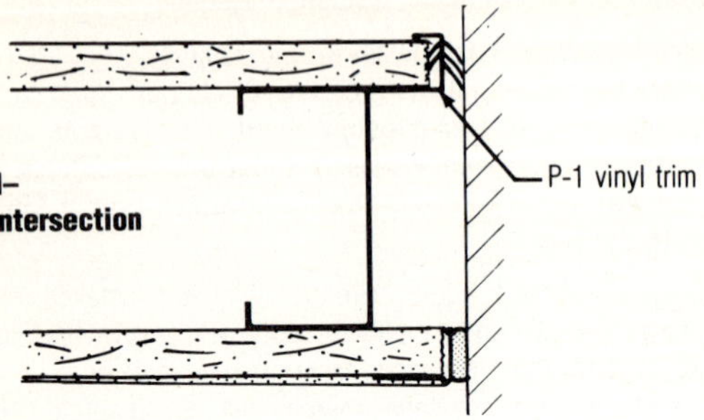

Detail–
wall intersection

Gypsum Board Edge Treatment—Where boards intersect dissimilar materials or structural elements, appropriate trim should be applied to the face-layer perimeter and SHEETROCK Acoustical Sealant applied to close the gap. P-1 Vinyl Trim may be used without sealant or joint treatment.

Partition-Structural Ceiling—Attach steel runner to structural ceiling to position partition. Cut steel stud ⅜″ min., ½″ max., less than floor-to-ceiling height. Attach gypsum board to stud at least 1½″ down from ceiling. Allow ⅜″ min. clearance atop gypsum boards; finish as required.

Partition-Radiant Heat Ceiling—Allow at least ⅛″ clear space between radiant-heated ceilings and walls or partition framing. Finish ceiling angle with P-2 Vinyl Trim or wood moulding fastened to wall members only.

Partition-Exterior Wall or Column—Attach steel stud to exterior wall or column to position partition. Attach gypsum board only to second steel stud erected vertically at max. 6″ from wall. Allow at least ⅜″ clearance between partition panel and wall. Caulk as required with SHEETROCK Acoustical Sealant.

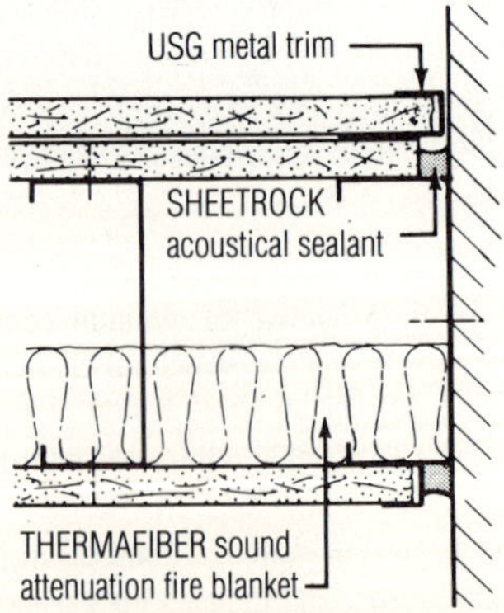

Partition–exterior wall

Perimeter relief at columns reduces possibility of cracking.

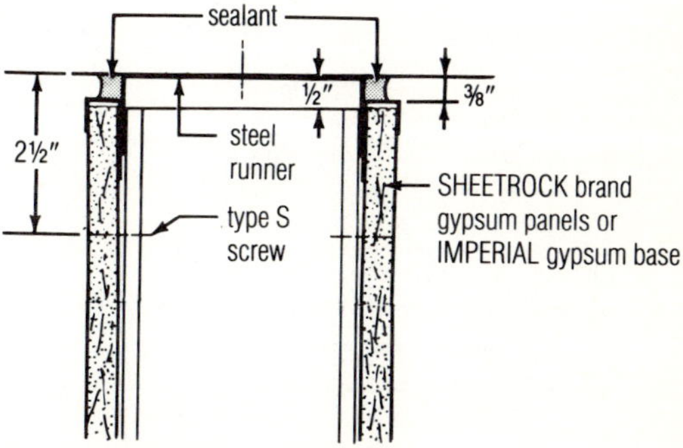

Partition–structural ceiling

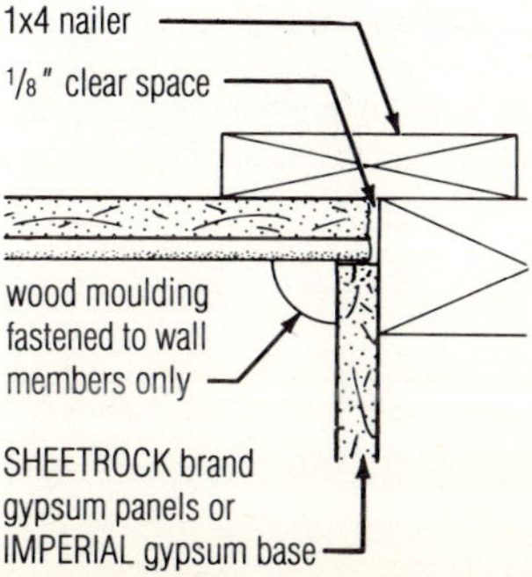

Partition-radiant heat ceiling

Furring-Exterior Wall—Allow ¼″ min. clearance between acoustical trim and intersecting exterior wall or column. Apply SHEETROCK Acoustical Sealant as required.

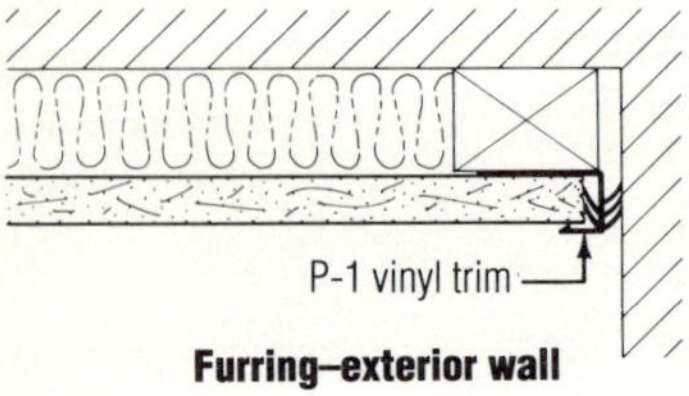

Furring–exterior wall

Ceiling-Exterior Wall—On suspended or furred ceilings, locate supports for gypsum board within 6″ of abutting surfaces but do not allow main runner or furring channels to be let into or come into contact with abutting masonry walls.

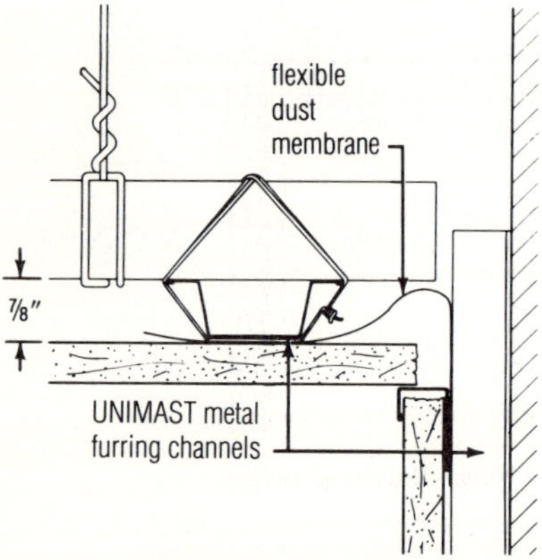

Ceiling–exterior wall

Partition-Column—Fur gypsum board away from concrete column using vertical steel studs. Attach stud in intersecting partition to stud within free-standing furring.

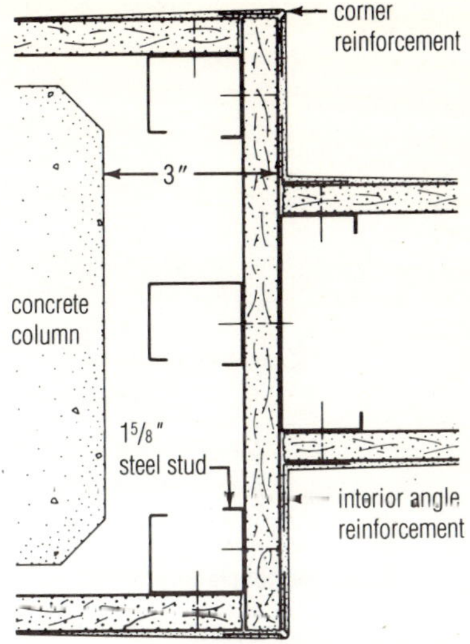

Partition–column

Floating Interior Angle Application

The floating interior angle method of applying gypsum board effectively reduces angle cracking and nail pops resulting from stresses at intersections of walls and ceilings. Fasteners are eliminated on at least one surface at all interior angles, both where walls and ceilings meet and where sidewalls intersect. Follow standard framing practices for corner fastening. Conventional framing and ordinary wood back-up or blocking must be provided where needed at vertical and horizontal interior angles. Apply gypsum board to ceilings first.

Ceilings

Use conventional single nail or screw application. Apply the first nails or screws approx. 7″ from the wall and at each joist. Use conventional fastening in the remainder of the ceiling area.

Sidewalls

Apply gypsum board on walls so that its uppermost edge (or end) is in firm contact with and provides support to the perimeter of the board already installed on the ceiling. Apply the first nails or screws approx. 8″ below the ceiling at each stud. At vertical angles omit corner fasteners for the first board applied at the angle. This panel edge will be overlapped and held in place by the edge of the abutting board. Nail or screw-attach the overlapping board in the conventional manner. Use conventional fastening for remainder of sidewall area.

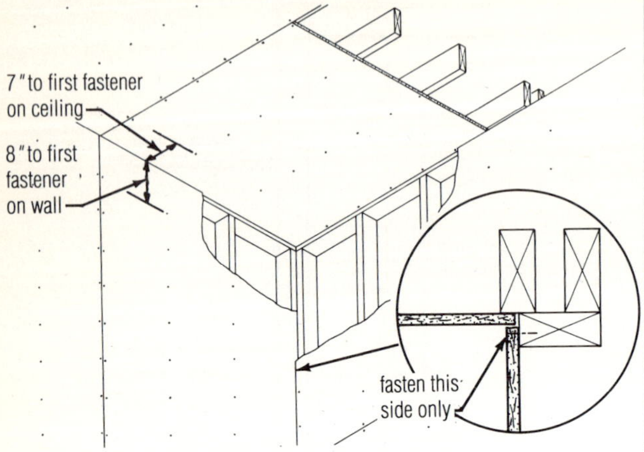

Detail–floating interior angle

Double Nailing

When double nailing is used with a floating interior angle, follow above spacing on first nail from intersection and use double nailing in rest of area. Conventional framing and ordinary wood back-up or blocking at vertical internal angles must be provided.

Fixture Installation

Electrical Fixtures

After electrical services have been roughed in and before gypsum board is installed, cut necessary openings in base and face layers of board to accept switches, outlet and fixture boxes, etc. Cut out openings with a keyhole saw, electric router with a drywall bit or with specially designed cutting tools which produce die-cut openings. (See "Tools and Equipment," Chapter 11.)

On TEXTONE Vinyl-Faced Gypsum Panels, holes made with a special outlet cutter should be cut from back of panel to avoid loosening vinyl around the cut. Erect panel in the usual manner.

Sealant—Where the partition is used as a sound barrier, do not install boxes back-to back or in the same stud cavity. Apply SHEETROCK Acoustical Sealant around all boxes to seal the cutout. See page 219 for typical sealant application. Electrical boxes having a drywall ring or device cover for use as a stop in caulking are recommended.

Fixture Attachment

Gypsum board partitions provide suitable anchorage for most types of fixtures normally found in residential and commercial construction. To ensure satisfactory job performance it is important to have an understanding of particular fixture attachment so that sound-control characteristics will be retained and attachment will be within the allowable load-carrying capacity of the assembly.

In wood-frame construction, fixtures are usually attached directly to the framing or to blocking or supports attached to the framing. Blocking or supports should be provided for plumbing fixtures, towel racks, grab bars and similar items. Fixture supports used with DUROCK Cement Board are shown in Chapter 5. Single or double-layer gypsum boards are not designed to support loads imposed by these items without additional support to carry the main part of the load.

The attachment of fixtures to sound-barrier partitions may impair the sound-control characteristics. Only lightweight fixtures should be attached to resilient wall surfaces constructed with RC-1 Resilient Channel unless special framing is provided (see cabinet attachment method, next page). Refrain from attaching fixtures to party walls so as to provide a direct flow path for sound. Gypsum boards used in the ceiling are not designed to support light fixtures or troffers, air vents or other equipment. Separate supports must be provided.

Fixture Attachment Types

Loading capacities of various fasteners and fixture attachments used with gypsum board partitions appear in load table on page 182. Fasteners and methods follow:

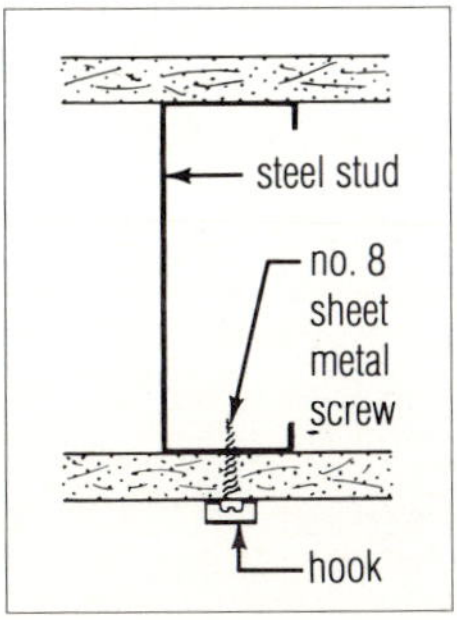

No. 8 Sheet Metal Screw—Driven into 25-ga. min. sheet metal plate or strip, laminated between face board and base board in laminated gypsum partitions. Also may be driven through gypsum board into a steel stud. Ideal for planned light fixture attachment.

Continuous Horizontal Bracing—Back-up for fixture attachment is provided with notched runner attached to steel studs with two ⅜″ pan head screws.

Bolt and Nested Channels—Bolt welded to nested 1½″ channels for use in mounting hanger brackets for heavy fixtures. Suitable for use in laminated gypsum partitions, provided that fixture attachments do not contact opposite coreboard.

Hollow Wall Anchors—¼″ hollow wall anchors installed in gypsum boards only. One advantage of this fastener is that threaded section remains in wall when screw is removed. Also, widespread spider support formed by the expanded anchor spreads load against wall material, increasing load capacity.

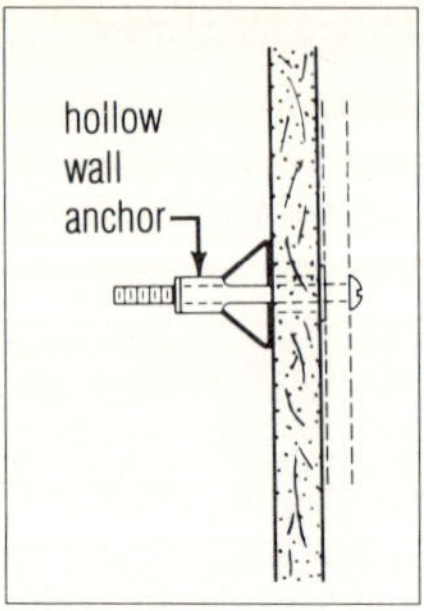

Toggle Bolt—¼″ toggle bolt installed in gypsum board only. One disadvantage of toggle bolt is that when bolt is removed, wing fastener on back will fall down into hollow wall. Another disadvantage is that a large hole is required to allow wings to pass through wall facings.

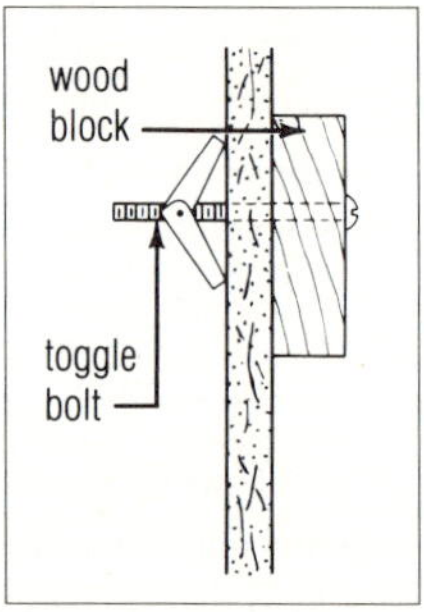

Bolt and 1½″ Channel—Bolt welded to single 1½″ channel and inserted in notches cut in steel stud for use in mounting hanger brackets for heavy fixtures. Suitable for use in gypsum board partitions with 3⅝″ steel studs.

Cabinet Attachment Method—Detailed on next page, allows kitchen, bathroom and other cabinets and fixtures (except lavatories and wall-mounted toilets) of moderate weight, and "Hollywood" style headboards on party walls using RC-1 Resilient Channel to be mounted without reducing the sound rating. Recommended only for residential and light commercial wood-frame construction. Suitable for loads including cabinet weight of 67½ lb. for studs spaced 16″ o.c. and 40 lb. for studs 24″ o.c. Loads are max. per lin. ft. of RC-1 Channel installed for cabinet attachment. Mounting cabinets back-to-back on a partition should be avoided since this practice creates a flanking path that increases sound transmission.

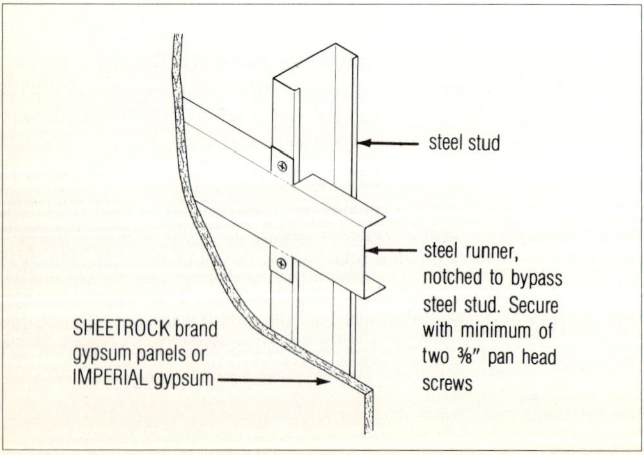

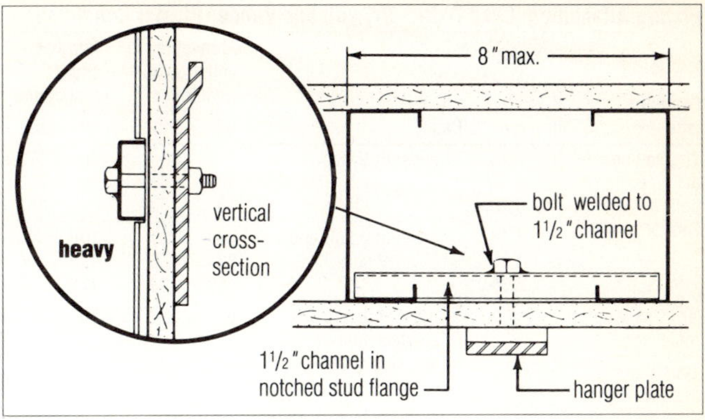

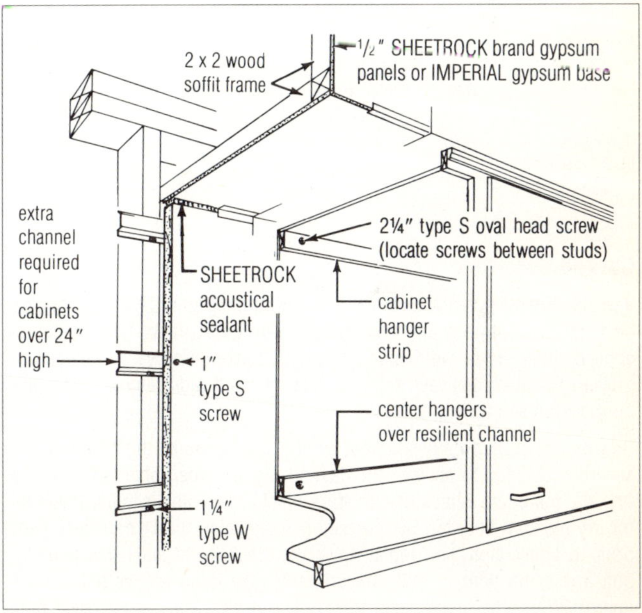

In this system, ⅝" gypsum board is installed with long dimension parallel to channels and fastened with 1" Type S screws spaced 12" o.c. along channels. Cabinets are attached to channels with 2¼" Type S screws spaced 12" o.c. and located between studs. Screws must be driven between studs. Screws that penetrate the stud cause a significant loss in the partition's sound rating.

Fixture Attachment Load Data—Drywall and Veneer Plaster Construction

Fastener type	Size in	Size mm	Base	Allowable withdrawal resistance lbf	Allowable withdrawal resistance N[1]	Allowable shear resistance lbf	Allowable shear resistance N[1]
Toggle bolt or hollow wall anchor	⅛	3.18	½″ gypsum board	20	89	40	178
	3/16	4.76		30	133	50	222
	¼	6.35		40	178	60	267
	⅛	3.18	½″ gypsum board and ST25 steel stud	70	311	100	445
	3/16	4.76		80	356	125	556
	¼	6.35		155	689	175	778
No. 8 sheet metal screw			½″ gypsum board and ST25 steel stud or 25-ga. steel insert	50	222	80	356
Type S bugle head screw				60	267	100	445
Type S-12 bugle head screw			½″ gypsum board and ST20 steel stud or 20-ga. steel insert	85	378	135	600
⅜″ Type S pan head screw			25-ga. steel to 25-ga. steel	70	311	120	534
Two bolts welded to steel insert	3/16	4.76	½″ gypsum board, plate and steel stud	175	778	200	890
	¼	6.35	½″ gypsum board, plate and steel stud	200	890	250	1112
Bolt welded to 1½″ chan.	¼	6.35	(see drawing, page 181)	200	890	250	1112

(1) Newtons

Curved Surfaces

Versatile SHEETROCK brand Gypsum Panels and IMPERIAL Gypsum Base can be formed to almost any cylindrically curved surface. Boards can be applied either dry or wet depending on the radius of curvature desired. To prevent flat areas between framing, shorter bend radii require closer than normal stud and furring spacing.

Boards are horizontally applied, gently bent around the framing, and securely fastened to achieve the desired radius. When boards are applied dry, the minimum radius of curvature meets many applications (see table for dry gypsum boards). By thoroughly moistening the face or back paper prior to application, and replacing in the stack for at least one hour, the board may be bent to still shorter radii (see table for wetted gypsum board). When the board dries thoroughly, it will regain its original hardness.

Installation

Framing— Cut one leg and web of top and bottom steel runner at 2″ intervals for the length of the arc. Allow 12″ of uncut steel runners at each end of arc. Bend runners to uniform curve of desired radius (90° max. arc). To support the cut leg of runner, clinch a 1″ x 25-ga. steel strip to inside of leg. Select the runner size to match the steel studs; for wood studs, use a 3½″ steel runner. Attach steel runners to structural elements at floor and ceiling with suitable fasteners as previously described.

Curved stairwell, faced with drywall, forms attractive design element in new shopping mall (above). Radius of curved gypsum board, joints treated, is shown in construction view (below).

Position studs vertically, with open side facing in same direction and engaging floor and ceiling runners. Begin and end each arc with a stud and space intermediate studs equally as measured on outside of arc. Secure steel studs to runners with ⅜″ Type S pan head screws; secure wood studs with suitable fasteners. On tangents, place studs 6″ o.c. leaving last stud free standing. Follow directions previously described for erecting balance of studs.

Panel Preparation—Select length and cut board to allow one unbroken panel to cover the curved surface and 12″ tangents at each end. Outside panel must be longer than inside panels to compensate for additional radius contributed by the studs. Cutouts for electrical boxes are not recommended in curved surfaces unless they can be made after boards are installed and thoroughly dry.

When wet board is required, evenly spray water on the surface which will be compressed when board is hung. Apply water with a conventional garden sprayer using the quantity shown in the table. Carefully stack boards with wet surfaces facing each other and cover stack with plastic sheet (polyethylene). Allow boards to set at least one hour before application.

Panel Application—Apply panels horizontally with the wrapped edge perpendicular to the studs. On the convex side of the partition, begin installation at one end of the curved surface and fasten panel to studs as it is wrapped around the curve. On the concave side of the partition, start fastening panel to the stud at the center of the curve and work outward to the ends of the panel. For single-layer panels, space screws 12″ o.c. Use 1″ Type S screws for steel studs and 1¼″ type W screws for wood studs.

For double-layer application, apply base layer horizontally and fasten to stud with screws spaced 16″ o.c. Center face layer panels over joints in the base layer and secure to studs with screws spaced 12″ o.c. Use 1″ Type S screws for base layer and 1⅝″ Type S screws for face layer. Allow panels to dry completely (approx. 24 hrs. under good drying conditions) before applying joint treatment.

Minimum Bending Radii of Dry Gypsum Board

Board thickness		Board applied with long dimension perpendicular to framing		Board applied with long dimension parallel to framing	
in	mm	ft	m	ft	m
½	12.7	20(1)	6.1	—	—
⅜	9.5	7.5	2.3	25	7.6
¼	6.4	5	1.5	15	4.6

(1) Bending two ¼″ pieces successively permits radii shown for ¼″ gypsum board.

Minimum Bending Radii of Wetted Gypsum Board(1)

Board thickness —in	Min. radius —ft	Length of arc —ft(2)	No. of studs on arc and tangents (3)	Approx. stud spacing —in (4)	Max. stud spacing —in (4)	Water required per panel side—oz (5)
¼	2	3.14	9	5.50	6	30
¼	2.5	3.93	10	5.93	6	30
⅜	3	4.71	9	7.85	8	35
⅜	3.5	5.50	11	7.22	8	35
½	4	6.28	8	11.70	12	45
½	4.5	7.07	9	11.40	12	45

(1) For gypsum board applied horizontally to a 4″ thick partition.

(2) Arc length = $\frac{3.14 \cdot R}{2}$ (for a 90° arc).

(3) No. studs = outside arc length/maximum spacing +1 (rounded up to next whole number).

(4) Stud spacing = outside arc length/no. of studs -1 (measured along outside of runner).

(5) Wet only the side of board that will be in tension. Water required per board side is based on 4′ x 8′ sheet.

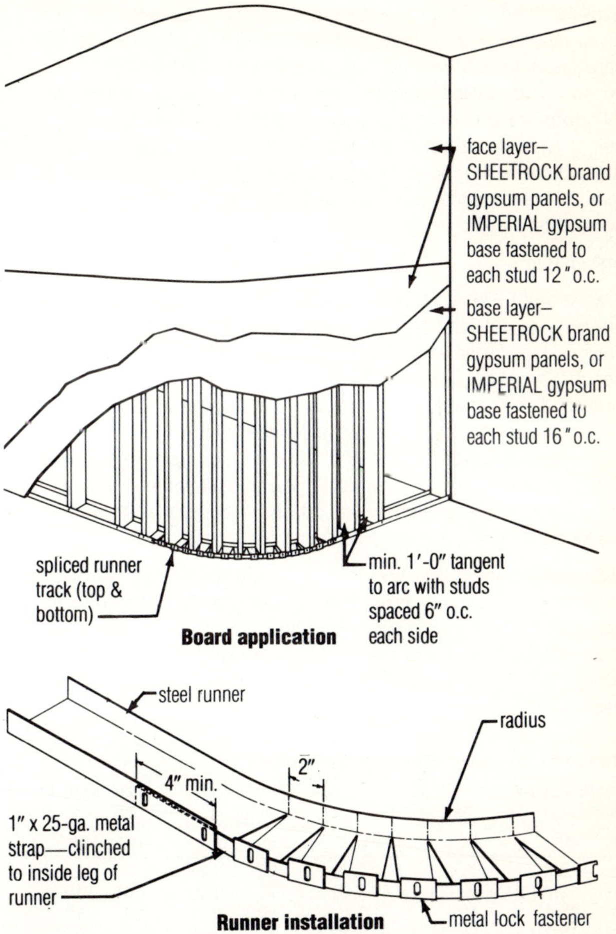

Board application

Runner installation

Arches

Arches of any radii are easily faced with gypsum panels or base and finished with a U.S. Gypsum Company joint system, or veneer plaster finish. Score or cut through back paper of panels at 1″ intervals to make them flexible. The board should previously have been cut to desired width and length of arch.

After board has been applied to arch framing with nails or screws, apply tape reinforcement (SHEETROCK Joint Tape for drywall panels or IMPERIAL Tape Type P or S for plaster base). Crease tape along center. Make scissor cuts half-way across tape and ¾″ apart to make tape flexible. Apply uncut half to curved surface, and fold cut half of tape onto wall surface. Finish as appropriate for drywall or veneer plaster construction.

Soffits

Gypsum board soffits provide a lightweight, fast and economical method of filling over cabinets or lockers and of housing overhead ducts, pipes or conduits. They are made with wood framing or with steel stud and runner supports, faced with screw-attached gypsum board. Braced soffits up to 24″ deep are constructed without supplementary vertical studs. Select components for the soffit size desired from table following. Unbraced soffits without horizontal studs are suitable for soffits up to 24″ x 24″. To retain fire protection, partitions and ceilings are finished with gypsum board before soffits are installed.

Installation

Braced Soffit—Attach steel runners to ceiling and sidewall, placing fasteners close to outside flange of runner. On stud walls, space fasteners to engage stud. Fasten vertical gypsum face board to web of face corner runner and flange of ceiling runner with Type S screws spaced 12″ o.c. Place screws in face corner runner at least 1″ from edge of board. Insert steel studs between face corner runner and sidewall runner and attach alternate studs to runners with screws. Attach bottom face board to studs and runners with Type S screws spaced 12″ o.c. Attach corner bead and finish. Where sound control is important, attach RC-1 Resilient Channel to framing before attaching gypsum board.

Unbraced Soffit—Attach steel studs to ceiling and sidewall, placing fasteners to engage wall and ceiling framing. Cut gypsum board to soffit depth and attach a soffit-length stud with Type S screws spaced 12″ o.c. Attach this preassembled unit to ceiling stud flange with screws spaced 12″ o.c. Attach bottom panel with Type S screws spaced 12″ o.c. Attach corner bead and finish.

Braced Soffit Design Maximum Dimensions(1)

Gypsum board thickness(2)		Steel stud size		Maximum width		Max. depth for max. width shown	
in	mm	in	mm	in	mm	in	mm
½	12.7	1⅝	41.3	60	1525	48	1220
½	12.7	2½, 3⅝	63.5, 92.1	72	1830	36	915
⅝	15.9	1⅝	41.3	60	1525	30	760
⅝	15.9	2½, 3⅝	63.5, 92.1	72	1830	18	455

(1) The construction is not designed to support loads other than its own dead weight and should not be used where it may be subjected to excessive abuse.
(2) Double-layer applications and ⅜″ board are not recommended for this construction.

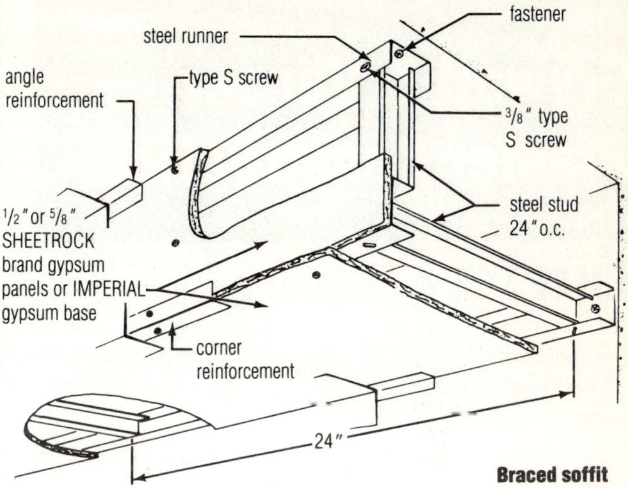

Braced soffit

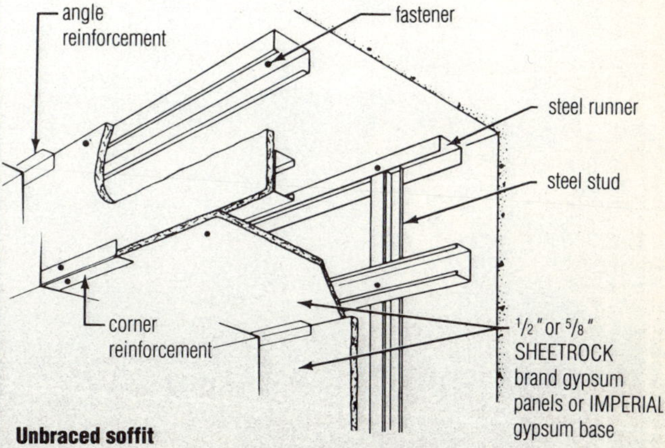

Unbraced soffit

Drywall & Veneer Plaster Construction

CHAPTER 4 Finishing

Levels of Gypsum Board Finishing

Contract documents traditionally have used nonspecific terms such as "industry standards" or "workmanlike finish" to describe how finished gypsum board walls and ceilings should look. This practice often has lead to misunderstanding about the degree of finishing sophistication required for any particular job.

A collective effort of four industry trade associations—Association of the Wall and Ceiling Industries-International (AWCI), Ceilings and Interior Systems Construction Association (CISCA), Gypsum Association (GA), and Painting and Decorating Contractors of America (PDCA)—has resulted in the adoption of industry-wide recommended specifications on levels of gypsum board finish. The work identifies five specific levels of finishing, enabling architects to more closely identify the sophistication required and allowing for better competitive bidding among contractors.

Key factors used in determining the sophistication level required include the location of the work to be done, type and angle of surface illumination (both natural and artificial lighting), type of paint or wallcovering to be used and method of application. Critical lighting conditions, gloss paints and thin wallcoverings require a high level of finish, while heavily textured and painted surfaces or surfaces that will be decorated with heavy-gauge wallcoverings require less sophistication.

The following level descriptions are taken from *GA-214-90 Levels of Gypsum Board Finish.*

Level 1— "Frequently specified in plenum areas above ceilings, in attics, in areas where the assembly would generally be concealed or in building corridors and other areas not normally open to public view. Accessories are optional at specifier discretion in corridors and other areas with pedestrian traffic. Some degree of sound and smoke control is obtained; this level of treatment is referred to as 'fire-taping' in some geographic areas. Where fire resistance rating is required for the gypsum board assembly, details of construction must be in accordance with reports of fire tests of assemblies that have met the fire rating requirement."

Level 1 may be achieved with SHEETROCK Joint Treatment products by applying the first (embedding) coat to joints as described on page 194 and to inside corners as described on page 196. Fire taping may exclude the double-back of the first coat over the tape.

Level 2— "Specified where water-resistant gypsum backing board (ASTM C630) is used as a substrate for tile; may be specified in garages, warehouse storage or other similar areas where surface appearance is not of primary concern."

Level 2 may be achieved with SHEETROCK Joint Treatment products by applying the first (embedding) and second (fill) coats to joints as described on pages 194-195, to inside corners as described on page 196, and one coat of joint compound applied over all fasteners, metal bead and trim.

Level 3— "Typically specified in appearance areas that are to receive heavy or medium texture (spray or hand applied) finishes before final painting, or where heavy grade wallcoverings are to be applied as the final decoration. This level of finish is not recommended where smooth painted surfaces or light to medium weight wallcoverings are specified."

Level 3 may be achieved with SHEETROCK Joint Treatment products by applying the first (embedding), second (fill) and third (finish) coats to joints as described on pages 194-195. Inside corners should be treated as described on page 196 and three coats of joint compound should be applied over all fasteners, metal bead and trim. If low shrinkage compounds are used, such as SHEETROCK Setting-Type Joint Compound (DURABOND), SHEETROCK Lightweight Setting-Type Joint Compound (EASY SAND), or SHEETROCK Lightweight All Purpose Joint Compound Ready-Mixed (PLUS 3), only two coats are required over fasteners, metal bead and trim. Surfaces should be primed with SHEETROCK First Coat or a flat latex paint with high solid content applied undiluted.

Level 4— "This level should be specified where light textures or wallcoverings are to be applied, or economy is of concern. In critical lighting areas, flat paints applied over light textures tend to reduce joint photographing. Gloss, semi-gloss, and enamel paints are not recommended over this level of finish. The weight, texture, and sheen level of wall coverings applied over this level of finish should be carefully evaluated. Joints and fasteners must be adequately concealed if the wallcovering material is lightweight, contains limited pattern, has a gloss finish, or any combination of these features is present. Unbacked vinyl wallcoverings are not recommended over this level of finish."

Level 4 may be achieved with SHEETROCK Joint Treatment products by following the same joint finishing and priming recommendations shown above for Level 3.

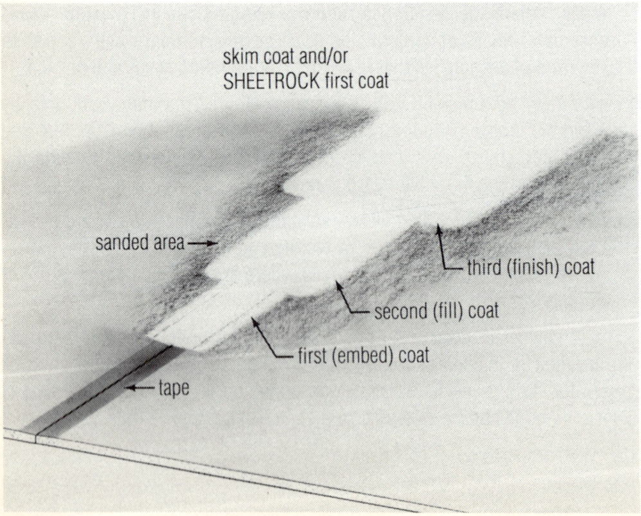

Level 5— "This level of finish is recommended where gloss, semi-gloss, enamel, or nontextured flat paints are specified or where severe lighting conditions occur. This highest quality finish is the most effective method to provide a uniform surface and minimize the possibility of joint photographing and of fasteners showing through the final decoration."

Level 5 may be achieved with SHEETROCK Joint Treatment products by following the same joint finishing recommendations as shown for Level 3 followed by application of a thin skim coat of joint compound over the entire surface. After the skim coat has dried, the entire surface should be primed with SHEETROCK First Coat or a flat latex paint with high solid content applied undiluted.

Joint Treatment for Drywall Construction

Application Conditions

In cold weather during joint finishing, temperatures within the building should be maintained within the range of 55° to 70°F. (13° to 21°C) and adequate ventilation should be provided. Also see "Quality Drywall Finishing in All Kinds of Weather," U.S. Gypsum Company folder J 75.

Check Working Surfaces

Gypsum panels must be firmly fastened to framing members without cutting the surface paper or fracturing the core. Make certain panel joints are aligned. When one panel is higher than another it becomes difficult to leave sufficient compound under the tape covering the high panel. Blisters, bond failure and cracks can easily develop in these areas.

Open spaces between panels of ¼" or more should be filled with compound at least 24 hours prior to embedding or first-coat work. SHEETROCK Setting-Type (DURABOND) and SHEETROCK Lightweight Setting-Type (EASY SAND) Joint Compounds, which are hardening types, are recommended for these large fills. With these setting-type compounds as a fill, joint treatment may begin as soon as the compound has hardened, eliminating the typical 24-hr. drying period. Good planning prior to hanging panels eliminates unnecessary joints.

Care of Equipment

Applicators must keep tools and equipment clean and in good repair to secure satisfactory results. With mechanical tools, parts must be replaced as soon as they show signs of wear.

Mixing joint compounds in dirty buckets or failure to wipe down bucket as material is used causes lumps, scratches and usually creates hard working material. With setting-type materials such as SHEETROCK Setting-Type (DURABOND) and SHEETROCK Lightweight Setting-Type (EASY SAND) Joint Compounds, a residue of dry compounds will shorten setting time of the new batch.

The hardening action of SHEETROCK Setting-Type (DURABOND) and SHEETROCK Lightweight Setting-Type (EASY SAND) Joint Compounds requires that all tools, mixing containers, bread pans, etc., used for application be thoroughly cleaned. Flush and clean these compounds from equipment with a conventional garden hose and brush before the setting action takes place. Immersion of equipment in water will not prevent hardening of the compound.

Mechanical tool application is not recommended with fast-setting SHEETROCK Setting-Type (DURABOND) and SHEETROCK Lightweight Setting-Type (EASY SAND) Joint Compounds.

Mixing Joint Compounds

1. Mix powder joint compounds in a clean 5-gal. container—preferably plastic for SHEETROCK Setting-Type (DURABOND) and SHEETROCK Lightweight Setting-Type (EASY SAND) Joint Compounds since plastic permits flexing container walls to break loose hardened compound. A commercial potato masher makes a convenient mixing tool. Power mixing saves considerable time, particularly where mixing in a central location is convenient. Power may be supplied by a ½" heavy-duty electric drill operating at 200 to 300 rpm (400 rpm max.). Drills operating at high speeds whip air into the compound, and also accelerate setting of setting-type compounds. (See page 432 for information on mixing paddles.) Keep mixing buckets and tools clean at all times. Containers having any residue of joint compounds in them may cause premature hardening, scratching and incompatibility problems.

2. Pour proper amount of clean drinkable water into container. Dirty water (such as that used to clean tools) will contaminate compound and cause erratic setting of SHEETROCK Setting-Type (DURABOND) and SHEETROCK Lightweight Setting-Type (EASY SAND) Compounds. The amounts for type of application and product used are shown in the directions on the package.

3. Sift joint compound into water, allowing complete wetting of the powder.

4. Mix as shown below:

a. For SHEETROCK Setting-Type (DURABOND) and SHEETROCK Lightweight Setting-Type (EASY SAND) Joint Compounds, follow mixing directions on the bag. Do not overmix; this may speed up hardening time. Note: Keep compound from being contaminated by any other materials such as other type joint compounds, dirty water or previously mixed SHEETROCK Setting-Type (DURABOND) and SHEETROCK Lightweight Setting-Type (EASY SAND) Joint Compounds. Contamination will affect the hardening time and properties of the compound. Do not remix if product has started to set.

Mix only as much SHEETROCK Setting-Type (DURABOND) and SHEETROCK Lightweight Setting-Type (EASY SAND) Joint Compound as can be used within time period shown on bag (usually about 2½ hours for 210, 1 hour for DURABOND 90, for example).

The compound will harden chemically after this time period, even under water. Do not attempt to hold wet mix or immerse joint compound-coated tools in water to hold back hardening. Retempering the compound is not recommended.

b. For SHEETROCK Drying-Type Powder Joint Compounds, stir until the powder is uniformly damp, then after approx. 15 min., remix vigorously until smooth. Note: Do not add extra water. Use specified amounts of water, as SHEETROCK Joint Compounds will retain their original mixed consistency over extended periods. On occasions, some slight liquid separation or settlement of compound may take place in the bucket, but a brief remix will restore the compound to its original consistency.

c. For SHEETROCK Ready-Mixed Compounds, mix contents lightly and use at package consistency for fasteners and corner beads. May be thinned for taping and finishing and for use with mechanical tools. Add water in half-pint increments to avoid overthinning. Remix lightly and test apply after each water addition. A potato masher-type mixer is recommended for Ready-Mixed Compounds. Drill-type mixers tend to whip air into the compounds.

Use cool to lukewarm (not hot) water. If compound should accidentally be overthinned, simply add additional Ready-Mixed Compound to thicken, then remix.

To hold the wet mix in a container for prolonged periods, cover the material with a wet cloth or a thin layer of water. When needed, pour off water and retemper as necessary.

Ready-Mixed Compound is sensitive to cold weather and must be protected from freezing. If material freezes in container, allow it to thaw at room temperature (do not force the thawing process). Usually it will again be usable, unless it has been subjected to several freeze-thaw cycles.

Ready-Mixed Compound can be used in tools and containers previously used for powder compound after normal cleaning.

Hand Tool Application

Prefilling "V" joint of SHEETROCK brand Gypsum Panels, SW Edge.

Prefilling Joints—This step is necessary with SW Edge board only. Fill the "V" groove between SHEETROCK brand Gypsum Panels, SW Edge with SHEETROCK Setting-Type (DURABOND 45 or 90) or SHEETROCK Lightweight Setting-Type (EASY SAND 45 or 90) Joint Compound. Apply compound directly over "V" groove with a flexible 5″ or 6″ joint finishing knife. Wipe off excess compound that is applied beyond the groove. Allow fill compound to harden.

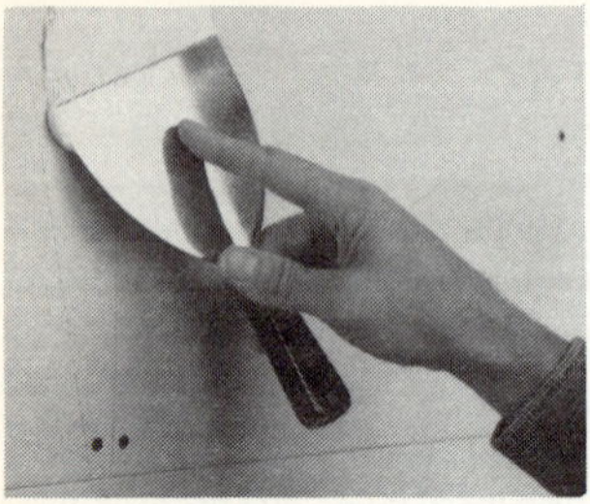
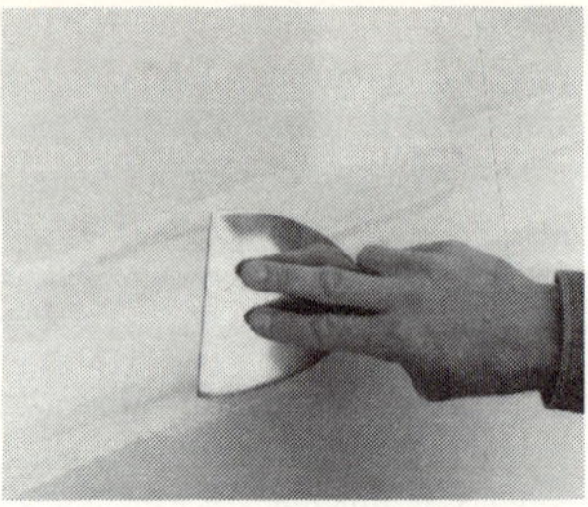

Embedding Tape—Drive home all fasteners that protrude above the gypsum panel surface. Using a broad steel finishing knife, apply a continuous coat of taping, all-purpose or setting-type joint compound to fill the channel formed by the tapered edges of the panels (above, left). Center and lightly press SHEETROCK Joint Tape into fresh joint compound. Working within a convenient arms-reach area, embed tape by holding knife at an angle to panel. Draw knife along joint with sufficient pressure to remove excess compound above and below tape and at edges (above, right). Leave sufficient compound under tape for proper bond but not over 1⁄32″ under edge. While embedding, apply a thin coat of joint compound over the tape (below, left). This thin coat reduces edge wrinkling or curling and makes the tape easier to conceal with following coats. Allow to dry completely. (See drying time guide on page 200.) Do not use topping compound for embedding tape.

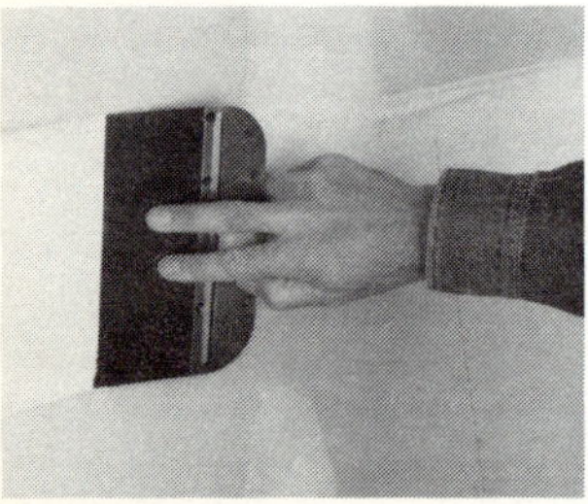
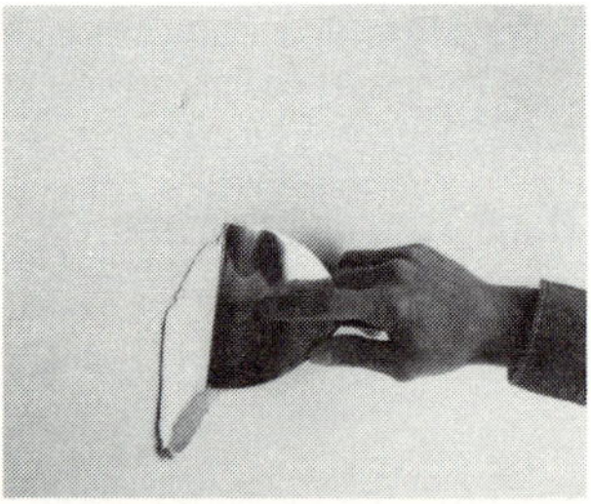

Spotting Fastener Heads—Use ready-mixed compounds at package consistency or powder compounds mixed per bag directions. Do not add excess water. Apply taping, all-purpose or setting-type joint compound over all fasteners (above, right) immediately before or after embedding tape. Fill only the fastener depression. Apply enough pressure on knife to level compound with panel surface. Allow each coat to dry. Repeat application until fastener depressions are flush with panel surface.

Filling Beads—Use ready-mixed compounds at package consistency or powder compounds mixed per bag directions. Apply all-purpose or setting-type compound at least 6″ wide over all corner beads (next page, top) and to trims that are to receive compound. Allow each coat to dry. Apply following coats approximately 2″ wider than preceding coats. For smoother finishing, the final coat of joint compound may be thinned slightly.

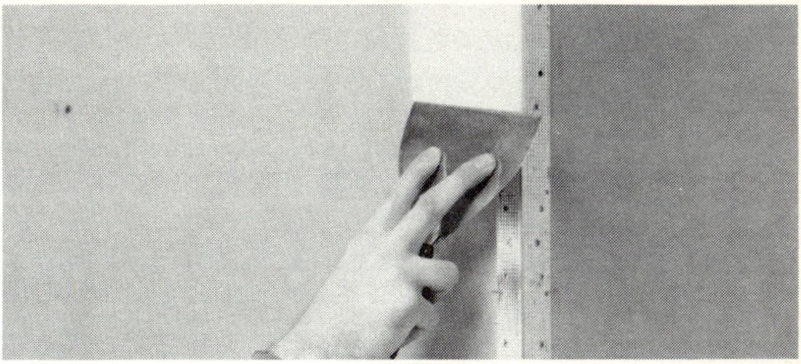

Fill Coat Application—After the tape embedding coat is dry, apply a topping or all-purpose compound fill (second) coat approx. 7″ to 10″ wide over taped joints (shown below), beads and trim. Feather edge of second coat approx. 2″ beyond edge of first coat. Spot fasteners with second coat. Allow to dry.

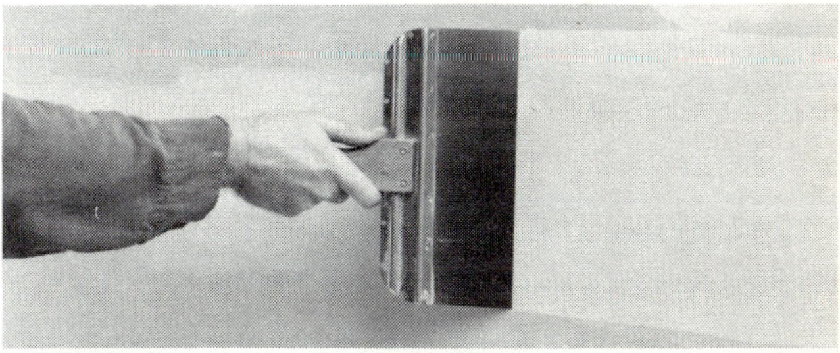

Fill coat application

Finish Coat Application—After second coat is dry, smooth tool marks and other protrusions with a finishing knife. Apply a thin finish (third) coat of ready-mixed topping or all-purpose compound over joints, fasteners, beads and trim. Finish compound may be applied at a slightly thinner consistency. Feather edges of third coats at least 2″ wider than second coats (shown below). SHEETROCK Lightweight All Purpose (PLUS 3) Joint Com-

Finish coat application

pound—Ready-Mixed, SHEETROCK Setting-Type (DURABOND) or Lightweight Setting-Type (EASY SAND) Joint Compounds, and SHEETROCK Lightweight All Purpose Joint Compound—Powder (AP LITE), require only two coats over metal corner beads and fasteners. Joints, fasteners, beads and trim should be finished as smooth as possible to minimize sanding. Go over the whole job to smooth and touch up with joint compound all scratches, craters, nicks and other imperfections in the dried finish coat.

End (Butt) Joints—Because ends of gypsum panels are flat and have no taper, end joints are difficult to conceal. Also, exposed paper on ends may cause visible ridging or beading. The following steps are recommended for joint treatment to minimize crowning and/or ridging of end joints:

1. Before attachment, bevel panel ends approx. ⅛″ at a 45° angle using a sharp utility knife. This keeps the paper ends apart and reduces expansion problems caused by the raw paper edge. Also, peel back and remove any loose paper from the end.

2. Gypsum panel ends should be loosely butted together. Ends should be separated slightly and not touching.

3. Apply compound and tape over the joint in the same manner as for tapered joints. Embed tape tightly to minimize joint thickness but leave sufficient compound under tape for continuous bond and blister prevention.

4. Finish the end joint to a width at least twice the width of a recessed edge joint. This will make the joint less apparent after decoration as the crown will be more gradual.

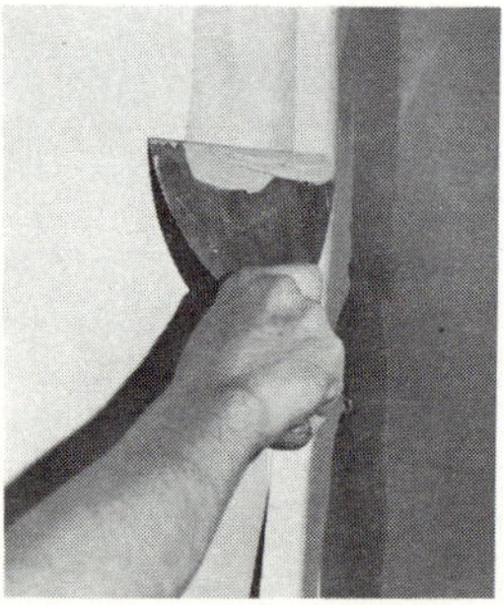

Finishing Inside Corners—Fold tape along center crease. Apply joint compound to both sides of corner and lightly press folded tape into angle. Tightly embed tape into both sides of angle with finishing knife (at right). Let dry. Next, apply a thin finish coat to one side of angle only. Allow to dry and apply finish coat to other side of angle.

Dry Sanding—Sand joint compounds to prepare gypsum drywall surfaces for decoration. Sand as necessary to remove excess joint compound from tool marks, lap marks and high crowned joints. Scratches, craters and nicks should be filled with joint compound, then sanded. Do not try to remove these depressions by sanding only.

Select sandpaper or abrasive-mesh cloth with grit as fine as possible. Excessively coarse sandpapers leave scratches that are visible after decoration. For conventional weight all-purpose compounds, use #120 grit or finer sandpaper (#200 grit or finer mesh cloth). For SHEETROCK Lightweight Ready-Mixed All Purpose (PLUS 3) Compound, SHEETROCK Lightweight All Purpose Powder Joint Compound (AP LITE) and topping compounds, use #150 grit or finer sandpaper (#220 grit or finer mesh cloth). Only sand surfaces coated with joint compound to avoid scuffing gypsum panel paper. Remove sanding dust before decorating.

Ventilate or use a dust collector to reduce dust in work areas. Use a NIOSH-approved respirator specified for mica and talc when air is dusty. Use of safety glasses is recommended.

Wet Sanding—Wet sanding or sponging finished joints, trim and fasteners is recommended rather than dry sanding to avoid creating dust. The best material to use for wet sanding is a high density, small celled, polyurethane sponge. This type of sponge material resembles high quality carpet padding. When only a touch-up is required, a general purpose sponge or smooth, soft cloth will work.

To wet sand, saturate the sponge with clean water containing no soap or additives. Water temperature should be cool to lukewarm, not hot. Wring out sponge only enough to eliminate dripping. To remove high spots, gently rub joints in a direction parallel to the joint compound surface; do not rub down into the joint compound. Use as few strokes as possible. Excessive rubbing will groove joints. Clean sponge frequently.

Mechanical Tool Application

Several types of mechanical and semi-mechanical tools are available. Tools used in the following sequence illustrate typical procedures.

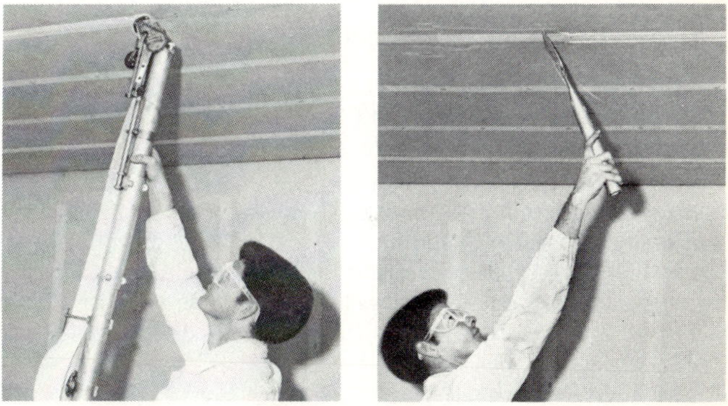

1. Using compound of suitable consistency, mechanically tape all joints; wipe down with broad knife. Allow to dry.

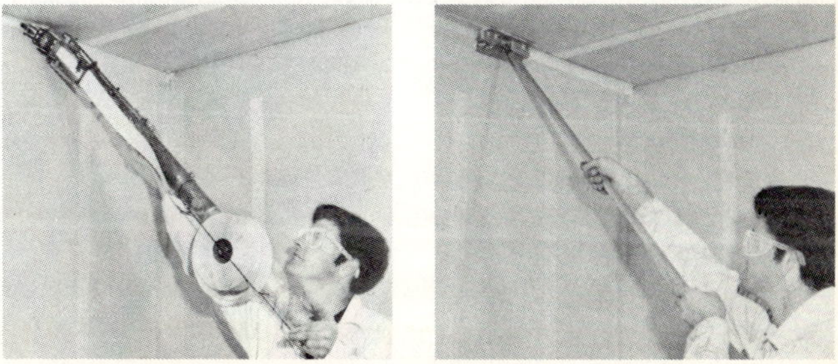

2. Mechanically tape interior angles. Finish both sides of angles with corner roller and corner finisher as shown above. Touch up with broad knife as necessary. Apply first coat to fastener heads and metal accessories. Allow to dry.

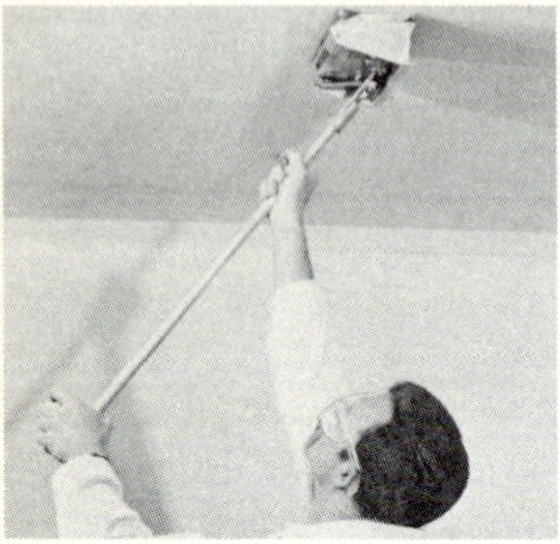

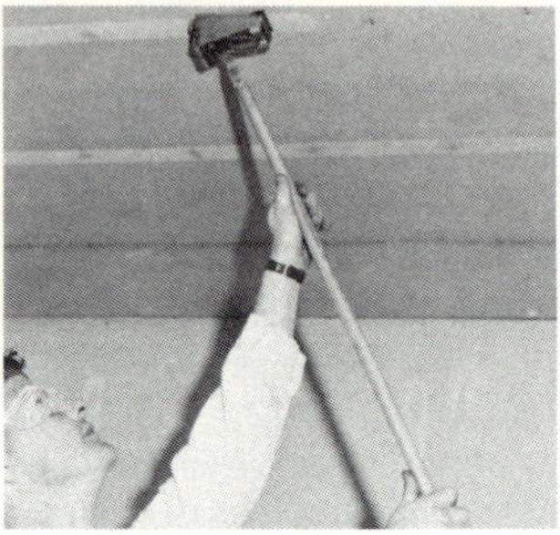

3. Apply fill coat of compound over tape on flat joints using hand finisher tool. Using compound of thicker consistency, spot fastener heads and apply second coat to metal accessories. Allow to dry.

4. Apply finish coat of compound to flat joints, feathering edges about 2″ beyond preceding coat. Apply finish coat to metal accessories and fastener heads. Allow to dry and smooth lightly as required. Remove all dust before decoration. Do not scuff face paper by oversanding.

Setting-Type Joint Compound Application

SHEETROCK Setting-Type (DURABOND) and SHEETROCK Lightweight Setting-Type (EASY SAND) Joint Compounds are chemical hardening products with varied working (setting) times for finishing interior gypsum panels and exterior gypsum ceiling boards. These specialized products provide short setting times for fast one-day finishing and extended times (up to 6 hours) to suit individual needs. Products with shorter working times have lower shrinkage. The following application guide will help you choose the proper product to meet your requirements.

Application Guide—Setting-Type Joint Compounds

Compound type	Approx. setting time-min.	Approx. working time-min.	Recommended application
20	20- 30	15	application needing very short working time
45	30- 80	20	prefill SW panels spot fastener heads embed metal beads
90	85-130	60	all applications
210	180-240	150	embed tape embed metal beads
300	240-360	210	application needing longer working time
DURABOND LC	check bag for setting and working times		application with IMPERIAL Type P Tape

For One-Day Finishing—Use the techniques shown for hand application; mechanical tool application is not recommended for SHEETROCK Setting-Type (DURABOND) and SHEETROCK Lightweight Setting-Type (EASY SAND) Joint Compounds because these compounds may harden in the tools, making them inoperable. In the following sequence, Steps 1 through 4 must be completed by mid-day. Planning and scheduling according to the setting times of the compounds is essential.

(Where SHEETROCK brand Gypsum Panels, SW Edge, are used, the first step is to fill the "V" grooves between panels.)

1. Embed SHEETROCK Joint Tape over all joints and angles.

2. Spot fastener heads throughout job.

3. Apply second (fill) coat over all joints and angles as soon as taping coat has set (hardened even though not dry).

4. Apply compound over metal corner reinforcement. For best results use compound that will set within 1½ to 2 hours. Use either previously mixed material or faster setting compound. Apply second coat to all fasteners.

5. After the second (fill) coat application has hardened, apply finishing coat of selected finishing compound to completely cover all joints, angles, corner bead and fasteners.

For SHEETROCK brand Exterior Gypsum Ceiling Board Surfaces—Use hand application techniques and a SHEETROCK Setting-Type (DURABOND) or SHEETROCK Lightweight Setting-Type (EASY SAND) Joint Compound to treat joints and fasteners in SHEETROCK brand Exterior Gypsum Ceiling Board applications. During periods of near-freezing temperatures, check weather forecast before beginning work. Minimum air, water, mix and surface temperatures of 45°F (7°C) must be ensured until compound is completely dry. Apply SHEETROCK Setting-Type (DURABOND) or SHEETROCK Lightweight Setting-Type (EASY SAND) Joint Compound in the following sequence:

1. Pre-fill joints of SHEETROCK brand Exterior Gypsum Ceiling Board with SHEETROCK Setting-Type (DURABOND) or SHEETROCK Lightweight Setting-Type (EASY SAND) Joint Compound. After pre-fill has set, tape all joints and angles in the ceiling with compound and SHEETROCK Joint Tape. When compound sets (hardens), immediately apply a fill coat of compound; allow to harden before finishing.

2. Apply SHEETROCK Setting-Type (DURABOND) or SHEETROCK Lightweight Setting-Type (EASY SAND) Joint Compound over flanges of SHEETROCK Zinc Control Joints, metal beads and trim. Spot all fastener heads.

3. After fill coat has set, apply compound finishing coat. Completely cover all joints, angles, beads, control joints and fasteners.

4. After the joint compound has dried, apply one coat of a good-quality latex flat exterior paint to equalize the joint and wallboard surfaces. Then follow with at least one coat of a balanced, good-quality alkyd or latex exterior finishing system as specified by the paint manufacturer.

For Use with SHEETROCK brand Gypsum Panels, Water-Resistant—In areas to be tiled, for tapered edge joints, embed SHEETROCK Joint Tape with SHEETROCK Setting-Type (DURABOND 45 or 90) or SHEETROCK Lightweight Setting-Type (EASY SAND 45 or 90) Joint Compound. When set, apply a fill coat of the same joint compound. Take care not to crown the joint. Wipe excess joint compound from the water-resistant panel surface before it sets. For butt joints and interior angles, embed SHEETROCK Joint Tape with SHEETROCK Setting-Type (DURABOND) or SHEETROCK Lightweight Setting-Type (EASY SAND) Joint Compound. A fill coat is not necessary. Again, take care not to crown the joint. For fasteners, spot fastener heads at least once with setting-type joint compound.

Fill and seal all openings around pipes, fittings and fixtures with a thinned down coat of a good quality tile adhesive. For best results, use tile adhesive both as a sealer and to set the tile. Thin to a paint-like viscosity and apply the thinned compound with a small brush onto the raw gypsum panel core at the cutouts and allow areas to dry thoroughly prior to application of tile. Before adhesive dries, wipe excess material from the surface of gypsum panels.

For areas not to be tiled, embed tape with SHEETROCK Setting-Type (DURABOND) or SHEETROCK Lightweight Setting-Type (EASY SAND) Joint Compound 45 or 90 in the conventional manner. Finish with at least two coats of a U.S. Gypsum Company joint compound to provide joint finishing for painting and wall papering.

Drying Time—Joint Compound Under Tape

This standard is based on evaporation of 10 lb. water per 250 ft. paper tape, corresponding to 1/16″ to 5/64″ wet compound thickness under the tape.

RH	RH = Relative Humidity			D = Days (24 hr.)		H = Hours		
98%	53 D	38 D	26 D	18 D	12 D	9 D	6 D	4½ D
97%	37 D	26 D	18 D	12 D	9 D	6 D	4½ D	3¼ D
96%	28 D	21 D	14 D	10 D	7 D	5 D	3½ D	2½ D
95%	25 D	17 D	12 D	8 D	6 D	4 D	2¾ D	2 D
94%	20 D	14 D	10 D	7 D	5 D	3¼ D	2¼ D	41 H
93%	18 D	12½ D	9 D	6 D	4 D	2¾ D	2 D	36 H
92%	15 D	11 D	8 D	5 D	3½ D	2½ D	44 H	32 H
91%	14 D	10 D	7 D	4¾ D	3¼ D	2¼ D	40 H	29 H
90%	13 D	9 D	6 D	4½ D	3 D	49 H	36 H	26 H
85%	10 D	6 D	4 D	3 D	2 D	34 H	25 H	18 H
80%	7 D	4¾ D	3¼ D	2¼ D	38 H	27 H	19 H	14 H
70%	4½ D	3½ D	2¼ D	38 H	26 H	19 H	14 H	10 H
60%	3½ D	2½ D	42 H	29 H	20 H	14 H	10 H	8 H
50%	3 D	2 D	36 H	24 H	17 H	12 H	9 H	6 H
40%	2½ D	44 H	29 H	20 H	14 H	10 H	7 H	5 H
30%	2¼ D	38 H	26 H	18 H	12 H	9 H	6 H	4½ H
20%	2 D	34 H	23 H	16 H	11 H	8 H	5½ H	4 H
10%	42 H	30 H	21 H	14 H	10 H	7 H	5 H	3½ H
0	38 H	28 H	19 H	13 H	9 H	6 H	4½ H	3 H
°F	**32°**	**40°**	**50°**	**60°**	**70°**	**80°**	**90°**	**100°**
°C	**0°**	**4°**	**10°**	**16°**	**21°**	**27°**	**32°**	**38°**

The drying times for thicker (or thinner) coats of wet compound between tape and panels will increase (or decrease) in proportion to the wet compound thickness.

The drying times shown in the chart apply when the exposed surface of tape is bare or nearly bare, and when adequate ventilation is provided. A heavy compound coat over tape lengthens drying time.

Finishing

Gypsum drywall provides smooth surfaces that readily accept paint, texture finishes and wallcoverings. For satisfactory finishing results, care must be taken to prepare surfaces properly to eliminate possible decorating problems commonly referred to as "joint banding" and "photographing." These problems are usually caused by differences between the porosities and surface textures of the gypsum panel face paper and the finished joint compound, and magnified by the use of gloss paints. Then, when viewed in direct natural lighting, the joints and fasteners in painted walls and ceilings may be visible.

Skim Coating

The best method to prepare any gypsum drywall surface for painting is to apply a skim coat of joint compound. This leaves a film thick enough to fill imperfections in the joint work, smooth the paper texture and provide a uniform surface for decorating. Skim coating is currently recommended in the industry when gloss paints are used. It is also the best technique to use when decorating with flat paints.

Skim Coat Application

Finish joints and fasteners in the conventional three-coat manner. After joints are dry, mix joint compound—preferably SHEETROCK All Purpose, SHEETROCK Lightweight All Purpose (PLUS 3) Ready Mixed or COVER COAT Compound—to a consistency approximating that used for hand taping. Using a trowel, broad knife, or long-nap texture roller, apply only sufficient amounts of joint compound to cover the drywall surface. Then immediately wipe the compound as tightly as possible over the panel surface using a trowel or broad knife.

Finishing and Decorating Tips

1. When sanding joint compound applied over joints, fasteners, trim and corner bead, take care to avoid roughening the panel face paper. Any paper roughened during sanding has raised fibers that are conspicuous after painting.

2. All surfaces (including applied joint compound) must be thoroughly dry and dust free before decorating.

3. After conventional finishing of gypsum panel joints and fasteners, apply a skim coat of joint compound over the entire surface. This is the best technique for minimizing surface defects under critical lighting conditions. Skim coating fills imperfections in joint work, smooths the paper texture and provides a uniform surface for decorating.

4. As an alternative to skim coating, apply SHEETROCK First Coat for priming to equalize joint and wallboard surfaces or apply undiluted flat latex paint with high solids content as a substitute. This procedure minimizes problems with concealment of joints and fasteners.

5. A ceiling or wall texture finish is an excellent method for masking imperfections and diffusing light across wall and ceiling surfaces.

6. Frequent job inspections forestall potential problems and help ensure project specifications are being met. Wall and ceiling surfaces should be inspected after the gypsum panels are installed, when the joints are being treated and after the joints are finished before the surface is decorated. These checks will reveal starved and crowned joints which always show up under critical lighting.

Priming

Surface Preparation—Proper preparation is essential for producing the best possible painted finish. Surfaces must be dry, clean, sound and free of oil, grease and efflorescence. Glossy surfaces must be dulled. Exposed metal should be primed with a good rust-inhibitive primer. New concrete should age 60 days or more before covering. Fill or level concrete with SHEETROCK Setting-Type (DURABOND) or SHEETROCK Lightweight Setting-Type (EASY SAND) Joint Compound or COVER COAT Compound. Treat drywall joints and nailheads with a U.S. Gypsum Company Joint System.

Also important for a superior paint job is the equalization of both the porosity and texture of the surface to be painted. The best way to achieve this is to skim coat the entire surface with SHEETROCK All Purpose, Lightweight All Purpose (PLUS 3) or Topping Joint Compound or COVER COAT Compound as described above. An excellent alternative to skim coating is priming the entire surface with SHEETROCK First Coat.

SHEETROCK First Coat Application

Mixing—Ready-mixed SHEETROCK First Coat should be stirred gently. Do not thin for brush or roller application. For spray application, if necessary, add water in half-pint increments up to a maximum 1 qt. of water per gallon. May be tinted.

If dry packaged product is used— follow directions on bag. Start with lukewarm water, 2½ gal. or 9.5 L for 25-lb. bag. For best results, use a heavy-duty ½″ electric drill mixer (550 rpm), a cage-type mixing paddle, and a clean, smooth-sided mixing container. Sift powder into water and power mix by triggering the drill on/off. Mix for one minute. Wipe sides of container with paint stick or spatula so that all powder goes into wet mix. Continue to mix until lump free. The initial mix should appear creamy yet very heavy in viscosity. Add 2 qt. (1.9 L) water and let soak 5-10 minutes, then remix. Add 2 qt. water and remix. Add another 2 qt. water and remix. SHEETROCK First Coat is now ready for brush or roller application. For spray application, add up to an additional 2 qt. water to obtain desired consistency. NOTE: overthinning may cause poor adhesion, lack of hide and unequal suction when dry. Do not exceed a total of 26 qt. (24.6 L) per 25-lb. bag. Use wet-mixed material within 7 days. May be tinted.

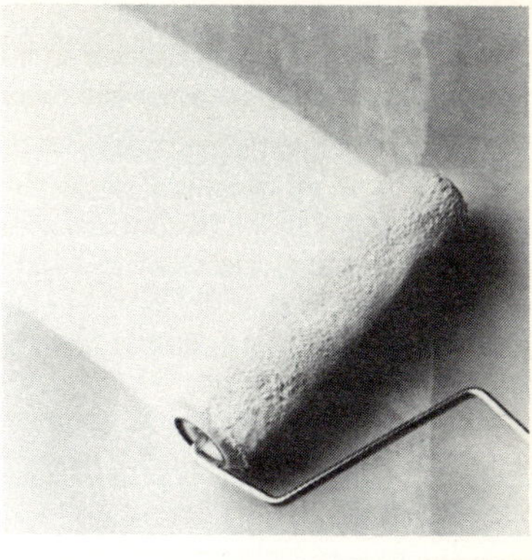

Specially formulated, fast-drying SHEETROCK First Coat equalizes surface texture and porosity to minimize decorating problems.

Application (walls and ceilings)—Apply a full coverage coat. Material dries to touch in under 30 min. Maintain minimum air, product mix and surface temperature of 55°F (13°C) during application and until surface is dry. Brush, roller, airless or conventional spraygun may be used.

Brush—Use a high-quality, professional paint brush.

Roller—Use a high-quality roller with ⅛″ to ¼″ nap on smooth and semi-smooth surfaces. For any surface, maximum nap length should not exceed ½″.

Conventional spraygun—Use Binks Model 2001 gun, pressurized external, with #565 fluid needle, #66 fluid nozzle and #65 PR air nozzle; or Binks model 18, pressurized internal, with #68 fluid needle, #68 fluid nozzle and #206 air nozzle; or Binks Model 18D gun, pressurized internal with #54-1209 fluid needle, #57 fluid nozzle and R-27 air nozzle; or similar equipment. Air hose is typically ⅜″ i.d. with ½″ fluid hose i.d.

Airless Spraygun—Use professional equipment that meets or exceeds the following when spraying through 50′ of ¼″ i.d. airless spray hose: output at least ¾ gal. per minute; pressure at least 2700 psi.; and accommodates a spray tip of 0.021″ at 2000 psi. Recommended equipment includes Graco ULTRA 1500, 1000 or 750 models with a suitable spray gun that will accommodate a RAC IV 519 (0.019) or RAC IV 521 (0.021) tip, a RAC IV Dripless Guard, and a 30-mesh filter.

Notes: Adjust atomizing air pressure and fluid flow rate so that a full coverage rate can be achieved by overlapping preceding application with one-quarter to one-half the fan width at a distance of 18″ from the surface. Air pressures and flow rates will vary with hose size and length and paint consistency.

SHEETROCK First Coat contains a high level of select pigments and fillers like conventional latex flat paints. When these paints are used in spray equipment previously used to spray PVA sealers which contain high levels of resin, clogging at the spray gun tip may result. The use of clean or new hoses is recommended to avoid this problem when spraying SHEETROCK First Coat.

Coverage—Approx. 300-500 sq. ft. per gal. of wet-mixed material depending upon factors such as application equipment and technique, condition of the substrate, amount of dilution and thickness and uniformity of coating.

Adding to Wall and Ceiling Textures—If slightly better spray properties, wet hide, improved bond, whiteness and surface hardness of texture are desired, SHEETROCK First Coat may be added to wet-mixed USG Ceiling and Wall Textures at a rate of up to 1-gal. (ready-mixed or fully mixed) SHEETROCK First Coat per 30, 32, 40 or 50-lb. bag of texture, or one 25-lb. bag of SHEETROCK First Coat may be mixed with 400 lb. (dry) ceiling and texture products. Reduce water quantity to account for addition of SHEETROCK First Coat based on 1:1 replacement basis. Surface priming recommendations on texture bag still apply.

Veneer Plaster Finishes

Veneer plaster finishes can be used in one or two-coat applications and can be given smooth or textured surfaces. Each method has its particular advantage.

Two-Coat Veneer Plaster Finish—Compared to many other finishes, two-coat veneer plaster provides a more durable, abrasion-resistant surface and can be finished to a truer plane than one-coat applications. These finishes can be used with steel or wood framing wherever the ultimate in appearance is desired. Ready for next-day decorating, assemblies with these monolithic gypsum surfaces offer excellent fire and sound ratings.

One-Coat Veneer Plaster Finish—Provides a hard monolithic surface at low cost. Complete application—from bare studs to decorated walls and ceilings—takes no more than 48 to 72 hr. Assemblies with one-coat veneer plaster application meet fire and sound requirements, and shorten construction schedules for added profit.

Job Environment

Maintain building temperature in comfortable working temperature above 55° F. Keep air circulation at minimum level prior to, during and following application until finish is dry.

If possible, maintain building temperature-humidity combination in the "normal drying" area of the graph. When dry conditions exist, relative humidity often can be increased by wetting down the floor periodically. During these periods, make every effort to reduce air movement by closing windows and deflecting heater blower and duct output away from the surfaces being plastered.

If building temperature-humidity combination is in the "rapid drying" area of the graph, special joint treatment measures must be taken. These include the use of SHEETROCK Setting Type (DURABOND) 45 or 90 Joint Compounds which are faster-setting and SHEETROCK Joint Tape.

Plaster Drying Conditions

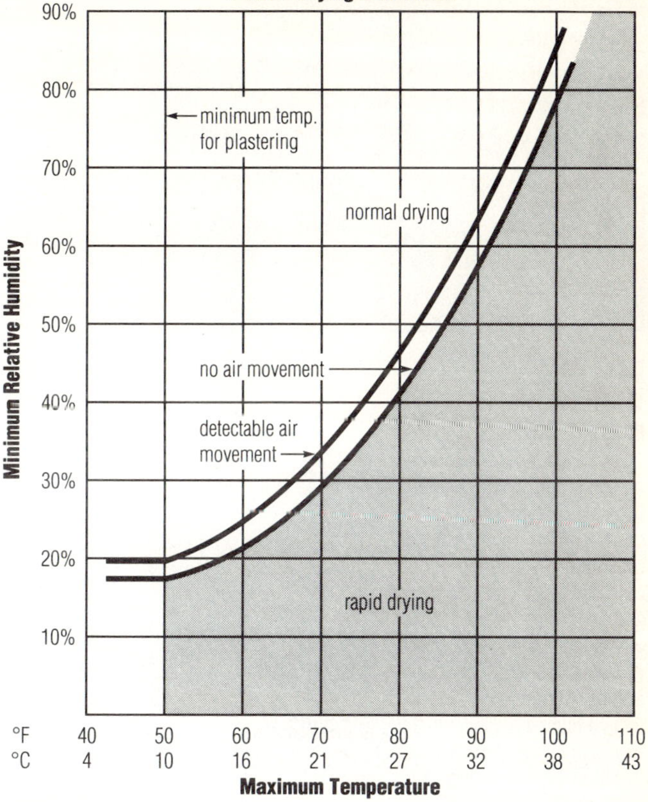

Grounds

Correct thickness of veneer plaster finish is one of the most important factors in obtaining good results. To ensure proper thickness, all corner beads, trim and expansion joints must be of the recommended type and be properly set.

Accessories must provide grounds for the following minimum plaster thicknesses:

1. Over veneer plaster gypsum base, one coat—1⁄16 in. (1.6 mm).
2. Over veneer plaster gypsum base, two coats—3⁄32 in. (2.4 mm).

Joint Treatment for Veneer Plaster Construction

For Wood-Framed Assemblies and Normal Drying Conditions—Align IMPERIAL Type P (pressure-sensitive) Tape over joint and press into place over entire length. Eliminate wrinkles and assure maximum adhesive bond by pressing entire length of tape with steel finishing knife or trowel. Press tape into corners with corner tool; do not overlap.

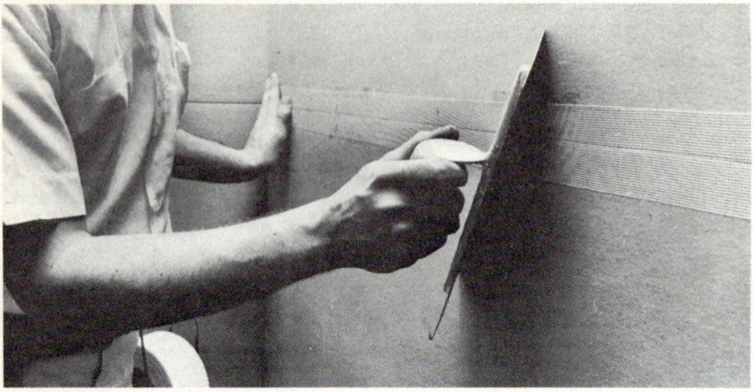

Simplified, wrinkle-free attachment of self-stick Type P IMPERIAL Tape speeds joint reinforcing, boosts production.

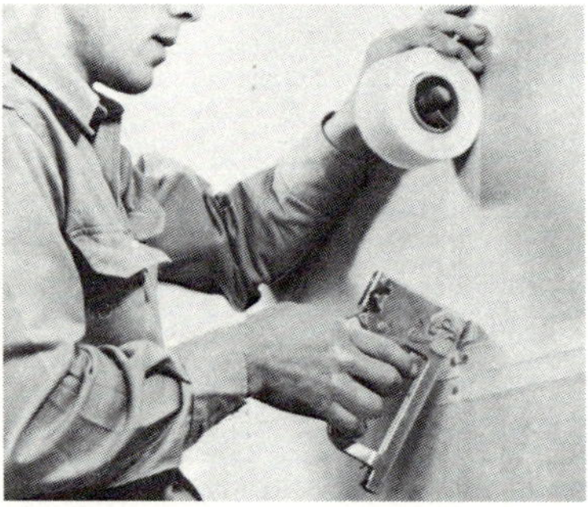

Or attach IMPERIAL Type S Tape with spring-driven hand stapler using ⅜″ staples. Use two staples at each end of tape; staple remainder at staggered 24″ intervals. At wall-ceiling angles, staple every 18″ to 24″ along ceiling edge only. For wall-to-wall interior angles, staple every 18″ to 24″ on one edge only, working from top to bottom. Position tape to bridge the joint at all interior corners without overlapping.

Embed tape and fill beads with a coat of veneer plaster being used, and allow to set—but not dry—prior to veneer plaster application. Slightly underfill in the bead by screeding along the bead with edge of trowel after setting the bead. (Best results are obtained by planning the finishing to permit continuous application from angle to angle.)

For Steel-Framed Assemblies and Rapid Drying Conditions—With steel framing and/or when building temperature-humidity conditions fall in the "rapid drying" area of the graph for steel or wood framing (see preceding page), use SHEETROCK Joint Tape embedded with a SHEETROCK Setting-Type (DURABOND) 45 or 90 Joint Compound.

Mix the compound in a clean 5-gal. container (plastic is preferred for setting-type compounds). Use a commercial potato masher or a ½″ heavy-duty 200 to 300-rpm electric drill with a cage-type paddle. Drill speed must not exceed 400 rpm. Use the amount of water shown on the bag and always sift the powder into the water to ensure complete wetting. Stir according to directions on bag. Note: Do not contaminate compound with other materials, dirty water or previous batches. Do not retemper batches.

Butter joints with compound using a trowel or steel finishing knife to force compound into the joints. Center SHEETROCK Joint Tape over joint and press it into the fresh compound with trowel held at a 45° angle. Draw trowel along joint with sufficient pressure to remove excess compound.

After tape is embedded, apply a thin coat of joint compound to reduce possibility of edge wrinkling or curling. Allow thin coat to harden, then apply a fill coat completely covering the tape and feathering 3″ to 4″ beyond edges of tape. Allow to harden before finishing. Plaster pre-fill is not required over SHEETROCK Setting-Type (DURABOND) Joint Compound.

Note: Under the following conditions a SHEETROCK Setting-Type (DURABOND) Joint Compound and SHEETROCK Joint Tape must be used: (1) where two-coat finish is applied over ½″ or ⅝″ base on 24″ o.c. framing; (2) where one-coat DIAMOND Interior Finish Plaster or IMPERIAL Finish is applied over ⅝″ base on 24″ o.c. framing; where IMPERIAL Gypsum Base and veneer plaster is used over steel framing.

Veneer Plaster Finish Application

Mixing and Proportioning

All veneer plaster finishes require the addition of water on the job. Water should be clean, fresh, suitable for human consumption, and free from mineral and organic substances that affect the plaster set. Water used for rinsing or cleaning is not suitable for mixing because it accelerates the plaster set.

Mechanical mixing is mandatory for veneer plaster finishes. Mix no more material than can be applied before set begins. Since veneer plaster finishes set more rapidly than most conventional plasters, always consult bag directions for specific setting times.

Veneer plaster finishes will produce mortar of maximum performance and workability when the correct equipment is used and mixing directions carefully followed. Proper mixing is one of the most important factors in producing mortar of maximum workability.

Use a cage-type mixer paddle driven by heavy-duty ½″ electric drill with a no-load rating of 900 to 1,000 rpm. Do not use propeller-type paddle or conventional mortar mixer. (For details of the cage-type mixing paddle and available electrical drills, see Technical Data Sheet P-502.)

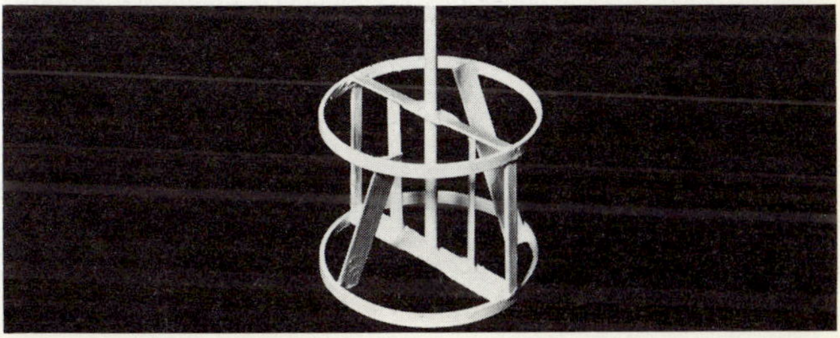

Cage-type mixing paddle is designed to draw material into and through paddle blades to disperse and blend ingredients by shear action rather than folding action of conventional mixers.

Mix plaster in 16 or 30-gal. smooth-sided container strong enough to withstand impacts that could cause gouging. Do not use brittle containers for mixing.

Correct mixing—rapid and with high shear action—is essential for proper dispersion of plaster ingredients. Slow mixing can reduce plasticity of material. Overmixing can shorten working time. Operated at correct speed, the cage-type design paddle mixes thoroughly without introducing excess air into the mix.

Mixing IMPERIAL Plasters

Water requirements for IMPERIAL Veneer Plasters:

IMPERIAL Basecoat—8 to 10 qt./80 lb. bag.

IMPERIAL Finish Plaster—11 to 13 qt./80 lb. bag.

Place water in a 12 to 16-gal. smooth-sided container. Start mixer, slowly add plaster and mix at least 2 min. to disperse lumps completely. Do not mix more than 5 min.

For sand float finish, add up to 20 lb. clean silica sand per 80-lb. bag of IMPERIAL Plaster to achieve desired texture. The use of more than 20 lb. of sand per bag will decrease hardness of surface. (Apply plaster in normal manner but omit final troweling. After surface has become firm, float to desired texture, using sponge, carpet or other float. Use water sparingly.)

Mixing DIAMOND Plasters

Water requirements for DIAMOND Veneer Plasters:

DIAMOND Veneer Basecoat—12 to 14 qt./80-lb. bag

DIAMOND Interior Finish Plaster —12 to 15 qt./50-lb. bag.

Place all but 1 or 2 qt. of water into mixing container; then with mixer operating, slowly add one bag of material. If a texture finish is desired, up to 50 lb. clean silica sand may be added per 50-lb. bag of DIAMOND Interior Finish Plaster. For electric cable heat systems, clean, sharp, fine silica sand must be added as follows: fill coat, 50 lb. but no less than 25 lb. per 50-lb. bag DIAMOND Interior Finish Plaster; finish coat, at least 12½ lb. per 50-lb. bag plaster. When material is wetted, add more water (1 to 2 qt.) to obtain desired consistency. Mix for minimum of two minutes, but no longer than five minutes.

When DIAMOND Finish is job-aggregated, one tablespoon cream of Tartar should be added for each bag of Finish to retard plaster and allow sufficient working time.

Application

Maintain temperature in all work areas at min. 55° to 60°F (13° to 16°C). Keep air circulation at minimum level during and after application until finish is dry.

IMPERIAL Plasters

Embed IMPERIAL Tape and fill beads with a tight, thin coat of IMPERIAL Plaster; allow to set, then plaster.

Plaster Finishing

IMPERIAL Plasters provide a wide range of finish options with three distinct application systems:

1) IMPERIAL Finish Plaster (one-coat).

2) IMPERIAL Basecoat with selected hand-applied or spray finish (two-coat).

3) IMPERIAL Basecoat MA with selected hand-applied or spray finish (two-coat).

IMPERIAL Finish Plaster (one-coat)—Scratch-in a tight, thin coat of IMPERIAL Finish Plaster over entire area, immediately doubling back with plaster from same batch to full thickness of 1⁄16″ to 3⁄32″. Fill all voids and imperfections. Final-trowel after surface has become firm, holding trowel flat and using water sparingly. Do not over-trowel.

For texture finished surfaces, with or without the addition of job added sand, final troweling is omitted. The surface is textured naturally as the material firms and water is removed into the base.

Best results are obtained by planning the plastering to permit continuous application from angle to angle. Where joining is unavoidable, use trowel to terminate unset plaster in sharp clean edge—do not feather out. Bring adjacent plaster up to terminated edge and leave level. Do not overlap. During finish troweling, use excess material to fill and bridge joining.

IMPERIAL Basecoat (two-coat)—Scratch-in a tight, thin coat of IMPERIAL Basecoat over entire area, immediately doubling back with plaster from same batch to full thickness of 1⁄16″ to 3⁄32″. Fill all voids and imperfections. Leave surface rough and open for proper bond of finish coat. Allow basecoat to set to provide proper suction for finish coat.

Finish coat materials are applied by scratching in and doubling back with the selected finish to achieve a smooth, dense surface for decoration, free of surface blemishes. For textured finishes, floating on textures with additional material is conducted once the surface has become firm. Final-trowel after surface has become firm, using water sparingly.

For spray-applied finish, mix Keenes Cement-lime-sand in proportion of 50 lbs. (23 kg) Keenes Cement to 100 lbs. (45 kg) dry double-hydrated lime, and up to 400 lb. (180 kg) but not less than 200 lb. (60 kg) clean, properly graded silica sand with sufficient water to form a smooth consistency for hand application. Apply this mix evenly over a properly prepared IMPERIAL Basecoat surface by first applying a well-ground-in scratch coat, then immediately double-back with sufficient material to cover the basecoat to a total thickness of 1⁄16″ to 1⁄8″. When the surface has become firm by water removal, float the surface to a uniform blemish-free flat texture. After floating, and while the application is still wet but totally firm, prepare additional finish material with the consistency adjusted for spray application. With either a hand-held hopper gun or machine-application equipment without catalyst, spray apply texture to provide a uniform texture appearance. Vary aggregate grading, aggregate proportion, number of passes over the surface, air pressure and nozzle orifice size as necessary to achieve the desired finish.

IMPERIAL Basecoat MA (two-coat)—Machine application of IMPERIAL Basecoat Plaster (MA) requires special equipment that provides for automatic catalyst injection. The machine should be operated in strict accordance with the manufacturer's directions. Successful results require advance job planning and operator training. The plaster base should be protected from overspray or contamination by lime or casein materials. Such materials adversely affect the bonding characteristics of plaster. Mask all areas to be protected from plaster overspray with plastic sheeting or paper secured with masking tape.

Mix one to three 3-lb. bags of IMPERIAL Catalyst in 3 gal. of clean water in a plastic pail. The amount of catalyst used will be determined by the desired setting time. Stir until material dissolves, let residue settle and pour solution into accelerator tank of the machine.

Mix plaster as previously described and pour into the machine through the 8x8-mesh screen hopper. Adjust water in plaster mix until 75% to 90% will pass the screen without shaking.

Caution: Clean mixing equipment after each batch. Clean hopper and screen free of set plaster to avoid acceleration. Machine should be completely cleaned after each four hours of use.

Adjust setting time by controlling catalyst flow at the machine. Test by spraying plaster on plaster base scrap.

With setting time adjusted to 30 min., spray taped joints and corner-bead flanges in a pattern wide enough to cover the tape and metal trim flanges. Immediately trowel level, completely embedding tape and metal component flanges. Leave no voids. Allow to set before plastering. This initial spraying is not required over SHEETROCK Setting-Type Joint Compound (DURABOND).

Adjust set for 20 to 30 min., spray IMPERIAL Basecoat Plaster (MA) over entire area to a thickness of 1⁄32", then immediately cross-spray to a total thickness of 1⁄16". Allow to set and dry sufficiently to provide proper suction for the finish coat.

For hand-applied finish, scratch-in and double-back with selected finish to fill out to a smooth, dense surface for decoration, free of surface blemishes, to a full plaster thickness of 1⁄8". Trowel after surface has become firm, using water sparingly. Texturing with desired finish is started once initial scratch and double-up application has become firm.

For spray-applied finish, mix and apply finish as described above for IMPERIAL Basecoat (two-coat).

Other IMPERIAL Basecoat Applications

Concrete Block—Surface must be porous and develop proper suction or be scored to provide adequate mechanical bond. Lightly spray walls with water to provide uniform suction. Fill and level all voids, depressions and joints with IMPERIAL Basecoat Plaster and allow to set; then apply subsequent coats as with gypsum base application, leaving final surface rough and open to provide proper bonding of the finish coat.

Monolithic Concrete—Prepare surface with a bonding agent applied according to manufacturer's directions. Fill all voids and depressions with IMPERIAL Basecoat Plaster and allow to set and partially dry. Then apply IMPERIAL Basecoat Plaster as with gypsum base or concrete block. *Important:* As no suction is provided over surfaces treated with a plaster bonding agent, it is essential that the applied basecoat surface be raked or broomed once the material has become firm for a rough and open surface to provide proper suction for the finish coat. Failure to do so may result in delamination of the finish material.

Integral Plaster Chalkboards

Plaster chalkboards offer maximum freedom in design. There is no limiting sheet size as is the case with fabricated boards; therefore, entire walls can be utilized as chalkboards. Maintenance is accomplished as easily as with conventional fabricated chalkboards. (Requirements for control joints in chalkboard surfaces are the same as for other gypsum surfaces.)

Chalkboard with Steel Stud-IMPERIAL Partitions

Follow directions for system construction. Locate floor and ceiling runners and position studs 16″ o.c. Attach IMPERIAL Gypsum Base using 1″ Type S Screws spaced 16″ o.c. When chalkboard area does not extend from floor to ceiling, use 701-A or 801-A Metal Trim to frame the IMPERIAL Base face layers that will be used as chalkboard. (All chalkboard surfaces must have two layers of IMPERIAL Gypsum Base.) Miter corners of the metal trim to form a neat joint. Attach chalkboard using 1⅝″ Type S Screws, driven through IMPERIAL base layer into the studs.

Plaster Application: use one or two-coat plaster for chalkboard surface. With one-coat work, apply IMPERIAL Finish Plaster to 1/16″ to 3/32″ thickness. Cover entire area with a tight, thin coat, then double back to full thickness. After surface has become firm, final-trowel to a smooth surface, using water sparingly.

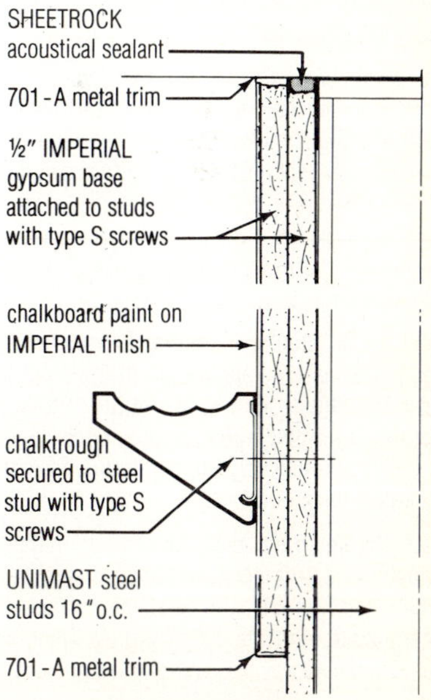

Plaster chalkboard

For two-coat application, apply IMPERIAL Basecoat to 1/16″ to 3/32″ thickness as described for single-coat application. Allow basecoat to set and partially dry; then apply IMPERIAL Finish or STRUCTO-GAUGE-lime smooth-trowel finish plaster. Leave surface very hard and polished.

Paint chalkboard, when dry, with one coat primer-sealer and two coats chalkboard paint.

Install chalktrough with 1 5/16″ Type S screws, driven through the two layers of IMPERIAL Gypsum Base and into the steel studs.

DIAMOND Plasters

DIAMOND Interior Finish Plaster

All finish materials and finish surfaces must be protected from contact with DIAMOND Interior Finish Plaster. This includes glass, ceramic materials, metal and wood. Apply wood, plastic or other exposed trim after plaster application.

DIAMOND Interior Finish Plaster should be applied to IMPERIAL Gypsum Base having unfaded blue face paper. However, under abnormal conditions where there is no alternative to using gypsum base faded from excessive exposure to sunlight or ultra-violet radiation, precautions should be taken to prevent delamination. Degraded gypsum base is indicated if face paper is not blue or grayish blue. When face paper color has become gray to tan, treat paper with an alum solution or bonding agent.

Degrading may occur when gypsum base has been installed long before the finish is applied.

For alum solution treatment, dissolve 3 lb. of alum in one gallon of water. Pour alum slowly into water and mix thoroughly. Allow solution to stand until any undissolved material has settled; then strain solution into tank-type sprayer (such as a lawn and garden sprayer). Spray solution on faded base so that it is wet but not soaked. One gallon of solution should treat 750 sq. ft. of base. Begin finish application before alum solution is completely dry. Note: Alum treatment shortens setting time of DIAMOND Interior Finish Plaster.

Begin application only after joints have been reinforced with glass fiber tape and preset with an application of DIAMOND Interior Finish Plaster or treated with SHEETROCK Joint Tape and SHEETROCK Setting-Type (DURABOND) Joint Compound. Apply a thin, tight scratch coat of this finish over entire working area. Immediately double back with material from same batch to a full 1/16″ to 3/32″ thickness.

Start the finish troweling as soon as material has become sufficiently firm to achieve a smooth trowel finish free from trowel marks, voids and other blemishes. Smooth and level the surface with trowel held flat; use water sparingly to lubricate. Final hard troweling should be accomplished prior to set as indicated by darkening of the surface.

A variety of textures ranging from sand float to heavy Spanish can be achieved with DIAMOND Interior Finish Plaster when job-aggregated with silica sand. Application is the same as for neat DIAMOND Interior Finish

Patterned swirl texture

Light to medium skip trowel texture

Sand-aggregated float finish

Heavy texture finish

Plaster except that once the surface has been leveled and sufficient take-up has occurred, begin floating material from the same batch with trowel, float, sponge or by other accepted local techniques.

DIAMOND Interior Finish Plaster also may be textured by skip-troweling. When applying in this manner, eliminate final troweling. When surface has become sufficiently firm, texture with material from same batch prior to set.

Painting or further decoration of DIAMOND Interior Finish Plaster is recommended and should be specified. However, in many residential applications, DIAMOND Interior Finish Plaster provides a uniform white color and may satisfy a job's specific acceptance specifications if skip-trowel and floated textured finishes are utilized. DIAMOND Interior Finish Plaster is formulated to allow quick drying and can be decorated, when thoroughly dry, with a latex base or breather-type paint. Under ideal conditions, painting can take place in as little as 24 hours which minimizes costly delays and speeds occupancy.

DIAMOND Veneer Basecoat Plaster

DIAMOND Interior Finish Plaster is intended for hand application over veneer basecoats, such as DIAMOND Veneer Basecoat Plaster, as well as direct to gypsum base as a one-coat system. Machine application is not recommended. When applied over veneer basecoat, be certain basecoat is not completely dry. Basecoat should be set and allowed to dry only partially to provide suitable suction.

DIAMOND Veneer Basecoat Plaster provides quality walls and ceilings for residential and commercial construction where the superior strength of IMPERIAL Basecoat Plaster is not essential. DIAMOND Veneer Basecoat Plaster produces a base that aesthetically enhances the finish by providing regulated suction, resulting in exceptional integral bond. Once basecoat is applied and has become firm, surface is raked or broomed to provide a rough and open surface for the finish coat.

Over Gypsum Base—Apply DIAMOND Veneer Basecoat Plaster from 1/16″ to 3/32″ thickness. When IMPERIAL Gypsum Base is used, reinforce all joints and interior angles with IMPERIAL Type P or Type S Tape. Embed tape and fill beads with DIAMOND Veneer Basecoat Plaster and allow to set, but not dry. After beads and joints have been properly prepared, apply a tight, thin coat of DIAMOND Veneer Basecoat Plaster over the entire area, immediately doubling back with plaster from the same batch to full thickness. Fill all voids and imperfections. Leave surface rough and open by cross raking with a fine wire rake, sponge or fine broom once the surface has become somewhat firm. Allow basecoat to set to provide proper suction for finish coat.

Over Concrete Block—Surface must be porous and develop proper suction or be scored to provide adequate mechanical bond. Lightly spray walls with water to provide uniform suction. Fill and level all voids, depressions and joints with DIAMOND Veneer Basecoat Plaster and allow to set; then apply subsequent coats as with gypsum base application, leaving final surface rough and open to provide proper bonding of the finish coat.

Over Monolithic Concrete—Prepare surface with a bonding agent applied according to manufacturer's directions. Fill all voids and depressions with DIAMOND Veneer Basecoat Plaster and allow to set and partially dry. Then apply DIAMOND Veneer Basecoat Plaster as with gypsum base or concrete block. Important: As no suction is provided over surfaces treated with a plaster bonding agent, it is essential that the applied basecoat surface be raked or broomed once the material has become firm for a rough and open surface to provide proper suction for the finish coat. Failure to do so may result in delamination of the finish material.

Radiant Heat Plaster System

Application-Radiant Heat Cable—After IMPERIAL Gypsum Base and joint reinforcement tape have been applied, install electric radiant heating cable in accordance with design requirements and cable manufacturer's specifications. Attach cable to ceiling in such a manner that it is kept taut and does not sag away from the base. All cable connectors and non-heating leads should be embedded into (countersunk), but not through, the gypsum base so they do not project below the heating wire.

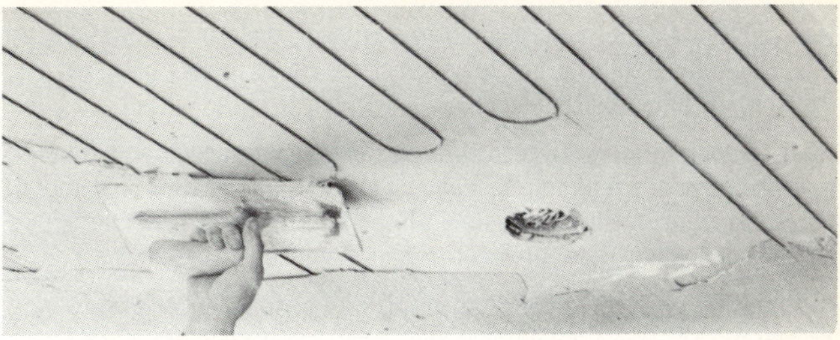

Trowel job-sanded DIAMOND Interior Finish Plaster over electric cable.

Fill Coat Application—Apply job-sanded DIAMOND Interior Finish Plaster in sufficient thickness to completely cover cable. Trowel plaster parallel to direction of cable but do not use cable as a screed. Level with a trowel, rod or darby to fill any low spots or to remove any high ridges, etc. Use a serrated darby or lightly broom the plaster surface prior to set to provide a key for the finish coat. Average thickness of fill coat should be 3⁄16″.

Finish Coat Application—Apply finish coat after fill coat has developed sufficient suction—in good drying weather, about two hours after the fill coat has set; in damp or cold weather, usually overnight unless good supplementary heat and ventilation are provided. Use job-sanded DIAMOND Interior Finish Plaster 1⁄16″ to 3⁄32″ thick, to bring total plaster thickness to 1⁄4″.

Scratch-in a tight thin coat over the entire area, immediately doubling back to full thickness. Fill all voids and imperfections. Scratch and double-back with the same mix of DIAMOND Interior Finish Plaster. When surface has become firm, hold trowel flat and final-trowel using water sparingly. Best results are obtained by continuous application of an entire ceiling. Always work to a wet edge to avoid dry joinings.

Texture Finish—When finish coat has become sufficiently firm, but unset, float surface to desired texture using a sponge, carpet, or other float. Use water sparingly. For heavier texture, additional material from the same batch may be applied to the firm surface to achieve a skip-trowel, Spanish, or other texture.

Simulated Acoustic Texture—Spray-apply max. 1⁄8″ thickness of IMPERIAL QT Texture Finish or similar product over a full 1⁄4″ thickness of sanded DIAMOND Interior Finish Plaster. Follow manufacturer's specifications.

Use of these finishes will slightly decrease heating system efficiency since simulated acoustical finishes are formulated with insulating-type aggregates.

Radiant Heat Systems to Monolithic Concrete

Surface Preparation—Concrete surface must be structurally sound and clean, free of dirt, dust, grease, wax, oil or other unsound conditions. Treat exposed metal with a rustproof primer. When corrosion due to high humidity and/or saline content of sand is possible, the use of zinc alloy accessories is recommended.

Remove form ridges to make surfaces reasonably uniform and level. Locate uneven ceiling areas and bad gravel pockets, which require filling prior to installing electric cable and filler.

After treating entire surface with a plaster bonding agent, according to manufacturer's directions, leveling may be done with fill-coat mix of DIAMOND Interior Finish Plaster. Minor leveling may be done with a SHEETROCK Setting-Type (DURABOND) Joint Compound.

Caution: Temperature of concrete ceiling with bonding agent applied must be above 32°F (0°C) before filler and finish applications are started, with air temperature above 55°F (13°C).

Radiant Heat Cable Application—After ceiling surface has been leveled, apply electric radiant heating cable according to design requirements and cable manufacturer's specifications. Attach cable to the ceiling so that it is kept taut and does not sag away from the ceiling. All cable connectors and non-heating leads must be securely attached to concrete ceiling.

Finishing—Mix and apply job-sanded fill coat DIAMOND Interior Finish Plaster according to directions in previous section. Apply 5⁄16″ basecoat parallel to direction of cable, completely covering cable and anchor devices.

Mix and apply finish coat after fill coat has developed sufficient suction. Use job-sanded DIAMOND Interior Finish Plaster 1⁄16″ to 3⁄32″ thick, to bring total plaster thickness to 3⁄8″. Apply finish coat in same manner described in previous section.

Do not energize heating cable until plaster is thoroughly dry. When either or both the completed radiant heat ceiling and room temperature are below 55°F (13°C), the thermostatic elevations should be made in 5°F (3°C) increments for each 24-hr. period until a room temperature of 55°F (13°C) is attained.

If completed radiant heat ceiling and room temperature are 55°F (13°C) or higher, thermostat may be set at desired temperature.

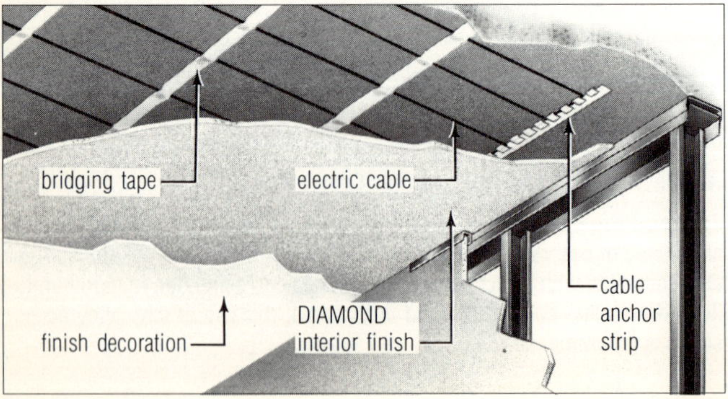

Concrete Coating Application

COVER COAT Compound

With ready-mixed COVER COAT Compound, drywall contractors are able to offer white, smooth or textured, ready-to-decorate surfaces on concrete ceilings and columns located above grade. Smooth application and excellent bonding strength make COVER COAT Compound ideal for filling small holes and crevices and for second and following covering applications with drywall methods and tools.

COVER COAT Compound should not be applied over moist surfaces or surfaces likely to become moist (by condensation or otherwise), on ceiling areas below grade, on surfaces which project outside the building, or on other areas which might be subject to moisture, freezing, efflorescence, pitting or popping, movement, or other abnormal condition.

Application

For best results, apply COVER COAT Compound before interior partitions are erected. Use the compound at package consistency to minimize shrinkage. If a thinner material for roller application is desired, the compound may be thinned by adding clean water (up to 1 to 1½ gal. per 61.7 lb. carton or pail) and mixing to desired consistency using a potato masher or low-speed drill type mixer. If applicator should inadvertently overthin, simply add additional compound to thicken and remix.

Protect COVER COAT Compound from freezing. During entire application, maintain temperature at or above 55°F (13°C), and provide heat and ventilation when necessary.

Prepare the concrete surfaces by removing any major portions or ridges. New concrete should age 60 days or more. Remove any form or parting oils, grease or efflorescence. Fill deep voids or offsets with SHEETROCK Setting-Type (DURABOND) or SHEETROCK Lightweight Setting-Type (EASY SAND) Joint Compound, then apply a skim coat of COVER COAT Compound over these areas after joint compound has hardened but not necessarily dried. All surfaces must be dry, clean and sound before applying COVER COAT Compound. Prime any exposed metal with a good, rust-inhibitive primer. Mix compound lightly and test-apply. Add small amounts of water if required.

Apply COVER COAT Compound over joints and ridges left by concrete forms with a flat finisher or knife. Fill in and/or level out small holes and lumps, ridges, lips, etc. with compound. Allow to dry.

Using two workers, apply first coat of compound to entire surface area of ceiling, beam, or column with flat finisher, long nap roller or regular knife. Keep moving in one direction, making sure that each application overlaps the previous one. Follow application with wide rubber squeegee or pole drywall blade, 24″ or wider, to smooth out fresh application, leaving a minimum of ridges and imperfections. Apply SHEETROCK No. 800 Corner Bead on angles and corners as required, embedding and covering both flanges with a smooth fill of COVER COAT Compound 3″ to 4″ wide. Allow to dry (under good drying conditions, 24 hr.).

Before second-coat application, sand and dust first coat. Apply second coat in manner described above or texture at this point if desired. Allow to dry. Sand to ultimate smoothness with fine sandpaper, if necessary. For texturing second coat, simply add water and/or sand. Use very fluid mix for fine texture, less fluid for coarse effects.

A very rough or uneven concrete surface may require three or more coats applied in the same manner.

COVER COAT Compound may be left undecorated. Not washable unpainted. If decoration is specified, follow directions on container.

More detailed directions, spray application and special-use information are available. Ask for U.S. Technical Data Sheet J-59.

Note: Check cracking may occur in excessively deep fills. For this reason successive coats are recommended for deep fills using a SHEETROCK Setting-Type (DURABOND) or SHEETROCK Lightweight Setting-Type (EASY SAND) Joint Compound for the first coat.

SHEETROCK Setting-Type Joint Compounds

A SHEETROCK Setting-Type (DURABOND) or SHEETROCK Lightweight Setting-Type (EASY SAND) Joint Compound is equally suitable for filling form offsets and voids left in interior concrete. As with COVER COAT Compound, SHEETROCK Setting-Type (DURABOND) and SHEETROCK Lightweight Setting-Type (EASY SAND) Joint Compounds should not be applied over moist surfaces or surfaces subject to moisture, or any abnormal condition.

Application

Grind off high plane differences in concrete level with adjacent area; remove any form oil, efflorescence or greasy deposits.

Prime exposed metal with a good rust-inhibitive primer.

Mix SHEETROCK Setting-Type (DURABOND) or SHEETROCK Lightweight Setting-Type (EASY SAND) Joint Compound according to bag directions.

Fill all offsets and voids with compound. Apply additional coats of SHEETROCK Setting-Type (DURABOND) or SHEETROCK Lightweight Setting-Type (EASY SAND) Joint Compound as required after each coat has set, but not necessarily dried.

Apply skim coat of SHEETROCK Setting-Type (DURABOND) or SHEETROCK Lightweight Setting-Type (EASY SAND) Joint Compound over entire surface. If an easier sanding surface is desired, apply final skim coat of COVER COAT Compound, SHEETROCK Lightweight Setting-Type (EASY SAND) Joint Compound or SHEETROCK All Purpose Ready-Mixed Joint Compound instead of SHEETROCK Setting-Type (DURABOND) Joint Compound.

Apply coat of SHEETROCK First Coat or a good quality, undiluted interior latex flat wall paint with high solid content over entire surface and allow to dry.

For textured ceiling, apply IMPERIAL QT Spray Texture Finish in uniform coat at rate up to 8 ft.^{2}/lb.

Acoustical Sealant Application

If gypsum board assemblies are to effectively reduce the transmission of sound, they must be airtight at all points. To achieve this, perimeters must be sealed with SHEETROCK Acoustical Sealant, a caulking material that remains resilient and will not shrink or crack. Also, penetrations for electrical outlets, medicine cabinets, plumbing, heating and air-conditioning ducts, telephone and intercom hookups and television antenna outlets must be effectively sealed.

Sealing or caulking for sound control is so important that it must be covered in the specifications, understood by the workers of all related trades, supervised by the foremen, and inspected carefully during construction.

Acoustical sealant application has proven to be the least expensive, most cost-effective way to seal assemblies and prevent sound leaks. SHEETROCK Acoustical Sealant is approved for use in all UL fire-rated assemblies without affecting fire ratings. All references herein to "caulk" or "caulking" indicate use of SHEETROCK Acoustical Sealant.

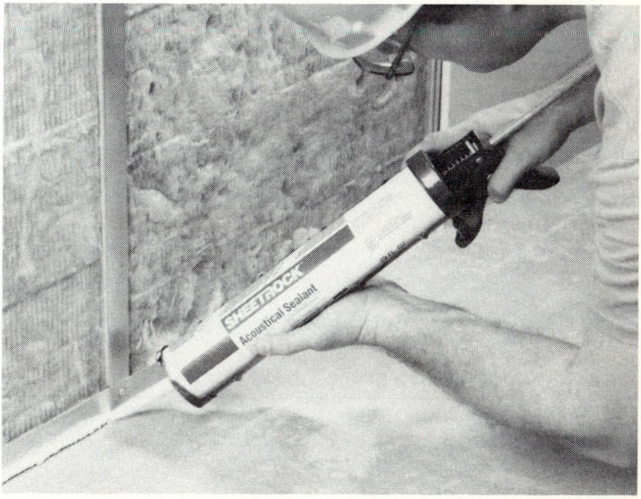

Tests conducted at the USG Research Center demonstrate that reliability of the perimeter seal is increased if perimeter relief does not exceed ⅛". When such a gap, around the base-layer perimeter, is caulked with a ¼" bead of sealant, installation of face panels compresses the sealant into firm contact with all adjacent surfaces to form a permanent airtight seal.

To be effective, sealant must be properly placed. Placement is as important as the amount used. The technical drawings on page 221 indicate correct and incorrect applications of acoustical sealant.

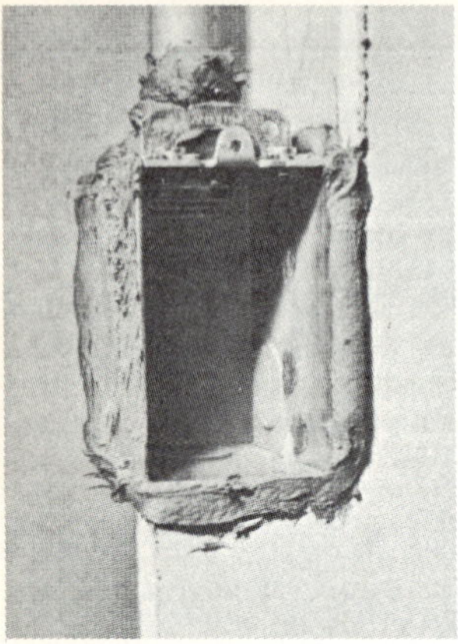

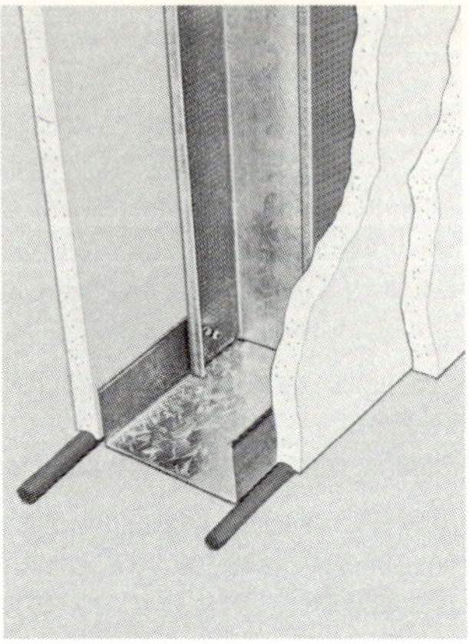

Proper caulking of outlet box (left), and double-layer partition (right).

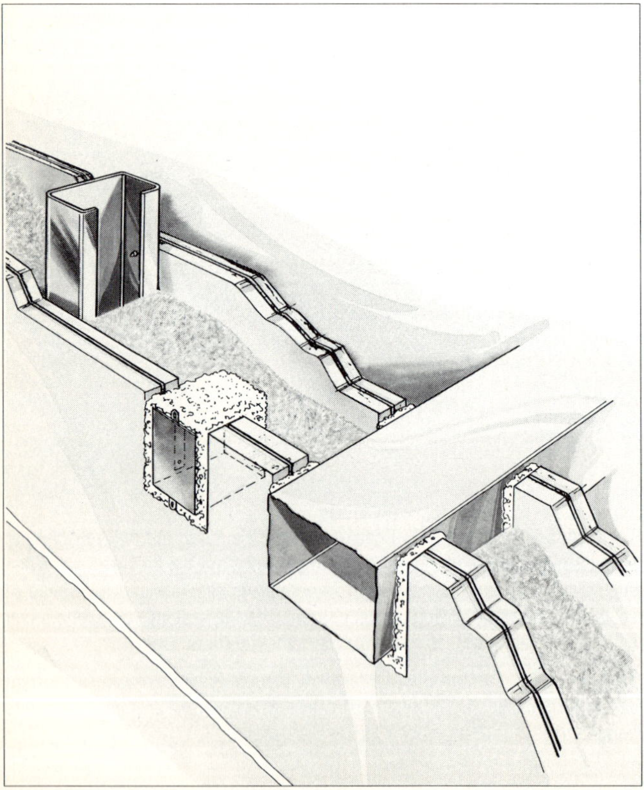

SHEETROCK Acoustical Sealant applied around ducts effectively seals the wall to reduce sound transmission.

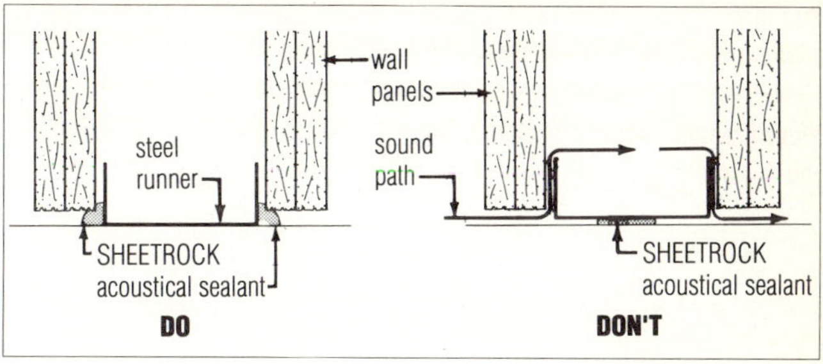

The assemblies tested consisted of 2½″ UNIMAST Steel Studs 24″ o.c., double-layer SHEETROCK brand Gypsum Panels, SW Edge, each side; 1½″ THERMAFIBER Sound Attenuation Fire Blankets between studs. Results of sealant conditions are shown below:

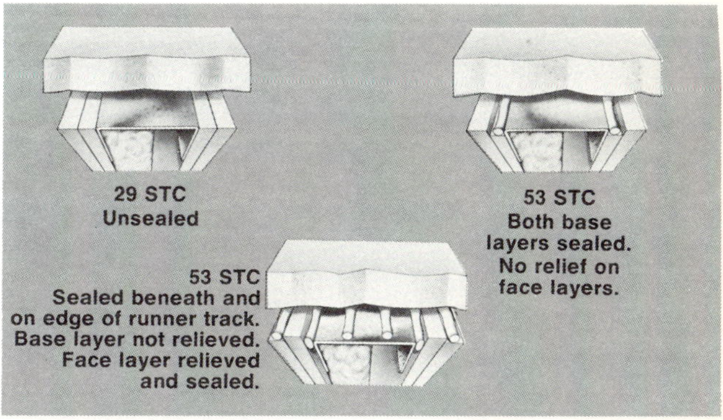

Installation

Partition Perimeter—Cut gypsum boards for loose fit around partition perimeter. Leave a groove no more than ⅛″ wide. Apply a ¼″ min. round bead of sealant to each side of runners, including those used at partition intersections with dissimilar wall construction. Immediately install boards, squeezing sealant into firm contact with adjacent surfaces. Fasten boards in normal manner. Gypsum panels may have joint treatment applied in normal manner over sealed joints, and gypsum base may be finished normally with veneer plaster. Or, panels may be finished with base or trim as desired.

For caulking application with metal trim over edge of boards where boards intersect dissimilar materials or cracking due to structural movement is anticipated, refer to "Perimeter Isolation" section on page 173.

Control Joints—Apply sealant beneath control joint to reduce path for sound transmission through joint.

Partition Intersections—Seal intersections with sound-isolating partitions that are extended to reduce sound-flanking paths.

Openings—Apply sealant around all cutouts such as at electrical boxes, plumbing, medicine cabinets, heating ducts and cold air returns to seal the opening. Caulk sides and backs of electrical boxes to seal them.

Door Frames—Apply a bead of sealant in the door frame just before inserting face panel.

Details–use of SHEETROCK acoustical sealant

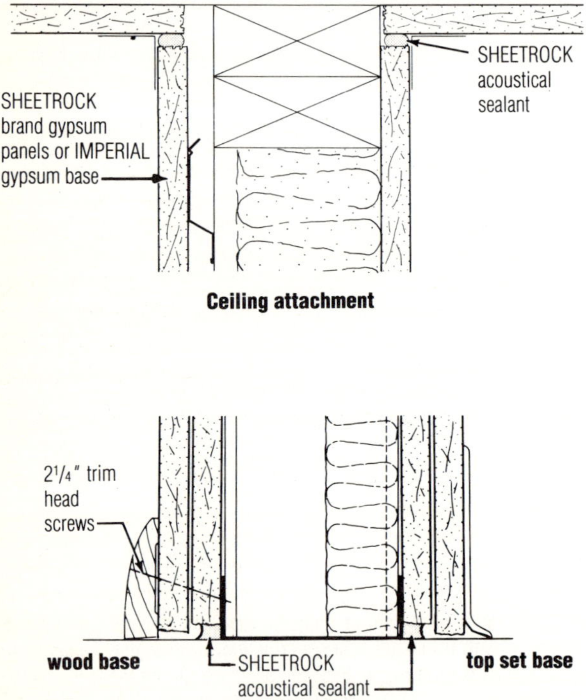

Floor attachment

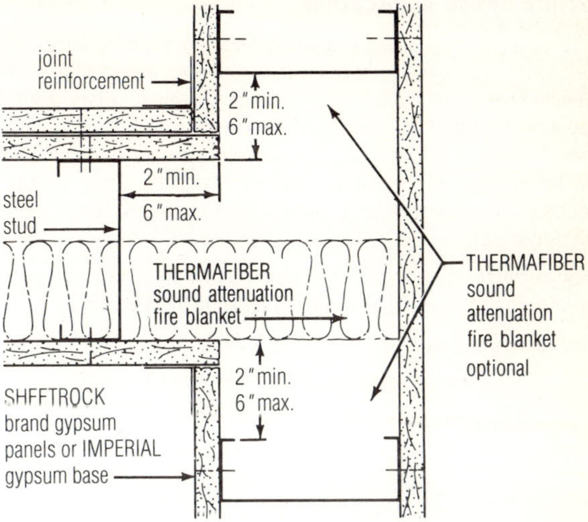

Sound-isolating partition intersection

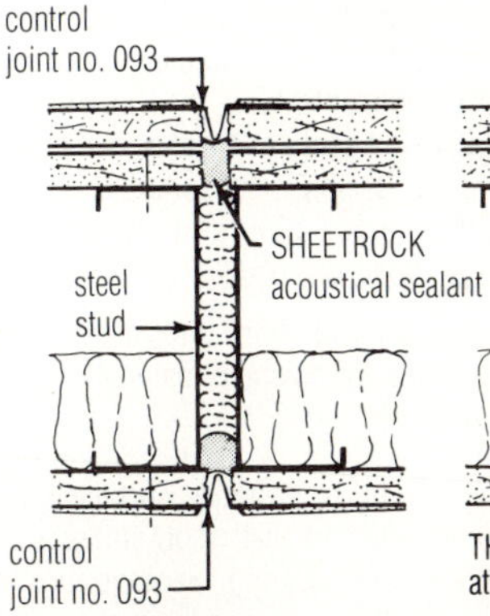

Partition control joint

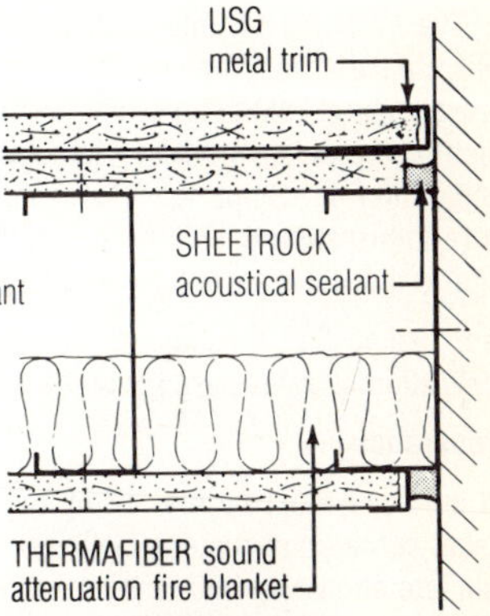

Partition wall intersection

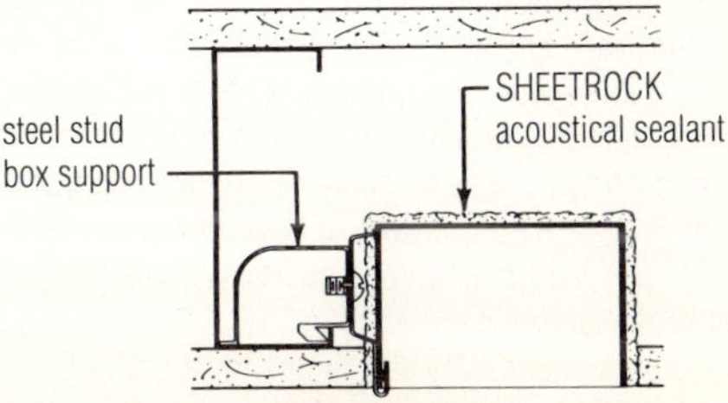

Outlet box

Texture Finish Application

Textured finishes for gypsum board surfaces are accepted for their decorative beauty and ability to obscure minor surface imperfections with economical spray application. U.S. Gypsum Company offers a full line of products to create fine, medium or coarse simulated acoustic texture finishes, as well as sand finishes. Also available is a sound-rated texture finish. Interesting wall patterns can be created by using texture finish products with stipple brushes, pattern devices, rollers, floats, trowels and finishing knives.

Note: Textured surfaces also can be created with veneer plaster finishes. See veneer plaster application section earlier in this Chapter.

General Limitations

1. Not recommended below grade or in high-humidity areas.

2. Heavy, water-based texturing materials may cause sag in gypsum panel ceilings under the following adverse conditions: high humidity, improper ventilation, panels applied parallel to framing and panels having insufficient thickness to span the distance between framing. The following table gives max. framing spacing for panels that are to be covered with water-based texturing materials.

3. The following surface preparation directions apply to new drywall and concrete surfaces. When redecorating an old, existing surface with a water-based texture, migrating stains or contaminants from the substrate may leach to the finished surface, resulting in discoloration and staining. See preparation directions for "Redecorating Ceilings" on pages 237-238 for more information on the proper surface preparation of existing surfaces prior to redecorating with a water-based texture.

See "Ceiling Sag Precautions" on page 362 for more information on the application of water-based textures and interior finishing materials.

Preparation

All surfaces must be dry, clean and sound. Dull glossy surfaces; prime metal with a good rust-inhibitive primer. Fill and seal wood surfaces. New concrete should age 60 days or more before covering. Remove form oils, grease, efflorescence. Grind down plane differences and remove grinding dust and sludge. Fill or level with SHEETROCK Setting-Type (DURABOND) or SHEETROCK Lightweight Setting-Type (EASY SAND) Joint Compound or COVER COAT Compound. Reinforce and conceal drywall joints using SHEETROCK Joint Tape and a U.S. Gypsum Company joint compound. Fill all fastener depressions with joint compound. Smooth surface scratches and scuffs. Correct plane irregularities, as these are accentuated by sharp, angular lighting. When prepared surfaces are dry and free of dust, apply a prime coat of SHEETROCK First Coat or a good quality, white, interior latex flat wall paint with high solids content. Paint should be applied undiluted and allowed to dry before decorating. NOTE: Application of a prime coat is to equalize the surface porosity and to provide a uniform color. Primers are not intended to reduce sag potential or to prevent migrating stains or contaminants from leaching to the finished surface.

Gypsum Panel Ceilings

	Application method (long edge relative to frame)	Max. framing spacing o.c. in	mm
	not recommended	—	—
12.7	perpendicular only	16	406
12.7*	perpendicular only*	24*	610*
15.9	perpendicular only	24	610

* ½″ SHEETROCK brand Interior Gypsum Ceiling Board provides the strength and sag resistance of ⅝″ standard board without the added thickness. Note: For adhesively laminated double-layer applications with ¾″ or more total thickness, 24 o.c. max.

Powder Texture Finishes

IMPERIAL QT Spray Texture Finish

Mixing—Use clean vessel equipped with variable speed power agitator. Sift texture finish into the recommended amount of water, agitating water during powder addition. Allow to soak for at least 15 min.—longer in cold water. Remix until a creamy (but aggregated) lump-free mix is obtained. Adjust spray consistency by adding small amounts of powder or water. Do not overthin, as poor adhesion, lack of hide and texture variation may result.

Optional Mixing Directions

a. **Adding joint compound to ceiling texture**—If slightly better spray properties, wet hide, improved bond and surface hardness are desired, one bag of Finish may be mixed with one bag of SHEETROCK Powder All Purpose Joint Compound. Water dilution should not exceed the total of specified amounts for each of the products used. Resultant texture will be slightly less aggregated and the appearance could be less white than the straight texture product. Surface priming recommendations still apply.

b. **Adding latex additive to ceiling texture**—If increased bond and surface hardness are desired, USG Latex Additive for IMPERIAL QT Texture may be added to the ceiling texture wet-mix at a rate of 1 to 2 pints per bag of Finish. The more latex additive used, up to 2 pints, the greater the bond and hardness of the dried surface. Surface-priming recommendations still apply.

c. **Adding paint to ceiling textures**—If better wet and dry hide, improved surface hardness, wider spray fan, and faster spray application are desired, SHEETROCK First Coat or a good-grade, compatible polyvinyl acetate, vinyl-acrylic, or acrylic-type paint in white, off-white or pastel colors only may be added to ceiling texture. For SHEETROCK First Coat, 25 lb. of SHEETROCK First Coat may be mixed with 400 lbs. (dry) of ceiling and wall texture products. Water dilution should not exceed the total of specified amounts for each of the products used. For polyvinyl acetate, vinyl-acrylic, or acrylic-type paint, wet-mix at a rate of 1 gal. per 32 or 40 lb. bag of texture by substituting 1 gal. paint for 1 gal. water. When adding 1 gal. paint to a fully diluted mix, the above properties remain appreciably the same but somewhat sparser aggregate surface may also result. Also, if aggregate is dentally brushed off, a lighter colored surface can result. Interior eggshell or semigloss paint products can be used. Compatibility of paint to be used with IMPERIAL QT products should be carefully checked before use. Exterior-grade latex paints also can be used if compatibility is satisfactory. Surface priming recommendations still apply.

Equipment—Use spray equipment similar to Grover 452-A 10-to-1 ratio, double-action pump with 7½″ stroke, equipped with 4′ pole gun having ⅜″ to ½″ round orifice, or Binks 7-E-2 hand gun with ⅜″ round opening. Use ¾″ to 1″ material hose, ⅜″ atomizing hose and ½″ air line from compressor to pump. Compressor must be adequate (85 cfm) for length and size of hose. Keep pressure as low as possible. Plaster mixers or hopper-type applicators also may be used.

Application—Apply at rate up to 8 sq. ft. per lb. Do not exceed recommended coverage, as subsurface defects, variations in base suction or color differences may show through, or lighter texture may result.

Surfaces with uneven suction may require two coats. Let first coat dry before applying second. Remove splatters immediately from woodwork and trim. Maintain 55°F (13°C) minimum air, water, product mix and surface temperature of the substrate during application and until surface is dry. Not washable but can be painted when redecoration is needed.

USG Spray Texture Finish

Mixing—Use a clean vessel equipped with variable-speed power agitator. In initial mix, stir powder into recommended amount of water per bag directions. Agitate during powder addition. Allow to soak for at least 15 min.—longer in cold water. Remix until a creamy, lump-free mix is obtained. Adjust spray consistency by adding small amounts of powder or water. Do not overthin, as poor adhesion, lack of hide and texture variation may result.

Equipment—Use equipment similar to Binks No. 18D gun with 53-R-21 nozzle combination, ½″ fluid hose, ⅜″ air hose, air-powered 4-to-1 ratio materials pump (minimum requirement) with double regulators, ½″ main line air hose and 7½ to 9-hp gasoline compressor.

Application—Apply with brush, roller or with suitable spray equipment at rate of up to 15 ft.^{2}/lb. for spatter and spatter/knockdown patterns, up to 25 ft.^{2}/lb. for orange peel and crow's feet pattern, and up to 40 ft.^{2}/lb. for fog coat and roller stipple patterns. Spray using 16″ to 20″ fan. Hold gun 16″ to 18″ from surface. Overlap preceding application with ½ to ⅔ of fan width. With 75′ to 125′ of ½″ hose, use 30 to 40 lb. fluid pressure and 50 to 60 lb. atomizing pressure. Then texture with roller or other tool. For finer textures and designs, use small brush, roller-stippler, fingers, whisk broom, crumpled paper, comb or similar items. Flatten raised portions of wet material or sand it when dry to provide further variation. Maintain 55°F (13°C) minimum air and surface temperature during application and until surface is dry. Avoid drafts while applying, but provide ventilation after application to aid drying. Do not use unvented gas or oil heaters. May be painted after overnight drying. Not washable unpainted.

SHEETROCK Wall and Ceiling Spray Texture (TUF TEX)

Mixing—Use a clean mixing vessel equipped with a variable speed power agitator. Using drinkable water and clean mixing equipment, slowly add dry powder texture to the recommended amount of water (see directions on bag) and mix to a heavy but lump-free consistency. Allow to soak for at least 15 min.—longer in cold water. Remix until mixture has a wet, smooth and creamy appearance. Adjust to desired spray consistency by adding small amounts of water to the wet mix. Do not overthin, as poor adhesion, lack of hide, or texture variation may result. If additional hardness and bond are desired, add up to 1 gal. of SHEETROCK First Coat flat latex paint or ¼ gal. (2 pts) of USG Latex Additive per 40 lb. bag of SHEETROCK Wall and Ceiling Spray Texture (TUF TEX). Note: Substitute volume of additives for equal volume of mixing water. Do not intermix with any materials other than those recommended.

Application—Using professional equipment, apply by machine or hand to create the desired effect. Application rates will vary depending on the texture pattern of choice. Generally, the application rate will vary from 20 ft²/lb (8.2 m²/kg) for fine patterns to 40 ft.²/lb. for full coverage-type applications. Maintain 55°F (13°C) minimum air and surface temperature during application and until surface is dry. Avoid drafts while applying product but provide ventilation after application to aid drying. Do not use unvented gas or oil heaters. May be painted after overnight drying. Unpainted surface is not washable or scrubbable.

USG Multi-Purpose Texture Finish

Mixing—Use a clean mixing vessel equipped with variable-speed power agitator. In initial mix, stir powder into recommended amount of water. See directions on bag. Agitate during powder addition. Allow to soak for at least 15 min.—longer in cold water. Remix until creamy, lump-free mix is obtained, then stir in up to 1 gal. water. To obtain suitable consistency for texturing as desired, do not use more than recommended amount of water. Do not overthin as poor adhesion, lack of hide and texture variation may result. Do not intermix with other materials.

Application—Apply with brush, roller or with suitable spray equipment at rate of up to 20 ft.²/lb.; then texture with roller or other tool. For finer textures and designs, use small brush, roller-stippler, fingers, whisk broom, crumpled paper, comb or similar items. Flatten raised portions of wet material or sand it when dry to provide further variation. Maintain 55°F (13°C) minimum air, water, product mix and surface temperature during application and until surface is dry. Avoid drafts while applying, but provide ventilation after application to aid drying. Do not use unvented gas or oil heaters. May be painted after overnight drying. Not washable unpainted.

USG Texture XII Drywall Surfacer

Mixing—Place recommended amount of water in suitable mixing container. Gradually add powder to water. Stir thoroughly with mechanical mixer until completely mixed and lump-free. Soak mix for 15 min.—longer in cold water; remix. Gradually add under agitation up to 1¾ gal. (6.6L) of water to reach desirable spraying consistency. Overthinning may result in poor adhesion, lack of hide, texture variation and inability to compensate for base suction variations. Do not exceed 3¼ gal. (12.3L) total water per bag. Use within 24 hours.

Equipment—Spray equipment: Binks Model 18 or 18D gun with #68 stainless-steel fluid nozzle, orifice size .110; air nozzle #101 carbide; atomizing pressure at gun 40-50 psi; air-hose size ⅜″ i.d. with ⅜″ swivel; fluid hose size ½″ i.d. with ½″ swivel; control-hose size 3⁄16″ with ¼″ swivel. Air-driven pump sizes: 4½:1 ratio for hose lengths up to 125′; 7½:1 for lengths up to 200′; 10:1 for lengths over 200′.

Application—Apply with spray gun using 24″ fan. Hold gun 18″ from surface and move parallel to surface. Avoid curved, sweeping strokes. Overlap preceding application with ½ to ⅔ of fan width. Apply a full coat in one direction, then immediately cross-spray in opposite direction. Use ½″ fluid hose with fluid-pressure variable depending on hose length. Coverage up to 35 ft.^{2}/lb. Air and surface temperatures should be 55°F (13°C) or higher during application and until surface is dry. Avoid drafts while applying, then provide adequate circulating ventilation to aid drying. Do not use unvented gas or oil heaters. May be painted after overnight drying. Not washable unpainted.

Sound-Rated Texture Finish

USG Acoustical Finish

Mixing—Read mixing and spray application directions completely before proceeding with mixing. Use a 7 cu. ft. or larger paddle-type plaster mixer with rubber-tipped blades (Anchor mixer) or a self-contained integral mixing/pumping spray texture tank with horizontal shaft and plaster or texture rig-type paddles mounted on a horizontal shaft. To ensure uniform product performance, mix a minimum of two bags. Add powder to water in quantity specified on bag. Mix approx. 5 minutes until lump-free, and a thick, foamy consistency is generated. (Initial mix will appear dry and heavy.) *Note:* If material is over-mixed, excessive foam will occur. Add more powder to break down foam and remix until proper foam level is reached. Additional mixing may be necessary during application to maintain foam consistency. Use wet-mixed material within 3-4 hours.

Spray Application—All pumps and hoses must be cleaned initially with water followed by approximately one gallon of SHEETROCK Ready-Mixed Joint Compound prior to spray application to prevent severe aggregate separation or clogging by the clean-out water.

For combined mixing/pumping units: Initially fill mixing hopper with necessary water to flush hoses. Pump all water from hopper, then drop joint compound into material reservoir of pump. Start pumping until compound feeds into hose. Immediately stop pump. Add water and powder in mixing hopper following mixing directions. When USG Acoustical Finish is properly mixed, pump out and discard joint compound. Turn on atomizing air, material valve and pump (in that order).

For pump units only: Add previously mixed finish to material hopper after pumping joint compound into hose. Then follow the start-up procedure as stated for combined mixing/pumping units.

Recommended spray pattern is 1½′ to 3′ in diameter. The spray gun should be held 2′ to 4′ from the surface, depending upon material density and atomizer pressure. Apply USG Acoustical Finish evenly, holding pole gun perpendicular to the surface being sprayed and slowly waving it from side to side until area is covered. Then immediately double back, crosshatching prior coat. Repeat same procedure as necessary until desired thickness is reached.

Elimination of spray lines and section seams is essential in producing an acceptable finish. Do not spray a portion of a ceiling in one day and the final portion on another day as a noticeable seam will result. If entire ceiling area cannot be sprayed to the final thickness in one day, spray the entire surface with a material coat of uniform thickness (min. ¼″). Complete to final thickness the following day using a cross-hatch application. Use natural breaks and boundaries to "frame" pattern edges and conceal seams. To measure average thickness, mark desired thickness on a blunt-tipped object (head of pencil or finishing nail) and insert into finish

For a different surface color or to redecorate (finish is not washable), use a good quality, flat latex paint (white or pastel) and spray apply over dried finish. There will be a minimal loss in NRC value.

USG Acoustical Finish provides textured, sound-rated finish.

Ready-Mixed Texture Finishes

USG QUIK & EASY Ready-to-Use Wall and Ceiling Texture

Product Preparation—Although USG QUIK & EASY Wall and Ceiling Texture is virtually ready to use, slight mixing will increase creaminess of the product. If remixing by drill is desired, use a heavy-duty drill; 300-600 RPM under load is best, with an open-blade mixer paddle. Heavier drills also may be used, but they tend to whip air bubbles into the mix. Plunge the mixer paddle up and down in the mix several times before switching on drill, then mix until smooth and uniform (about 1 to 2 min.).

Thinning—Experiment with small amount of mixed material prior to use, adjusting water proportions to match product viscosity to individual requirements. For thickest finish (heavy trowel, roller or brush application), use as is. For thinnest finish (light roller or spray application), add no more than 1½ qts. of water per 3.5 gal. (13.2L) pail, or 2 qts. of water per 4.5 gal. (17L) pail.

Coverage

1. For heavy trowel, brush or roller application, use spread rate up to: 90 ft.^{2}/3.5-gal. (13.2L) pail; 115 ft.^{2}/4.5 gal (17L) pail.

2. For thin application, use spread rate up to: 250 ft.^{2}/3.5 gal. (13.2L) pail; 320 ft.^{2}/4.5 gal. (17L) pail.

3. Actual coverage may vary depending on factors such as the condition of the substrate surface, dilution, application techniques and uniformity of the coating.

Aggregate Additions

Vermiculite—Up to 3¼ lbs. (1.5 kg) per 3.5 gal. (13.2L); up to 4 lbs. (1.8 kg) per 4.5 gal. (17.2L).

Perlite—Up to 2¾ lbs. (1.3 kg) per 3.5 gal. (13.2L); up to 3½ lbs. (1.6 kg) per 4.5 gal. (17.2L).

White Silica Sand—Up to 15 lbs. (6.8 kg) per 3.5 gal. (13.2L); up to 19¼ lbs. (9 kg) per 4.5 gal. (17L).

Note: Water is not to exceed 1¼ gal./3.5 gal. (4.7L/13.2L) pail; 1½ gal./4.5 gal (6L/17L) pail.

Application—Apply finish with trowel, roller or brush; troweling provides best results. Protect from freezing in containers during application and in place until dry. Maintain 55°F (13°C) min. air and surface temperature during application and until texture is dry. Avoid drafts, but provide good circulating ventilation. Do not use unvented gas or oil heaters.

For spray application, use spray equipment similar to Grover 452-A 10-to-1 ratio, double-action pump with 7½″ stroke, equipped with 4′ pole gun having ⅜″ to ½″ round orifice, or Binks 7-E-2 hand gun with ⅜″ round opening. Use ¾″ to 1″ material hose, ⅜″ atomizing hose and ½″ air line from compressor to pump. Compressor must be adequate (85 cfm) for length and size of hose. Keep pressure as low as possible. Plaster mixers or hopper-type applicators also may be used.

Do not exceed recommended coverage, as subsurface defects, variations in base suction or color differences may show through, or lighter texture may result. Remove splatters immediately from woodwork and trim. Maintain 55°F (13°C) minimum air and surface temperature during application and until surface is dry.

All walls must be painted after texturing except for noncontact areas. Not washable unpainted.

USG Ready-Mixed Texture Compound

Mixing and Thinning—Product is ready to use but slight mixing will increase creaminess of the product. To mix, transfer contents into a suitable vessel. Use an open-blade mixer paddle on a heavy-duty drill, preferably at 300-600 RPM under load. Higher speeds tend to whip air bubbles into the product. Plunge the mixer paddle up and down in the mix about 10 times before switching on drill. Mix, adding water as recommended, until smooth and uniform, always keeping paddle completely immersed to avoid whipping in air bubbles.

Experiment with small amount of mixed material prior to use, adjusting water proportions to match product viscosity to individual requirements. For brush or roller/crow foot pattern, dilute with 1-1½ gal. (3.8-5.7L) of water per 50 lb. (22.7 kg).

If large areas are being textured, prevent shade differences by sorting out enough boxes bearing identical manufacturing codes to cover entire area. Also, be sure to add exactly the same amount of water to each 50-lb. (22.7 kg) batch of ready-mixed texture finish.

Application—Apply finishes according to product directions. Finishes must be evenly spread and free from runs, sags and other blemishes. Allow each coat to dry before applying following coat. Not washable unpainted.

Coverage—Up to 400 ft²/50 lb. (37.2m²/22.7 kg) container. Coverage shown here should only be considered an estimate. Actual coverage can vary depending on factors such as the condition of the substrate surface, the amount of dilution of the product, spray techniques and the uniformity of the coating.

Creating Texture Patterns

Texture finishes and compounds offer opportunities for a variety of patterns and appearances. The number of patterns that can be created is limitless, but several patterns are particularly popular. Here are some commonly used patterns and information about how to achieve them.

Fog and Spatter

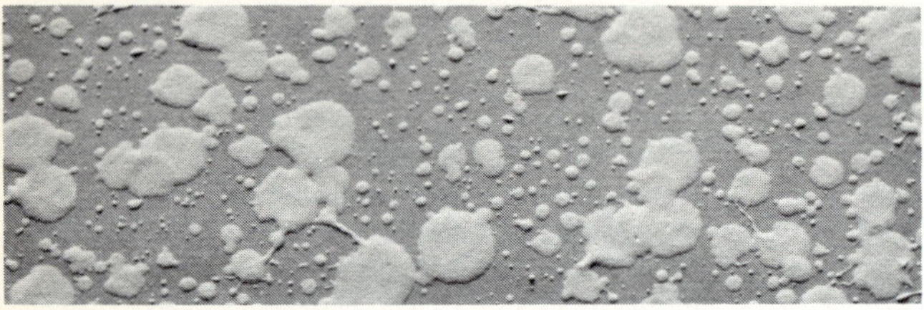

Material for Fog and Spatter—

- USG Spray Texture Finish
- SHEETROCK Wall and Ceiling Spray Texture (TUF TEX)
- USG Multi-Purpose Texture Finish

Application—Spray.

Equipment—Binks 18D gun or equivalent, equipped with a #53 fluid nozzle and R-21 fan cap.

Procedure—Mix products to a thin, latex-paint consistency. For a good fog coat, which is always the first application, atomizing air should be approximately 60 psi and material feed pressure approx. half the atomizing pressure.

When spraying, apply in long even strokes with no wrist action, holding gun perpendicular to surface and approx. 18″ from surface. Apply material as uniformly as possible avoiding lap marks. After fog coat has been applied, allow about 10-15 min. for surface to partially dry, then apply spattering by removing the R-21 fan cap and reducing atomizing air to approx. 15 psi and material feed to approx. 10 psi. While applying spatter coat, move spray gun in a rapid random fashion standing about 6′ from surface. Size of spatters depends on pressures used. Amount (or density) of spatters on surface depends on personal preference.

Orange Peel

Material for Orange Peel—

- USG Multi-Purpose Texture Finish
- SHEETROCK Wall and Ceiling Spray Texture (TUF TEX)
- USG Spray Texture Finish

Application—Spray.

Equipment—Same as for fog coat except atomizing air pressure should be 40 psi and material feed pressure approx. 20 psi. When applying, follow same procedure as fog coat, but use slightly more material to give a good orange peel pattern. Degree of orange peel pattern depends on amount of material applied to surface.

Knock-Down and Skip-Trowel

Material for Knock-Down and Skip-Trowel—

- USG Multi-Purpose Texture Finish
- SHEETROCK Wall and Ceiling Spray Texture (TUF TEX)
- USG Spray Texture Finish

Application—Spray.

Equipment—Pole gun, hopper, or Binks 18D or Binks 7E2 gun.

Knock-Down Procedure—Apply as spatter as described except use material at heavy latex-paint consistency. After spattering surface, wait about 10-15 min., then very lightly flatten only tops of spatters with flat blade or flat hand trowel. Again, size of spatters depends on pressures used.

Skip-Trowel Procedure—Apply as spatter coat but at very low pressures to allow for large spatters on surface. Wait approx. 10 to 15 min., then use blade as in the knock-down procedure, but applying more pressure. A common job practice is to add some clean silica sand to emphasize skip-trowel effect.

Roller Texture

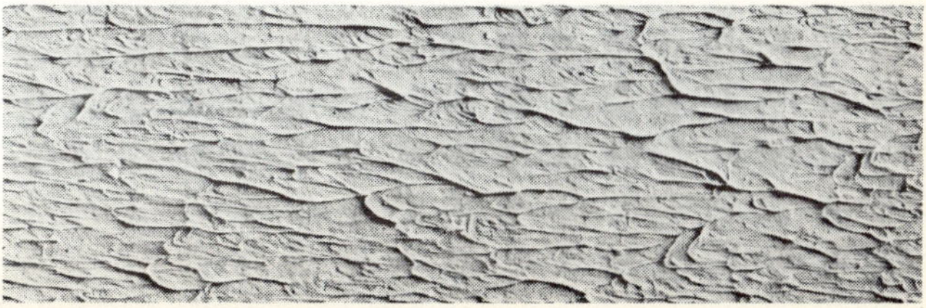

Material for Roller Texture—

- USG Multi-Purpose Texture Finish
- SHEETROCK Wall and Ceiling Spray Texture (TUF TEX)
- USG Ready-Mixed Texture Compound
- USG QUIK & EASY Ready-To-Use Wall and Ceiling Texture
- SHEETROCK Lightweight All Purpose Joint Compound (PLUS 3)— Ready-Mixed
- SHEETROCK Topping or All Purpose Joint Compound—Ready-Mixed
- SHEETROCK Topping or All Purpose Joint Compound—Powder

Application—Hand.

Equipment—Paint pan or large pail, paint roller ¼″ to 1″ nap. Short nap produces lower stipple—fine pattern. Long nap produces higher stipple—course pattern.

Procedure for Hand Texturing—Mix product to latex-paint consistency, depending on desired texture. Completely wet out roller with material, then apply to surface as evenly as possible, covering entire surface. Let partially dry to a "dull wet" appearance, then roll again for desired texture effect.

Crow's Foot

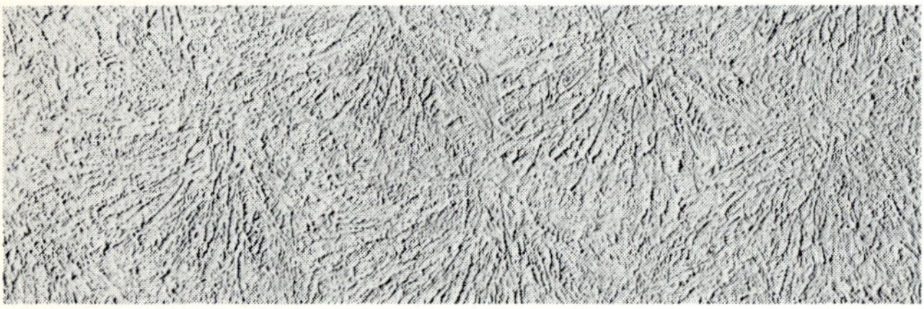

Use same material, equipment and application as for roller texture, then use texture brush instead of paint roller to texture surface.

Procedure is the same as for roller texture, except that after material has partially dried to dull wet finish, stamp surface with texture brush prewetted with texture material for desired stipple finish.

Swirl Finish

Material for Swirl Finish—

- USG Multi-Purpose Texture Finish
- SHEETROCK Wall and Ceiling Spray Texture (TUF TEX)
- USG Ready-Mixed Texture Compound
- USG QUIK & EASY Ready-To-Use Wall and Ceiling Texture
- SHEETROCK Lightweight All Purpose Joint Compound (PLUS 3)—Ready-Mixed
- SHEETROCK Topping or All Purpose Joint Compound—Ready-Mixed
- SHEETROCK Topping or All Purpose Joint Compound—Powder

Application—Hand.

Equipment—Same as for roller texture, plus wallpaper-type brush.

Procedure—Apply as a roller texture. Let surface dry to dull wet finish, then use wallpaper brush to achieve desired swirl texture, rotating brush in circular motion on the wet surface.

Brocade or Travertine

Material for Brocade or Travertine—

- USG Multi-Purpose Texture Finish
- SHEETROCK Wall and Ceiling Spray Texture (TUF TEX)
- USG Ready-Mixed Texture Compound
- USG QUIK & EASY Ready-To-Use Wall and Ceiling Texture
- SHEETROCK Lightweight All Purpose Joint Compound (PLUS 3)—Ready-Mixed,
- SHEETROCK Topping or All Purpose Joint Compound—Ready-Mixed
- SHEETROCK Topping or All Purpose Joint Compound—Powder

Application—Hand.

Equipment—Same as for roller texture.

Procedure—Same as for crow's foot. After crow's foot texture has been achieved, wait 10-15 min., then knock down tips only by lightly drawing a flat blade across surface.

Other Hand-Textured Effects

Textured effects cited above are only a few of the many imaginative textures possible. Other effects can be achieved using different texturing tools.

A string-wrapped roller produces an attractive striated stone effect while cross-rolling gives an additional interesting squared pattern. For finer designs and textures, use a small brush, roller-strippler, whisk broom, crumpled paper, comb, sponge or similar items. Flattening raised portions of wet material or sanding when dry provides further variations. Material also may be scored to represent block, tile or cut stone outlines.

Glitter Effects

Sparkle, particular under artificial light, can be added to ceilings for unusual and interesting effects with the use of glitter. Glitter comes in 1⁄32″ to 1⁄16″ cuts in silver, gold, blue, red, fuscia or green, and is applied with hand-cranked or air-powered guns. Application usually consists of embedding glitter in the freshly applied ceiling texture while surface is still wet. Depending on effect desired, figure one lb. of glitter per 500 ft.^{2} of ceiling.

Resurfacing

Where ceilings or sidewalls are so badly disfigured that an entirely fresh surface is desirable, they may be resurfaced using a layer of 1⁄4″ or 3⁄8″ SHEETROCK brand Gypsum Panels or predecorated TEXTONE Vinyl-Faced Gypsum Panels (walls only). Ceilings may also be redecorated with texture finishes. For resurfacing masonry walls, see application of gypsum board to wall furring, described in Chapter 3.

Preparation—Remove all trim (this may not be necessary if ¼″ panels are used). To remove trim easily, drive all nails completely through the trim with a pin punch. Remove all loose surfacing material. Fill small holes with joint compound or patching plaster. Patch large holes to the surrounding level with single or multiple layers of gypsum board nailed to framing and shimmed out as required.

Electrical outlet boxes for switches, wall receptacles and fixtures should be extended outward to compensate for the added gypsum panel thickness.

Locate joists and studs by probing or with a magnetic "stud finder." Snap a chalk line to mark their full length and mark their location on the adjacent wall or ceiling. Where great irregularities of surface exist, apply furring strips not over 16″ o.c., using wood shingles to shim out to a true, even plane.

Installation—Apply SHEETROCK brand Gypsum Panels with long dimension placed horizontally or vertically. Fasten with gypsum board nails, spaced 7″ o.c. on ceilings, 8″ on walls. Nails must be long enough to penetrate into framing members at least ⅝″. Nail TEXTONE Vinyl-Faced Gypsum Panels over existing walls with matching color nails using a plastic-headed hammer.

Gypsum panels may be adhesively applied over sound, existing walls with laminating adhesive for regular or irregular surfaces or liquid contact adhesive for flat, smooth surfaces (see directions in Chapter 3), or with SHEETROCK Setting-Type (DURABOND) or SHEETROCK Lightweight Setting-Type (EASY SAND) Joint Compound.

Finish SHEETROCK brand Gypsum Panels with metal corner reinforcement and joint treatment as necessary, and replace all trim.

Redecorating Ceilings

Redecorating cracked, discolored or damaged ceilings with texture can make old ceilings look like new. Spray-applied texture finishes cover small cracks and other imperfections and provide beautiful surfaces. Redecorating over large-aggregate texture surfaces is especially effective since these surfaces normally are not easily cleaned, rolled or brush-painted. Yet they are easily spray-painted with texture. These modernized ceilings add value and beauty. Best of all, most jobs can be done in one day without removing rugs, furniture or light fixtures.

Preparation—Surface cracks larger than hairline size should be treated with a drywall joint compound and tape, and thoroughly dried, prior to redecorating. Tobacco smoke stains require predecorating attention and treatment with special paint. Remove grease stains using mild detergent. Seal water-stained surfaces with primer specifically recommended by the manufacturer. Remove soot or dirt by "air dusting" surfaces. Wash mildew-contaminated surfaces with a solution of 1 qt. household bleach such as Clorox (sodium hypochlorite) to 3 qt. water. Cover all furniture, rugs, etc. with drop cloths and wear gloves and protective clothing as well as eye protection. For heavy mildew deposits, two applications of the bleach solution may be necessary. On textured ceilings, heavy coats of bleach are not recommended. Mist-coat surface with bleach solution using an aerating device such as a trigger-type household sprayer. No rinsing of the bleach solution is necessary since this would rewet the texture and cause serious bond problems. Let bleach dry thoroughly, then respray surface with IMPERIAL QT Spray Texture Finish.

On previously painted mildew-contaminated surfaces, apply bleach solution with a scrub brush. When dry, rinse the painted surface to remove bleach, dry, and then spray-apply desired texture finish.

Caution. Treatment for mildew will not necessarily prevent its recurrence if humidity, temperature and moisture conditions are favorable for further mildew growth.

Redecorating with Texture

After properly preparing the surface as described above, redecorate with a texture following the guidelines below.

Painted surfaces—On dark painted ceilings, prime with SHEETROCK First Coat or white latex or white alkyd flat ceiling paint to provide acceptable base color that subsequent texture coating can hide. Dull semi-gloss surfaces completely by sanding lightly. Liquid sandpapers are not recommended.

Previously textured surfaces—Priming a previously textured ceiling with a paint primer is not necessary. If required, one or two coats of SHEETROCK First Coat or a good-quality, undiluted, white interior latex wall paint is sufficient prior to redecorating with a coarse texture only.

Wallpaper or vinyl wall covering—Remove material and prime ceiling surface with appropriate primer prior to texturing.

Plaster ceilings—Surface must be in paintable condition. Prior to texturing, cover with primer-sealer specifically recommended by paint manufacturer.

Mask surfaces by covering floors and walls with .85 to 1-mil-thick polyethylene film, available in 8′ widths, folded and rolled in half for easier handling. Spread polyethylene film on floor, making sure that all areas are completely covered. Next, apply wide masking tape to wall-ceiling intersection, fastening only top of tape to wall and leaving bottom hanging free. Fasten one edge of folded poly film to loose edge of tape, then unfold film to full 8′ width. Press tape into firm contact with both wall and film.

Cover furniture, cabinets, light fixtures—anything that will remain in the room during spraying operation. Building owner usually is responsible for having ceiling light fixtures lowered so they can be quickly and completely covered.

Mixing—Use clean mixing vessel equipped with variable-speed power agitator. Sift texture material into the recommended amount of water, agitating water during powder addition. Stir until a creamy (but aggregated) lump-free mix is obtained. Adjust spray consistency by adding small amounts of powder or water. Do not overthin, as poor adhesion, hide and texture variation may result. Do not intermix with other materials.

Equipment—Use spray equipment similar to Grover 452-A 10-1 ratio, double-action pump with 7½″ stroke, equipped with 4′ pole gun having ⅜″ to ½″ round orifice, or Binks 7-E-2 hand gun with ⅜″ round opening. Use ¾″ to 1″ material hose, ⅜″ atomizing hose and ½″ air line from compressor to pump. Compressor must be adequate (85cfm) for length and size of hose. Keep pressure as low as possible. Plaster mixers or hopper-type applicators also may be used.

Application—Apply IMPERIAL QT Spray Texture Finishes at a rate of up to 8 ft^2/lb. Do not exceed recommended coverage, as subsurface defects, variations in base suction or color differences may show through, or lighter texture may result. Maintain 55°F (13°C) minimum air and surface temperature during application and until surface is dry.

Redecorating a Textured Surface with Paint

After properly preparing the surface as described on pages 237 and 238, redecorate a previously textured surface with paint following the guidelines below.

Brush application of paint over a textured ceiling is not recommended. In redecorating by hand, use a long-nap paint roller with ½″ to ¾″ nap. Any good-quality interior latex or vinyl acrylic paint in flat, egg-shell or semigloss can be used. Slight dilution of paints with water, particularly high-viscosity types, may be necessary for smoother, easier spreading. Apply paint by rolling in one direction, immediately followed by cross rolling. Use light pressure and avoid over-rolling and saturating the surface to minimize loosening of surface aggregate.

Whether spraying or rolling, avoid drafts while applying, but provide adequate circulation and ventilation to aid drying.

Precautions

Ventilate or use a dust collector to avoid creating dust in the workplace. A NIOSH-approved respirator should be used if the air is dusty. The use of safety glasses is recommended. Do not take internally. Keep out of reach of children.

Cement Board Construction

CHAPTER 5 Interior & Exterior Systems

Cement Panel Products

DUROCK Cement Board and DUROCK Exterior Cement Board Systems are designed to provide fire-resistant and water damage-resistant assemblies for both interior and exterior construction. The systems may be used in load-bearing and non-load bearing applications with either wood or steel framing spaced max. 16″ o.c. All systems utilize conventional construction methods and equipment.

The principal product used in each system is the DUROCK Cement Board, a panel product formed in a continuous process that consists of an aggregated portland cement core reinforced with polymer-coated, glass-fiber mesh embedded in both surfaces.

Cement Board Sizes and Packaging

	Thickness		Width		Length		Ship. Units
Type	Standard	Custom[1]	Standard	Custom[1]	Standard	Custom[1]	(pcs)[2]
Cement Board	½″	⅝″	32″, 36″	48″	5′	3′ to 8′	40
	½″	⅝″	32″		8′	4′ to 8′	30
Underlayment	5⁄16″		48″		4′	4′ to 8′	40
Exterior Board	½″	⅝″	48″	32″	8′	4′ to 10′	20

(1) Minimum quantity required for custom sizes.
(2) Stretch wrapped and shipped in packaging units as shown.

DUROCK Cement Board—Especially sized and formulated for use in interior and exterior areas that may be subject to water or high moisture/humidity conditions, such as bath tub or shower enclosures, bathroom floors, bath and kitchen counter tops and steam rooms. Excellent residential panel for soffits, walls, privacy fences, and chimney enclosures. DUROCK Cement Board is a superior substrate for ceramic tile, slate and quarry tile on all interior surfaces. Also UL-listed as a wall shield and floor protector for room heaters and stoves. Panels are manufactured in three widths for minimum cutting and easy handling and installation in tub and shower areas. Available in various lengths and thicknesses. Use larger panel sizes for larger projects such as commercial kitchens and gang showers. Aggregated portland cement core resists water penetration and will not deteriorate when wet. DUROCK Cement Board exceeds the ANSI Standards for cementitious backer units. See National Evaluation Service Report No. 259 for fire-resistant designs and/or conditions of use. Reports are subject to reexamination, revisions and possible changes.

DUROCK Cement Board Limitations

1. Systems using DUROCK Cement Board are designed for positive or negative uniform loads up to 30 psf with studs spaced max. 16″ o.c.

2. Maximum stud spacing: 16″ o.c. (24″ o.c. for cavity shaft wall assembly); maximum allowable deflection, based on stud properties only, L/360. Maximum fastener spacing: 8″ o.c. for wood and steel framing; 6″ o.c. for ceiling applications.

3. Maximum dead load for ceiling system is 7.5 psf.

4. Steel framing must be 20-ga. or heavier.

5. Do not attach panels with drywall screws or drywall nails.

DUROCK Underlayment—A strong, thin substrate designed for use under ceramic or thin-cut stone tile floors and counter tops. DUROCK Underlayment is similar in composition to DUROCK Cement Board, but comes in convenient 4′ x 4′ panels and is only 5⁄16″ thick, reducing the variation in level between an overlaid ceramic tile floor and abutting carpet or wood floors. The reduced thickness also eliminates the need to cut down entry doors in threshold applications and allows easier installation of kitchen appliances such as dishwashers. Not for use over exposed framing; requires underlaid substrate material such as plywood for structural stability. May be applied directly over old substrate on counter tops to save time.

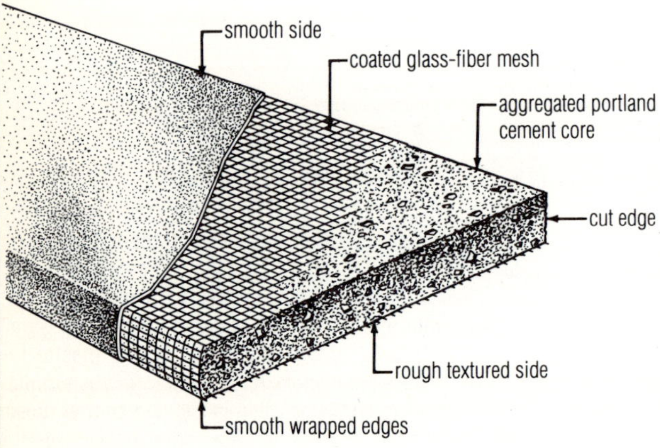

DUROCK Exterior Cement Board—A superior base for DUROCK Exterior Finish, thin brick, ceramic tile, exposed aggregate or exterior insulation and finish systems (EIFS) on building exteriors. DUROCK Exterior Cement Board is especially formulated to withstand moisture and wind conditions and freeze-thaw cycles in exterior applications up to 40 psf. Panels are ½″ thick and measure 4′ x 8′ for convenient handling and rapid installation. Assembly can be adapted readily to factory or jobsite panelization. Easily installed with nails or screws. Finished systems weigh considerably less than masonry or precast or poured concrete, reducing structural foundation requirements and in-place costs. See National Evaluation Service Report No. 396 for allowable values and/or conditions of use. Report is subject to reexamination, revisions and possible changes.

DUROCK Cement Board and Exterior Cement Board

Typical physical properties

Property	ASTM Test Ref.	Cement board value	Under-layment value	Exterior board value
Flexural strength-psi	C947-81	750	1250	1000
Indentation strength—psi 1″ dia. disc @ 0.02″ indent.	D2394	2300	2300	2300
Uniform Load—psf	—	30 max.	—	40 max.
Water absorption-% by wt. 24 hrs.	C473-84	10	10	10
Nail pull resistance-lb. 0.4″ head diameter (wet or dry)	C473-84	125	—	125
Weight-psf	C473-84	3	2	3
Freeze/thaw resistance-procedure A number of cycles with no deterioration	C666-84	100	100	100
Surface burning characteristics-flame/smoke	E84-84	5/0	5/0	5/0
Thermal "R"/k value	C177	0.26/1.92	—	0.26/1.92
Standard method for evaluating ceramic floor tile installation systems	C627	Residential	Residential	—
Min. bending radius[1]—ft.	—	8	—	8

(1) Requires special framing. Details available on request.

Thermal Properties

Thickness (in)	Product	"R"
3	THERMAFIBER SAFB	11.1
3½	THERMAFIBER FS-15	13.0
½	SHEETROCK brand Gypsum Panels	0.45
⅝	SHEETROCK brand Gypsum Panels	0.56
½	DUROCK Exterior Cement Board	0.26
1	DUROCK Insulation Board	3.8

DUROCK Exterior Cement Board Limitations

1. Maximum stud spacing: 16″ o.c. for exterior wall assemblies, 24″ o.c. for cavity shaft wall assembly; maximum allowable deflection, based on stud properties only, L/360 for tile and thin brick finishes, L/240 for DUROCK Exterior Finish and DUROCK EIFS.

2. DUROCK Exterior Systems may be used on sills sloped a minimum of 10 degrees and up to 2′0″ deep provided that the framing is 16″ o.c., ½″ minimum structural sheathing is placed behind the DUROCK Exterior Cement Board, and the installation is adequately caulked, flashed and waterproofed.

3. DUROCK Exterior Cement Board may not be used as a structural sheathing—for racking resistance, separate bracing must be specified.

4. These systems are designed for positive or negative uniform loads up to 40 psf with studs spaced 16″ o.c. max.

DUROCK Accessory Products

DUROCK steel screw

DUROCK wood screw

DUROCK Wood and Steel Screws—Developed especially for use with DUROCK Cement Boards. All DUROCK Screws are made with a special anticorrosive coating that is superior to cadmium plating or zinc. Wafer head design with countersinking ribs allows flush seating while preventing strip-outs. Increased bearing surface provides greater pull-through resistance. Neat, flush appearance makes application of finishing material easier. *For 14 to 20-ga. steel framing,* use 1¼″ and 1⅝″ DUROCK Steel Screws. *For wood framing,* use 1¼″, 1⅝″ or 2¼″ DUROCK Wood Screws. Packaging: 1¼″ screws, 5,000 pieces per carton, or twenty 150-piece boxes (shown above) per carton; 1⅝″ screws, 4,000 pieces per carton; 2¼″ screws, 2,000 pieces per carton.

Nails—1½″ 11-ga. hot-dipped galvanized roofing nails may be used for attachment of DUROCK Cement Board, DUROCK Underlayment or DUROCK Exterior Cement Board to wood framing.

Staples—¼″ x ⅞″ galvanized staples may be used only for DUROCK Underlayment attachment to plywood substrate.

DUROCK Interior Tape—A specially designed tape for use with DUROCK Cement Board. Tape is 2″ wide, polymer-coated, open glass-fiber mesh. Packaging: 2″ (nom.) x 75′ rolls; 24 rolls per carton.

DUROCK Exterior Tape—Specifically designed for use with DUROCK Exterior Cement Board. Tape is 4″ wide, polymer-coated, open glass-fiber mesh for embedding in DUROCK Latex Fortified Mortar or DUROCK Exterior Basecoat. Packaging: 4″ (nom.) x 150′; 4 rolls per carton.

DUROCK Insulation Board—An expanded polystyrene (EPS) board, aged for stability, available in 1″, 1½″ or 2″ thicknesses. Panels are 2′ x 4′ and have nominal density of not less than 1.0 pcf per ASTM C-579; R-3.8/in. thickness. Flame spread less than 25 and smoke developed less than 450 per ASTM E-84. Packaging: 1″—24 pcs., 192 sq. ft., 16 lb. net wt.; 1½″—16 pcs., 128 sq. ft., 16 lb. net wt.; 2″—12 pcs., 96 sq. ft., 16 lb. net wt.

DUROCK Reinforcing Mesh—A light green, open-weave, glass fiber fabric treated for alkaline resistance. Packaging: 38″ x 50-yd. rolls, four per carton.

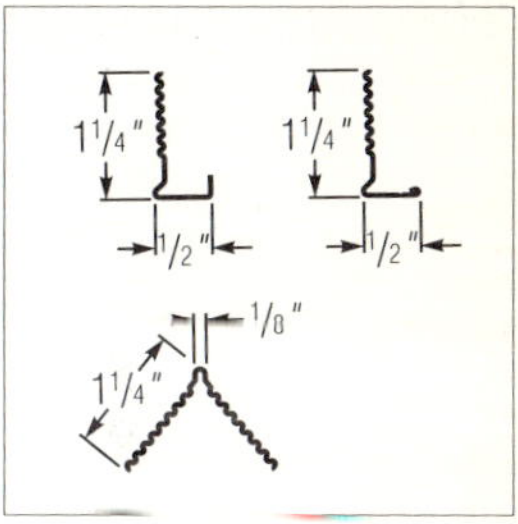

DUROCK Trim and Bead—Provide superior edge protection and corner reinforcement for DUROCK Exterior Cement Board. DUROCK J-Trim 1¼″ x ½″ x 8′; DUROCK-L-Trim, 1¼″ x ½″ x 8′; DUROCK Corner Bead, 1¼″ x 1¼″ x 8′ zinc, 50 pcs./carton.

DUROCK Control Joint—Formed from corrosion-resistant material. Plastic tape protects ¼″ wide, 7⁄16″ deep opening and is removed after application of finish. Supplied in 8′ length. 25 pcs./carton.

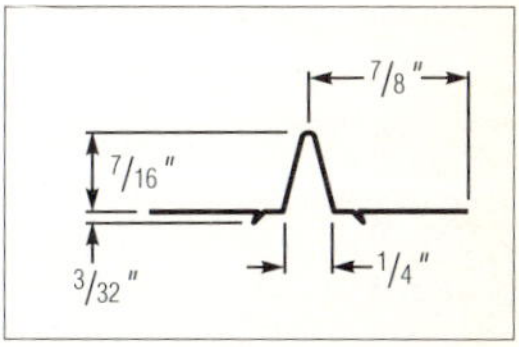

DUROCK Basecoat, Mortar, Grout and Finish

DUROCK Exterior Basecoat—A ready-to-mix portland cement mortar containing dry latex polymers. For use with DUROCK Exterior Cement Board and DUROCK Exterior Finish. Gray in color. Provides a smooth, uniform surface for finish; reinforces exterior board joints. Packaging: 50-lb. bags. Approximate coverage: 80-90 ft.^{2}/50 lb. for 1⁄16″ thickness.

DUROCK Latex Fortified Mortar—A ready-to-mix, thin-set mortar containing dry latex polymers. A superior mortar for use with DUROCK Cement Board or Exterior Cement Board for setting ceramic tile, quarry tile or thin brick. White or gray. Packaging: 50-lb. bags. Approximate coverage: 40-50

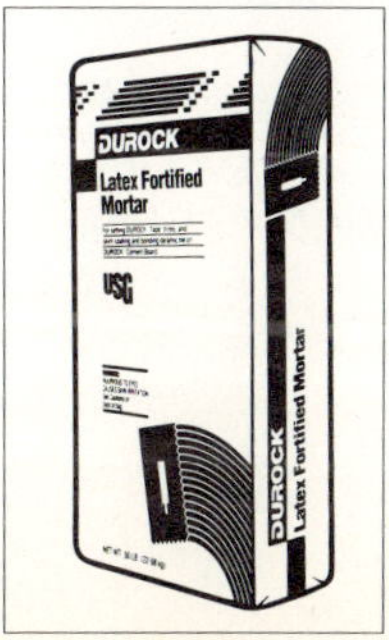

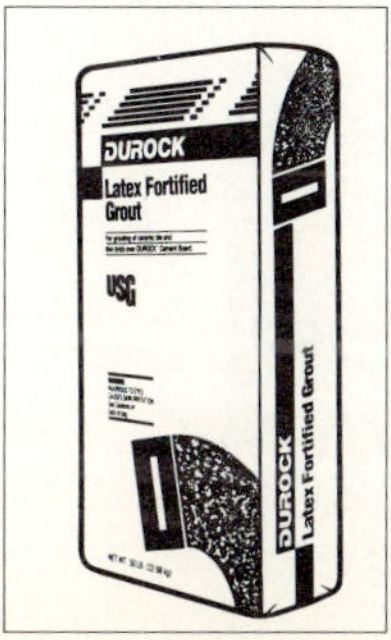

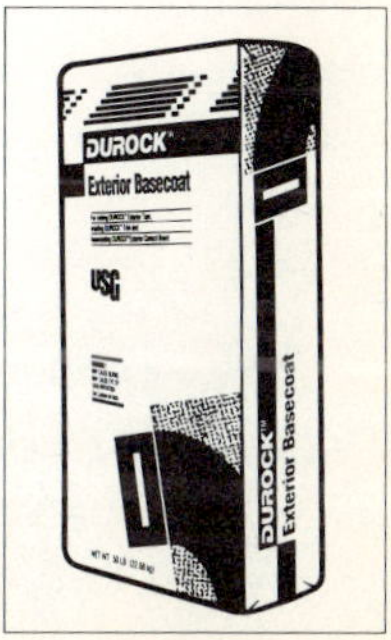

ft.²/50 lb. for ⅛″ thick skim coat; 80-90 ft.²/50 lb. for bond coat (back buttering not included).

DUROCK Latex Fortified Grout—A ready-to-mix grout containing dry latex polymers. For filling joints between tiles or thin brick over DUROCK Cement Board or Exterior Cement Board. White or gray. Packaging: 50-lb. bags. Approximate coverage: 60-80 ft.²/50 lb. for ½″ x 6″ x 6″ tile with ⅜″ wide joints; for thin brick systems, 30-45 ft.²/50 lb. for ½″ x 2¼″ x 8″ thin brick with ⅜″ wide joints.

DUROCK Exterior Finish—A ready-mixed acrylic polymer emulsion coating for long-lasting beauty on exterior walls. Used in conjunction with DUROCK Exterior Cement Board and applied directly over DUROCK Exterior Basecoat material. Durable, textured stucco finish resists impacts and cracks; dries to touch in a few hours. Available in Fine or Coarse finishes and 33 standard colors. Custom colors are available on special order. Numerous textures can be created using conventional stucco tools and application techniques.

Durable texture coating is based on a 100% acrylic polymer emulsion. DUROCK Exterior Finish is flexible and accommodates thermal expansion and contraction without cracking or delaminating under normal conditions, including repeated freeze-thaw cycles. Installation time can be as little as

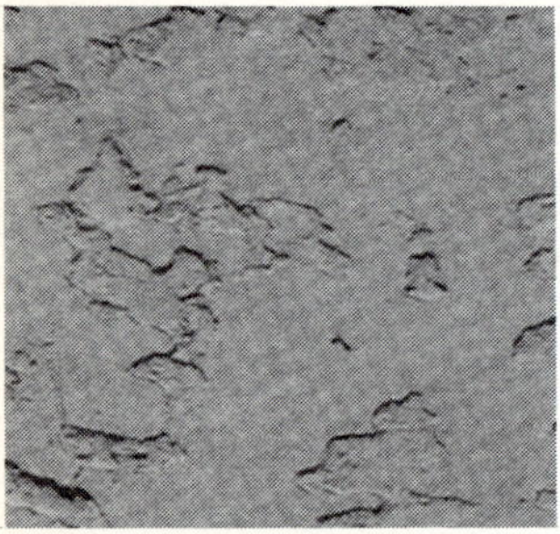

Fine Finish—Spanish Trowel on a smooth first coat. Before surface moisture leaves, apply a second coat using random, overlapping trowel strokes.

Coarse Finish—Float Trowel on finish allowing aggregate to act as thickness gauge. Remove excess material. Float surface to achieve desired texture with plastic or wood float.

Fine Finish—Lace Trowel on first coat. Before surface moisture leaves, skip-trowel on second coat with trowel held flat. If desired, knock down finish with trowel.

Fine Finish—Sandstone Trowel on finish allowing aggregate to act as thickness gauge. Using a plastic or wood float, float surface to achieve desired texture.

Spray Application—Finish may be spray float surface to achieve desired texture. applied for a variety of desired textures.

two days with board erection, joint treatment and basecoat application done on the first day.

This system affords exceptional design freedom. Choice of 33 standard colors and two textures, plus numerous specialty colors, offer many texture possibilities. Color and texture can be varied to contrast or complement glass and other exterior materials. Can be used in combination with ceramic tile, thin brick and stone aggregate finishes on the same base. Rapid system installation and fast curing time allow quicker occupancy, reduced labor and installed costs. (Refer to Technical Folder SA-700 for complete information on color selections.)

DUROCK Over-Coat—An unaggregated polymer finish coating for application over existing DUROCK Finishes, stucco and stucco-like surfaces. Ideal for blending patch repairs with existing stucco surface or for altering or freshening colors of existing stucco finish. Comes in 33 standard colors. Approx. coverage: 175-200 ft.^{2}/gal.

Job Preparation and Design Considerations

Estimating

Estimate material requirements for cement board applications in much the same manner used for estimating gypsum drywall applications. There are, however, certain important differences.

Measure surface area "solid," ignoring cutouts for doors and windows, then adjust as necessary. Take into account different thicknesses required to match drywall surfaces for interior projects—typically ½" for walls, ⅝" or ½" for ceilings and 5/16" underlayment for floors and counter tops.

Fastener spacing requirements are 8" for walls and 6" for ceilings—considerably more frequent than for drywall construction. Plan for fasteners accordingly. Average usage—1,600 pcs./1,000 sq. ft.

A water barrier, not a vapor retarder, is required behind cement board in many applications. Water barrier must be Grade D building paper, No. 15

asphalt felt or equivalent. Ensure that total square footage of moisture barrier exceeds total square footage of cement board to take into account material overlaps required to prevent moisture penetration.

To estimate basecoats, mortars, grouts and adhesives, consult packaging and coverage information above for those products.

All materials should be delivered and stored in their original unopened packages and stored in an enclosed shelter providing protection from damage and exposure to the elements. Even though the stability and durability of DUROCK Cement Board is unaffected by the elements, moisture and temperature variations may have an effect on the bonding effectiveness of basecoats and adhesives.

Water Barriers and Vapor Retarders

Cement boards are vapor permeable but provide resistance to water penetration. If a vapor retarder or waterproof construction is specified, a separate barrier must be applied over or behind the cement board.

Various humidity and temperature conditions may require a vapor retarder. A qualified engineer should be consulted to determine the proper location of the retarder to prevent moisture condensation within the wall.

Window and Door Openings—All window and door openings must be properly flashed and caulked. Grout exterior steel door frames with portland cement mortar.

Air and Water Infiltration—Flashing and sealants as shown in construction details must be provided to resist air and water infiltration. DUROCK

Exterior Cement Board is not a water barrier. A water barrier must be installed over the studs with a 2″ overlap or stapled to the back of the DUROCK Exterior Cement Board before it is applied. DUROCK Exterior Cement Board must be clean and dry before application of exterior finishes. Also, for ceramic tile and thin brick, a ⅛″ min. thick leveling/skim coat of DUROCK Latex Fortified Mortar must be applied to the exterior surface of the DUROCK Exterior Cement Board and allowed to set 24 hours. Accessories for exterior finishes should be made of corrosion-resistant materials.

Control Joints

Certain interior wall surface constructions should be isolated with surface control joints (sometimes referred to by the industry as expansion joints) or other means where: (a) a wall abuts a structural element or dissimilar wall or ceiling; (b) construction changes within the plane of the wall; (c) interior surfaces exceed 16' in either direction. Surface control joint width should comply with architectural practices. Location and design of building control joints must be detailed by the architect. (See U.S. Gypsum Company Technical Folders SA-700 and SA-932 for further information.)

In large wall areas where it is desirable to minimize the number of joints or where other design considerations dictate the use of a longer or wider board, use board ½″ thick, 48″ wide, 8′ long, or ½″ x 32″ x 8′.

High Moisture Areas

Pool enclosures—Cement board systems may be used for the walls and ceilings around indoor swimming pools. In areas of high moisture and chlorine content, adequate consideration should be given for ventilation to protect against deterioration of metal hangers and framing members.

Steam rooms and saunas—Where temperatures exceed 120°F for extended periods, use dry-set or latex-portland cement mortar; do not use organic adhesive.

Leaching and Efflorescence

Latex leaching and efflorescence are natural phenomena which occur with the use of latex modified mortars and grouts through no fault in the products. To help protect against their occurrence, follow current industry guidelines and recommendations.

Wall Furring Applications

For new construction or renovation, contact your local United States Gypsum Company representative for specifications and details on Z-furring and DWC-furring systems.

Interior Applications

Framing

Framing for DUROCK Cement Board attachment must not exceed 16″ o.c. (24″ o.c. for UL Design U459). Studs of freestanding furred walls must be secured to exterior wall with furring brackets or laterally braced with horizontal studs or runners spaced 4′ o.c. max. If necessary for tub or shower

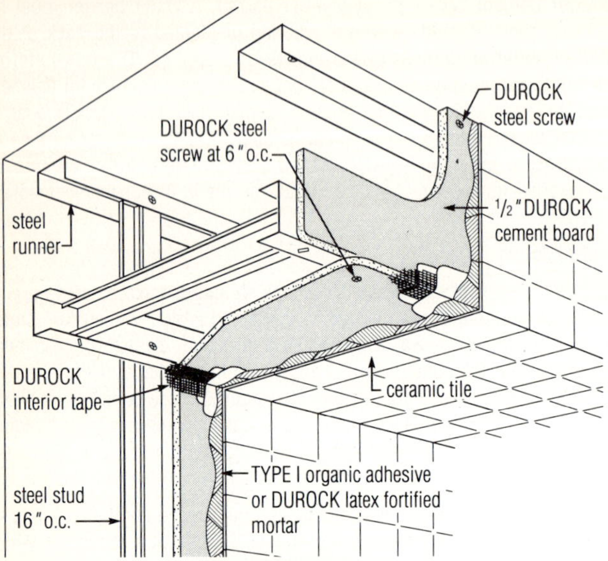

Steel soffit framing

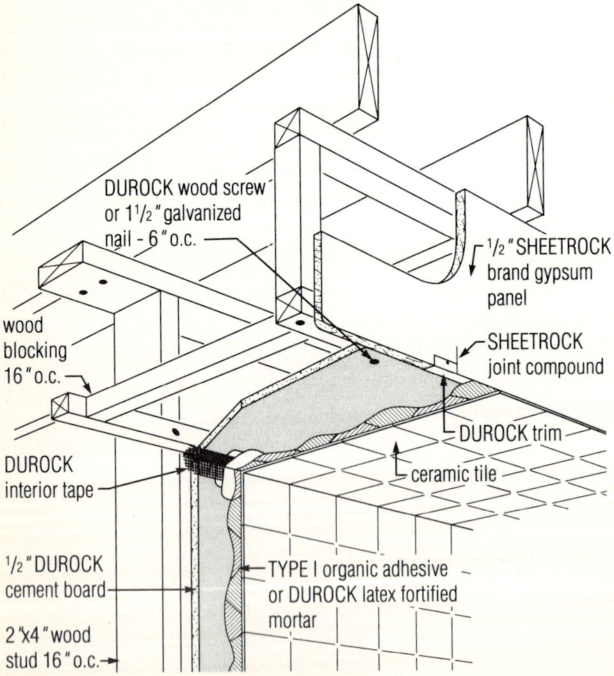

Wood soffit framing

surround applications, fur out studs so inside face of receptor is flush with cement board face. Install appropriate blocking or headers to support tub and other plumbing fixtures and to receive soap dishes, grab bars, towel racks and other accessories and hardware.

Ceiling framing must be capable of supporting the total ceiling dead load, including insulation, ceramic tile, bonding materials and cement board with deflection not exceeding L/360.

Floor framing must be covered with min. 5/8″ exterior grade plywood subfloor. Apply 3/8″ bead of multipurpose adhesive to center of top flange of joists. Place plywood panels with long dimension across or parallel to wood or steel joists spaced max. 16″ o.c. Fasten to steel joists with 1 15/16″ pilot point Type S-12 screws spaced 16″ o.c. Fasten plywood to wood joists with suitable nails or screws spaced max. 12″ o.c. or as required by code.

For counter tops, install min. 3/4″ exterior grade plywood base across wood cabinet supports spaced 16″ o.c. Position ends and edges over supports.

Fixture Attachment

To ensure satisfactory job performance, fixture attachment should provide sufficient load-carrying capacity within the assembly. Framing and bracing must be capable of supporting the partition elements and fixture additions within L/360 allowable deflection limit. Install bracing and blocking flush with the face of the framing to keep the stud faces smooth and free of protrusions.

Heavy gauge metal straps mounted on the studs are not recommended supports because they cause bowing in the board and interfere with the flat, smooth application of DUROCK Cement Board and ceramic tile. When heavy anchor plates must be used, fur out studs with a metal strap or wood shim to provide an even base for the cement board. If required, the board may be ground or drilled to provide relief for projecting bolts and screwheads.

Fixture Attachment Load Table

Fastener type	size		Base assembly	Allowable withdrawal resistance		Allowable shear resistance	
	in	mm		lb	N(1)	lb	N(1)
toggle bolt or hollow wall anchor	1/8	3.18	1/2″ cement board & steel stud	70	311	100	445
	3/16	4.76		80	356	125	556
	1/4	6.35		155	689	175	778
3/8″ Type S-12 pan head screw	3/8	9.5	20-ga. steel to 20-ga. steel	53	236	133	680
two bolts welded to steel insert	3/16	4.76	1/2″ cement board, plate and steel stud	175	778	200	890
	1/4	6.35		200	890	250	1112
bolt welded to 1 1/2″ chan.	1/4	6.35	see plumber's bracket page 181	200	890	250	1112

(1) Newtons

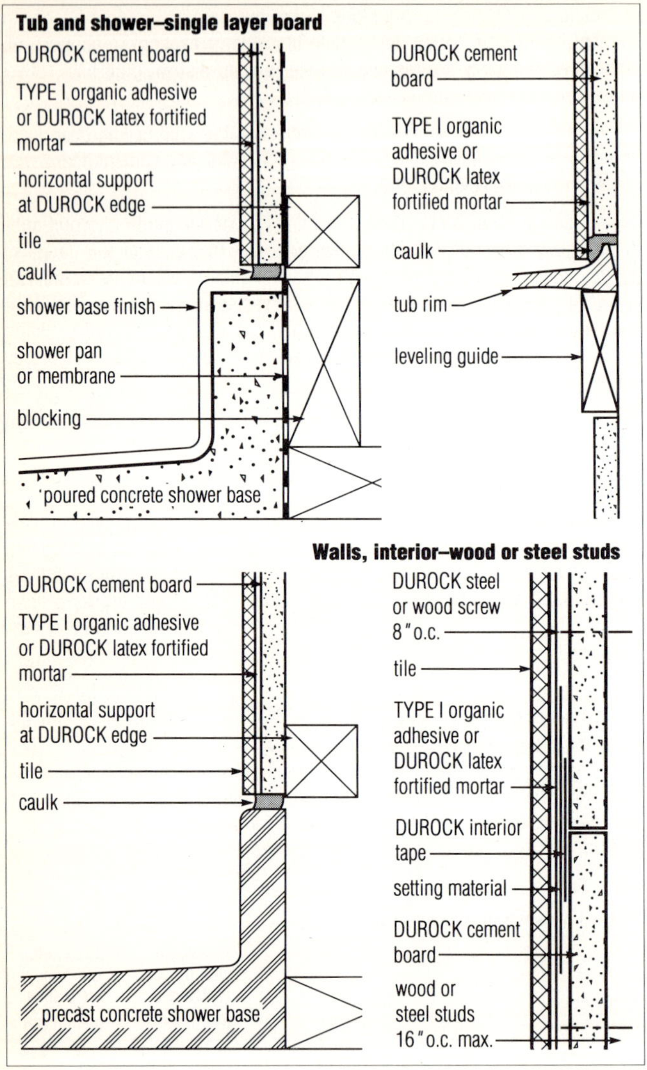

Panel Fabrication

Cutting, shaping and applying DUROCK Cement Board panels is handled in much the same fashion as drywall. Using a utility knife and a straightedge, cut through the glass-fiber mesh scrim on both sides of the board and snap as you would drywall. Cutouts for penetrations must be made on both sides of the board first, then they may be tapped out with a hammer. To ensure that the cuts occur at the right locations on both sides of the sheet, it often is best to drive a nail through the board at the center of the penetration location and measure accurately from that point with a tape measure. A wood rasp is useful for shaping cutouts and board edges. Tools required for laminating and attachment include notched trowels and a power screw-gun.

Panel Application

For floors—Laminate 5/16″ DUROCK Underlayment to subfloor using Type 1 ceramic tile adhesive, latex fortified mortar or thin-set mortar mixed with acrylic latex additive applied to subfloor with 1/4″ square-notched trowel for thin set, 5/32″ V-notched trowel for mastic. Place underlayment with joints staggered from subfloor joints. Fit ends and edges closely but not forced together. Fasten to subfloor with 1¼″ DUROCK Wood Screws or 1½″ galvanized roofing nails spaced 8″ o.c. in both directions with perimeter fasteners at least 3/8″ and less than 5/8″ from ends and edges; or with 1/4″ x 7/8″ galvanized staples spaced 4″ o.c. in both directions.

½″ DUROCK Cement Board may be used instead of DUROCK Underlayment. Use the same procedure except fastening with staples is not permitted.

Tile/carpet underlayment

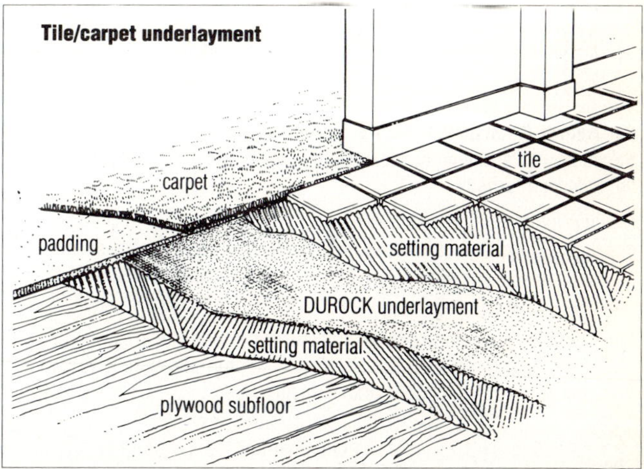

For walls—After tub, shower pan or receptor is installed, place temporary 1/4″ spacer strips around lip of fixture. Cut DUROCK Cement Board to required sizes and make necessary cut-outs. Fit ends and edges closely but not forced together. Install board abutting top of spacer strip. Stagger end joints in successive courses. Fasten boards to wood studs spaced max. 16″ o.c. and bottom plates with 1¼″ DUROCK Wood Screws or 1½″ galvanized roofing nails spaced 8″ o.c. Fasten boards to steel studs spaced max., 16″ o.c. and bottom runners with 1¼″ DUROCK Steel Screws spaced 8″ o.c. with perimeter fasteners at least 3/8″ and less than 5/8″ from ends and edges. In double-layer walls where cement board is installed over base-layer gypsum boards, apply a water barrier (not a vapor retarder) over gypsum board.

For counter tops—Laminate 5⁄16″ DUROCK Underlayment to plywood using ceramic tile mastic, latex fortified mortar or thin-set mortar mixed with acrylic latex additive applied to plywood with 1⁄4″ square-notched trowel for thin set, 5⁄32″ V-notched trowel for mastic. Fasten to plywood with 1¼″ DUROCK Wood Screws or 1½″ galvanized roofing nails spaced 8″ o.c. in both directions and around edges; or with 1⁄4″ x 7⁄8″ galvanized staples spaced 4″ o.c. in both directions and around edges.

½″ DUROCK Cement Board may be used instead of DUROCK Underlayment. Use the same procedure except fastening with staples is not permitted.

Counter tops

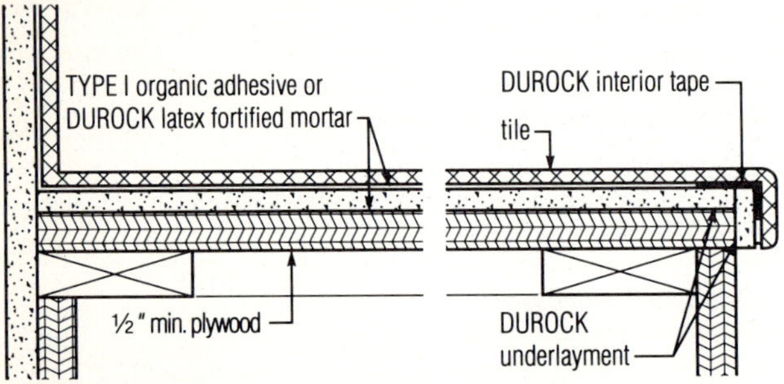

For Hearth Extensions—To use panels in hearth extension designs, use the guidelines specified by local building code, fireplace manufacturer, and the following formula:

$$\frac{\text{k-value DUROCK}}{\text{k-value specified}} \times \text{Hearth extension thickness (specified)} = \text{Thickness of DUROCK Panels (not less than hearth extension specified)}$$

For example, if the fireplace manufacturer or code requires one layer of .75-in. millboard with a k-value of .84, use the formula as follows to determine the required layers of DUROCK Panels:

$$\frac{1.92}{.84} \times .75 \text{ in.} = 1.71 \text{ in. of DUROCK Board or four layers}$$

Joint Treatment

Fill joints with tile-setting mortar or adhesive and then immediately embed tape and level joints. As an alternative, apply DUROCK Interior Tape over the joints and then apply tile-setting mortar or adhesive, forcing it through the tape to completely fill and level the joints. This may require several passes to accomplish.

Panel Finishing

Install tile or thin brick and grout in accordance with ANSI A108.4 for Type I organic adhesive or ANSI A108.5 for dry-set or latex portland cement mortar and ANSI A108.10 for grouts. Before tile application begins, the moisture content of the DUROCK Cement Board should be allowed to adjust as closely as possible to the level it will reach in service. Avoid extreme changes in environmental conditions during the curing of the tile setting material. Provide adequate ventilation to carry off excess moisture.

For small areas where the DUROCK Cement Board will not be tiled, such as a board extending beyond the tiled area and abutting another surface, treat joints as follows. Seal tile backer board with thinned ceramic tile mastic. Mix four parts adhesive with one part water. Embed SHEETROCK Joint Tape over joints and treat fasteners with SHEETROCK Setting-Type Joint Compound (DURABOND 45 or 90) or SHEETROCK Lightweight Setting Type Joint Compound (EASY SAND 45 or 90) applied in conventional manner. Flat trowel SHEETROCK Setting-Type or Lightweight Setting Type Joint Compound over board to cover fasteners and fill voids to a smooth surface. Finish joints with at least two coats SHEETROCK Ready-Mixed Joint Compound. Do not apply ready-mixed joint compound over unsealed board.

UL-Listed Wall Shields and Floor Protectors

Description

Space-saving DUROCK Cement Board (min. ½″ thick) and Exterior Cement Board are listed by Underwriters Laboratories, Inc. for use with UL-listed solid-fuel room heaters and fireplace stoves. Used as a wall shield, DUROCK

Cement Board reduces by two-thirds the manufacturer-specified clearance between room heater or stove and a combustible wall surface. Note: Clearance shall not be less than 12″ in any application.

Used as a floor protector, DUROCK Cement Board may be used in place of ⅜″ thick millboard having a thermal conductivity of $k \leq 0.84$ Btu in/(ft²h °F) in the minimum dimensions specified by the room heater/stove manufacturer.

Installation

For wall shield—Cut ½″ DUROCK Cement Board panel to size with a carbide tip or utility knife scoring tool or circular saw using a carbide-tip blade. Smooth edges with wood rasp. Attach furring strips to wall framing with 2¼″ galvanized roofing nails with ¾″ minimum framing penetration. Attach DUROCK Cement Board through furring to wall framing with 2¾″ galvanized roofing nails with ¾″ minimum framing penetration.

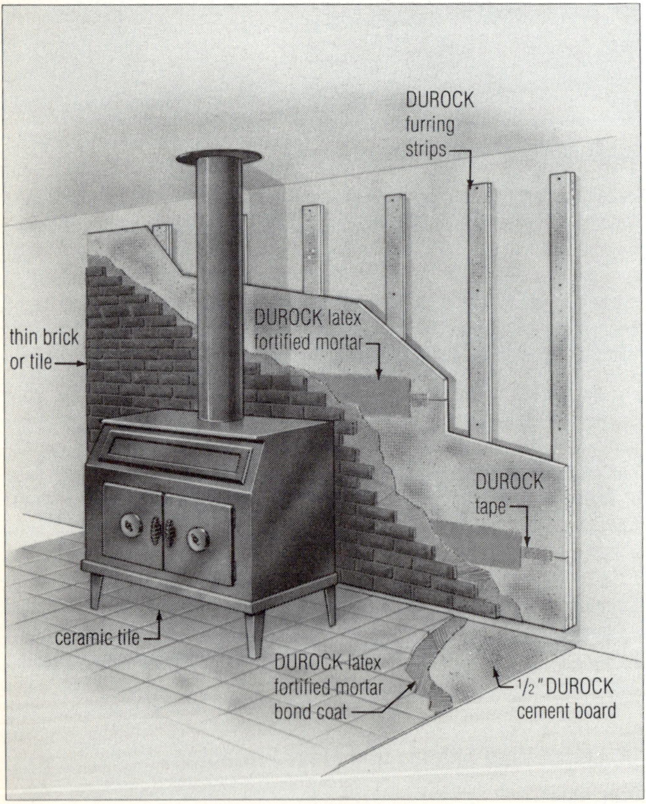

For floor protector—Apply ⅛″ to ¼″ thick latex-fortified portland cement to solid surface—never on top of carpeting or padding. Attach ½″ DUROCK Cement Board with 1⅝″ DUROCK Wood Screws or 1½″ galvanized roofing nails at 8″ o.c. both directions and with ¾″ minimum flooring penetration.

Finishing—Fill joints with latex-fortified portland cement mortar and then immediately embed DUROCK Interior Tape and level the joints. As an alternative, apply tape over the joints and then apply tile-setting mortar or adhesive, forcing it through the tape to completely fill and level the joints. This may require several passes to accomplish. Optional: Finish with thin brick or ceramic quarry tile set in a bed of DUROCK Latex-Fortified Mortar. Grout tiles.

Minimum wall shield/floor protector extensions—The technical drawings below show the minimum wall shield or floor protector extension beyond the room heater or stove.

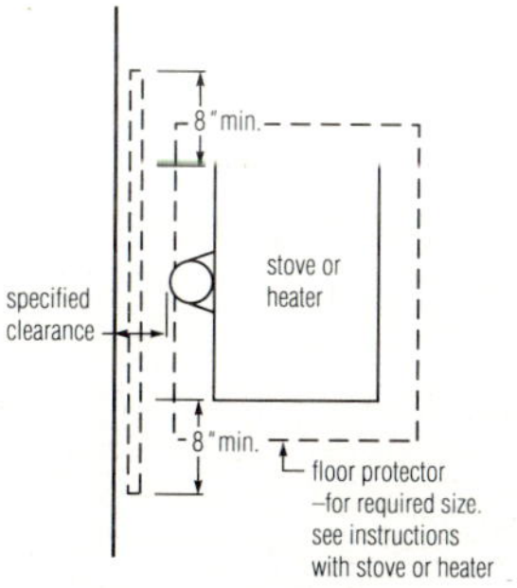

Along-the-wall installation – plan view

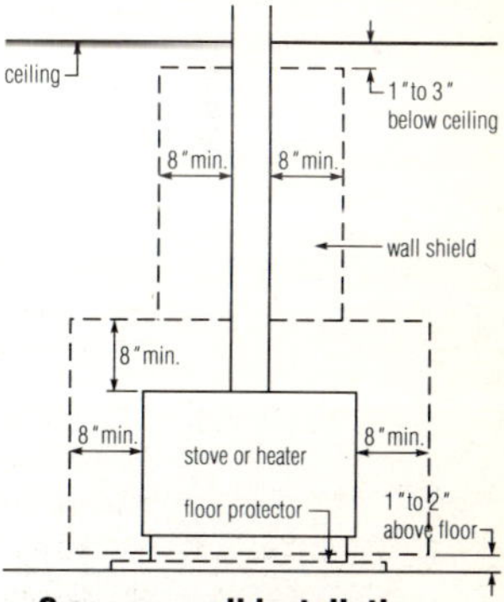

Corner or wall installation – front view

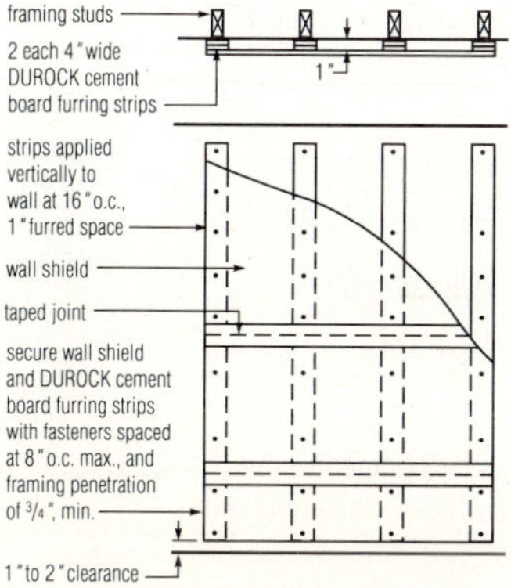

Furred wall shield attachment

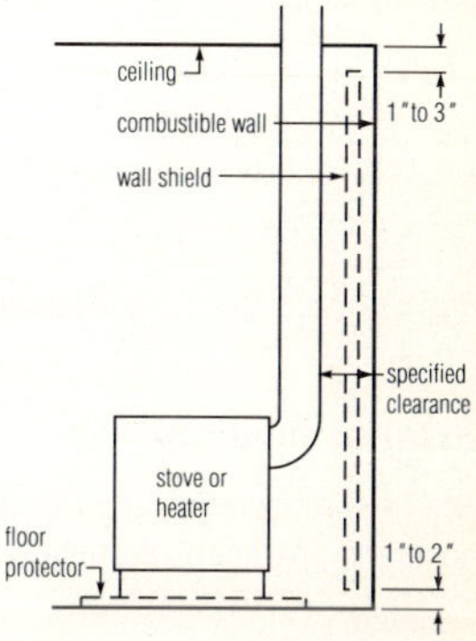

Rear flue stove – side view

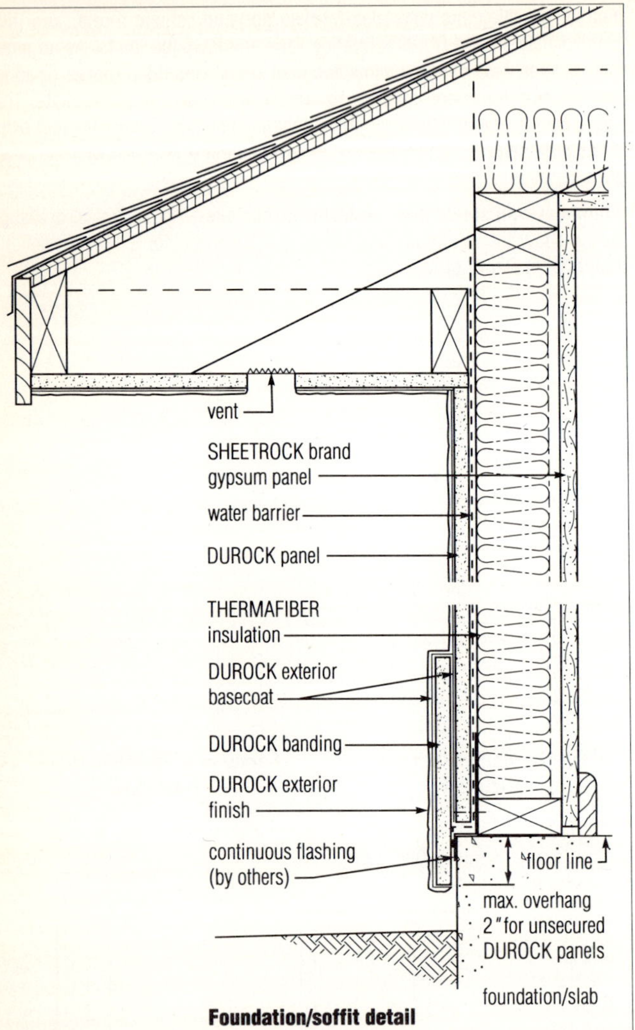

Foundation/soffit detail

Exterior Applications

See "Job Preparation and Design Considerations" earlier in this Chapter for general preparation information and special considerations.

Framing

Steel or wood framing must be designed not to exceed L/360 deflection for tile and thin brick finishes, L/240 for DUROCK Exterior Finish and exterior insulation finish system. Steel framing must be 20-ga. or heavier with a corrosion-resistant metal coating equivalent to G60 hot-dipped galvanized.

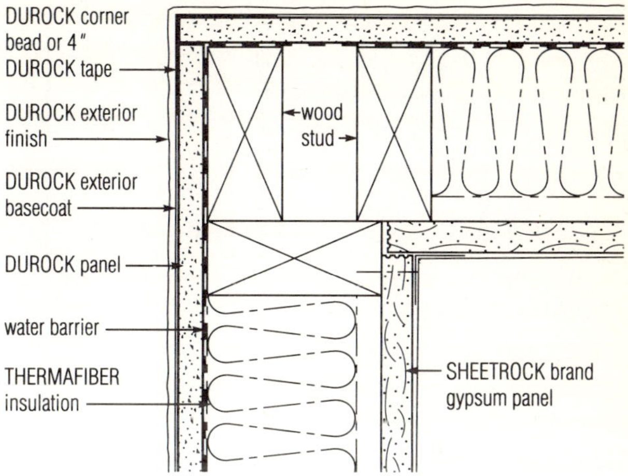

Corner

Space wood and steel framing a maximum of 16″ o.c. Adequate diagonal bracing meeting design requirements must be installed flush with the framing members prior to application of cement board. The studs of freestanding furred walls must be secured to exterior wall with wall furring brackets or laterally braced with horizontal studs or runners spaced 4′ o.c. max.

Water barrier membrane—For steel framing, secure membrane with tape or adhesive and immediately apply DUROCK Panels or staple membrane to back of cement board with ½″ crown, ¼″ to ⅜″ leg staples. Extend membrane 2″ to 3″ beyond board edges so that membrane will lap in a shingle-like manner upon board erection to prevent water penetration into stud cavity. For wood framing, plywood or gypsum sheathing, staple membrane to structure and immediately apply cement board.

Panel Application

Apply DUROCK Panels with rough side toward exterior and with ends and edges over supports. Fit ends and edges closely, but not forced together. For cement board with staple-attached membrane, apply in a shingle-like manner beginning at bottom of wall. Stagger end joints in successive courses.

Fasten DUROCK Panels to framing with DUROCK Screws: 1¼″ and 1⅝″ DUROCK Steel Screws for 14 to 20-ga. steel framing; 1¼″, 1⅝″ or 2¼″ DUROCK Wood Screws for wood framing as specified. Drive fasteners in field of cement board first, working towards ends and edges. Hold cement board in firm contact with framing while driving fasteners. Space fasteners max. 8″ o.c. for walls, 6″ o.c. for ceilings, with perimeter fasteners at least ⅜″ and less than ⅝″ from ends and edges. Drive nails and screws so bottom of heads are flush with surface of cement board, to provide firm panel contact with framing. Do not overdrive fasteners.

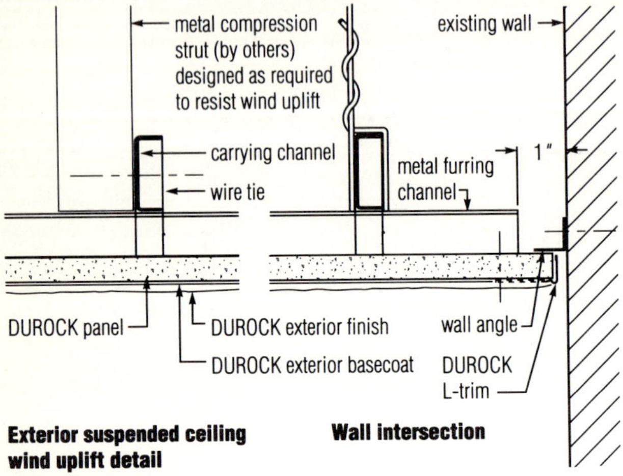

Exterior suspended ceiling wind uplift detail

Wall intersection

Soffits and Ceilings—DUROCK Exterior Cement Board Systems finished with ceramic tile, thin brick and DUROCK Exterior Finish may be used on properly vented soffits and ceilings with DUROCK Screws spaced 6″ o.c. max. A qualified structural engineer should evaluate design including uplift bracing.

DUROCK Cement Board Systems provide an exceptional, virtually maintenance-free suspended ceiling when attached to DONN RIGID X suspension system or metal furring channel wire-tied to carrying channel. Consult your U.S. Gypsum Company representative for details concerning a particular application.

Trim Accessories

When specified, DUROCK Trim Accessories are applied using DUROCK Screws, hot-dipped galvanized roofing nails, or stainless staples. Space nails or screws a max. 16″ o.c. and staples 6″ to 9″ o.c. in each flange as required to hold flange flat against panel. For DUROCK Exterior Finish System, treat trim accessories with basecoat and level with adjacent board areas. Fill all voids and depressions with basecoat and feather mortar edges. The treated joints and trim areas must be allowed to cure for a minimum of four hours before application of basecoat.

Joint Treatment

For tile and thin brick finishes, fill joints with DUROCK Latex Fortified Mortar. Embed DUROCK Exterior Tape centered over all joints and corners but not overlapped. For DUROCK Exterior Finish System, fill joints with DUROCK Exterior Basecoat and then immediately embed 4″ tape and level the joints. As an alternative, apply DUROCK Exterior Tape over joint and then apply DUROCK Latex Fortified Mortar or DUROCK Exterior Basecoat, forcing it through tape to completely fill and level joints. This may require several passes. For exposed aggregate finishes, apply DUROCK Exterior Tape and immediately apply the required thickness of coating over entire surface and embed stone aggregate according to manufacturer's directions.

Suspended ceiling detail

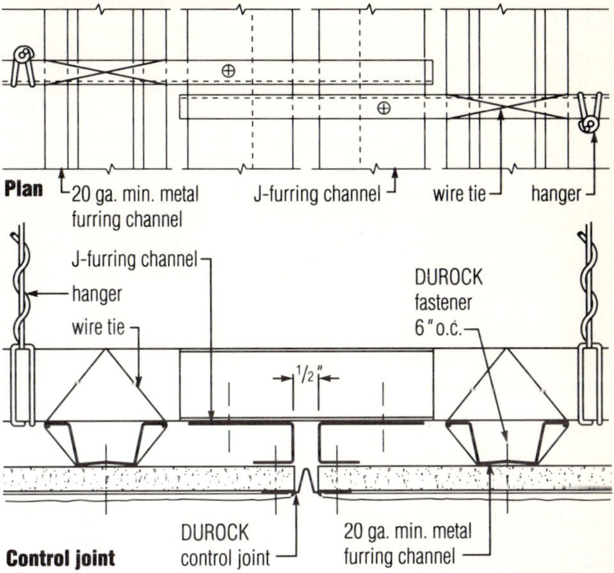

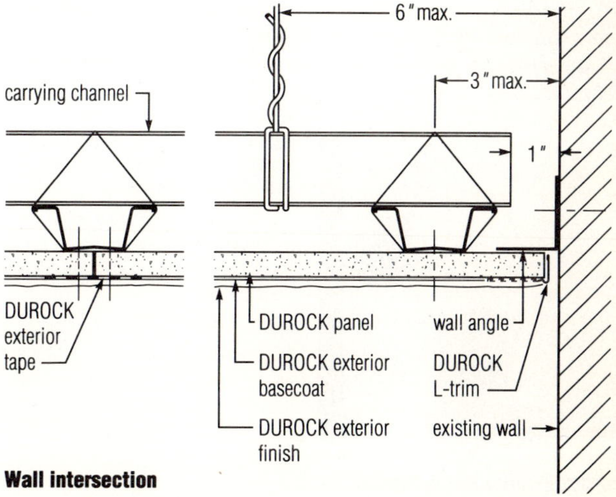

DUROCK Exterior Finish Application

Mixing Finish Products

Mixing of mortar, grout or exterior basecoat or exterior finish can be accomplished by hand using a potato masher or similar device, however power tools speed the process and assure a more uniform product. Mixing paddles powered by a ½″ heavy-duty electric drill with a speed of max. 400 rpm. are both popular and readily available. They come in a variety of sizes and styles. A cage type mixer is recommended. When mixing, use drinkable water and follow specific instructions on bag. Remix occasionally during use. Do not mix more than can be applied during the working time period. Larger mixing equipment may be used if multiple applicators are involved.

Application of these materials requires standard or notched trowels, depending on the application. Steel trowels should not be used with DUROCK Exterior Finish as discoloration may result.

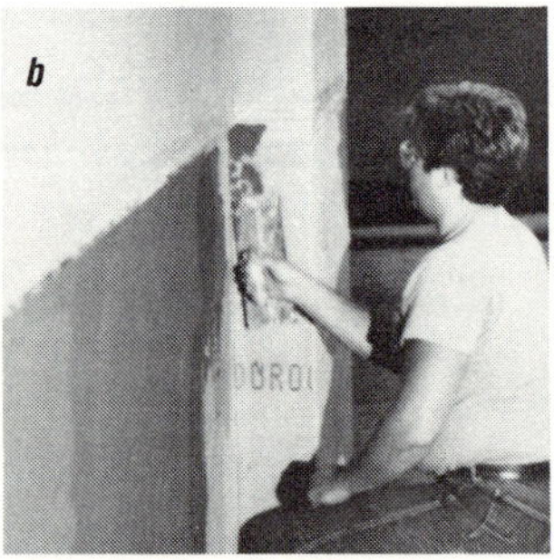

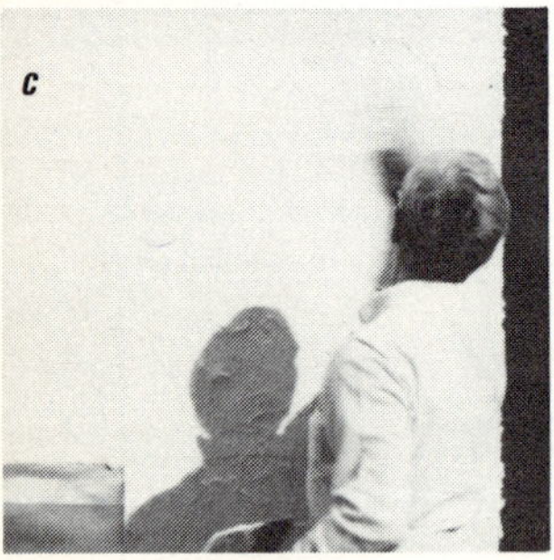

***a.** Cement board is applied to framing with DUROCK SCREWS or hot-dipped galvanized roofing nails.*

***b.** Basecoat is applied over entire area, including treated joints.*

***c.** Exterior finish is applied after basecoat has cured for 24 hours.*

DUROCK Exterior Finish System

Basecoat—Apply a 1⁄16″ minimum thick, uniform layer of DUROCK Exterior Basecoat over the entire surface after joints and trim have cured a minimum of four hours. Leave surface smooth and flat. Under rapid drying conditions, dampen surface as necessary to improve workability. Allow basecoat to cure 24 hours before application of DUROCK Exterior Finish Coat.

Finish Coat—Trowel-apply DUROCK Exterior Finish in a 1/16″ minimum thick, uniform layer over all base-coated surfaces. Do not add sand or other additives to create heavier textures (material is not designed for texture heavier than 3/16″).

If necessary, adjust consistency and working properties by adding up to 8 oz. clean water per 67-lb. pail of finish material. Add the same amount of water to all subsequent pails to ensure color uniformity. Mix well for uniform consistency. Texture as required, using plastic or wood floats. (Note: steel floats may cause discoloration.)

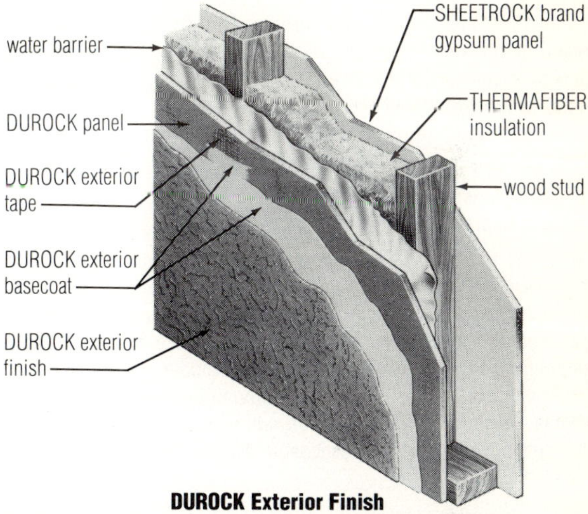

DUROCK Exterior Finish

DUROCK Exterior Finish Applied to Concrete or Masonry—Prior to application of basecoat, new poured-in-place concrete construction should be allowed to cure for 28 days and all structural cracks and large surface voids must be filled and leveled. For new construction, the joints between the concrete masonry units should be struck flush. For existing construction the joints must be pre-filled with DUROCK Exterior Basecoat. For poured-in-place concrete, any ridges caused by form separation shall be leveled. Small surface voids (i.e. cracks, spalled areas) must be prefilled with DUROCK Exterior Basecoat and allowed to cure for 24 hrs. prior to the actual basecoat application. Several coats may be required. Allow basecoat to cure a minimum of four hours between each coat. Apply DUROCK Exterior Basecoat and DUROCK Exterior Finish as stated above.

DUROCK Exterior Insulation & Finish System (EIFS)

The DUROCK Exterior Insulation & Finish System (EIFS) combines the thermal advantages of expanded polystyrene insulation with the superior strength and integrity of cement board. DUROCK Exterior Cement Board offers a nondeteriorating and extremely stable substrate that is superior to gypsum sheathing while DUROCK Insulation Board provides excellent thermal resistance. The system is completed with DUROCK Basecoat and Exterior Finish.

Apply DUROCK Exterior Basecoat directly to the entire back surface of the insulation board using a ⅜″ notched trowel. Back wrap insulation board a minimum of 2″ at all terminations. Then immediately apply DUROCK Insulation Board horizontally in a running bond pattern, starting at the bottom and supported vertically by permanent or temporary supports. Slide insulation board diagonally into place. Joints in the insulation board must be offset from joints in the cement board substrate. Cut insulation board to fit openings and corners. Edges of the insulation board must not align with corners of windows, doors or other wall penetrations.

The insulation board must butt tightly without any basecoat material between the boards. Apply firm pressure over the entire insulation board surface to ensure uniform contact. Use a straightedge at least 6′ long to check plane level of the adjacent insulation boards. Cut slivers of material to fit and fill gaps. Allow adhesive to cure 24 hours before sanding or applying reinforcing mesh and basecoat.

Sand out-of-plane edges of insulation board to produce a level surface. Do not attempt to prefill low areas with DUROCK Exterior Basecoat.

Trowel apply to the insulation board a 1/16″ thick, uniform layer of DUROCK Exterior Basecoat. Immediately embed DUROCK Reinforcing Mesh into the wet basecoat material and smooth with a trowel until mesh is fully embedded and pattern of mesh is not visible beneath the surface. Sections of mesh should be lapped min. 2″ on all sides. Smooth out wrinkles by working from center to edges. Wrap mesh around corners 8″ min. from both directions. Reinforce wall penetrations with 9″ x 12″ pieces of mesh applied at 45° angle to corners. Allow basecoat to dry min. 24 hr.

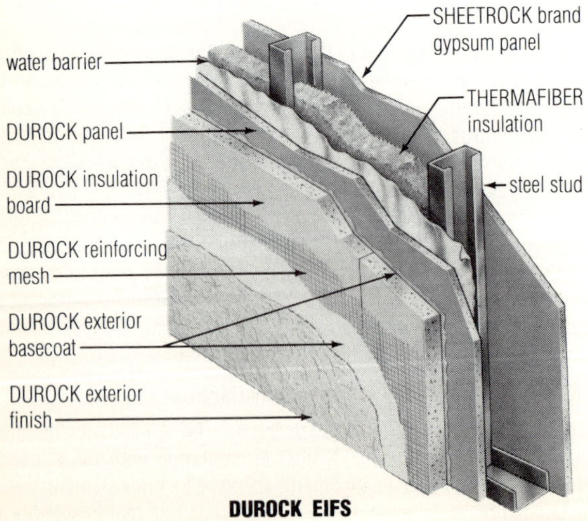

DUROCK EIFS

Trowel-apply DUROCK Exterior Finish in the same manner as for Exterior Finish System above. (Refer to U.S. Gypsum Company Technical Folder CB-269 DUROCK Exterior Insulation & Finish System for construction details and architectural specifications.)

Other Finishing Applications

Thin Brick and Ceramic Tile Systems—Apply a ⅛″ min. thick leveling/skim coat of DUROCK Latex Fortified Mortar over DUROCK Panel surfaces. Leave surface smooth and flat. Allow to set 24 hours before application of bond coat for setting tile and thin brick.

Ceramic tile and thin brick on walls may not exceed ⅞″ thickness, 18″ x 18″ size, and 10 psf. Install thin brick or ceramic tile in accordance with ANSI 108.5 specifications and manufacturer's directions. Using the notched trowel required for the thickness of thin brick or tile being installed, apply DUROCK Latex Fortified Mortar to obtain uniform setting bed. Back-butter the thin brick or ceramic tile for 100% mortar contact. Install units by firmly pressing them into freshly applied mortar. Use a sliding and twisting motion to embed units and obtain a 100% mortar contact. Beat-in ceramic tile in accordance with accepted practice. Apply DUROCK Latex Fortified Grout after mortar has set firmly for 24 hours. Mix and apply grout according to directions on package. Force maximum amount of grout into joints. Tool and compress grout into joints to provide neat and uniform appearance. Clean grout from finished surfaces and cure installation as required by ANSI A108.10 specification.

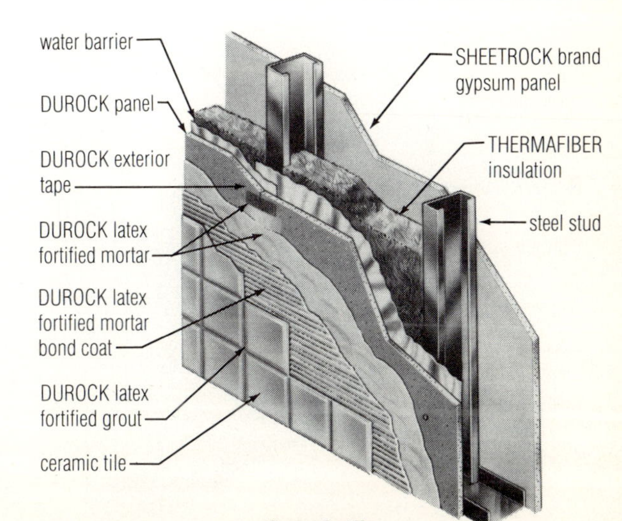

Ceramic tile

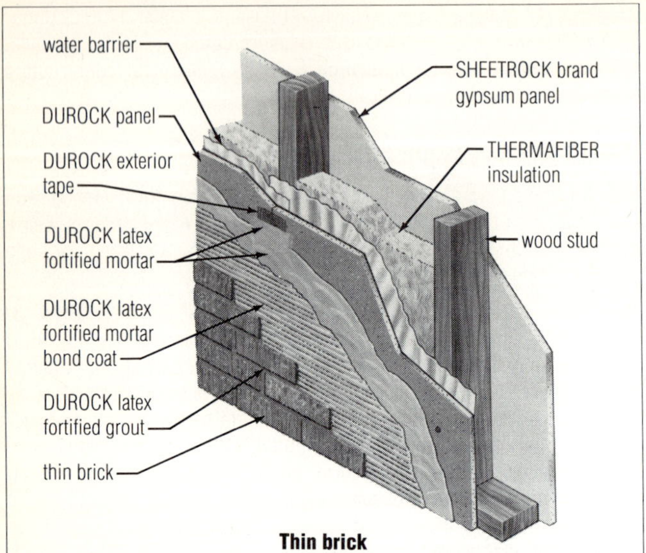

Thin brick

Stone Aggregate/Epoxy Matrix System—DUROCK Panel is a suitable substrate for many epoxy matrix stone aggregate products. Contact epoxy matrix suppliers regarding suitability of their products for this use. Mix and apply epoxy material directly to the taped DUROCK Panel surface according to manufacturer's directions. Follow immediately with specified aggregate application.

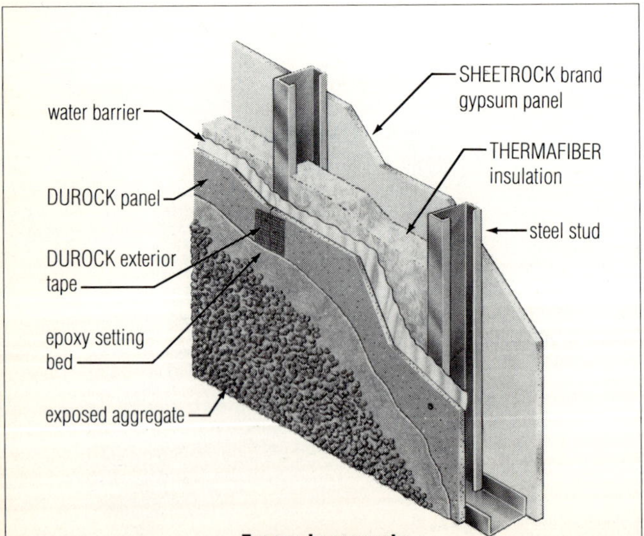

Exposed aggregate

Layered Details—To create bands, quoins, dentils, lintels and other layered details, cut DUROCK Panel to specified size and shape. Laminate to basecoated DUROCK Panel following same application procedure as with ceramic tile.

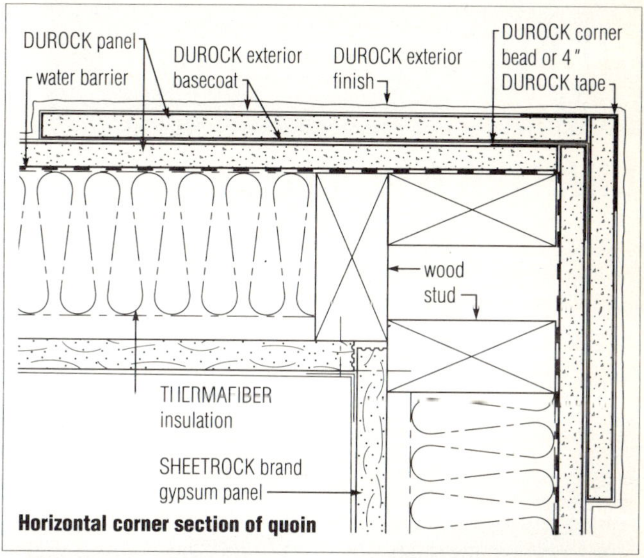

Horizontal corner section of quoin

Layering DUROCK Exterior Cement Board provides design options such as dentils, lintels, and bands that are easily fabricated and installed.

Conventional Lath & Plaster Construction

CHAPTER 6 Products

Performance is the Key

A completed plaster job can be no better than the basecoat or finish materials used and the base to which they are applied. U.S. Gypsum Company plastering products have gained their superiority on one basis: *performance.*

This record of performance extends through the complete U.S. Gypsum Company line—broadest in the industry—of plaster and lime products, plaster bases and accessories, designed to work together in a wide range of wall and ceiling systems.

The basic materials recommended by U.S. Gypsum Company for quality plaster walls and ceilings are described on the following pages. All meet the essential requirements of function, economy and speed of installation.

The U.S. Gypsum Company trademark on a product is assurance of consistent high quality and proven performance to meet your construction needs.

U.S. Gypsum Company technical sales representatives are ready to consult with contractors, architects and dealers on plastering materials, systems and special job conditions. They may be reached by contacting the nearest U.S. Gypsum Company sales office (see inside back cover) or the Company's headquarters in Chicago, Illinois, at (312) 606-4000.

Plaster Bases

Proper use of U. S. Gypsum Company plaster bases and plasters provides the secure bond necessary in order to develop surface strength and resistance to abuse and cracking. These characteristics are common to both metal lath and gypsum plaster bases.

Gypsum Plaster Bases

ROCKLATH Plaster Base is a gypsum lath that provides a rigid, fire-resistant base for the economical application of gypsum plasters. ROCKLATH Base requires less basecoat plaster than does metal lath.

The gypsum core of this lath is faced with multilayer laminated paper formulated by a U. S. Gypsum Company process to provide proper absorp-

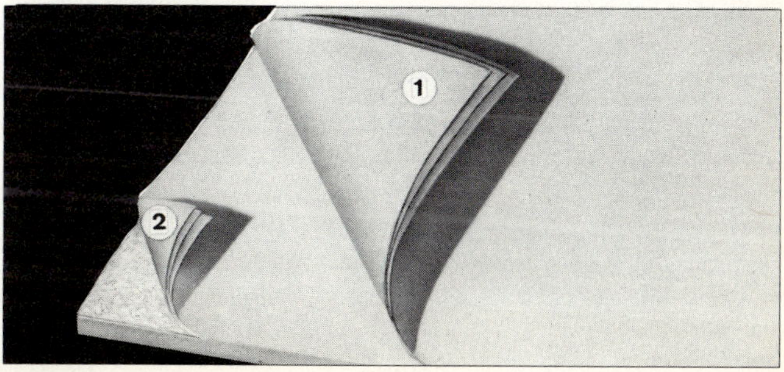

ROCKLATH multi-ply face paper

layers (1) are highly absorbent to draw moisture from the plaster mix uniformly and quickly so that the plaster takes on anti-slump strength before it can slide; the inner layers (2) are chemically treated to form a barrier against moisture penetration, thus reduce softening of the gypsum core and consequent sag after the board is in place. Face paper is folded around the long edges and the ends are square-cut.

ROCKLATH Base complies with ASTM C37. Other features are:

Fire Resistance—When used with gypsum plaster, gypsum plaster bases provide assemblies with fire ratings of up to 2 hrs. for partitions, ceilings and column fireproofing.

Strength—When securely attached, gypsum plaster bases add lateral stability to the assembly.

Sound Resistance—Partitions faced with gypsum plaster bases and plaster on both sides have excellent resistance to sound transmission; resilient attachment further improves ratings, makes assemblies suitable for party walls.

Bonding—Gypsum plaster bonds to these gypsum plaster bases with a safety factor far higher than required to meet usual construction standards.

Durability—Not harmfully affected by decay, dry rot or normal moisture; will not attract vermin.

Gypsum Plaster Base General Limitations

1. Maximum frame spacing is dependent on thickness and type of lath used.

2. To be used with gypsum plaster only. Bond between lime or portland cement plaster and ROCKLATH Base is inadequate.

3. Should not be used in areas that are exposed to excessive moisture for extended periods or as a backing for ceramic tile or other similar surfacing materials commonly used in wet areas; UNIMAST Galvanized Metal Lath and portland cement-lime plaster are recommended.

4. ROCKLATH Base is unsuitable for veneer plasters and finishes.

Note: Gypsum basecoat plasters have slightly greater dimensional stability than gypsum lath. Therefore, the stability of the lath would govern in design problems. Refer to Appendix for coefficients of expansion, and data on drying shrinkage.

Available in ⅜″ and ½″ thicknesses and in two different products for specific uses:

Regular ROCKLATH Plaster Base—Rigid board used for nail or staple application to wood; screw attachment to UNIMAST Steel Studs and Metal Furring Channels.

ROCKLATH FIRECODE Plaster Base—A gypsum lath that combines all the advantages of Regular ROCKLATH Base with additional resistance to fire exposure—the result of a specially formulated core containing special mineral materials.

Note: See Appendix for Thermal Resistance Values (R).

Specifications—ROCKLATH Plaster Bases

ROCKLATH product	Thickness		Width		Length		pc/ bdl	Approx. wt.	
	in	mm	in	mm	in	mm		lb/ft²	kg/m²
Regular	⅜	5.9	16	406	48	1.2	6	1.4	6.8
Regular	½	12.7	16	406	48	1.2	4	1.8	8.8
Regular	½	12.7	24	610	96	2.4	2	1.8	8.8
FIRECODE	⅜	5.9	16	406	48	1.2	6	1.4	6.8

Metal Lath

The metal lath and accessory products identified here are manufactured by Unimast Inc. and marketed exclusively by United States Gypsum Company.

UNIMAST Metal Lath—Mesh material formed from sheet steel that has been slit and expanded to form a multitude of small openings. It is made in Diamond Mesh and Riblath types and in two different weights for most types. Manufactured from steel protected by a coating of black asphaltum paint. Diamond Mesh and ⅜″ Riblath are also available in galvanized steel. Comply with Federal Specification QQ-L-101C.

Ends of bundles of UNIMAST Metal Lath are spray painted in different colors for various weights, thus simplifying stocking and handling. All 3.4-lb. lath is painted *red*; 2.5-lb. diamond mesh and 2.75-lb. ⅛″ riblath are *white*; 1.75 diamond mesh is not end-painted.

In addition, UNIMAST Metal Lath offers these features:

Strength—Metal lath embedded within the plaster provides reinforcement.

Flexibility—Readily shaped to ornamental contours to a degree not possible with other plaster bases.

Fire Resistance—When used with gypsum plaster, metal lath provides excellent fire-resistant construction; up to two hrs. for partitions and four hrs. for ceilings and column fireproofing.

Security—Metal lath and plaster surfaces are extremely difficult to penetrate; provide excellent protection against break-through.

Available in following types and styles:

UNIMAST Junior Diamond Mesh Lath

UNIMAST Self-Furring Diamond Mesh Lath

UNIMAST Paper-back Lath

UNIMAST 4-Mesh Z-Riblath

UNIMAST ⅜″ Riblath

Metal Lath General Limitations

1. Metal lath products should not be used with magnesium oxychloride cement stuccos or stuccos containing calcium chloride additives.

2. In ceiling assemblies, certain precautions concerning construction, insulation and ventilation are necessary for good performance. A min. of ½ sq. in. net-free vent area is recommended per sq. ft. of horizontal surface in plenum or other space.

UNIMAST Junior Diamond Mesh Lath—A small diamond mesh metal plaster base (approx. 11,000 meshes per sq. yd.). A general all-purpose lath, best for ornamental, contour plastering. The small mesh openings conserve plaster and reduce droppings. Also available in *self-furring* type having ¼″ dimple indentations spaced 1½″ o.c. each way, for use as exterior stucco base, column fireproofing and for replastering over old surfaces.

UNIMAST Paper-back Metal Lath—Asphalt paper-backed Junior Diamond Mesh Lath available painted or galvanized. Regular or Self-Furring type. Asphalt-impregnated paper is factory-bonded to the back. Paper is vapor permeable. Meets Federal Specification UU-B-790a, Type I, Grade D, Style 2.

Paper-backed painted lath is recommended for lath and plaster back-up of interior tile work and other inside work.

Paper-backed galvanized lath is a recommended base and reinforcement for some exterior wall construction, including stucco and other machine or hand-applied exterior surfacing materials.

UNIMAST 4-Mesh Z-Riblath—A "flat rib" type of lath with smaller mesh openings, suitable for "double-up" plastering. More rigid than diamond mesh, excellent as nail-on lath, and for tie-on work on flat ceilings. Not recommended for contour lathing.

UNIMAST ⅜″ Riblath—A herringbone mesh pattern with ⅜″ V-shaped ribs running lengthwise of the sheet at 4½″ intervals, with inverted intermediate ³⁄₁₆″ ribs. The heavy ribs provide exceptional rigidity. Used when supports are spaced more than 16″ o.c. and not more than 24″, and for 2″ solid studless metal lath and plaster partitions. Also used as a centering lath for concrete floor and roof slabs. Unsuitable for contour plastering. Min. ground thickness must be 1″.

Specifications—UNIMAST Metal Lath

Product	Approx. wt.		Finish		Size				Color code	Packaging (bundle)		
					Width		Length				Quantity	
	lb/yd²	kg/mm²	paint	galv.	in	mm	in	m		Pcs	yd²	m²
	1.75	0.9	x	x	27	686	96	2.4	None	10	20	16.7
Diamond Mesh – Regular	2.5	1.4	x	x	27	686	96	2.4	White	10	20	16.7
	3.4	1.8	x	x	27	686	96	2.4	Red	10	20	16.7
– Self-Furring	2.5	1.4	x	x	27	686	96	2.4	White	10	20	16.7
	3.4	1.8	x	x	27	686	96	2.4	Red	10	20	16.7
	1.75	0.9	x	x	27	686	96	2.4	None	10	20	16.7
– Asphalt Paper-backed	2.5(1)(2)	1.4	x	x	27	686	96	2.4	White	10	20	16.7
	3.4(1)(2)	1.8	x	x	27	686	96	2.4	Red	10	20	16.7
⅜″ (9.5mm) Riblath	3.4	1.8	x	x	27	686	96	2.4	Red	10	20	16.7
4-Mesh Z-Riblath	2.75	1.5	x		27	686	96	2.4	White	10	20	16.7
	3.4	1.8	x		27	686	96	2.4	Red	10	20	16.7

(1) Add 0.3 lbs/yd for weight of paper. (2) Regular or Self-Furring.

Trim Accessories

UNIMAST Corner Beads should be used on all external plaster corners to provide protection, true and straight corners, and grounds for plastering; UNIMAST Casing Beads, around wall openings and at intersections of plaster with other finishes. Both products are made from galvanized steel.

Limitation: Galvanized steel accessories are recommended for interior use only. For exterior application and where corrosion due to high humidity and/or saline content of aggregates is possible, the use of zinc alloy accessories is required. Should not be used with magnesium oxychloride cement stucco or portland cement stucco containing calcium chloride additives.

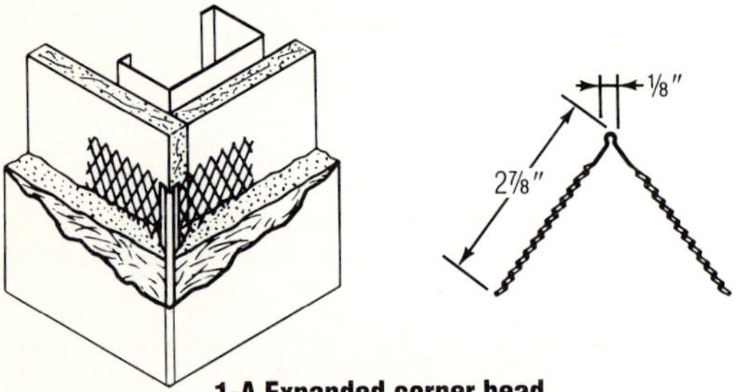

1-A Expanded corner bead

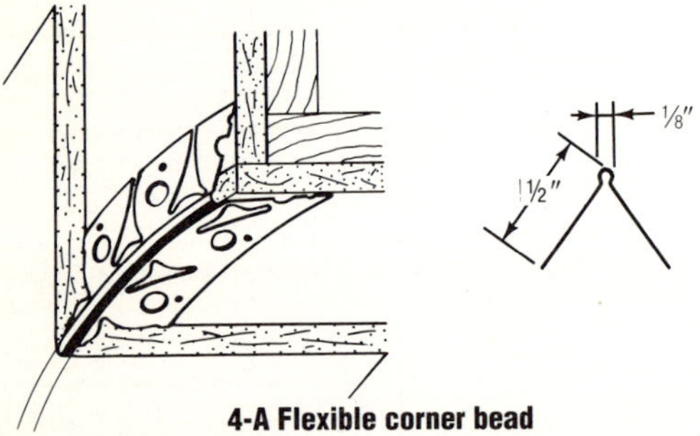

4-A Flexible corner bead

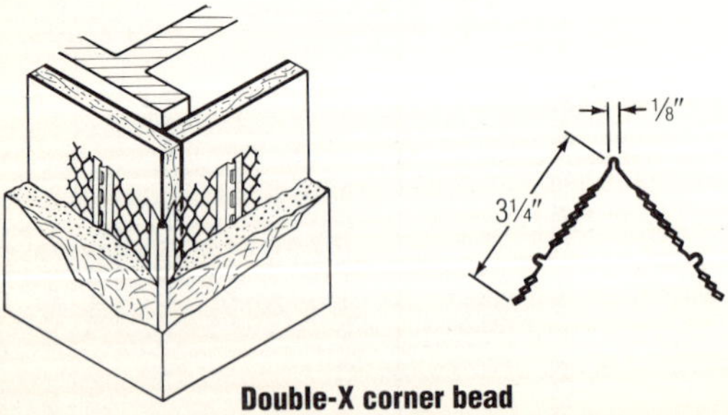

Double-X corner bead

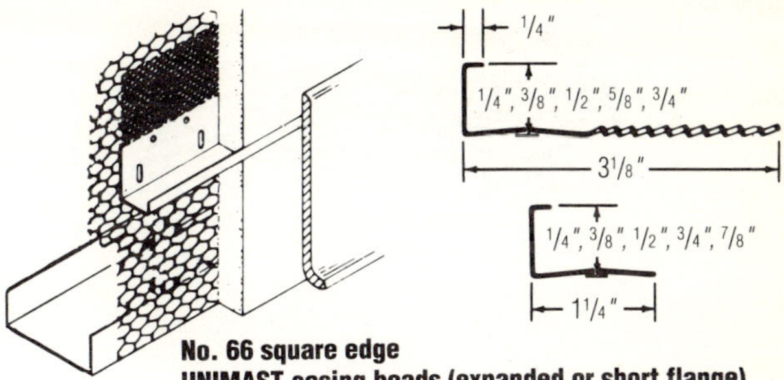

No. 66 square edge
UNIMAST casing beads (expanded or short flange)

Corner and Casing Beads

UNIMAST 1-A Expanded Corner Bead—A general-purpose corner bead, economical and most generally used. Has wide expanded flanges that are easily flexed. Preferred for irregular corners. Provides increased reinforcement close to nose of bead.

UNIMAST 4-A Flexible Corner Bead—A special utility, solid punch-pattern bead. By snipping flanges, this bead may be bent to any curved design where required.

UNIMAST Double-X Corner Bead—Has full 3¼″ flanges easily adjusted for plaster depth on columns. Ideal for finishing corners of structural tile and rough masonry. Has perforated stiffening ribs along expanded flange.

UNIMAST Casing Beads—Used as a plaster stop and exposed to eliminate the need for wood trim around window and door openings; also recommended at junction or intersection of plaster and other wall or ceiling finishes, and as a screed. May be used with UNIMAST Metal Lath, ROCKLATH Plaster Base, or masonry construction. In order to ensure proper grounds for plastering, ¾″ casing beads are recommended for use with metal lath, ⅝″ expanded flange beads with all masonry units, ⅞″ beads when solid flange is applied under gypsum plaster base, ½″ beads when flange is applied over gypsum base. Also available in ¼″ and ⅜″ ground heights for special application. Available in galvanized steel or zinc alloy for exterior applications.

Cornerite and Striplath

These products are strips of painted or galvanized Diamond Mesh Lath used as reinforcement. **Cornerite,** bent lengthwise in the center to form a 100° angle, should be used in all internal plaster angles where metal lath is not lapped or carried around; over gypsum lath, anchored to the lath, and over internal angles of masonry constructions. Also used in the "floating angle" method of applying gypsum lath to wood framing in order to reduce plaster cracking. **Striplath** is a similar flat strip, used as a plaster reinforcement over joints of gypsum lath and where dissimilar bases join; also used to span pipe chases, and as reinforcement of headers over openings.

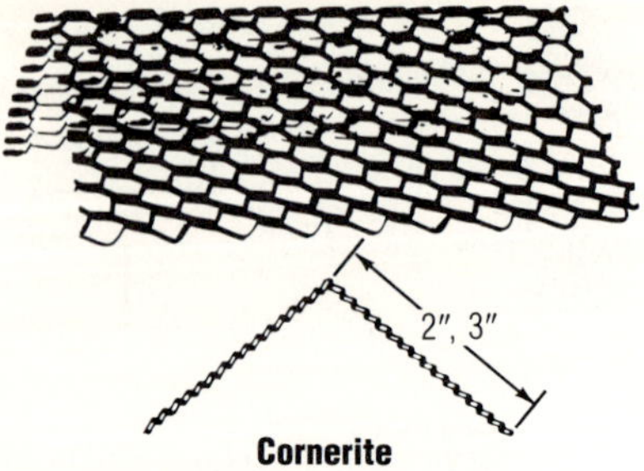

Cornerite

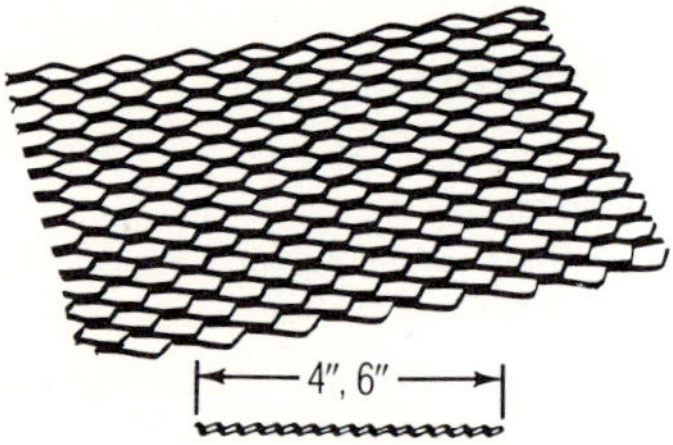

Striplath

Control Joints

SHEETROCK Zinc Control Joint—Designed to relieve stresses of both expansion and contraction in large plastered areas. Made from roll-formed zinc alloy, it is resistant to corrosion in both interior and exterior uses with gypsum or portland cement plaster. An open slot, ¼" wide and ½" deep, is protected with plastic tape that is removed after plastering is completed. The short flanges are perforated for keying and attachment by wire-tying to metal lath or by stapling to gypsum lath. Thus the plaster is key-locked to the control joint, which not only provides plastering grounds but can also be used to create decorative panel designs. Sizes and grounds: No. 50, ½" ; No. 75, ¾"; No. 100, 1" (for exterior stucco).

Max. Spacing—SHEETROCK Control Joints for Interior Assemblies

		Max. single dimension		Max. single area	
System	**Location**	**ft**	**mm**	**ft²**	**m²**
Metal Lath & Plaster	Partition	30	9144	—	—
	Ceiling	50(1)	15240	2500	230
		30(2)	9144	900	83.6
Gypsum Lath & Plaster	Partition	30	9144	—	—
	Ceiling	50(1)	15240	2500	230
		30(2)	9144	900	83.6

(1) With perimeter relief. (2) Without perimeter relief.

For exterior application where wind pressure exceeds 20 psf, back control joints with 2″ wide butyl tape applied to the sheathing. Install joints with flanges under self-furring lath and attach temporarily with Bostitch 9/16″ "G" staples or equal, spaced 6″ apart on each side. Positive attachment of flange to framing through the lath is required using fasteners 12″ o.c. Break supporting members, sheathing and metal lath behind control joints. When vertical and horizontal joints intersect, vertical joint should be continuous; horizontal joint should abut it. Apply sealant at all splices, intersections and terminals.

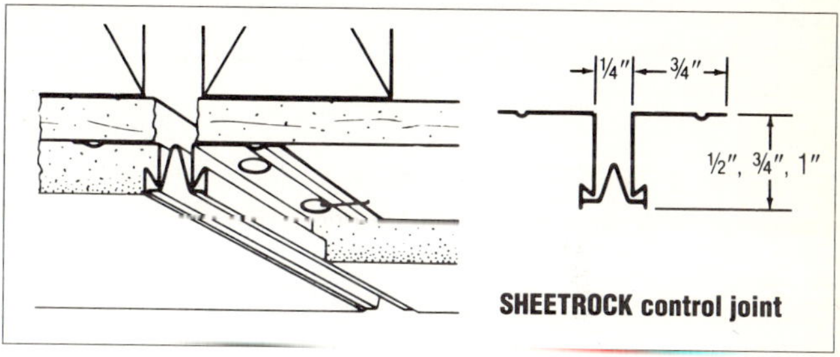

SHEETROCK Zinc Control Joint

Control Joint Limitations

1. Where sound and/or fire ratings are prime considerations, adequate protection must be provided behind the control joint.

2. Functions only with transverse stresses.

3. Should not be used with magnesium oxychloride cement stucco or portland cement stucco containing calcium chloride additives.

UNIMAST Double-V Expansion Joint—Provides stress relief to control cracking in large plastered areas. Made with expanded flanges of corrosion-resistant galvanized steel or zinc for exterior use in ½″ or ¾″ grounds.

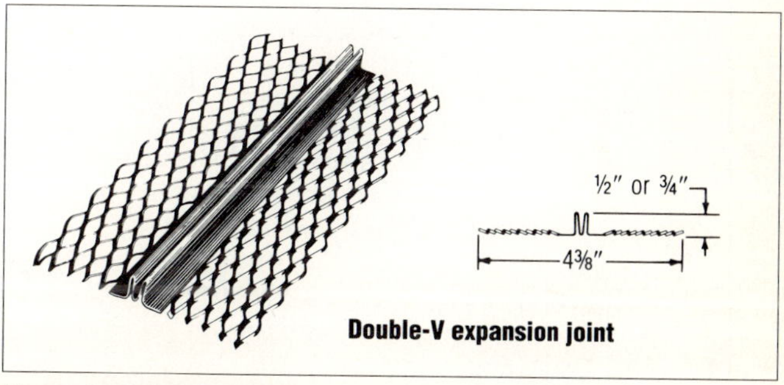

UNIMAST Double-V Expansion Joint

Specifications—Trim Accessories

Product	Depth or grounds in	Depth or grounds mm	Flange width in	Flange width mm	Finish	Length[1] 7′	8′	9′	10′	12′	Pcs. per ctn.	Approx. weight lb/1000 ft.	Approx. weight kg/100m
UNIMAST 1-A Expanded Corner Bead			2⅞	73.0	Galv.	x					40	195	29
					Galv. or Zinc Alloy		x		x		30	195	29
UNIMAST 4-A Flexible Corner Bead			1½	38.1	Galv.		x		x		30	179	27
UNIMAST Double-X Corner Bead			3¼	82.6	Galv.		x	x	x	x	30	253	38
UNIMAST #66 Square Expanded Flange Casing Bead[2]	¼	6.3	3⅛	79.4	Galv. or Zinc Alloy				x		30	50	7
	⅜	9.5	3⅛	79.4	Galv. or Zinc Alloy				x		30	56	8
	½	12.7	3⅛	79.4	Galv. or Zinc Alloy		x		x		8′-40,10′-30	213	32
	⅝	15.9	3⅛	79.4	Galv. or Zinc Alloy				x		30	241	36
	¾	19.1	3⅛	79.4	Galv. or Zinc Alloy		x		x		8′-40,10′-30	255	38
	⅞	22.2	3⅛	79.4	Galv. or Zinc Alloy				x		30	—	—
	1	25.4	3⅛	79.4	Galv. or Zinc Alloy				x		30	—	—
	1¼	31.8	3⅛	79.4	Galv. or Zinc Alloy				x		30	—	—
UNIMAST #66 Square Short Flange Casing Bead	¼	6.3	1¼	31.7	Galv. or Zinc Alloy				x		30	44	6
	⅜	9.5	1¼	31.7	Galv. or Zinc Alloy				x		30	47	7
	½	12.7	1¼	31.7	Galv. or Zinc Alloy				x		30	178	26
	¾	19.1	1¼	31.7	Galv. or Zinc Alloy				x		30	202	30
	⅞	22.2	1¼	31.7	Galv. or Zinc Alloy				x		30	212	32
UNIMAST Cornerite			2	50.8	Paint or Galv.		x				75	90	13
			3	76.2	Paint or Galv.		x				75	132	20
UNIMAST Striplath			4	101.6	Paint or Galv.		x				75	90	13
			6	152.4	Paint or Galv.		x				75	—	—
SHEETROCK Zinc Control Joint #50	½	12.7	¾	19.1	Zinc Alloy		x				25	172	26
#75	¾	19.1	¾	19.1	Zinc Alloy		x				25	192	29
#100	1	25.4	¾	19.1	Zinc Alloy		x				25	216	32
UNIMAST Double-V Expansion Joint	½	12.7	2³⁄₁₆	76.7	Galv.		x				24	308	46
	¾	19.1	2³⁄₁₆	6.7	Galv.		x				24	379	56

Clip and Screws

A specially formed steel clip and self-drilling screws are available for positive attachment and rapid erection of gypsum plaster bases and metal lath.

BRIDJOINT Field Clip B-1—Used to support and align end joints that do not fall opposite structural members; designed for use with ⅜″ ROCKLATH Base. Approximate usage is 350 clips per 100 yd^2 ($84m^2$) of lath, based on 16″ o.c. stud spacing. Shipped 500 pc./package; 19 lb. (8.6kg)/1000 pcs.

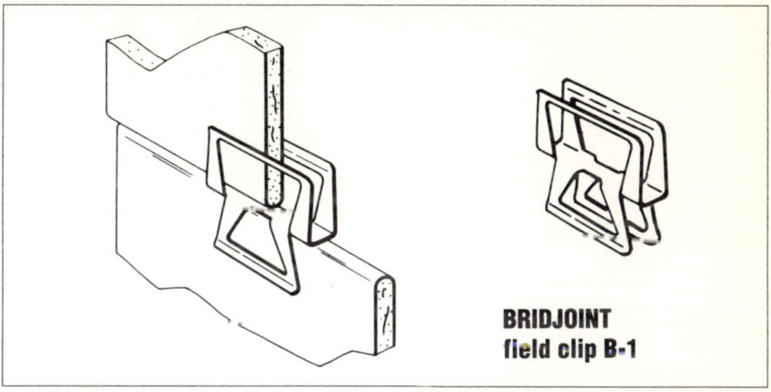

BRIDJOINT Field Clip B-1

Screws—The result of continuing development aimed at producing the best possible attachment of gypsum boards to steel, wood or gypsum supports simply and quickly. A complete line of self-drilling, self-tapping steel screws is available to improve construction systems and simplify installation methods. All screws are highly corrosion-resistant and have a Phillips head recess for rapid installation with a special bit and power-driven screwgun. (For complete data and Selector Guide for Screws, see Chapter 1.)

Bugle head, high and low threads, slotted point of screws.

Economical SUPER-TITE Screws with specially designed drill point for steel studs.

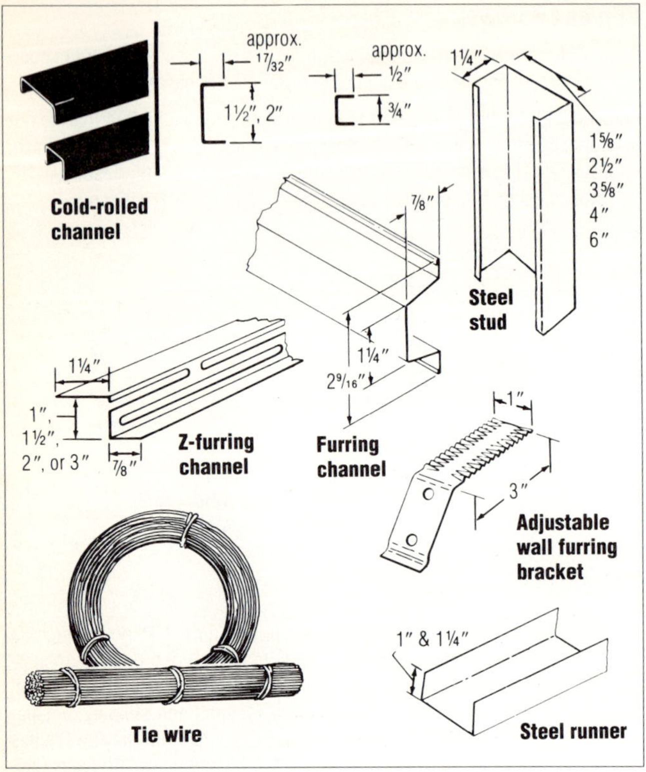

Framing Components

United States Gypsum Company sells and distributes steel studs, runners and accessories manufactured by UNIMAST Incorporated. Steel framing members offer the advantages of light weight, low material cost and quick erection, superior strength and versatility in meeting job requirements. All are noncombustible, made from galvanized steel.

UNIMAST Steel Studs and Runners—Channel-shape and roll-formed from galvanized or corrosion-resistant steel. Used in non-load and load-bearing interior partition and exterior curtain wall systems. Limited chaseways are provided by punchouts in the web. Assemblies using these studs are low in cost with excellent sound and fire-resistance characteristics. Available in various styles and widths to meet functional requirements.

For complete data on UNIMAST framing component sizes and availability, see section in Chapter 1. For installation, see Chapter 2.

UNIMAST Cold-rolled Channels—Formed from 16-ga. steel, black asphaltum painted or galvanized. Used for furring, suspended ceilings, partitions and ornamental lathing. Sizes: ¾″, 1½″, 2″ .

Specifications—Structural Accessories(1)

Product	Size		Lengths		Shipping	Approx. weight			
						Unit			
	in	mm	ft	m	unit	lb	kg	lb/1000 ft	kg/100 m
UNIMAST Cold-Rolled Channel(2)(3)	¾	19.1	10, 16, 20	3.05 4.88, 6.10	20 pc	62, 99, 124	28, 44, 56	300	45
	1½	38.1	10, 16, 20	3.05, 4.88, 6.10	10 pc	50, 79, 99	22, 35, 44	500	74
	2	50.9	16, 20	4.88, 6.10	10 pc	62, 100	28, 45	629	94
UNIMAST Tie Wire/Hanger Wire(3)	8-ga 18-ga	4.1 1.2	1429 16620	436 5066	Coil Coil	100 100	45 45	— —	— —
	8-ga 18-ga 18-ga	4.1 1.2 1.2	1429 8310 4155	436 2533 1266	Bdl. Hank Hank	100 50 25	45 23 11	— — —	— — —

(1) See Chapter 1 for other structural accessories. (2) Painted. (3) Galvanized.

UNIMAST Metal Furring Channels—Roll-formed, hat-shaped section of galvanized steel. This 25-ga. channel may be attached with furring clips or tie wire to the 1½″ main carrying channel and spaced 16″ o.c. for economical screw attachment of ROCKLATH Base as a base for either adhesively applied acoustical tile or a basecoat plaster; 4′ span max. Also available made from 20-ga. galvanized steel for heavier loads and longer spans. The furring channel also provides noncombustible furring for exterior walls; may be spaced up to 24″ o.c. Face width 1¼″, depth ⅞″. (See Chapter 1 for data on SHEETROCK Z-Furring Channels.)

UNIMAST Adjustable Wall Furring Brackets—Used for attaching ¾″ furring channels to exterior masonry walls. Made of galvanized steel with corrugated edges. Brackets are attached to masonry and act as supports for horizontal channels 24″ o.c. in braced furring systems.

UNIMAST Tie Wire—18-ga. galvanized soft annealed wire for tying metal lath to channels and furring to runner channels.

Hanger Wire—8-ga. for suspended ceiling channel runners when spaced not more than 4′ o.c.

STRUCTOCORE Security Wall Components—Steel sheets that provide continuous reinforcement for the cementitious fireproofing materials used in the STRUCTOCORE Security Wall System (see page 319 for installation procedures). Components include STRUCTOCORE Forming Sheets in sizes shown in table below; 1¾″ x 1¾″ floor-ceiling angle; 4″ x 1½″ vertical abutment angle; 4″ x 4″ intersection angle; vertical abutment angle; and 5½″ x 8′ flat accessory plate.

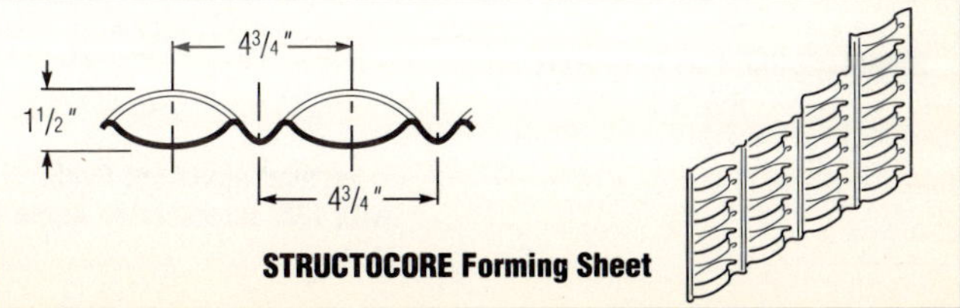

STRUCTOCORE Forming Sheet

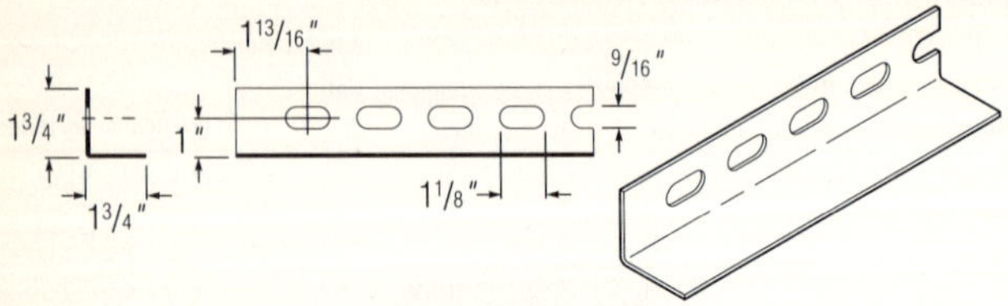

Floor–Ceiling Angle
1 3/4" x 1 3/4" Leg

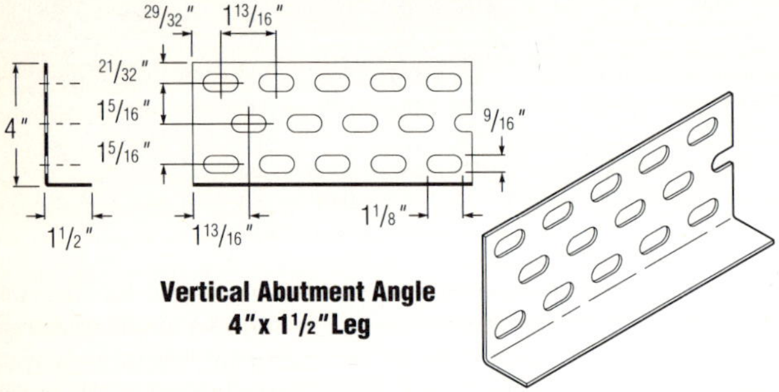

Vertical Abutment Angle
4" x 1 1/2" Leg

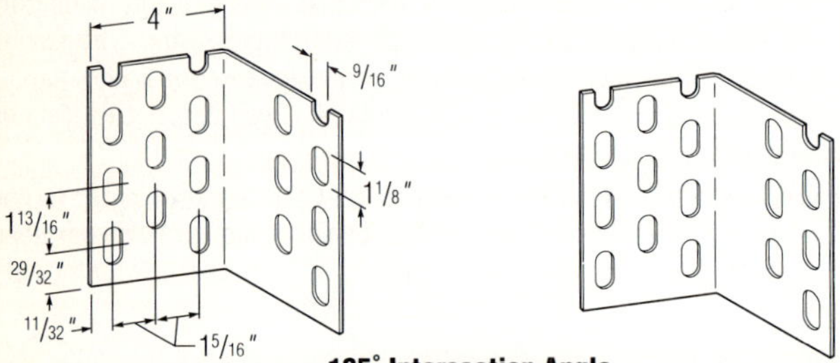

135° Intersection Angle
4" x 4" Leg

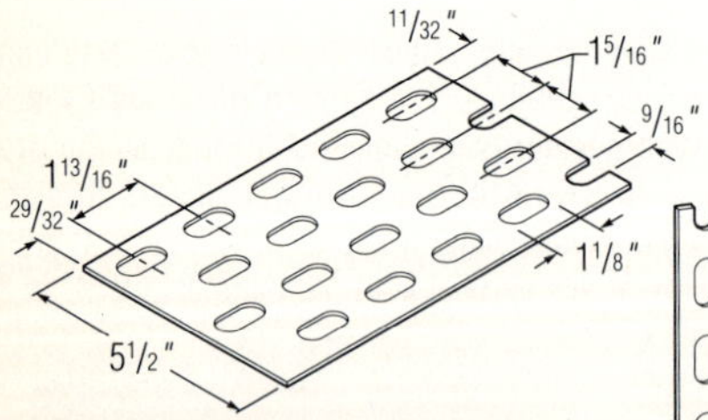

STRUCTOCORE Flat Accessory Plate
5 1/2" x 8'

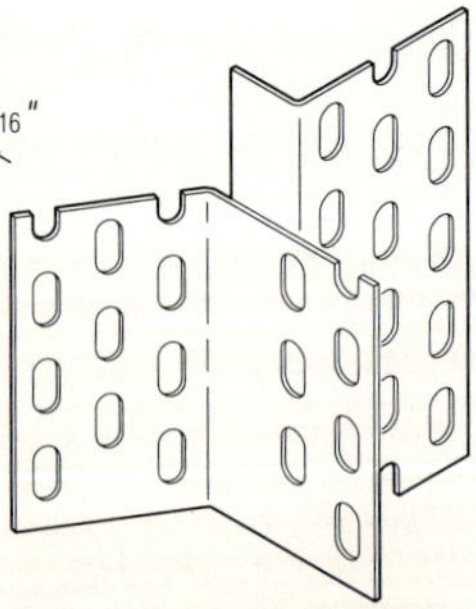

Vertical Abutment Angle
with 135° Intersection Angle

Design Properties—STRUCTOCORE Steel Sheets

Gauge	Approximate lbs./ft²	Size (nom.)	Yield strength	Finish
18	2.10	4′ x 8′, 4′ x 10′ 3′ x 8′, 3′ x 10′ 4′ x 4′	33 ksi	galvanized steel (G-60)
16	2.60	4′ x 8′, 4′ x 10′ 3′ x 8′, 3′ x 10′ 4′ x 4′	33 ksi	galvanized steel (G-60)
14	3.18	4′ x 8′, 4′ x 10′ 3′ x 8′, 3′ x 10′ 4′ x 4′	33 ksi	galvanized steel (G-60)
12	4.44	4′ x 8′, 4′ x 10′ 3′ x 8′, 3′ x 10′ 4′ x 4′	33 ksi	galvanized steel (G-60)

Plasters

The main ingredient of all gypsum plasters is gypsum rock—hydrous calcium sulfate—which has a water content of about 20% in chemical combination. During processing, about ¾ of this chemically combined water is removed from the gypsum rock by means of a controlled calcination process. When water is added at the job, the material crystallizes (sets), reverting to its original chemical composition.

United States Gypsum Company plasters are specifically formulated to control setting time and other important characteristics. These depend upon the intended use and method of application, the climatic conditions of the area and job conditions.

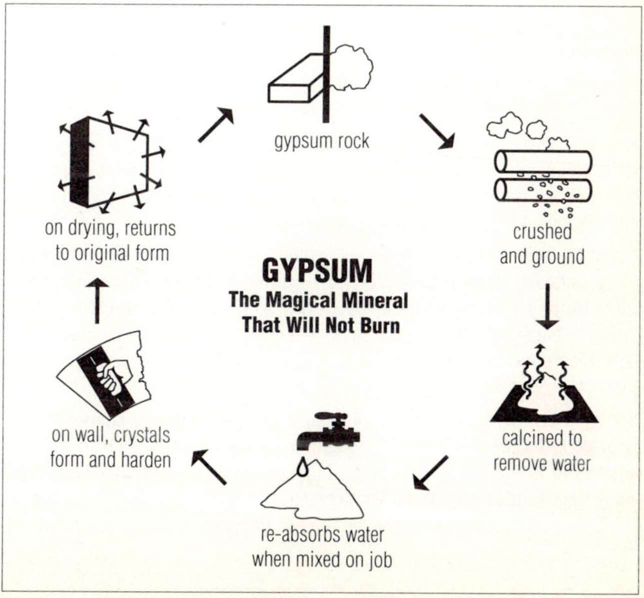

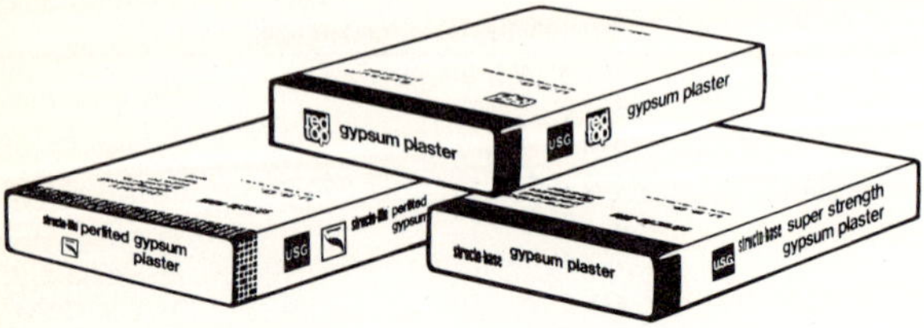

Basecoat Plasters

For the beauty and durability of which plaster is capable, certain requirements should be followed as to the number of coats applied. Three-coat work is necessary on all metal lath and is desirable on all gypsum lath; two-coat work is acceptable on unperforated gypsum lath and on the interior face of rough concrete block, clay tile or porous brick. Outside masonry should be furred prior to plastering to prevent seepage and condensation.

In preparing for plastering, consideration should be given to the selection of materials not only for compatibility but also for the quality of the structure to be plastered. It is wise to upgrade plastering specifications when possible.

Red Top Gypsum Plaster—Preferred for its low cost and excellent workability; must be job-aggregated. Three types: *Regular*—for sand aggregate, hand application; *LW*—for lightweight aggregate, hand application (not recommended over metal lath when smooth-trowel lime finish is used); *Machine Application*—for sand or lightweight aggregate. Perlite aggregate is not recommended when vertical lift exceeds 30′ or hose length is over 150′. Meets ASTM C28. Available in 100-lb. bags.

Red Top Two-Purpose Plaster—Suitable for machine or hand application; reduces inventory requirements. Must be job-aggregated; perlite aggregate not recommended when vertical lift exceeds 30′ or hose length is over 150′. Meets ASTM C28. Available in 100-lb. bags.

Structo-Base Gypsum Plaster—Develops higher strength than conventional plasters. Used in Structocore Security Wall applications. Also ideal for handball courts, hospital corridors, schools, etc. requiring high-strength basecoat. Superior as sanded scratch and brown coat over metal lath. Meets ASTM C28. Available in 100-lb. bags. Available in Regular and Machine Application.

Structo-Lite Gypsum Plaster—Contains mill-mixed perlite aggregate and is preferred in cold weather when aggregate may freeze, or when suitable aggregate is not readily available. Lighter weight and greater insulation value than sanded basecoats. Three types (not available in all areas; consult U. S. Gypsum Company representative): *Regular*—for gypsum or metal

Gypsum Basecoat Plasters—Coverage and Technical Data

		Ratio: aggregate (vol.)/ basecoat (wt.)		Approx. compressive strength dry[1]		Approx coverage per ton of gypsum basecoat[2]					
						gypsum lath		metal lath		unit masonry	
Plaster product	**Mix**	**ft³/100 lb**	**m³/ton[5]**	**lb/in² (psi)**	**MPa[4]**	**yd²/ton**	**m²/ton[5]**	**yd²/ton**	**m²/ton**	**yd²/ton**	**m²/ton[5]**
RED TOP Gypsum and Two-Purpose Plasters	sand	2.0	1.24	875	6.00	180	165	114	104	140	129
	sand	2.5	1.55	750	5.17	206	190	131	121	160	147
	sand	3.0	1.86	650	4.48	232	213	148	136	181	167
	perlite	2.0	1.24	700	4.82	176	162	112	103	137	126
	perlite	3.0	1.86	525	3.62	224	206	143	132	174	160
	vermiculite	2.0	1.24	465	3.21	171	157	109	100	133	123
	vermiculite	3.0	1.86	290	2.00	215	198	137	126	168	154
STRUCTO-BASE	sand	2.0	1.24	2800 min.[6]	19.30	154	142	99	91	120	110
	sand	2.5	1.55	1900 min.[6]	13.10	185	170	118	109	144	132
	sand	3.0	1.86	1400 min.[6]	9.65	214	197	136	125	167	154
STRUCTO-LITE	regular	—	—	700	4.82	140	129	89[3]	82[3]	109	100
RED TOP Wood Fiber	neat	—	—	1750	12.06	85	78	54	49	66	60
	sand	1.0	0.62	1400	9.65	135	124	86	79	105	97

(1) Average laboratory results when tested in accordance with ASTM C472. Figures may vary slightly for products from individual plants. (2) Grounds (including finish coat): gypsum lath (face of lath), metal lath ¾″ (back of lath), unit masonry ⅝″. (3) Lightweight aggregated plasters are not recommended over metal lath when the finish coat is to be smooth troweled. (4) Megapascals (MN/m²). (5) Metric ton. (6) Laboratory evaluations for sanded plaster are based on use of graded Ottawa silica sand.

lath; *Masonry*—for high-suction unit masonry; *Type R or S* for specific UL-listed assemblies. Not recommended over metal lath when smooth-trowel lime finish is used or machine application when vertical lift exceeds 30′ or hose length is over 150′. Meets ASTM C28. Available in 80-lb. bags.

RED TOP Wood Fiber Plaster—Contains selected wood fiber and can be used with the addition of water only. When used over masonry bases or for machine application, 1 cu. ft. of sand per 100 lb. of plaster must be added. When used as a scratch or brown coat, 1 cu. ft. of sand can be added.

RED TOP Wood Fiber Plaster can be applied to all standard lath and masonry surfaces and is recommended as a scratch coat for metal lath.

Wood Fiber Plaster, neat, weighs approx. ¼ less than a sanded gypsum basecoat and generally provides greater fire resistance than normally sanded gypsum plaster at a slightly higher cost. Complies with ASTM C28. Available in 100-lb. bags.

Basecoat Plaster Limitations

1. Where sound isolation is prime consideration, use sand aggregate only.

2. Do not use where water or excessive moisture is present. May be applied to exterior soffits equipped with suitable drips and casings and protected from direct exposure to rain and moisture.

3. Not recommended for masonry or concrete walls or ceilings coated with bituminous compounds or waterproofing agents. Interior of exterior walls shall be furred and lathed prior to plastering to prevent seepage and condensation.

4. The only United States Gypsum Company plaster recommended for embedding electric heat cables is job-sanded DIAMOND Interior Finish Plaster applied directly to properly prepared monolithic concrete or IMPERIAL Gypsum Base (see page 214 for more information). If IMPERIAL Gypsum Base and job-sanded DIAMOND Interior Finish Plaster are used for a radiant heat system, the cable-sheath operating temperatures must never exceed 125°F.

5. Basecoats containing job-mixed lightweight aggregate or STRUCTO-LITE Gypsum Plaster must be finished with an aggregated finish plaster.

Portland Cement-Lime Plaster—This mix is used for interior applications where high-moisture conditions exist, or for exterior stucco. Prepared as follows:

Job-mixed Stucco—Mix BONDCRETE or MORTASEAL Mason's Lime with portland cement and sand in accordance with ASTM C926, Type L basecoat, Type FL finish coat. Suggested proportions: *scratch coat*—1 bag portland cement, 1 bag lime, 8 cu. ft. sand; *brown coat*—1 bag Portland cement, 1 bag lime, 10 cu. ft. sand; *finish*—1 bag portland cement, 1½ bags lime, 9 cu. ft. sand.

Prepared Finish—ORIENTAL Exterior Finish Stucco (see page 291).

Portland Cement Plaster Limitations

1. Scratch, brown and finish coats of portland cement plasters require curing with water after set.

2. Must not be applied directly to smooth, dense surfaces, gypsum lath or gypsum block. Self-furring metal lath must be secured to such surfaces before plaster is applied.

3. Control joints should be provided to compensate for shrinkage during drying.

4. A Keenes cement-lime putty finish must never be used over a portland cement basecoat.

Finish Plasters

Conventional plaster walls are finished with gauging plasters and finishing limes or with prepared finishes. United States Gypsum Company provides a range of products with a variety of characteristics. See page 471 in the Appendix for a comparison of various finish plasters.

Finish Plaster General Limitations

1. A smooth trowel finish should not be used over lightweight aggregate gypsum basecoat applied over metal lath. Only sand float finishes are recommended over metal lath.

2. Where the gypsum basecoat is STRUCTO-LITE plaster or contains lightweight aggregate (perlite or vermiculite) and a smooth trowel finish is used over any plaster base except metal lath, the finish coat should be RED TOP Gauging Plaster and lime: a) with addition of ½ cu. ft. of perlite fines, or, b) with addition of 50 lb. of No. 1 white silica sand per 100 lb. gauging plaster, or, c) use a Quality (RED TOP, CHAMPION or STAR) factory aggregated gauging plaster.

3. Gypsum or lime-based finishes, including Keenes cement, should not be used directly over a portland cement basecoat or over concrete block or other masonry surfaces.

4. Smooth-trowel high-strength finishes, such as STRUCTO-GAUGE Gauging Plaster and Keenes Cement, must not be used over STRUCTO-LITE Plaster or a basecoat with a lightweight aggregate.

5. Gauged-lime putty and RED TOP Finish applied over conventional basecoat plasters must age 30 days, be thoroughly dry and properly sealed before decorating. Quick-drying vinyl acrylic latex or alkali-resistant alkyd primer-sealers are recommended.

6. Primers containing polyvinyl acetate (PVA) are not recommended and should not be specified for use over wet plaster of any kind, over lime-gauging or lime-containing plasters. The PVA film is subject to rewetting and will almost certainly result in bond loss and subsequent paint delamination. In view of these precautions, strictly follow the specific lime-locking product recommendations of paint manufacturers for painting lime-gauging putty finishes and for lime-containing veneer plaster finishes.

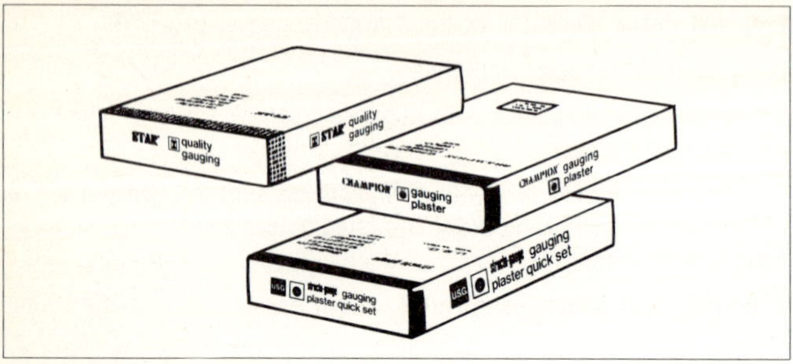

Gauging Plasters

Lime, when used alone as a finish plaster, does not set, is subject to shrinkage when drying, and lacks a hard finish. Gauging plaster is blended into the lime putty in the proper proportions to provide controlled set, early hardness and strength, and to prevent shrinkage cracks.

Gauging plasters are carefully ground and screened to proper particle sizes to make the plasters quick-soaking and easily blended with lime putty.

High-strength RED TOP Keenes Cement and STRUCTO-GAUGE Gauging Plaster are to be used only over sanded, wood fiber or veneer basecoat plasters. Over lightweight aggregated basecoats, use white or regular gauging plaster that is properly aggregated.

RED TOP Gauging Plaster—Blends easily with lime putty for durable smooth-trowel or sand-float finishes in residential construction. Provides high strength, hardness and abrasion resistance superior to many other surfaces. Easily painted or decorated. Applied over a gypsum or veneer plaster basecoat. Available in *Regular*, unaggregated; *Quality*, with perlite fines for lightweight-aggregated basecoats. Two types: Quick Set (30-40 min); Slow Set (50-70 min.). Meets ASTM C28. Available in 100-lb. bags.

CHAMPION and STAR Gauging Plasters—Selected for their whiteness and provide for smooth-trowel or sand-float lime-putty finishes. Effectively resist cracking, provide hardness and abrasion resistance required for normal interior walls and ceilings. Applied over a gypsum or veneer plaster basecoat. Job-aggregated, sand-float finish may be job colored. Available in *Regular*, unaggregated; *Quality*, with perlite fines for lightweight-aggregated basecoats. When mixed with recommended proportions of lime putty, CHAMPION Plaster sets in 20-30 min.; STAR Plaster in 40-60 min. Meet ASTM C28. Available in 50 and 100-lb. bags.

STRUCTO-GAUGE Gauging Plaster—Mixed with lime putty, produces high-strength, durable white smooth-trowel finish for high-traffic areas. Excellent hardness and abrasion resistance to withstand abuse. Faster and easier to apply than Keenes Cement. Used over high-strength sanded, veneer plaster or wood fiber gypsum basecoats. Not for use over lightweight-aggregated or portland cement basecoats or masonry. Two types: *Slow Set* (60-75 min.) for regular sanded basecoats; *Quick Set* (30-40 min.) for low-suction veneer basecoats. Meets or exceeds ASTM C28. Under equivalent application conditions will provide a harder finish than RED TOP, CHAMPION OR STAR Gauging Plasters. Available in 100-lb. bags.

RED TOP Keenes Cement—Only retemperable gauging plaster; provides best gauging for lime-sand float finishes; also suitable for job color. Can be used as a smooth-trowel finish, offering strong, hard surfaces when densified by extensive troweling through set. Permits mixing large batches for job-colored finishes. Requires high-strength gypsum basecoat. Two types: *Regular* (3-6 hr. set); *Quick Trowel* (1-2 hr. set.) Meets ASTM C61. Available in 100-lb. bags.

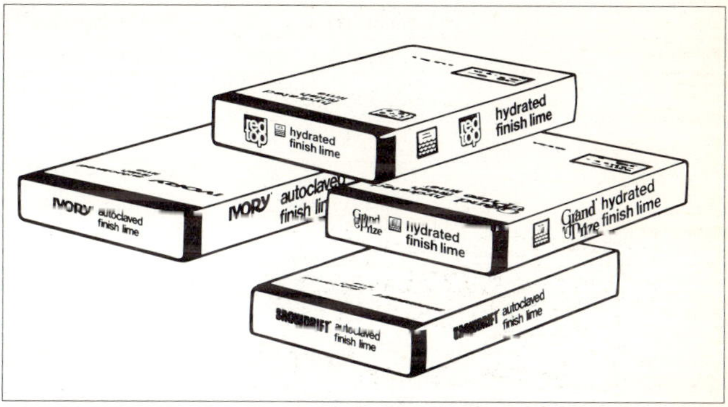

Finish Limes

The purpose of finish lime is to provide bulk, plasticity and ease of spread for the finish coat. There are two types of finish lime: (1) double hydrate (Type S), (2) normal or single hydrate (Type N). Each requires different preparation in order to produce a good finish-lime putty.

IVORY and SNOWDRIFT Finish Limes—Autoclaved (double-hydrate)—immediately develop high plasticity when mixed with water and do not require overnight soaking. Virtually eliminate the possibility of future expansion within the finish coat because of unhydrated magnesium oxides. These limes are easy to apply and have excellent spreading qualities. Comply with ASTM C206, Type S. Available in 50-lb. bags.

RED TOP and GRAND PRIZE Finish Limes—Single-hydrate—economical, easy working, uniform, white and plastic. Require soaking at least 16 hrs. to develop proper plasticity and the degree of hydration for use. They comply with ASTM C6, Type N. Available in 50-lb. bags.

Prepared Finishes

IMPERIAL Finish Plaster—Provides the ultimate in surface hardness and abrasion resistance. Available for hand application—provides a smooth-trowel or float or spray-texture finish ready for decoration. Used as finish coat in STRUCTOCORE Security Wall Systems. Complies with ASTM C584. Available in 80-lb. bags.

DIAMOND Interior Finish Plaster—Offers a strong, hard white surface for construction where the extra hardness of IMPERIAL Finish Plaster is not required. Extremely adaptable to textured finishes. Complies with ASTM C587. Available in 50-lb. bags.

RED TOP Finish—Mill-mixed gauged interior finish requiring addition of water only. Has stabilized set, excellent troweling characteristics. Two formulations available: *Regular Set*, for use over conventional sanded gypsum basecoat, and *Quick Set*, for use over IMPERIAL and DIAMOND Basecoats. Not for use over lightweight aggregate gypsum basecoat. Available in 50-lb. bags.

Finish Plasters Coverage(1)

Product	Ratio of mix (dry wt.)			Approx. coverage(2)	
	lime	gauging	sand(3)	yd^2/ton	m^2/t(4)
IMPERIAL Finish Plaster	—	—	—	360	330
DIAMOND Interior Finish Plaster	—	—	—	550	510
RED TOP Finish	—	—	—	390	360
CHAMPION, STAR and RED TOP Gauging	2 2	1 1	— 8	390 280	360 260
RED TOP Keenes Cement	2 1 1	1 2 2	8 8 —	270 270 370	250 250 340
STRUCTO-GAUGE Gauging	1 2	1 1	—	380 430	350 400

(1) Over conventional basecoat plasters; over veneer basecoats, coverage is increased. (2) 1/16″ (1.6mm) thickness. (3) Natural, uniformly graded, clean slice sand. (4) Metric ton.

Ornamental Plasters

USG Moulding Plaster—Used for specialized work such as ornamental trim or running cornices. The plaster grain is very fine, ideal for sharp detail when used neat for cast work. Controlled set provides uniform workability. For running cornice work, add a small amount (max. 50%) of lime putty to add plasticity and to act as a lubricant for the template. Provides approx. 1.5 cu. ft. per 100 lbs. Complies with ASTM C28. Available in 50- and 100-lb. bags.

HYDROCAL White Gypsum Cement—Has exceptional strength and is recommended for ornamental work and for castings made using latex, urethane, and other mould materials where high green strength minimizes breakage. In 50- and 100-lb. bags.

HYDROCAL FGR-95 Gypsum Cement—Unique high-strength product used with glass fibers and/or glass fiber mat for fabricating lightweight fire resistant decorated shapes, architectural elements, column covers, cornices, and trims. Adapts to most patterns, accepts most coatings. Safe and nontoxic; zero flame spread, zero smoke contribution material. In 50- and 100-lb. bags.

For additional information, contact Dept. 983-2, Industrial Gypsum Division, United States Gypsum Company, Chicago, Illinois.

Ornamental Plasters Approx. Yield

Product	Bag size lb	Approx. volume dry kg	ft³/100 lb	m³/t(1)
USG Moulding Plaster	50 & 100	22.7 & 45.4	1.5	0.94
HYDROCAL White Cement	50 & 100	22.7 & 45.4	1.3	0.81
HYDROCAL FGR-95 Gypsum Cement	100	45.4	1.0	0.62

(1) Metric ton

Special Additives

RED TOP Retarder and Red Top Accelerator—Available for use with plaster when required by job or climate conditions. When used in excess, setting and drying problems can arise. Avoid use of too much retarder which can weaken the plaster finish. Available in 1½-lb. (Retarder) and 2-lb. (Accelerator) packages.

Prepared Exterior Finish

Oriental Exterior Finish Stucco—A white, water-resistant finish for exterior portland cement-lime basecoats. Mill-prepared; requires water only. Easily hand or spray-applied as float, splatter-dash and other texture finishes; not designed for smooth-trowel finish. One ton covers 150-200 sq. yd. in ⅛" thickness. Available in 15 additional colors (Southwest only). In 100-lb. bags.

Masons and Stucco Lime

BONDCRETE Air-Entraining Masons and Stucco Lime—A fine-grind, white, high-purity dolomitic lime, pressure-hydrated for immediate use in portland cement-lime plaster for interior and exterior use. Recommended for use in scratch, brown and finish coats. It produces a stucco that is free-flowing for spray application or easily spread with light trowel pressure. Excellent water retention allows the cement-lime plaster to resist suction and allow sufficient time for finishing. Not recommended for use with gypsum gauging plaster in gauged lime-putty finish. BONDCRETE Lime meets ASTM C207, Type SA. Available in durable, 3-ply, weather-resistant 50-lb. bags.

MORTASEAL Autoclaved Masons Lime—A specially hydrated, high-purity, immediate-use dolomitic lime for use where non-air-entrained products are desired. Lacking air-entraining additive, it gives mortar greater compressive strength; exceeds minimum values of all ASTM C270 mortar types in both 7 and 28-day tests. High water retention reduces need for retempering mortar during use. Reduces water permeance. No soaking required. Meets ASTM C207, Type S for mason's lime. Available in durable, 3-ply, weather-resistant 50-lb. bags.

Conventional Lath & Plaster Construction

CHAPTER 7 Product Application

General Planning Procedures

Two ingredients are required for a quality plaster job—quality products and skilled workers employing correct lathing and plastering procedures.

U.S. Gypsum Company plaster bases, plasters and plastering accessories are top quality, job-proven products designed to work together. But without proper planning and correct installation by the contractor, these products cannot be expected to produce the desired results.

This Chapter deals with the basic recommendations and installation procedures to follow in completing the best possible job. It describes wood and steel framing, applications of conventional plaster bases in fire and sound-rated assemblies, and includes frame-spacing and fastener-selector charts.

Good lathing and plastering practices can give the contractor greater profit through fewer callbacks, less waste along with lower job costs, and high quality results that produce quicker sales and a favorable business reputation.

Planning the Job

Advance planning by the plastering contractor can mean savings in time and materials cost and a better-appearing job.

Two areas of planning deserve special attention. In high-rise work it is essential to determine availability and charges for use of hoisting equipment on the job well in advance of the time when it will be needed. Failure to do so can result in costly delays while the hoist is tied up by other trades.

In all types of jobs it is wise to plan for clean-up as the work proceeds, not when the job is finished. Contractors who have adopted the practice affirm that it reduces job costs. They have discovered that it is easier, faster and cheaper to remove drop cloths of roofing paper than to scrape up set plaster. And stuffing electrical boxes with paper before plastering begins is far less costly than digging out set plaster when they are accidentally filled. When machines are shut down, they should be hosed off and thoroughly cleaned—made ready for a fresh start. The benefits from these good working practices add up to faster completions, less down time and equipment maintenance, and more profit.

Estimating Materials

Accurate take-off and estimating quantities of materials are an essential part of job planning. Underestimating causes expensive job delays while quantities are refigured and orders placed. Overestimating invariably results in damage or loss of at least part of the surplus materials.

All lime-base finish plasters are subject to staining from local environmental and/or atmospheric conditions. Painting should be included in the estimate but may be omitted for a cost savings if the plaster is satisfactory.

The tables in Chapter 6 contain data needed for accurate estimating: packaging, coverage of various cementitious materials, and number of accessories needed per 100 sq. yd. of finished surface. Similar data on UNIMAST Steel Components and screws can be found in Chapter 1.

General Job Conditions

Handling and Storage

All successful plaster jobs require adequate equipment: power mixers, mortar boards, scaffolding and tools. Ample scaffolding should be provided to permit continuous application of both basecoat and finish plasters for a complete section of wall or ceiling. Obtain clean water for washing all mixing tools.

Lath and plaster products should be ordered for delivery to the job just prior to application. Materials stored on the job for longer periods are exposed to damage and abuse.

Rather than ship all plaster to the job at one time, fresh plaster should be delivered as needed. Plaster stored for long periods is subject to variable moisture conditions and aging that can produce variations in setting time and performance problems.

Store plastering products inside, in a dry location and away from heavy-traffic areas. Stack plaster bags on planks or platforms away from damp floors and walls. Store gypsum plaster bases flat on a clean dry floor; vertical storage may damage edges or deform board. Protect metal corner beads, casing beads and trim from being bent or damaged. All materials used on the job should remain in their wrappings or containers until used.

Warehouse stocks of plaster products should be rotated to ensure a supply of fresh materials and to prevent damage to plaster through aging and contact with moisture.

Environmental Conditions

1. When outdoor temperatures are less than 55°F (13°C), the temperature of the building must be maintained above 55° both day and night for an adequate period prior to the erection of gypsum plaster base, the application of plaster, while the plastering is being done, and until the plaster is dry. The heat should be well distributed in all areas, with deflection or protective screens used to prevent concentrated or irregular heat on plaster areas near the source.

2. Ventilation must be provided to properly dry the plaster during and subsequent to its application. In glazed buildings, this should be accomplished by keeping windows open sufficiently to provide air circulation; in areas lacking normal ventilation, moisture-laden air must be mechanically removed.

3. To develop proper performance characteristics, the drying rate of plastering materials must be strictly controlled during and after application. Plaster should not be allowed to dry too slowly or too fast. If possible, maintain building temperature-humidity combination in the "normal drying" area of the graph on next page. Excessive ventilation or air movement should be avoided to allow plaster to properly set.

Plaster Drying Conditions

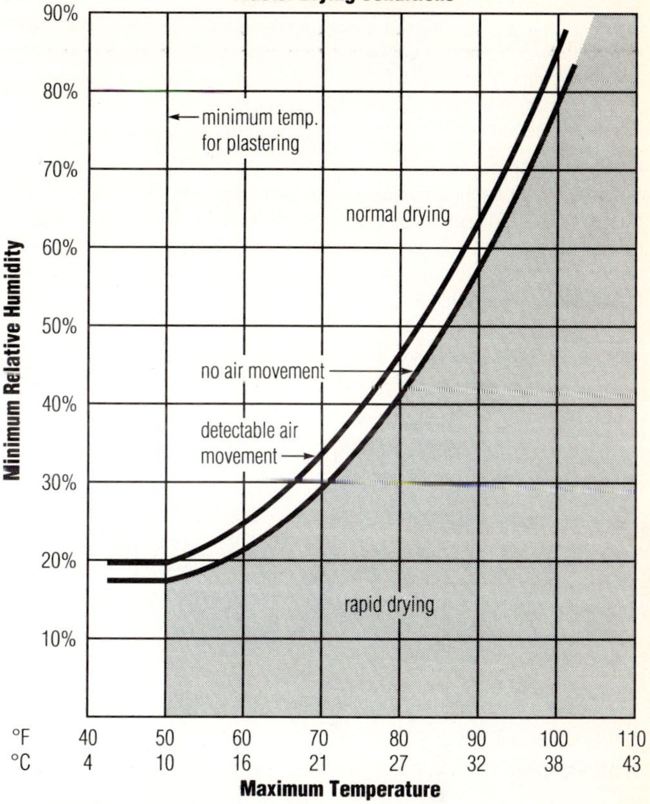

Framing Installation

Requirements for framing with wood and UNIMAST Steel Studs are the same for veneer plaster and drywall construction covered in Chapter 2 of this Handbook. Maximum frame spacing for plaster base is as follows:

Frame Spacing—Gypsum Base (Ceilings & Sidewalls)

	Base thickness		Max. frame spacing o.c.	
Type framing	**in**	**mm**	**in**	**mm**
Wood	⅜	9.5	16	406
	½	12.7	24	610
UNIMAST Steel Stud	⅜	9.5	16	406
	⅜	9.5	24(1)	610
	½	12.7	24	610
¾ (19.1 mm) Channel	⅜	9.5	16	406
UNIMAST Metal Furring	⅜	9.5	16	406
	½	12.7	24	610

(1) Three-coat plastering.

Reinforcing

Openings in a gypsum lath-and-plaster system, such as door frames, borrowed lights, etc., cause a concentration of stresses in the plaster, typically at intersection of head and jamb. The use of additional reinforcement (channels, runners, Striplath, self-furring diamond mesh lath) is recommended at the weakened area to distribute concentrated stresses.

Wood or metal inserts used as reinforcing or for attachment of cabinets and shelving on nonresilient surfaces should always be applied behind the plaster base to prevent unnecessary damage to the plaster surface. Heavy fixtures such as water closets and lavatories should be supported by separate carriers and not by the lath and plaster surface. (See "Fixture Attachment" on page 326.)

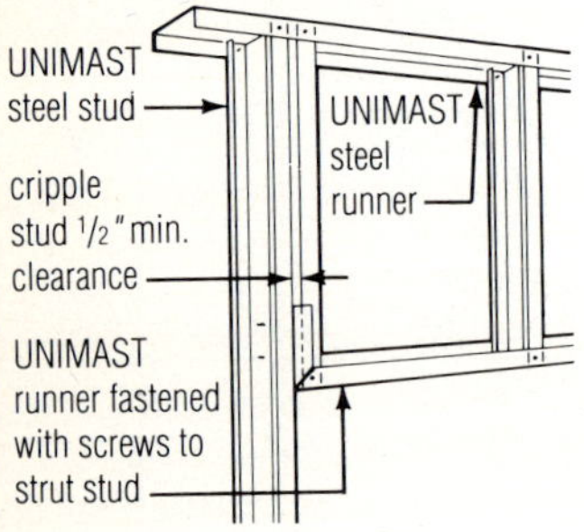

Reinforcing of frame

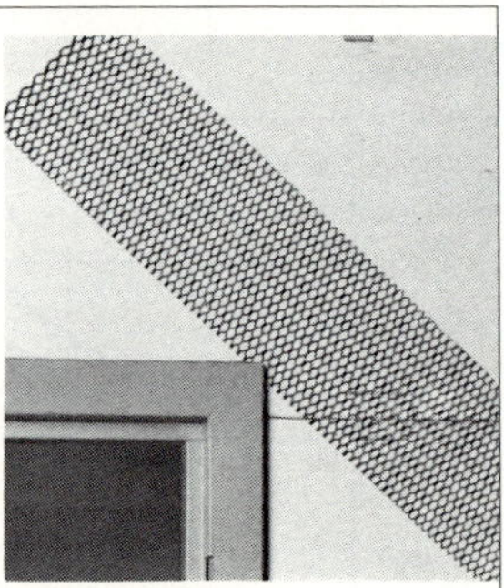

Reinforcing at door

Wall Furring

Exterior wall furring provides a way of spacing the plaster base and plaster away from masonry walls to produce an air space, a chase for services and space for insulation. By furring, uneven walls can be changed to true, even surfaces. Plaster base can be quickly attached, and the uniform plaster base saves plastering material and labor.

Exterior masonry walls should be furred out and a vapor retarder provided. Several systems are available; each provides structural and cost advantages for special furring conditions.

A properly designed wall furring system should provide:

1. Condensation control.
2. Protection from moisture seepage.
3. Insulation and vapor retarder.
4. Some isolation from structural movement. Exterior walls are subject constantly to changing dimensions due to temperature changes and wind loads.

Wood Strip-ROCKLATH Base Furring

For masonry wall furring, ROCKLATH Plaster Base and gypsum plaster over wood furring strips is an economical assembly. The wood furring is usually nom. 1x2 or 2x2 strips spaced 16″ o.c. for ⅜″ lath, 24″ o.c. max. for ½″ lath. Apply furring vertically and securely attach to the masonry. If necessary, use small wooden wedges to shim strips to a level surface.

Installation—Apply 16″ x 48″ ROCKLATH Plaster Base at right angles to furring strips with end joints occurring between strips. For ⅜″ ROCKLATH use 1⅛″ nails, for ½″ thickness use 1¼″ nails. When ROCKLATH Base has been installed, reinforce inside corners with UNIMAST Cornerite.

Wood Furring—direct attachment

Steel Stud-ROCKLATH Base Furring

This free-standing furring assembly consists of ROCKLATH Plaster Base screw-attached to UNIMAST Steel Studs and finished with gypsum plaster. The assembly offers a maximum of free space for encasing pipes, ducts or conduits.

With a 6-mil polyethylene film installed under the ROCKLATH Base, the assembly provides an effective vapor retarder.

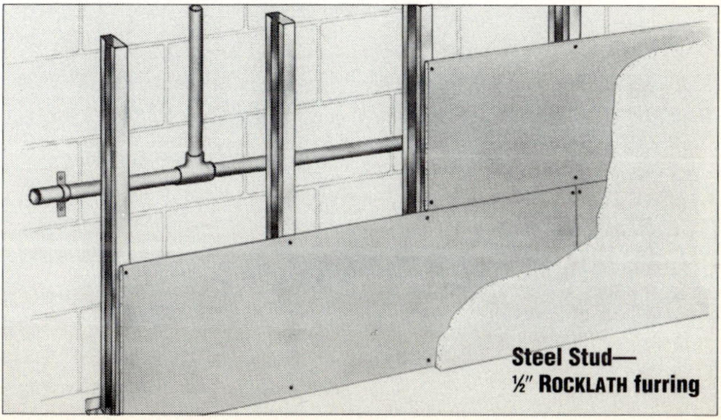

Steel Stud—½″ ROCKLATH furring

Installation—Align floor and ceiling runners parallel to wall and positioned to provide required chase space. Attach to concrete slabs with concrete stub nails or power-driven anchors 24″ o.c., to suspended ceilings with toggle bolts or hollow metal fasteners 16″ o.c., or to wood framing with 1″ Type S screws 16″ o.c.

Position steel studs vertically in runners, 16″ o.c. for ⅜″ lath, 24″ o.c. for ½″ lath, and with all flanges in same direction. Anchor all studs that are adjacent to doors and window frames, partition intersections and corners to floor and ceiling runner flanges with screws. Attach ⅜″ ROCKLATH Base to studs with two 1″ Type S screws at each stud and B-1 Clips at floated lath ends. Attach ½″ ROCKLATH Base with lath ends supported by framing using four screws per stud. Apply ½″ sanded basecoat plaster, lime putty finish.

UNIMAST Steel Stud-ROCKLATH Furring—Max. Height

Steel stud			Spacing cc		Max. height(2)	
Style	Width in	mm	16″ (406mm)	24″(1) (610mm)	ft-in	mm
158ST25	1⅝	41	x		8′3″	2515
				x	7′3″	2210
212ST25	2½	64	x		10′9″	3275
				x	9″6″	2895
358ST25	3⅝	92	x		13′9″	4190
				x	12′0″	3660
400ST25	4	102	x		14′9″	4495
				x	13′0″	3960
600ST25	6	152	x		18′9″	5720
				x	15′0″	4570

(1) 24″ spacing limited to ⅜″ ROCKLATH Base and 3-coat sanded plaster to ½″ grounds or ½″ ROCKLATH Base and sanded plaster. (2) Based on L/360 deflection or bending stress and no intermediate support. Studs over 12′ high require midheight anchor to exterior wall.

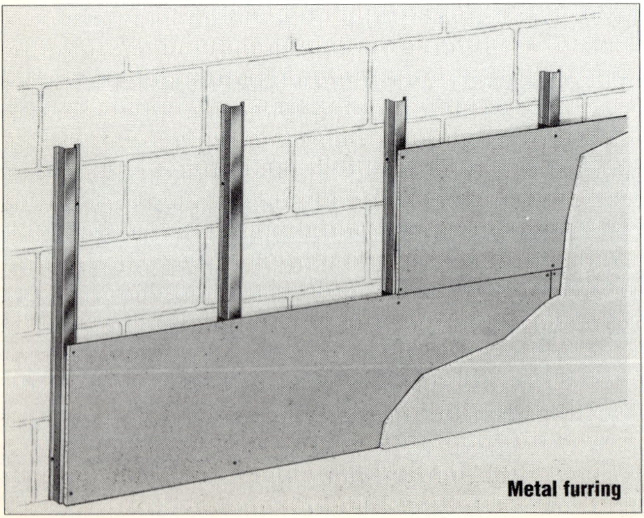

Metal furring

Metal Channel-ROCKLATH Base Furring

For direct attachment with metal furring channels, ROCKLATH Plaster Base is screwed to UNIMAST Furring Channels that are attached directly to an exterior masonry wall. When a 6-mil polyethylene film is included, the system provides an excellent vapor retarder.

Installation—Apply channels vertically to masonry not more than 16″ o.c. Fasten each channel with hammered or power-activated stud fasteners. If there is a possibility of water penetration, install an asphalt felt protection strip between the furring channel and wall surfaces.

Frame Spacing and Attachment

For Furred Ceilings—Fasten ¾″ UNIMAST Cold-rolled Channel or ⅜″ pencil rods directly to bottoms of framing members. On concrete joists, 8-ga. galvanized wire can be put in place before the concrete is poured. Space fur-

Frame and Fastener Spacing—ROCKLATH Plaster Base

Type framing	Base thickness in	Base thickness mm	Fastener[1]	Max. frame spacing in	Max. frame spacing mm	Max. fastener spacing in	Max. fastener spacing mm
Wood	⅜	9.5	Nails 13 ga., 1⅛″ long, 19/64″ flat head, blued	16	406	5	127
			Staples—16-ga. galv. flattened wire, flat crown 7/16″ wide, 1″ divergent legs				
	½	12.7	Nails 13-ga., 1¼″ long, 19/64″ flat head, blued	24	610	4	102
			Staples 16-ga. galv. flattened wire, flat crown 7/16″ wide, 1″ divergent legs				
UNIMAST Steel Stud (ST)	⅜	9.5	1″ Type S Screws	16	406	12	305
	½	12.7		24	610	6	152
UNIMAST Metal Furring	⅜	9.5	1″ Type S Screws	16	406	12	305
	½	12.7		24	610	6	152

(1) Metric fastener dimensions: 19/64″ = 7.5 mm; ⅜″ = 9.5 mm; 7/16″ = 11.1 mm; 19 = 25.4 mm; 1⅛″ = 28.6 mm; 1¼″ = 31.8 mm.

Max. Frame Spacing—Metal Lath[1]

Product	Weight lbs/yd²	Weight kg/m²	Spacing in	Spacing mm
Diamond Mesh[2]	2.5	1.4	12[3]	305[3]
	3.4	1.8	16	406
⅜″ (9.5mm) Riblath	3.4	1.8	19	610
4-Mesh Z-Riblath	2.75	1.5	16[4]	406[4]
	3.4	1.8	19[5]	483

(1) For spacing on fire-rated constructions, see test reports. (2) 2.5-lb. lath should not be used for ceilings. (3) 16″ o.c. permitted with wood framing and 2″ solid partition. (4) Spacing of metal ceiling grills 12″ o.c. (5) 24″ spacing with solid partition.

Support Area—Hangers

Hanger size and type	Max. ceiling area per hanger ft²	m²	Allowable tensile load lbs[3]
9-ga. galvanized wire	12.5	1.2	340
8-ga. galvanized wire	16	1.5	408
3/16″ (4.8 mm) mild steel rod[1][2]	20	1.9	546
1/4″ (6.4 mm) mild steel rod[1][2]	22.5	2.1	972
3/16″ x 1″ (4.8mm x 25.4mm) mild steel flat[1][2]	25	2.3	3712

(1) Where severe moisture conditions may occur, rods galvanized or painted with rust-inhibitive paint or galvanized straps are recommended. (2) Not manufactured by United States Gypsum Company. (3) Based on minimum yield 33,000 psi.

Max. Spacing—Main Runner—Carrying Channels

Main runner c. r. channel size		Max c. to c. spacing of main runners		Max. spacing of hangers along runners	
in	mm	ft	mm	ft	mm
3/4	19.1	3	914	2	610
3/4	19.1	2 1/4	686	3[1]	914
1 1/2	38.1	4	1219	3	914
1 1/2	38.1	3 1/2	1067	3 1/2	1067
1 1/2	38.1	3	914	4	1219
2	50.8	4	1219	5	1524
2	50.8	2 1/2	762	6	1829
2	50.8	2	610	7	2134

(1) For concrete joist construction only—where 8-ga. wire may be inserted in joist before concrete is poured.

Max. Spacing—Cross-Furring Members

Cross-furring size	Max. c. to c. spacing of cross-furring in	mm	Main runner or support spacing ft	mm
3/4″ (19.1 mm) C. R. Channel	24	610	3	914
3/4″ (19.1 mm) C. R. Channel	19	483	3 1/2	1067
3/4″ (19.1 mm) C. R. Channel	16	406	4	1219
1″ (25.4 mm) H. R. Channel	24	610	4	1219
1″ (25.4 mm) H. R. Channel	19	483	4 1/2	1372
1″ (25.4 mm) H. R. Channel	12	305	5	1524
3/8″ (9.5 mm) Pencil Rod[1]	19	483	2	610
3/8″ (9.5 mm) Pencil Rod[1]	12	305	2 1/2	762

(1) Primary usage is on furred ceiling members.

ring members as shown in the above cross-furring member spacing table. For joists spaced about 25″ o.c., attachment of 3/4″ channels may be on alternate joists; if greater than 25″ o.c. but not more than 48″ o.c., place attachment at every joist.

On steel joists or beams, place 3/4″ cold-rolled channels at right angles to joists; attach with 3 strands 18-ga. galvanized wire.

For Suspended Ceilings—Space 8-ga. wire hangers not over 4′ o.c. in direction of 1 1/2″ carrying channels and not over 3′ o.c. at right angles to direction of carrying channels. If hanger wires are 3′ o.c. in direction of 1½″ channels, then channels at right angles may be 4′ o.c. Place hangers within 6″ of ends of carrying channel runs and of boundary walls, girders or similar interruptions of ceiling continuity. Position and level carrying channel and saddle tie securely with hanger wire.

Position ¾″ UNIMAST Cold-rolled Channel (cross-furring) across carrying channels, spacing them 12″ to 24″ depending on type of metal lath to be used, and saddle tie carrying channels with three strands of 18-ga. tie wire.

Apply UNIMAST 3.4-lb. Diamond Mesh Lath, ⅛″ Z-Riblath or ⅜″ Riblath as specified with long dimension of sheets across the supports. Details on lathing procedures and control joints follow later in this Chapter.

Plaster Base Application

Plaster bases may be classified as gypsum base, metal lath base or masonry base. These materials provide a level surface for plastering and add reinforcement to the plaster. As such, they must be rigid enough to accept plaster and produce a secure bond between plaster and base—both necessary to develop strength and resistance to abuse and cracking.

To ensure adequate rigidity of plaster constructions, recommendations for the spacing of supports and fasteners must be strictly followed.

Apply plaster bases to ceilings first and then to partitions, starting at the top and working down to the floor line.

ROCKLATH Plaster Base—An ideal high-suction rigid base for gypsum plasters, should be applied face out with long dimension across supports and with end joints staggered between courses. Cut lath accurately so it slips easily into place without forcing and fits neatly around electrical outlets, openings, etc. Install any lengthwise raw cut edges at bottom strip or wall-ceiling angle. Apply Cornerite to all interior angles and staple to the lath only.

UNIMAST Metal Lath—Should be applied with long dimension across supports and with end joints staggered between courses. Apply Riblath with the rib against supports. Lap ends of metal lath 1″ and sides at least ½″. Lap Riblath by nesting outside ribs. If end laps occur between supports, they should be laced or tied with 18-ga. tie wire. Secure lath to all supports at intervals not exceeding 6″. At all interior angles, metal lath should be formed into corners and carried out onto abutting surface.

Clay Tile and Brick—Frequently used for plaster bases. Care should be taken to make sure that surfaces are sufficiently porous to provide suction for the plaster and are scored for added mechanical bonding. Smooth-surfaced clay tile that is glazed or semi-glazed does not offer sufficient bond for plaster.

Concrete Block—A satisfactory base for plaster. The surface should be porous, for proper suction, or face-scored for adequate mechanical bond. Units must be properly cured to minimize dimensional changes during and subsequent to plastering.

Monolithic Concrete—Ceilings, walls, beams and columns should have a complete and uniform application of a high-quality bonding agent before plastering. This surface treatment produces an adhesive bond suitable for direct application of gypsum plasters.

Plastering Direct to Exterior Masonry Walls—Not recommended. Exterior walls are subject to water seepage and moisture condensation that may wet the plaster and damage interior decoration.

Bituminous Waterproofing Compounds—Do not provide a good plaster base. Gypsum plasters should not be applied to surfaces treated with these compounds.

Rigid Foam Insulations—Such insulations have not proven to be satisfactory bases for direct application of gypsum plaster because of rigid foam insulation's low suction characteristics and low structural strength which may result in cracking of the plaster.

U.S. Gypsum Company does not recommend direct application of plaster to rigid foam insulation. However, some rigid foam insulation manufacturers have specific directions for application when direct plastering is to be used, as well as detailed specifications for plaster mixes and methods of application to be employed.

U.S. Gypsum Company has designed various furring systems (covered earlier in this Chapter) that avoid the need to apply plaster to these unsuitable surfaces and do provide high-quality plaster finishes over the inside of exterior walls.

Fastener Application

Correct fastener selection and adherence to fastener spacing are extremely important to good plastering performance and absolutely essential in meeting the requirements of specific fire-rated constructions.

Screw attachment of ⅜" ROCKLATH Base to steel studs.

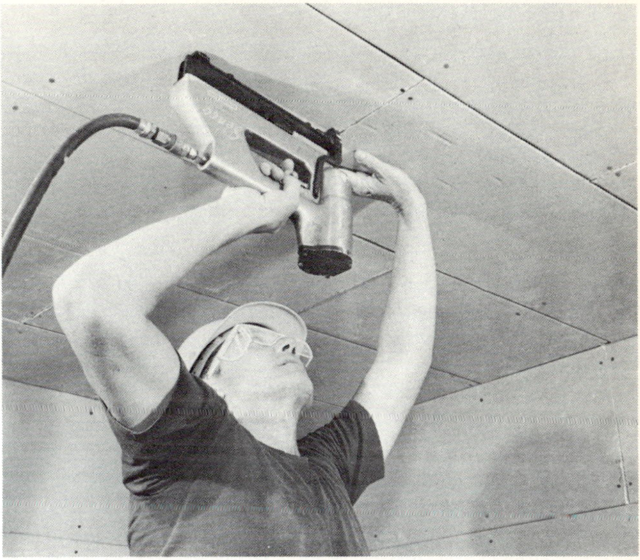

Power nailer used to attach ROCKLATH Plaster Base to ceiling.

Gypsum Plaster Bases—Attached to framing with screws, nails, staples or special clips. Nails, screws and staples should be driven so that the fastener head or crown bears tightly against the base but does not cut the face paper. To prevent core fracturing, they should be driven at least ⅜″ away from ends and edges. Staples should be of flattened wire driven so the crown is parallel to wood framing. Screws should be used to attach gypsum plaster bases to UNIMAST Steel Studs, Furring Channels or RC-1 Resilient Channels.

For screw attachment of single-layer ⅜″ ROCKLATH Base to UNIMAST Steel Studs or Furring Channel, 1″ Type S bugle head screws are used. Attach metal door frames to steel studs, using ⅜″ Type S-12 pan head screws; for steel studs to metal runners up to 20-ga., ⅜″ Type S pan head screws. For steel to steel attachment from 20 to 14-ga., use ⅝″ Type S-12 low-profile head. For attaching base to 20 to 14-ga. studs, use Type S-12 screws. See Chapter 1 for complete data on screws and page 430 for information on electric screw guns.

Nail Application—Begin from center of base and proceed toward outer ends or edges. When nailing, apply pressure adjacent to nail being driven to ensure base is secured tightly on framing member. Position nails on adjacent ends or edges opposite each other and at least ⅜″ from ends and edges. Drive nails with shank perpendicular to plaster base. The nail heads should be driven flush with paper surface but not break paper.

UNIMAST Metal Lath—Attach to cold-rolled channel framing with tie wire (min. 18 ga.) and to wood framing with fasteners engaging two strands or a rib and providing at least ¾″ penetration.

Gypsum Lath-UNIMAST Steel Studs

Screw Attachment—For 16″ x 48″ ROCKLATH Base, fasten ⅜″ thickness to studs spaced 16″ o.c. with two 1″ Type S screws per stud spaced 2″ from edge of lath and support lath ends with BRIDJOINT B-1 Clips at top, center and bottom of each butt joint. For ½″ x 16″ x 48″ lath, fasten to studs 24″ o.c. with 1″ Type S screws, three screws per stud. Type S-12 screws are required for heavier gauges. Fasten ½″ x 24″ x 96″ long-length lath the same as regular ½″ lath. Drive screws with an electric screwgun.

Gypsum Lath-Wood Framing (direct)

Nail ROCKLATH Plaster Base with face out and long dimension across framing members. Stagger end joints in successive courses with ends of lath falling between framing members. Butt all joints together and cut lath to fit neatly around electrical outlets and other openings. For ⅜″ ROCKLATH Plaster Base, when end joints occur between framing members, secure ends to abutting and adjacent panels with BRIDJOINT B-1 Clips.

For ⅜″ ROCKLATH Plaster Base with 16″ o.c. stud spacing, use four fasteners, 5″ o.c. per 16″ width of lath. For ½″ ROCKLATH Plaster Base and max. support spacing of 24″ o.c., use five fasteners, 4″ o.c. per 16″ width of lath. Place fasteners at least ⅜″ from edges and ends of lath. Make all interior plaster angles the floating type and space first fasteners at least 10″ from corner. Reinforce angle with Cornerite stapled to lath surface.

Metal Lath-Wood Framing (direct)

Apply metal lath with long dimension of sheet across supports. Lap ends of lath at least 1″ and if laps occur between supports, lace or tie with 18-ga. tie wire. Attach with fasteners 6″ o.c. so fastener engages two strands or a rib and provides at least ¾″ penetration.

On walls, place metal lath so that the lower sheets overlap upper sheets and, where possible, stagger ends of lath in adjacent courses.

At all interior angles, form lath into corners and carry out onto abutting surface. Secure lath to joists with 1″ galvanized nails, to studs with nails or staples providing min. ¾″ penetration.

Control Joint Application

Lath and plaster surfaces will not resist stresses imposed by structural movement. Additionally, plaster assemblies are subject to dimensional changes caused by fluctuations in temperature and humidity. (See thermal and hygrometric coefficients of expansion in Appendix.) Such surfaces should be isolated from the following structural elements by SHEETROCK Zinc Control Joints, casing beads or other means where:

a. A partition or ceiling abuts any structural element other than the floor, a dissimilar wall or partition assembly, or other vertical penetration.

b. The construction changes within the plane of a partition, or ceiling and wings of "L-," "U-" and "T-" shaped ceilings are joined.

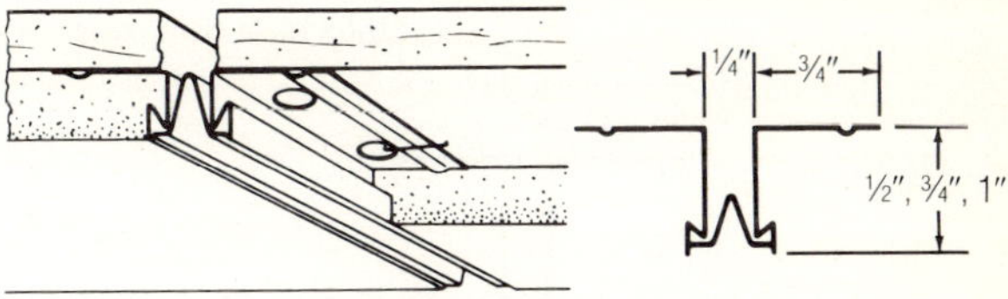

SHEETROCK control joint

In long partition runs, control joints should be provided at max. 30′ o.c. Door frames extending from floor to ceiling may serve as control joints. For less-than-ceiling-height doors, control joints extending from center or both corners of frame to ceiling is an effective application. If control joints are not used, additional reinforcement is required at corners to distribute concentrated stresses. (Door frame details appear later in this Chapter.) In exterior wall furring systems, control joints must be provided at the same locations where control joints in the exterior walls are located and at max. 30′ o.c.

Control joints will not accommodate transverse shear displacement on opposing sides of joint. A joint detail comprising casing beads each side of joint opening is typically used to accommodate expansion, contraction and shear. Such joints require special detailing by designer to control sound and fire ratings where applicable as well as dust and air movement. In exterior walls, particular attention is required to resist wind, driving rain, etc., by adequate flashing, backer rod, sealants and gaskets as required.

Large interior ceiling areas with perimeter relief should have control joints spaced at max. 50′ o.c. in either direction; without perimeter relief, 30′ o.c. maximum in either direction. The continuity of both lath and plaster must be broken at the control joints. Control joints should be positioned to intersect light fixtures, heating vents, or diffusers, etc., which already break ceiling continuity and are points of stress concentration.

Max. Spacing—SHEETROCK Zinc Control Joints for Interior Assemblies

		Max. single dimension		Max. single area	
System	Location	ft	mm	ft²	m²
Metal Lath & Plaster	Partition	30	9144	—	—
	Ceiling	50(1)	15240	2500	230
		30(2)	9144	900	83.6
Gypsum Lath & Plaster	Partition	30	9144	—	—
	Ceiling	50(1)	15240	2500	230
		30(2)	9144	900	83.6

(1) With perimeter relief. (2) Without perimeter relief.

Installation—Provide a break in the lath at location of control joint. At this location install double framing members, one on each side of the break and ½″ to ¾″ apart. Place control joints over all control or relief joints within

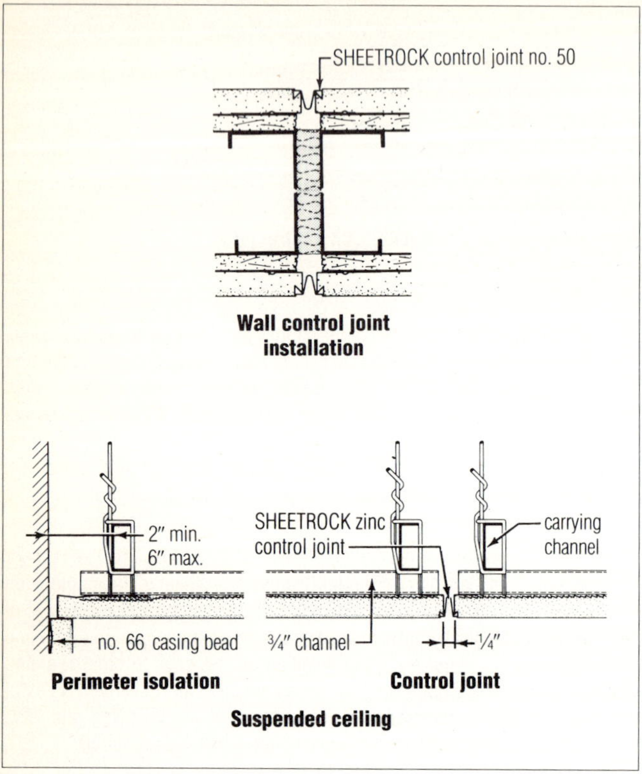

structural frame of building. Staple or wire tie perforated flanges of control joint to lath. Plaster flush to grounds. Remove factory-applied protective tape after completion of finished surface.

SHEETROCK Zinc Control Joints must be properly insulated or otherwise protected when used in fire-rated assemblies.

Basecoat Plaster Application

For the beauty and durability of which plaster is capable, certain requirements should be followed regarding the number of coats applied. Three-coat work is necessary on all metal lath and on edge-supported gypsum lath used in ceilings; three-coat work is desirable on all gypsum lath but two-coat work is acceptable when gypsum lath is properly supported and on masonry plaster bases (rough concrete block, clay tile, porous brick).

In preparing for plastering, consideration should be given to the selection of materials not only for compatibility but for the quality of the structure to be plastered. It is wise to upgrade plastering specifications when possible.

The architect's specifications and the plaster base used will determine the plastering method, either two-coat or three-coat.

Two-Coat Plastering with Conventional Plasters— Generally accepted for plaster application over gypsum lath and masonry. The base (first) coat should be applied with sufficient material and pressure to form a good bond to the base and to cover well; then be doubled back to bring plaster out to grounds, straightened to true surface with rod and darby without use of additional water, and left rough to receive finish (second) coat.

Three-Coat Plastering—Required over metal lath and edge-supported gypsum lath used in ceilings. Preferred for other bases because it develops a harder, stronger basecoat. The scratch (first) coat should be applied with sufficient material and pressure to form good full keys on metal lath, and a good bond on other bases, and then cross-raked. The brown (second) coat should be applied after scratch (first) coat has set firm and hard, brought out to grounds and straightened to true surface with rod and darby without use of additional water, and left rough to receive finish (third) coat.

To obtain the full hardness, high strength and superior performance available in gypsum basecoat plasters, water, aggregates and setting time must be carefully controlled. In addition, proper mixing and drying of the plaster are required to obtain these superior functional characteristics.

Grounds

The thickness of conventional basecoat plaster is one of the most important elements of a good plaster job. To ensure proper thickness, grounds should be properly set and followed.

Grounds may be defined as wooden strips, corner beads (plumbed and aligned), or metal casing beads applied at perimeter of all openings and other locations.

In addition to these, and especially on walls with no openings and on ceilings, plaster screeds should be installed to ensure plumb and level surfaces. Plaster screeds are continuous strips of plaster, approximately 4″ wide, applied either vertically or horizontally and plumb with the finish wall line, allowing for 1/16″ finish coat.

Grounds

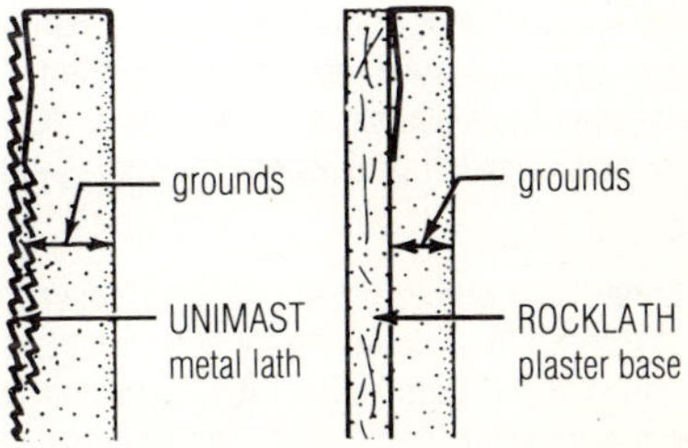

Grounds should be set to obtain the following minimum plaster thicknesses:

1. Over gypsum lath .... 1/2″ (12.7mm)
2. Over brick, clay tile or other masonry .... 5/8″ (15.9mm)
3. Over metal lath, measured from face of lath .... 5/8″ (15.9mm)

Mixing

Use of the proper type of mechanical mixer ensures that the plaster aggregate and water are evenly mixed. Keep the mixer continually clean—a very important precaution because partially set material is a powerful accelerator.

Proportioning is *weight* of gypsum to *volume* of aggregate.

A No. 2 shovel used to add sand to the plaster mix generally carries 15 lb., or approx. one-sixth cubic foot. Thus, a 100:1 mix would use 6 shovels of sand to 100 lb. of gypsum plaster, a 100:2 mix 12 to 13, and a 100:3 mix 18 to 19 shovels of sand.

Perlite is generally packaged in bags containing 3 or 4 cu. ft. for easy proportioning.

Prepare only one hour's supply of plaster at one time and do not remix if plaster has started to set. All such plaster should be discarded.

Water—All gypsum plasters require addition of water on job. Water should be clean, fresh, suitable for domestic consumption, and free from mineral and organic substances which affect plaster set. Water used earlier for rinsing or cleaning containers and tools should not be used as it accelerates plaster set.

Only enough water should be used to provide a plaster of workable consistency. Too much water in machine-applied plasters (in excess of 10% more than for hand-applied mixes) or over-aggregated plasters will cause weak, soft walls and ceilings. Excessive water reduces plaster strength and hardness.

Aggregates—Added to conventional gypsum plasters to extend coverage, reduce shrinkage, and lower cost. Aggregates recommended are: sand, which is denser, stronger and provides greater sound transmission loss than lightweight aggregates; and perlite, a lightweight aggregate that generally offers better fire resistance, insulation values and reduced weight. For sand-float finishes the aggregate should be a fine silica sand.

All aggregates used should have proper gradation of size as outlined in ASTM C35. Improperly sized aggregates will produce weak walls. Sand should be clean and free of dirt, clay and foreign matter that might affect the setting time of plaster. Perlite-aggregated plasters should not be machine-applied when vertical lift is over 30′ or hose length exceeds 150′. Maximum recommended proportions for aggregates are shown in table on next page.

Setting Time—The proper setting time for conventional basecoat plasters is generally from 2 to 4 hours after mixing, and this should be checked for close conformity on both the scratch coat and brown coat operations. Normally, plaster shipped to the job will fall in this range. If conditions exist that affect normal setting time, retarders or accelerators may be used.

The danger of "quick set" plaster is insufficient time to get the plaster from mixer to walls without retempering on mortar board, and such retempering will produce a plaster of lower than normal strength. The correction for "quick set" is to add the minimum required amount of retarder in solution with water in the mixer.

Max. Aggregate Quantity—Gypsum Plasters

			Under smooth trowel finishes				Under texture finishes			
			sand[2]		perlite[3]		sand[2]		perlite[3]	
Base	No. coats[1]	Coat	ft³/100 lb	m³/t	ft³/100 lb	m³/t	ft³/100 lb	m³/t	ft³/100 lb	m³/t
Gypsum Lath	3	Scratch	2	1.24	2	1.24	2	1.24	2	1.24
		Brown	3	1.86	2	1.24	3	1.86	3[4]	1.86[4]
	2	Basecoat[5]	2.5	1.55	2	1.24	2.5	1.55	2	1.24
Metal Lath	3	Scratch	2	1.24	—	—	2	1.24	2	1.24
		Brown	3	1.86	—	—	3	1.86	2	1.24
Unit Masonry	3	Scratch	3	1.86	3	1.86	3	1.86	3	1.86
		Brown	3	1.86	3	1.86	3	1.86	3	1.86
	2	Basecoat[5]	3	1.86	3	1.86	3	1.86	3	1.86
STRUCTO-CORE	3	Scratch	2	1.24	—	—	—	—	—	—
Steel Sheet		Brown	2	1.24	—	—	—	—	—	—

(1) Includes finish coat. (2) Approx. 6 No. 2 shovels of sand equal 1 cu. ft. (0.028 m³). (3) In a construction with metal lath as the plaster base, perlite aggregate is not recommended for use in the basecoat plaster, except under a float finish. For a smooth trowel finish over a perlite aggregated basecoat on any plaster base except metal lath, add ½ ft.³ of fine silica sand per 100 lb. of gauging plaster, or use aggregated gauging. (4) Only if applied 1″ thick, otherwise 2 ft.³. (5) Basecoat applied scratch and double-back.

If plaster does not set for 5 to 6 hr., no harm will be done to the resulting plaster surfaces, but a "slow set" of the plaster (generally one taking more than 6 hr.) should be avoided by adding accelerator at the mixer, as such plaster may be subject to a "dryout," particularly in hot, dry weather, and will have a lower than normal strength when finally set.

Only RED TOP Accelerator and RED TOP Retarder should be used. Other materials may not be compatible.

Heating and Ventilation—Plaster must not be applied to surfaces that contain frost. A min. temperature of 55°F (13°C) should be maintained for adequate period prior to, during, and after application of plaster. In cold, damp or rainy weather, properly regulated heat should be provided but precautions must be taken against rapid drying before set has occurred. This prevents "dryouts."

As soon as set occurs in conventional plasters, free circulation of air should be provided to carry off excess moisture. Heating should be continued to ensure as rapid drying as possible. In hot, dry weather, protect plaster from wind and from drying unevenly or too rapidly before set has taken place and after set to avoid excessive shrinkage. If windows or curtain walls are not in place, exterior openings in the building should be screened.

Finish Plaster Application

Finish plasters applied to basecoats provide the surface for final wall or ceiling decoration. Finish coats should be applied only to properly prepared basecoats.

Trowel finish

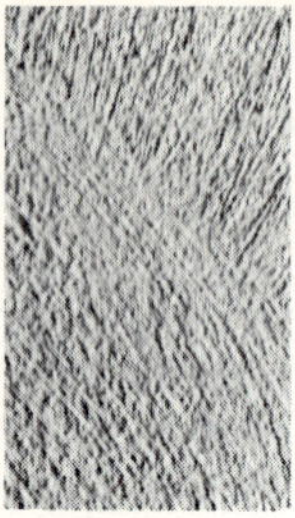

Float finish

Spray finish

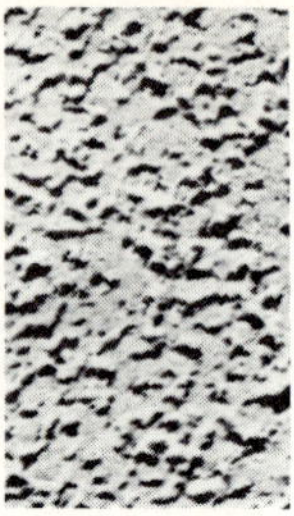

Texture finish

Trowel Finishes—Used where a smooth, easily maintained surface is desired, often as a base for paint or wall coverings. The degree of hardness, porosity, and polish is determined by the materials and application techniques used. When a smooth-trowel gauged lime-putty finish is used over a basecoat containing lightweight aggregate on any plaster base except metal lath, three options are available. Either add at least 50 lb. of fine silica sand or ½ cu. ft. of perlite fines per 100 lb. of gauging plaster or use a mill-aggregated "Quality" gauging plaster.

Application—To avoid blistering, allow basecoat to dry sufficiently or use a quick-set gauging plaster. Use 50-lb. bag of IVORY or SNOWDRIFT Lime with 5½ to 6 gal. water. Machine-mix for immediate use. For a medium-hard finish, mix 100 lb. STRUCTO-GAUGE Gauging Plaster or 200 lb. CHAMPION, STAR or RED TOP Gauging Plaster to each 200 lb. dry lime (approx. 400 lb. putty). *For extremely hard finish*, mix one part STRUCTO-GAUGE Plaster to one part lime.

Scratch in thoroughly, then immediately double-back to a thickness of not more than 1/16" and trowel to a smooth, dense surface ready for decoration.

Float or Spray Texture Finishes—Provide attractive, durable finishes where surface textures are desired. They are recommended for use over all types of gypsum basecoats and are the most desirable finishes from the standpoint of crack resistance. The surface texture is easily controlled and can be produced by spray application, or a variety of hand tools.

Application, Sand Float Finish—Machine-mix finish in proportion of 100 lb. RED TOP Keenes Cement, 50 lb. dry hydrated lime, approx. 400 lb. sand and water to produce a mixture with smooth, plastic consistency.

Scratch in thoroughly over dry basecoat, then immediately double-back to a thickness of not more than ⅛". Hand float to produce a uniform texture free of blemishes. Use water sparingly during floating.

Application, Machine Spray Texture Finish—Acceptable finishes can be achieved using either a hand held hopper gun or other machines specifically designed for spray applying plaster.

The aggregate size, number of passes over the surface, air pressure and nozzle orifice can be varied to achieve the desired texture. When spraying, it is best to spray first in one direction and then in another direction crossing the first direction at right angles.

Before beginning spray application, test the pattern and make the necessary adjustments to give the desired appearance. There are many things that affect the pattern including the following:

1. *Orifice size*—the smaller the orifice or nozzle tip, the finer the spray.

2. *Air pressure*—if no other changes are made, the higher the air pressure, the finer the spray.

3. *Liquid state*—the liquid state of the material should be like medium thick cream. The condition is obtained by taking the regular mix and adding more water until it comes to the desired consistency. It is good practice to screen the mix through a screen that will pass no particles larger than the aggregate being sprayed.

Basecoat must be free of ridges or other surface imperfections. Spraying texture finish direct to basecoat is not normally recommended. The preferred method is to hand-apply a scratch coat before spray-applying finish. Finish materials for this method include gauging plaster, either regular or high-strength; Keenes cement job mixed with lime and silica sand; or various single component prepared finishes designated for two component systems.

With finish coat material mixed for hand application, apply a well ground-in scratch coat over properly set and partially dry brown coat. After scratch coat is applied, double back with sufficient material to cover the basecoat completely. When the surface has become firm by water removal, float it to a uniform, blemish-free flat texture. After scratch and double-back set, and while material is in a wet state, spray material should be prepared using the same proportions as the finish material and mixed to proper fluidity for achieving the final texture finish. Spray texture to a uniform thickness and appearance.

Other Texture Finishes—Many pleasing and distinctive textures are possible using various techniques in finishing. Finishes may range from an extremely fine stipple to a rough, heavy or coarse texture. Variety is limited only by the imagination of the designer or the ingenuity of the applicator.

Finish Plaster Limitations

Certain precautions must be observed when applying finish coat plasters over various basecoats:

1. A smooth-trowel finish should not be used over lightweight aggregate gypsum basecoat applied over metal lath. A sand-float finish is recommended.

2. Where the gypsum basecoat over any plaster base except metal lath is STRUCTO-LITE or contains lightweight aggregate and a smooth-trowel finish is used, the finish coat should be RED TOP Gauging Plaster and lime, with addition of ½ cu. ft. of perlite fines or 50 lb. of No. 1 white silica sand per 100 lb. gauging plaster or mill-aggregated "Quality" Gauging Plaster.

3. Gypsum or lime-base finishes, including Keenes Cement, should not be used directly over a portland cement basecoat or over concrete block or other masonry surfaces.

4. In smooth trowel finishes, gauging plasters providing an extremely hard surface, such as STRUCTO-GAUGE and Keenes Cement, must not be used over STRUCTO-LITE Plaster or a basecoat with a lightweight aggregate.

5. Lime putty cannot be used without the addition of gauging plasters. When used alone as a finish plaster, lime does not set, is subject to shrinkage when drying and lacks hard finish.

Gauging Plasters

RED TOP, CHAMPION and STAR Gauging Plasters—Gauging plaster (see following pages for full description) is blended into the lime putty in the proper proportions to provide controlled set, early hardness and strength, and to prevent shrinkage cracks.

Mixing—Add gauging plaster to lime putty in proportion of 1 part dry gauging plaster by weight to 2 parts dry lime by weight or 1 part dry gauging by volume to 3 parts lime putty by volume. To mix, form a ring of lime putty on mixing board. The volume of putty used depends on wall or ceiling area to be covered. A hod of lime putty weighs approx. 100 lb., a 12-qt. bucket of lime putty about 35 lb. (50 lb. dry lime equals 100 lb. lime putty). After forming the putty ring, pour clean water into center of ring in correct proportions: 6 qt. water to 100 lb. lime putty; 2 qt. water to each 12-qt. bucket of lime putty. Next, sift *Slow or Quick Set* gauging plaster into water; 25 lb. gauging plaster to one hod of lime putty. Thoroughly wet gauging plaster and blend materials thoroughly to prevent gauging "streaks" and provide uniform density.

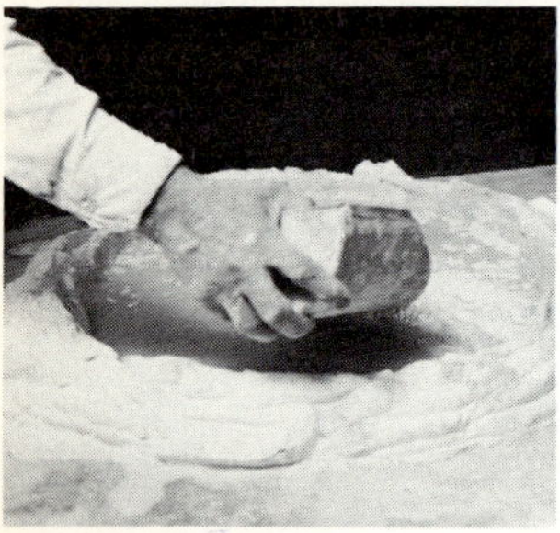

Sift gauging plaster into water.

Mix to blend thoroughly.

To protect against finish coat check or map-cracking, add ½ cu. ft. perlite fines or 50 lb. fine silica sand to every 100 lb. of gauging used. This addition is necessary when applying smooth troweled finishes over lightweight aggregate basecoats. Mill-aggregated "Quality" gauging plasters are generally available and eliminate the need for on-the-job measuring.

Application—Apply the gauged lime putty over a partially dry basecoat. Scratch in a thin coat, well ground into the basecoat, and double-back with a second coat, filling imperfections. After basecoat has absorbed most of excess water from finish, trowel to densify surface. As final set takes place, water-trowel surface to provide a dense, smooth surface.

STRUCTO-GAUGE Gauging Plaster—This high-strength gypsum finishing plaster is used with lime putty to produce an easily applied finish of extreme hardness.

Finish hardness may be altered by adjusting the proportions of lime putty and STRUCTO-GAUGE Plaster. Since the material cannot be retempered, use of regular Keenes Cement is recommended when retempering is a factor. STRUCTO-GAUGE Plaster is not recommended for use where excessive or continued moisture conditions exist. Application must be over high-strength sanded or wood-fibered gypsum basecoats.

Mixing—For a hard finish, mix proportions of 100 lb. dry hydrated lime (200 lb. of lime putty) to 100 lb. STRUCTO-GAUGE Plaster. For a medium hard finish, mix proportions of 200 lb. dry hydrated lime (400 lb. of lime putty) to 100 lb. STRUCTO-GAUGE Plaster. For best results, machine-mix.

Application—Apply like regular finish over a partially dry gypsum basecoat. Scratch in thin coat, well ground into base, and immediately double-back with second coat, filling imperfections. Water-trowel to a smooth, hard finish, free of all blemishes. Continue troweling until final set has taken place. Clean tools and equipment after each mix.

RED TOP Keenes Cement—A high-strength white gypsum plaster used with finish lime putty for extremely hard, dense surfaces. It is the only gypsum plaster that can be retempered. Made in two types: *Regular* (3-6 hr. set) and *Quick Trowel* (1-2 hr. set). Quick-troweling Keenes must be used with a min. of 25 lb. dry hydrated lime per 100 lb. of Keenes.

Keenes Cement-Lime Finish is similar in many respects to a lime gauged finish except that Keenes Cement, instead of gauging plaster, is used in varying proportions depending on the hardness required and is generally used as a float finish. If a smooth-trowel hard finish is desirable, use STRUCTO-GAUGE Gauging Plaster. Keenes Cement is for interior use over sanded or wood-fibered gypsum basecoats. Do not apply smooth-trowel finish over lightweight aggregate basecoats.

Application, Sand Float Finish—Commonly used, hard float finish which may be satisfactorily colored. Mix in proportions of 100 lb. Keenes Cement to 50 lb. dry lime and 400 lb. sand, with or without lime-proof colors. Apply in same manner as for Keenes Cement-Lime Finish but instead of final troweling, use a wood, cork, sponge or felt-covered float to bring sand particles to surface to produce a pleasing, durable sand finish.

Application, Keenes Cement-Lime Finish—For medium hard finish, mix proportion of 100 lb. lime putty (50 lb. of dry hydrated lime) to 100 lb. Keenes Cement. For a hard finish, use 50 lb. lime putty (25 lb. dry hydrated lime) to 100 lb. Keenes Cement. Apply finish coat over a set high-strength basecoat that has been broomed and is partially dry. Spray with water if surface is too dry but do not soak. Scratch-in thin coat and then double-back with second coat to a true surface. Trowel with water to a smooth, glossy finish, free from blemishes, until finish has set.

(See page 290 for lime-gauging ratios and coverage.)

Gauging Plasters—Technical Data

Product	Set time with lime putty (min.)
CHAMPION (white)	20-30
STAR (white)	40-60
RED TOP Gauging (local-gray)	slow set 50-75 quick set 30-40
RED TOP Keenes Cement (white)	regular 180-360 quick set 60-120
STRUCTO-GAUGE (white or light gray)	slow set 60-75 quick set 30-40

Finish Limes

The two types of finish lime are: *Type S* (also called autoclaved, pressure or double hydrate); *Type N* (also called normal or single hydrate). Both produce a good finish lime putty, but their preparation differs. Weather precautions:

In Cold Weather—A few precautions will result in improved quality and easier working. Where weather and water are cold, lime develops better plasticity when soaked overnight. Best conditions are a warm room and water temperature above 50°F (10°C).

It is important to note that in cold weather the lime putty-gauging mixture requires a longer time to set. Therefore, gauging content should be increased or quick-set gauging added to offset the slower setting time.

Proper heat and ventilation are extremely important. Windows should be opened slightly so that moisture-bearing air moves out of the building. Fast drying after setting is essential to a hard finish.

Many cold-weather problems with finish lime are a direct result of improper basecoat conditions. Finish should go over a set, fairly dry basecoat. The basecoat will dry slowly in winter, so heat and ventilation are needed. The water retentivity of lime putty, plus a cold, "green" base, does not provide enough suction to remove excess moisture. Blistering and cracking can occur due to slow set, or the lime may work back through set of the gauging.

In Hot Weather—Precautions include proper soaking of lime putty. When the sun is hot, hydrated lime requires ½ to 1 gal. more water per 50 lb. The water should be cool. Soaking of putty in shade prevents undue water evaporation and helps to prevent curdling and loss of spreading properties. Avoid soaking for periods longer than two or three days.

For application of lime putty-gauging finish plaster, make sure that basecoat is set and partially dry. If applied over a dried-out basecoat, water will be drawn from the finish coat, resulting in severe check-cracking. Spray the basecoat before finish coat application and trowel coat until final set.

IVORY and SNOWDRIFT Finish Limes—Autoclaved (double-hydrate). *Mixing*—Machine equipment must be clean. Place 5½ to 6 gal. clear water per 50-lb. bag of lime in mixer. Using a motor-driven, propeller-type mixer, the complete mixing of lime putty takes 2 to 3 min. and results in a high-quality, easy-working putty. Machine-mixed putty is plastic and coverage is increased from 10% to 15%. With a paddle-type mixer, the mixing time is about 15 min. *Hand Mixing*—For immediate use, place 5½ to 6 gal. water per 50 lb. IVORY or SNOWDRIFT Finish Lime in mixing box. Add finish lime to water and hoe sufficiently to eliminate lumps. Screen putty through 8-mesh screen before using. *Overnight Soak*—Place water hose in bottom of a level soaking box. Sift lime through screen into box. When full, run water *slowly*, but continuously, until a small amount of excess water is visible over top of lime. If excess water remains on the surface the following morning, absorb excess water by screening in additional IVORY or SNOWDRIFT Lime, allow to soak a few minutes, then blend into putty by hoeing. For use, if necessary, screen through 8-mesh hardware cloth and mix with gauging plaster that meets job requirement. *Application*—follows directions for Gauging Plasters.

RED TOP and GRAND PRIZE Finish Limes—Single or Normal Hydrate. *Machine Mixing*—Produces a smoother, more plastic putty, easier to use and with better coverage. Use approx. 6 gal. water to each 50-lb. bag of RED TOP or GRAND PRIZE Finish Lime. *Hand Mixing*—Slowly sift RED TOP or GRAND PRIZE Lime into water in soaking box. Allow material to take up water for about 20 or 30 min. and then hoe briskly to mix thoroughly.

Let machine or hand mix soak for min. of 16 hrs. to develop full workability and plasticity. For use, screen through 8-mesh hardware cloth and mix with gauging plaster that meets job requirements. Application follows directions for Gauging Plasters.

Prepared Finishes

United States Gypsum Company's prepared finishes—IMPERIAL Finish Plaster, DIAMOND Interior Finish Plaster, RED TOP Finish—shorten construction time and provide hard, abrasion-resistant surfaces. Allow basecoat plaster to set but not completely dry before application of prepared finishes. If basecoat plaster has dried, it is required to mist the surface before applying finish.

Mixing—Prepared finishes require the addition of water on the job. Water should be clean, fresh, suitable for human consumption, and free from mineral and organic substances that affect the plaster set. Water used for rinsing or cleaning is not suitable for mixing because it accelerates the plaster set.

Mechanical mixing is mandatory for prepared finishes. Mix no more material than can be applied before set begins. Since prepared finishes set more rapidly than the most conventional plasters, always consult bag directions for specific setting times. Prepared finishes will produce mortar of maximum performance and workability when the correct equipment is used and mixing directions carefully followed. Proper mixing is one of the most important factors in producing mortar of maximum workability.

Use a cage-type mixer paddle driven by heavy-duty ½" electric drill with a no-load rating of 900 to 1,000 rpm. Do not use propeller-type paddle or conventional mortar mixer. (For details of the cage-type mixing paddle and available electrical drills, see U.S. Gypsum Company Data Sheet P-502.)

Mix plaster in 16- or 30-gal. smooth-sided container strong enough to withstand impacts that could cause gouging. Do not use brittle containers for mixing.

Correct mixing—rapid and with high shear action—is essential for proper dispersion of plaster ingredients. Slow mixing can reduce plasticity of material. Overmixing can shorten working time. Operated at correct speed, the cage-type design paddle mixes thoroughly without introducing excess air into the mix.

IMPERIAL Finish Plaster—Scratch-in a tight, thin coat of IMPERIAL Finish Plaster over entire area, immediately doubling back with plaster from same batch to full thickness of 1/16″ to 3/32″. Fill all voids and imperfections. Final-trowel after surface has become firm, holding trowel flat and using water sparingly. Do not over-trowel.

Best results for IMPERIAL Finish Plaster are obtained by planning the plastering to permit continuous application from angle to angle. Where joining is unavoidable, use trowel to terminate unset plaster in sharp, clean edge—do not feather out. Bring adjacent plaster up to terminated edge and leave level. Do not overlap. During finish troweling, use excess material to fill and bridge joining.

DIAMOND Interior Finish Plaster—Scratch-in a tight, thin coat of DIAMOND Interior Finish Plaster over entire area, immediately doubling back with plaster from same batch to full thickness of 1/16″ to 3/32″. Fill all voids and imperfections. Final-trowel after surface has become firm, holding trowel flat and using water sparingly. Do not over-trowel.

A variety of textures ranging from sand float to heavy Spanish can be achieved with DIAMOND Interior Finish Plaster when job-aggregated with silica sand. (When DIAMOND Interior Finish Plaster is job-aggregated, one tablespoon cream of Tartar should be added for each bag of finish to retard plaster and allow sufficient working time.) Application is the same as for neat DIAMOND Interior Finish Plaster except that once the surface has been leveled and sufficient take-up has occurred, begin floating material from the same batch with trowel, float, sponge or by other accepted local techniques.

DIAMOND Interior Finish Plaster also may be textured by skip-troweling. When applying in this manner, eliminate final troweling. When surface has become sufficiently firm, texture with material from same batch prior to set.

RED TOP Finish—Scratch-in a tight, thin coat of RED TOP Finish over entire area, immediately doubling back with plaster from same batch to full thickness of not more than 1/16″. Final-trowel to a smooth, dense surface ready for decoration.

Ornamental Plasters

For complete information on the use of ornamental plasters, contact Dept. 983-2, Industrial Gypsum Division, United States Gypsum Company, Chicago, Illinois.

Special Additives

RED TOP Retarder and RED TOP Accelerator are available for use with plaster when required by job or climatic conditions. When used in excess, setting and drying problems can arise. Avoid use of too much retarder which can weaken the plaster finish.

Proper drying of conventional plaster surfaces is important to their ultimate performance. Rapid drying after the plaster has set is good because it produces high strength. Drying too fast, before the plaster has set may leave insufficient water for the chemical reaction needed to properly set plaster. For standard plasters, slow drying should be avoided because it may cause "sweatout" and reduce strength. Proper ventilation should be provided to remove moisture-laden air.

Prepared Exterior Finish

ORIENTAL Exterior Finish Stucco requires addition of water only. Do not add waterproofing or antifreeze compounds, sand or other materials. All tools and equipment must be clean.

Measure water accurately. To ensure color uniformity, use the same amount of water for each batch mixed.

Do not overtrowel, as this may cause color to concentrate unevenly on the surface.

Mixing—For mechanical mixing, add 100 lb. ORIENTAL Finish to 3 gal. water and mix for approx. 3 min. If more than one mixer load can be applied within three hours, mix entire amount, dump into box and blend to uniform color.

For hand mixing, add 100 lb. ORIENTAL Finish to 3 gal. water and allow to soak for approx. 15 min. Then hoe thoroughly to smooth consistency. If more than one batch can be applied within three hours, blend several batches together.

Hand Application—Apply ORIENTAL Finish only over a portland cement-lime and sand basecoat that is level but left rough under the darby, or broomed and properly cured.

Water-spray basecoat to provide a uniformly damp surface. Remove loose and projecting particles from basecoat. Then apply a thin coat well-ground into the base and completely covering it. Double-back and fill out to uniform thickness of about ⅛″. Cover total area in one operation to eliminate joinings. Trowel three or four times before texturing.

For a float finish, use a cork, wood, carpet or sponge-rubber float and work surface to an even texture, free from blemishes, as the material stiffens and starts to set. Floating must be done without use of additional water.

Machine Application—ORIENTAL Exterior Finish Stucco may be applied in a single or two-coat operation. Excessive mixing water should not be used as this may result in "lime bloom." For one-coat application, spray-apply material to basecoat to uniform thickness of approx. ⅛". Work from wet edges to complete an entire unbroken area in one continuous operation to eliminate joinings. For two-coat application, apply first coat with a trowel as in hand application. After troweling uniformly level, apply textured second coat, machine-spraying to total thickness of approx. ⅛".

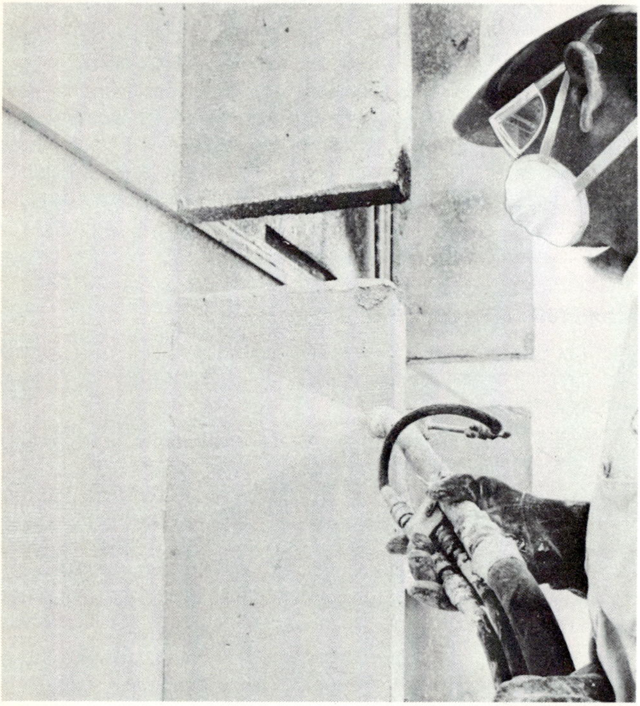

Solid keying of ORIENTAL Exterior Stucco in machine-spray application ensures high strength of finish.

Masons and Stucco Lime

Portland Cement-Lime Plaster—Used for interiors where high moisture conditions exist such as in steam rooms, dairies, showers, etc. It also is used as basecoat for exterior stucco. (Portland cement is not manufactured by U.S. Gypsum Company.)

Surface Preparation—Monolithic concrete surfaces should be brush-cleaned of all dust, loose particles and other foreign matter. Remove all laitance and efflorescence by washing with a 10% solution of commercial muriatic acid and water, then rinsing well with clear water. Remove grease or form oil by wiping with naphtha spirits. A high-quality plaster bonding agent should be applied to all surfaces according to manufacturer's directions.

Limitations

1. Brown coat requires curing with water after set.

2. Must not be applied directly to smooth, dense surfaces, gypsum lath or gypsum block. Self-furring metal lath must be secured to such surfaces before plaster is applied.

3. Control joints should be provided to compensate for shrinkage during drying.

4. A Keenes Cement-Lime Putty Finish must never be used over a portland cement basecoat.

Job-Mixed Stucco—Mix mason's lime such as BONDCRETE or MORTASEAL Lime with portland cement and sand in accordance with ASTM C926 and in the following proportions:

Job-Mixed Stucco Proportions

	Mix					
	Portland cement		BONDCRETE or MORTASEAL lime		sand[2]	
Coat[1]	lb	kg	lb	kg	ft³	m³
Scratch	94	43	40-50[1]	18-23[1]	5-6[1]	0.14-0.17
Brown	94	43	50	23	6-7[1]	0.17-0.20
Finish	94	43	100	45	7-10	0.20-0.28

(1) Upper end of range for use over concrete block where greater water retention and plasticity are required; lower end of range for use over metal reinforcing mesh with exterior sheathing or building paper. (2) Quantity used varies, depending on shape and size of local sand particles.

STRUCTOCORE Security Wall System

STRUCTOCORE Security Wall Systems consist of specially formed steel sheets that provide continuous reinforcement for monolithic, high-strength, fire-resistant plaster finish applications. STRUCTOCORE Security Wall Systems are ideal for use in lieu of reinforced concrete or concrete block.

The system is constructed from STRUCTOCORE Steel Sheets, spray-applied STRUCTO-BASE Gypsum Basecoat Plaster (Machine Application), and finished with IMPERIAL Finish Plaster.

STRUCTOCORE Sheets are welded or screwed to 18-ga. continuous steel angles that are securely anchored to floor and ceiling. High-strength STRUCTO-BASE Gypsum Basecoat Plaster (Machine Application) is sprayed to both sides of sheets to a minimum 3½″ thickness. IMPERIAL Finish Plaster (nominal 3000-psi compressive strength) is hand troweled over basecoat producing a hard, durable, smooth or textured finish at low cost.

STRUCTOCORE Steel Sheets are screw-attached for fast installation and then STRUCTO-BASE Plaster is spray-applied.

Framing

Layout—Partition layout for the STRUCTOCORE Security Wall System is accomplished in a conventional manner except measured from the center line of the partition.

Panel Angles—Measure the opening and determine the wall center line. Align and set trailing end of floor angle about ½″ from the wall or vertical structural member. Align outstanding legs ¾″ off one side of the center line. Securely anchor 1¾″ x 1¾″ 18-ga. STRUCTOCORE Floor-Ceiling Angles to the floor plate with fasteners not exceeding 16″ o.c. Repeat process at header.

Where angle run is greater than length of angles, space angle ends ⅛″ apart and continue layout. Field cut last angle to about ½″ from existing or adjacent walls. In order to increase attachment area and ease of installation, STRUCTOCORE Vertical Abutment Angles (4″ x 1½″ 18-ga.) may be used in place of the Floor-Ceiling Angles (1¾″ x 1¾″ 18-ga.). With 4″ leg in plane of wall, attach 4″ x 1½″ 18-ga. STRUCTOCORE Vertical Abutment Angles to the side walls. Cut angles to a min. ⅛″ short of the floor, ceiling and header angles at ends.

Sheet Layout—STRUCTOCORE Forming Sheets can be installed individually or assembled in multiples and raised into position. The width is perpendicular to the ribs; the length is parallel to the ribs. Sheets can be erected from bottom up or hung from the top angle. Hanging the sheets allows for self plumbing and ease of installation.

Start with a full STRUCTOCORE Sheet on the upper course and position the ribs vertically. Care should be employed in order to level the first row of sheets. Allow ⅛″ minimum between the web of the Vertical Abutment Angle and the forming edge. Screw-attach or weld the STRUCTOCORE Sheets to leg of the Floor-Ceiling Angle at every other rib. For walls over 12′0″ high, attach at every rib. Screw-attach the trailing edge of the sheet to the Vertical Abutment Angle 7″ o.c. Overlapping of courses may be increased to accommodate wall height and length in order to reduce cutting of sheets. If sheets must be cut, start second course sheets with a partial-width sheet. Overall rib length can be up to 1¼″ less than clear height. Nest or interlock sheets in increments of three rows of corrugations (2¾″ minimum) and one course horizontally and connect horizontally with wedges, screws or welds. An additional screw should be driven when a screw has stripped. Temporary clamping should be employed during fastening. This can easily be accomplished with the STRUCTOCORE Shoe Horn Splice (which can be removed or left in place after fastening), locking-type pliers or C-clamps. Provide temporary bracing (strong backs) to plumb the forming until after the plaster application scratch coat fully sets.

Remove free oil and loose rust from the forming to permit adequate plaster application. Verify wall straightness and conformance with specified tolerances for surfaces to receive direct application. Sequence plaster installation properly with the installation and protection of other work.

Basecoat Plaster Application

Spray-apply scratch coat to a minimum ⅜″ thickness. Before set, trim or remove high spots or surplus material from angles and floor line. Allow scratch coat to set and dry thoroughly (approximately 24 to 48 hours) before application of subsequent brown coat.

After scratch coat sets, install plaster dots (approximately 1″ x 4″) plumb with finish wall line, and subsequent screeds, allowing 1⁄16″ for finish coat. Screed composition shall consist of STRUCTO-BASE Gypsum Basecoat Plaster (Machine Application) with RED TOP Accelerator added, if necessary, to expedite the setting reaction. Build up plaster screeds (grounds) to provide necessary plaster thickness of wall surface. Screeds shall be approximately 4″ wide to avoid warping and out-of-plumb areas.

Spray-apply brown coat, laying on proper thickness with one pass and overlapping successive strokes. Rod brown coat flush with screeds, leaving walls ready for finish coat.

Plaster flush with metal frames and other built-in metal items or accessories that act as a plaster ground, unless otherwise shown. Where plaster is not terminated at metal by casing beads, cut basecoat free from metal before plaster sets.

Finish Plaster Application

Apply finish coat 48 hours after brown coat. Apply a tight scratch coat over entire working area; immediately double back with material from the same batch to a nominal 1⁄16″ thickness.

Start finish troweling as soon as material has become sufficiently firm to achieve a smooth surface free from marks, voids and other blemishes.

Where plaster is not terminated at metal by casing beads, groove finish coat at the junctures with metal.

Lightweight STRUCTOCORE Security Walls provide strength, save space.

Replastering Old Plaster Surfaces

In plastering over old plaster surfaces, certain precautions should be exercised to ensure a satisfactory result. Generally, the old surface is lime mortar plaster on wood lath, is badly cracked, and usually has been covered with canvas and/or multiple coats of paint.

The following suggestions for lathing and plastering over such old surfaces are listed in order of preference for best results:

1. If the old plaster and lath are removed, 3⁄8″ ROCKLATH Plaster Base may be applied to the framing and plastered in the same manner as for new work, following all applicable specifications.

2. If the old lath and plaster are left in place, the following methods may be used, after determining that the framing is of adequate size to carry the additional weight of a new plaster finish (average 8 lb./ft.^{2}).

(a) Apply 1″ x 3″ furring strips 16″ o.c. with 12d 9-ga. nails, 3¼″ long or of sufficient length to achieve 1¾″ min. penetration into framing. Then apply ⅜″ ROCKLATH Base and plaster in same manner as specified for new work.

(b) Apply 3.4-lb. self-furring diamond mesh metal lath over old surface by nailing through into framing, using 2″ 11-ga. 7⁄16″-head barbed-shank galvanized roofing nails, 6″ o.c. Wire tie side and end laps. Apply plaster in three coats. RED TOP Gypsum Plaster can be used with max. 2 cu. ft. of sand for scratch coat, max. 3 cu. ft. of sand for brown coat, or with 2½ cu. ft. of sand for scratch and brown coats. Lightweight aggregate should not be used in replastering when using metal lath.

3. If the old plaster is removed and wood lath left in place, all loose laths should be renailed and the lath repeatedly sprayed with water over a period of several hours in order to wet thoroughly. Then replaster as specified in 2(b). Note: If wood lath is not thoroughly nailed and wetted, cracking of the plaster may occur. Finish coat may be smooth trowel or sand float, as desired, mixed and applied per applicable specifications.

Integral Plaster Chalkboards

Plaster chalkboards offer definite design advantages. There is no limiting sheet size as is the case with fabricated boards; therefore, entire walls can be utilized as chalkboards. Maintenance is accomplished as easily as with conventionally fabricated chalkboards. Plaster chalkboards may be used with most plastered partition systems.

Chalkboard with UNIMAST Steel Stud Partitions—Follow general directions for system construction. Fasten floor and ceiling runner tracks in place and set studs 16″ o.c. Toggle bolts installed after plastering may be used for

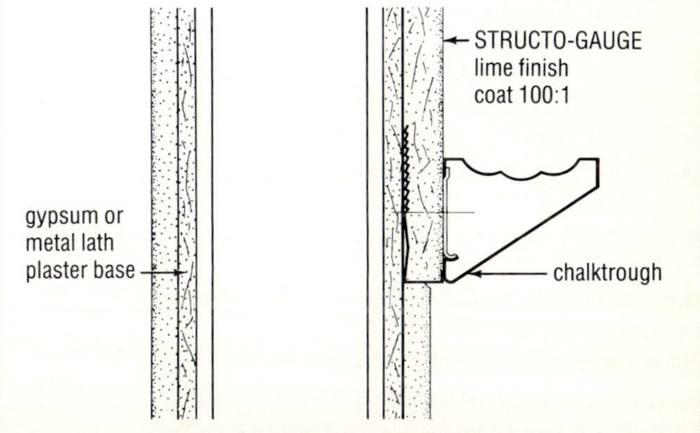

Cross-section thru chalkboard

chalk and eraser-trough clips. Attach ROCKLATH Plaster Base or metal lath and install UNIMAST No. 66 Expanded Flange Casing Bead, stapling to plaster base or wire-tying to metal lath, 8″ o.c. Plumb and level all casing bead installations.

Mix STRUCTO-BASE Plaster in proportion of 2 cu. ft. sand per 100 lb. plaster; apply to ½″ grounds. Finish with STRUCTO-GAUGE Gauging-lime mixed 100 lb. gauging per 100 lb. lime; apply to max. 1⁄16″ thickness.

Paint chalkboard when dry, with one coat primer-sealer and two coats of quality chalkboard paint.

Door Frames

Hollow metal door frames are shop-fabricated of 16-ga. and 18-ga. primed steel. Floor anchor plates of 16-ga. steel, with two anchor holes to prevent rotation, are welded to trim flanges to dampen door impact vibrations. Jamb anchor clips should be formed of 18-ga. steel, welded in the jamb and head.

Frames used with various plaster systems must be rigidly secured to the floor and partition construction to prevent twisting or other movement. If door frames are free to twist upon impact, cracking of plaster will result and eventually the frames will loosen. In addition to the framing specifications described, door closers are recommended on all oversize doors where the weight, including hardware, is over 50 lb.

Grouting of Door Frames—Always recommended, and *required* where heavy or oversize doors are used. As a grout, use a 100:2 RED TOP Gypsum Plaster-sand mix, adding enough water so that the material is stiff but workable.

Under no condition should the lath and plaster terminate against the trim of the door frame. Grouting of exterior door frames with gypsum plasters is not recommended.

Control Joints—Also help prevent cracking of plaster at door frames. To break continuity of framing for control joint location, install door frame and place friction-fit cripple studs next to frame uprights. Allow ¼″ clearance for SHEETROCK Zinc Control Joints Nos. 50, 75 and 100. Continue with plaster base application using required control joint at break in framing above door frame.

Door Frames with Studless Metal Lath Partitions—Follow general directions for fabricating door frames. Use four jamb anchors on each jamb and wire tie to support frame. Use temporary bracing to hold frame until plaster has set.

Door Frames with Stud-Metal Lath Solid Partitions—Fabricate as previously described with four jamb anchors welded to trim returns. Anchor frame to floor with power-driven fasteners.

Insert studs into steel door frame. Nest studs in notches of jamb anchor clips and wire tie. Install a ⅜″ round rod or a ⅛″ x 1¼″ flat bar across head

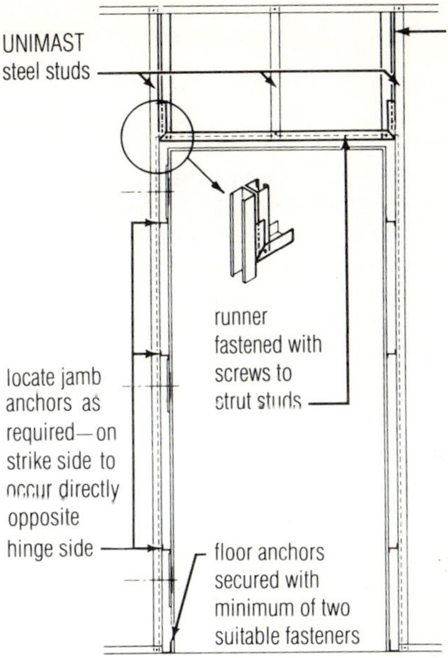

Note: See U.S. Gypsum technical literature for reinforcing requirements of doors over 50 lb. and up to 200 lb.

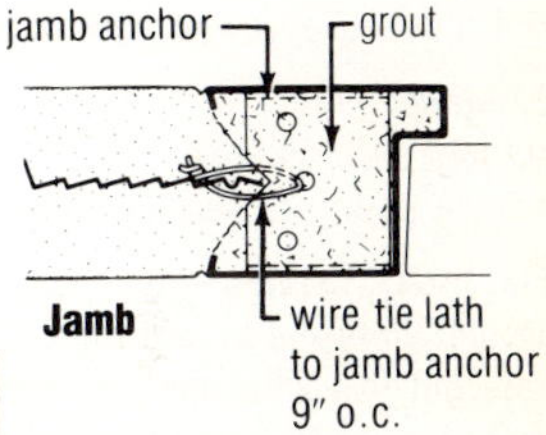

Jamb

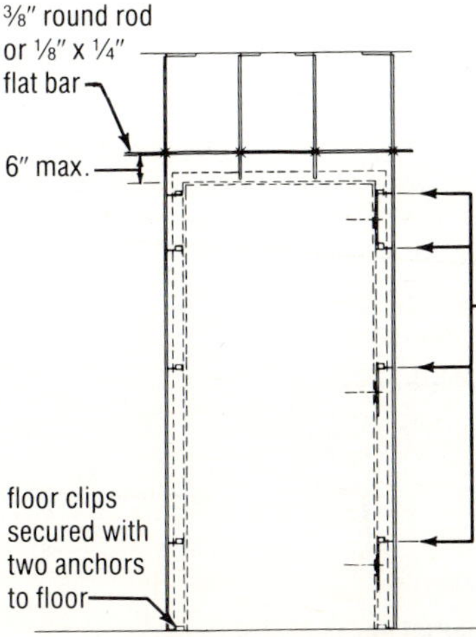

Elevation cross section

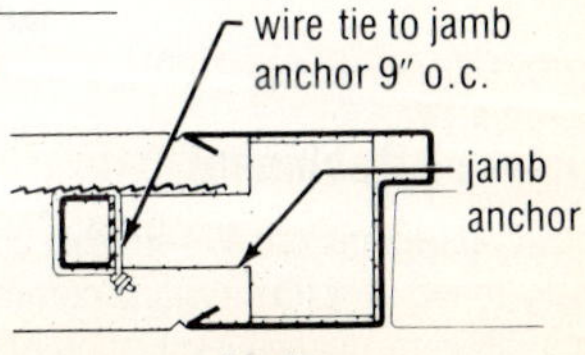

Jamb

Note: See frame elevation (studless metal lath partitions) for location of jamb anchors

of door, extending to engage first stud beyond frame. Wire-tie bar at each channel intersection.

Grout steel door frames solid with mortar when scratch coat of plaster is applied.

Caulking Procedures

Where a plaster partition is used as a sound barrier, SHEETROCK Acoustical Sealant should be used to seal all cutouts and all intersections with the adjoining structure. Caulking at runners and around the partition perimeter between gypsum lath and/or plaster and the structure is required to achieve sound transmission loss values on the job that approximate those determined by test. Caulking has proven to be the least expensive way to obtain better sound control.

The surfaces to be caulked should be clean, dry and free of all foreign matter. Using an air-pressure-activated or hand caulking gun, apply SHEETROCK Acoustical Sealant in beads about ⅜″ round.

Partition Perimeters—When gypsum lath is used, leave a space approx. ¼″ wide between lath and floor, ceiling and dissimilar walls. Appropriate metal edge-trim or casing beads applied to the lath may be used to create this space. Fill groove with SHEETROCK Acoustical Sealant.

When conventional plaster is applied to metal lath, rake out plaster to form a ⅜″ groove at partition perimeter, and fill groove with acoustical sealant. Finish over groove with base or trim as desired.

Openings—Apply a ⅜″ min. round bead of acoustical sealant around all cutouts such as at electrical boxes, medicine cabinets, heating ducts and cold air returns to seal the opening.

Electrical Fixtures—Apply caulking to the backs of electrical boxes and around all boxes to seal the cutout. Avoid cutting holes back to back and adjacent to each other. Electrical boxes having a plaster ring or device cover for use as a stop for caulking are recommended.

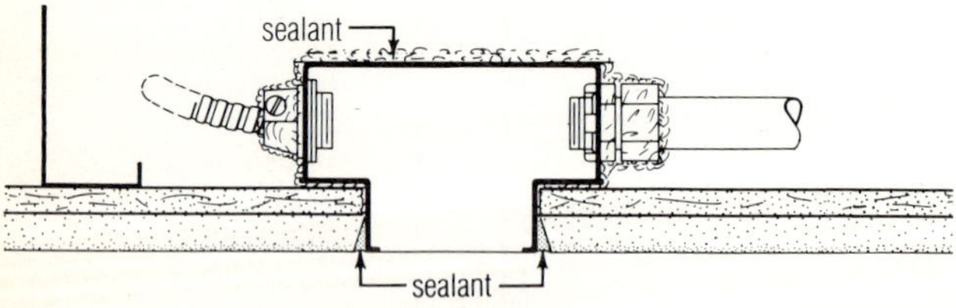

Outlet box

Fixture Attachment

Plaster partitions provide suitable anchorage for most types of fixtures normally found in residential and commercial construction. To ensure satisfactory job performance, evaluation of load requirements of unusual or heavy fixtures and preconstruction planning are needed so that attachments will be within the load-carrying capacity of the construction.

The carrying capacity of a given attachment depends upon the strength of the plaster used. Plaster having a compressive strength of at least 900 lb/in^2 was used to develop the data shown in the fastener load table on page 329.

The attachment of fixtures to sound-barrier partitions may impair the sound-control characteristics desired. Refrain from attaching fixtures to party walls so as to avoid a direct path for sound flow. Plastered ceilings are not designed to support light fixtures or troffers, air vents or other equipment. Separate supports should be provided.

In wood-frame construction, fixtures are usually attached directly to the framing or to blocking supports attached to the framing. Blocking or supports should be provided for plumbing fixtures, towel racks, grab bars and similar items. Lath and plaster membranes are not designed to support loads imposed by these items without additional support to carry the main part of the load.

To provide information for proper construction, an investigation of loading capacities of various fasteners and fixture attachments used with plaster partitions was conducted at the USG Corporation Research Center. These fasteners and attachments were tested:

Picture Hooks—A flattened wire hook attached to the wall with a nail driven diagonally downward. Depending on size, the capacity varies from 5 to 50 lbs. per hook. Suitable for hanging pictures, mirrors and other lightweight fixtures from all plaster partitions.

Fiber and Plastic Expansion Plugs—A sheet metal or wood screw driven into a fiber or plastic plug. Annular ribs are provided on outside of plastic plug to ensure a positive grip in wall. As screw is inserted, rear end of plug expands and holds assembly in place. Suitable for attaching lightweight fixtures in all partitions.

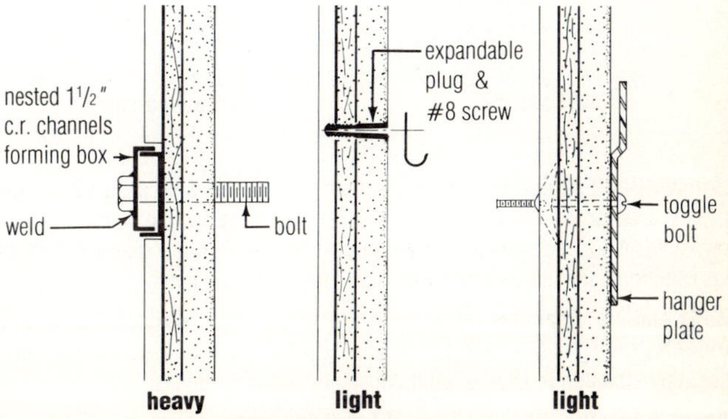

Fixture attachments

Toggle Bolts—Installed in lath and plaster only. Disadvantages of toggle bolt are that when bolt is removed, wing fastener on back will fall down into a hollow wall and a large hole is required to allow wings to pass through wall facings.

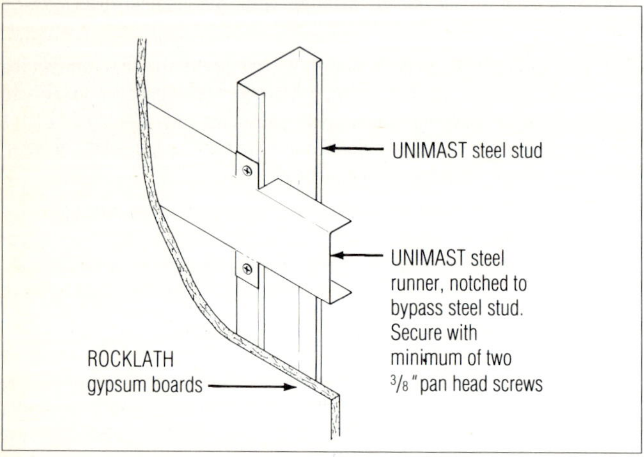

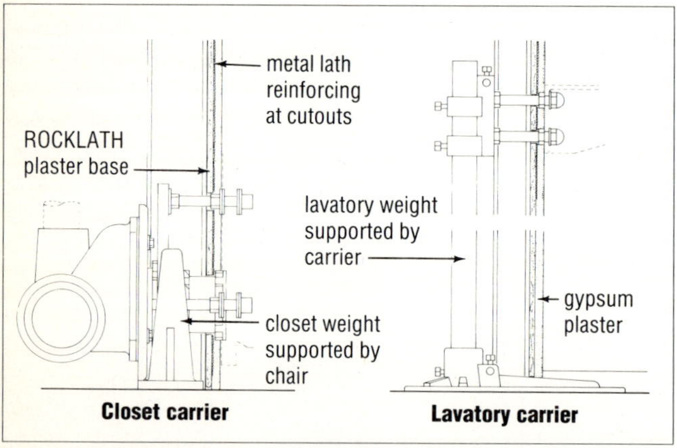

Hollow Wall Fasteners—Installed in lath and plaster only. One advantage of this type fastener is that threaded section remains in wall when screw is removed. Also, widespread spider support formed by the expanded anchor spreads load against wall material, increasing load capacity.

Bolts and 1½″ Channels—Two 5⁄16″ bolts welded to 1″ channels for use in mounting hanger brackets for heavy fixtures. Two nested channels are securely attached to back of studs in steel-framed partitions.

Angle Brackets—Standard 10″ x 12″ shelving brackets spaced 24″ o.c. and fastened to wall with three-hole anchorage. Fastened to steel studs with sheet metal screws or to lath and plaster with toggle bolts or hollow wall fasteners.

Continuous Horizontal Bracing—Back-up for fixture attachment is provided with notched runner attached to steel studs with two ⅜″ pan head screws.

Slotted Standards—With adjustable shelf brackets, are fastened 24″ o.c. to steel studs with sheet metal screws or to lath and plaster with toggle bolts or hollow wall fasteners. Normal standard spacing: 24″ o.c. for 24″ stud spacing, and 32″ o.c. for 16″ stud spacing. Limited to six shelves per partition height.

Separate Supports—Individual carriers or chairs placed in the core wall, recommended where heavy bathroom fixtures such as lavatories and water closets without floor supports are required.

Fixture Attachment Load Data Lath and Plaster Construction

Fastener				Allowable withdrawal resistance		Allowable shear resistance	
Type	Size		Substrate	lbf	N[1]	lbf	N[1]
Plug and screw	#6 #8 #12		metal or gypsum lath and plaster	10 20 30	45 89 133	40 50 60	178 222 267
Toggle bolt or hollow wall fastener	⅛″ ³⁄₁₆″ ¼″	3.2mm 4.8mm 6.4mm	metal or gypsum lath and plaster	75 125 175	334 556 778	50 140 150	222 623 667
Plumber's bracket bolted to 1½″ channel	⁵⁄₁₆	7.9mm	see drawing, page 181			300	1334

(1) Newtons.

Insulating Blankets

See page 172 for data on THERMAFIBER Insulating and Sound Attenuation Fire Blankets for use in sound-rated assemblies.

General Construction

CHAPTER 8 System Design Considerations

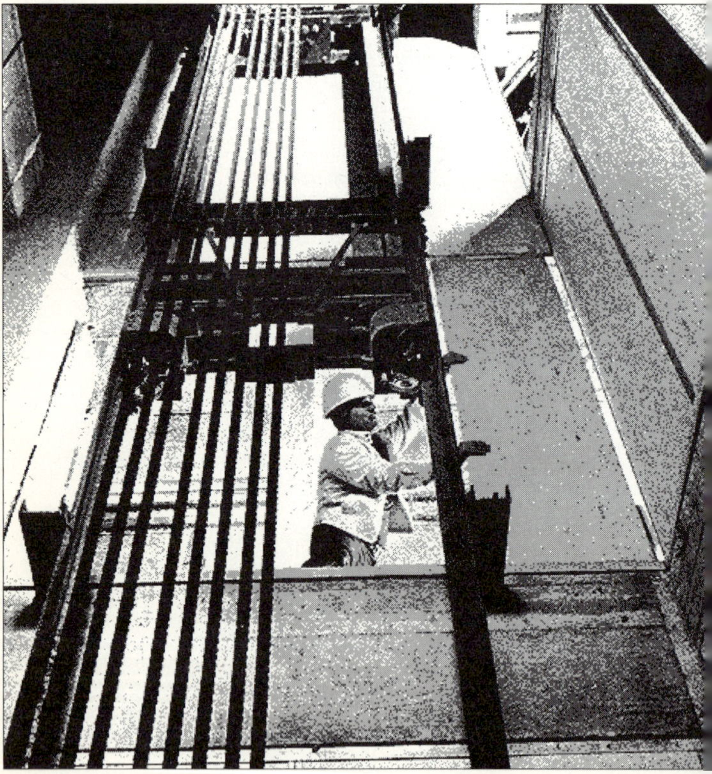

System Technical Data

United States Gypsum Company leads the industry in developing high-performance systems to meet specialized requirements for modern building design and in documenting their performance at recognized testing laboratories. These systems provide fire resistance, sound control, structural capacity and aesthetics for improved function and utility while reducing construction time and cost. All are constructed of quality products and released only after thorough testing and field trial.

In most instances, fire resistance and often sound-attenuation performance applies equally to systems constructed with gypsum panels and gypsum bases. Gypsum base with veneer plaster finish provides an acceptable alternative to gypsum panels. Therefore, the term "gypsum board" is used throughout this chapter to refer to both types of products. Only where performance differs greatly are the products treated separately.

Structural Criteria

Design of any structure must take into account the load conditions that will exist and the resulting stresses and movements. Load-bearing walls must be designed to carry the weight of the structure, its components and other loads that occur once the building is occupied.

The amount of axial load that structural members can bear will vary with the amount of lateral load (pressure from wind or other horizontal forces) that the final assembly may incur.

Manufacturers of structural components, particularly steel framing (studs, runners, joists) provide tables that identify the maximum allowable loads for various components under specific conditions. These tables typically start at 0 psf lateral loads and increase in 5 or 10 psf increments to about 40 psf. Interior partitions are typically rated for 5 psf lateral loads.

Interior non-bearing partitions are not designed to carry axial loads. Limiting heights are based on stress or deflection limits for given lateral load. Height limitations depend on the gauge of the steel used, dimensions of the stud, stud spacing and the allowable deflection limit.

Curtain walls are not regarded as load-bearing walls and are not designed to carry axial loads. However, finished curtain wall assemblies do need to withstand wind loads within certain stress or deflection limits. Limiting height tables from the framing manufacturer should be consulted.

Load-span capacity of steel studs are based on the following factors as applicable:

1. AISI Specifications for the Design of Cold-formed Steel Structural Members.
2. Yield strength.
3. Structural and physical properties of members.
4. Bending stress.
5. Axial stress.
6. Shear stress.
7. Allowable deflection.
8. Web crippling at supports.
9. Lateral bracing ratio.

Stud Selection

Selection of a stud gauge and size must take into account a number of factors. The key consideration is whether the assembly is for a load-bearing, non-load-bearing or curtain wall application. Other variables include anticipated wall height, weight and dimensions of mounted fixtures, fire rating desired, sound attenuation needed, anticipated wind loads, insulation requirements, deflection allowance and desired impact resistance.

In general, stronger or heavier studs are needed to accommodate taller walls. Stronger studs also reduce deflection and vibration from impacts such as slamming doors. Wider studs may be required to handle insulation requirements. Fire-rated constructions typically have been designed, tested and approved based on using the lightest gauge, shallowest stud depth and maximum stud spacing as indicated in the assembly description. Stud gauge and depth may generally be increased without affecting the fire-resistance rating of the assembly.

Strength and performance characteristics can be achieved in a variety of ways. Wall strength can be increased by using heavier gauge material, stronger stud designs, narrower stud spacing or larger web dimensions. Studs typically are selected to maintain cost control and design integrity. Increased strength requirements generally are met by first increasing metal gauge or stud style before increasing stud dimensions.

Unimast Inc. manufactures two different style studs with a 3⅝″ web dimension:

- ST style studs are designed for non-load bearing interior drywall partition applications. They have a minimum 1¼″ flange width on both sides. The web design incorporates a "keyhole" cutout for bracing and for electrical, communication and plumbing lines.
- SJ style studs are designed with ribs formed lengthwise in the steel web for added strength. Flange width is 1¾″ on one side and slightly less on the other side to permit nesting. Cutouts in the web are oval shaped to accommodate bracing, utility service and mechanical attachments.

For specific stud design and assembly information, consult U.S. Gypsum Company Technical Folder SA-923 and Unimast Inc. Catalog UN-30.

Fire and Sound Tests

Fire and sound test data aid in comparing and selecting materials and constructions. In addition, these data frequently are essential for securing building code or agency approval. The USG Construction Selector SA-100 provides tested fire resistance and acoustical performance for various systems.

Fire resistance refers to the ability of an assembly to serve as a barrier to fire and to confine its spread to the area of origin. Spread of fire from one area to another occurs because (a) the barrier collapses, (b) openings in the barrier allow passage of flame or hot gases or (c) sufficient heat is conducted through an assembly to exceed specified temperature limitations. These characteristics form the basis for judging when an assembly no longer serves as a barrier in a test.

A *fire-resistance rating* denotes the length of time a given assembly can withstand fire and give protection from it under precisely controlled laboratory conditions. All tests are conducted under standard conditions and test methods as described in the *Standard, Fire Tests of Building Construction and Materials*, ASTM E119. The standard is also known as ANSI/UL 263 and NFPA 251. The ratings are expressed in hours and apply to floors, beams, columns and walls.

For assemblies tested at U.L., ratings are specific to the designs tested. Unless described in the design, insulation may not be added under the assumption that the rating either will remain the same or improve. Addition of insulation in the concealed space between the ceiling membrane and the roof structure may reduce the hourly rating of an assembly by causing premature disruption of the ceiling membrane and/or higher temperatures on structural components under fire exposure conditions. For steel stud walls, the dimensions and gauge of studs specified for an individual design are minimums. Ratings still apply to the design when constructed of steel studs of a heavier gauge and/or larger dimensions than the stud specified.

Sound control refers to the ability to attenuate sound passing through a partition.

The *Sound Transmission Class (STC)* is a widely used rating of sound attenuation performance—accurate for speech sounds but not for music, mechanical equipment noise or any sound with substantial low-frequency energy. Tested per ASTM E90; rated per ASTM E413.

The *Music, Mechanical Equipment Transmission Class (MTC)* system, developed by U.S. Gypsum Company, measures music and machinery/mechanical equipment sounds—provides a more accurate rating of attenuation from these low-frequency energy sources. The MTC rating does not replace STC, but complements it.

The *Impact Insulation Class (IIC)* is a numerical evaluation of a floor-ceiling assembly's effectiveness in retarding the transmission of impact sound, also determined from laboratory testing. Tested per ASTM E492; rated per ASTM E989.

The *Noise Reduction Coefficient (NRC)* is a measure of sound absorption. This is an important consideration for controlling acoustics within a confined area. It does not generally apply to the performance of a structural system.

Fire and sound tests are conducted on U.S. Gypsum Company products assembled in a specific manner to meet requirements of established test procedures. Substitution of materials other than those tested or deviation from the specified construction may adversely affect performance and result in failure. For complete information on test components and construction, see the test report.

Additional information about fire and sound testing can be found in the Appendix.

Typical Fire Systems

A large number of systems have been designed and tested for fire resistance. The systems vary greatly in both design and performance. Nevertheless, certain basic system designs are commonly used. As a frame of reference, several typical designs and their accompanying fire ratings are shown below for wood frame and steel frame assemblies.

Wood Frame Partitions

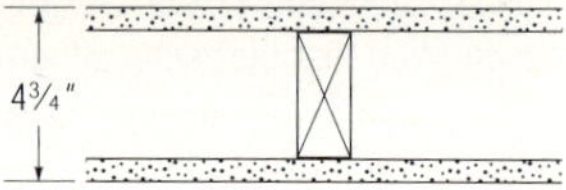

1-hr. Rating

UL Design U305

Drywall System

Studs:	Wood 2x4 (nom.).
Stud spacing:	16″ o.c.
Gypsum panel:	⅝″ SHEETROCK brand Gypsum Panel, FIRECODE Core, or ⅝″ SHEETROCK brand Gypsum Panel, Water-Resistant, FIRECODE Core, each side.
Application:	Horizontal.
Attachment:	1⅞″ cement-coated nails spaced 7″ o.c.
Joints:	Exposed or taped and treated.
Insulation:	3″ batts THERMAFIBER SAFB (optional).
Perimeter:	May be caulked with SHEETROCK Acoustical Sealant.

Veneer Plaster System

Studs:	Wood 2x4 (nom.).
Stud spacing:	16″ o.c.
Gypsum panel:	⅝″ IMPERIAL Gypsum Base, FIRECODE Core, each side.
Application:	Horizontal.
Attachment:	1⅞″ cement-coated nails spaced 7″ o.c.
Joints:	Taped.
Finish:	3⁄32″ DIAMOND or IMPERIAL plaster finish both sides.
Insulation:	3″ batts THERMAFIBER SAFB (optional).
Perimeter:	May be caulked with SHEETROCK Acoustical Sealant.

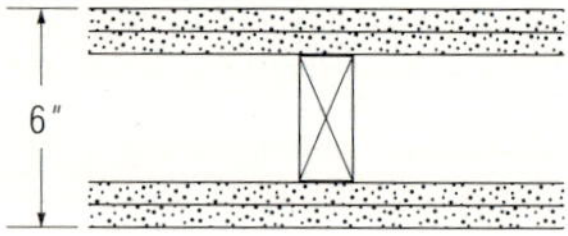

2-hr. Rating

UL Design U301

Drywall System

Studs:	Wood 2x4 (nom.).
Stud spacing:	16″ o.c.
Gypsum panel:	Two layers of ⅝″ SHEETROCK brand Gypsum Panel, FIRECODE Core, or ⅝″ SHEETROCK brand Gypsum Panel, Water-Resistant, FIRECODE Core, each side.
Application:	Horizontal or vertical—joints of face layer staggered over joints of base layer.
Attachment:	Base layer—1⅞″ cement-coated nails spaced 6″ o.c. Face layer—2⅜″ nails 8″ o.c.
Joints:	Exposed or taped and treated.
Perimeter:	May be caulked with SHEETROCK Acoustical Sealant.

Veneer Plaster System

Studs:	Wood 2x4 (nom.).
Stud spacing:	16" o.c.
Gypsum panel:	Two layers of 5/8" IMPERIAL Gypsum Base, FIRECODE Core.
Application:	Horizontal or vertical—joints of face layer staggered over joints of base layer.
Attachment:	Base layer—1 7/8" cement-coated nails spaced 6" o.c. Face layer—2 3/8" nails 8" o.c.
Joints:	Taped.
Finish:	3/32" DIAMOND or IMPERIAL plaster finish both sides.
Perimeter:	May be caulked with SHEETROCK Acoustical Sealant.

Steel Frame Partitions

1-hr. Rating
UL Design U465
Drywall System

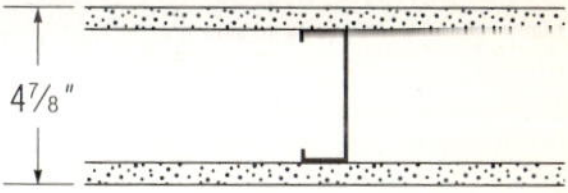

Studs:	Steel 3 5/8" x 25-ga. (min.).
Stud spacing:	24" o.c.
Gypsum panel:	5/8" SHEETROCK brand Gypsum Panel, FIRECODE Core, or 5/8" SHEETROCK brand Gypsum Panel, Water-Resistant, FIRECODE Core, each side.
Application:	Vertical.
Attachment:	Type S screws 8" o.c.
Joints:	Taped and treated.
Insulation:	3" batts THERMAFIBER SAFB (optional).
Perimeter:	May be caulked with SHEETROCK Acoustical Sealant.

Veneer Plaster System

Studs:	Steel 3 5/8" x 25-ga. (min.).
Stud spacing:	24" o.c.
Gypsum panel:	5/8" IMPERIAL Gypsum Base, FIRECODE Core, each side.
Application:	Vertical.
Attachment:	Type S screws 8" o.c.
Joints:	Taped.
Finish:	3/32" DIAMOND or IMPERIAL plaster finish both sides.
Insulation:	3" batts THERMAFIBER SAFB (optional).
Perimeter:	May be caulked with SHEETROCK Acoustical Sealant.

2-hr. Rating
UL Design U411 or U412
Drywall System

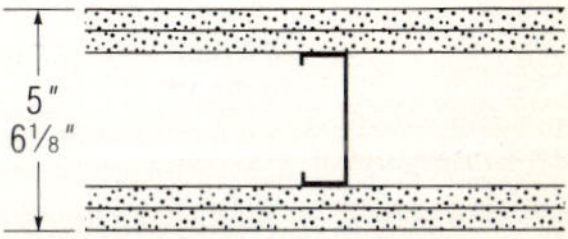

Studs:	Steel 2 1/2" x 25-ga.
Stud spacing:	24" o.c.
Gypsum panel:	Two layers of 5/8" SHEETROCK brand Gypsum Panel, FIRECODE Core, or 1/2" SHEETROCK brand Gypsum Panel, FIRECODE C Core, each side.

Application:	Vertical—joints of face layer staggered over joints of base layer.
Attachment:	Base layer—1″ Type S screws 8″ o.c. Face layer—laminated with joint compound or attached with 1⅝″ Type S screws 12″ o.c.
Joints:	U411, exposed or taped and treated; U412, outer layer taped and treated.
Perimeter:	May be caulked with SHEETROCK Acoustical Sealant.

Veneer Plaster System

Studs:	Steel 2½″ x 25-ga.
Stud spacing:	24″ o.c.
Gypsum panel:	Two layers of ⅝″ IMPERIAL Gypsum Base, FIRECODE Core, or ½″ IMPERIAL Gypsum Base, FIRECODE C Core.
Application:	Vertical—joints of face layer staggered over joints of base layer.
Attachment:	Base layer—1″ Type S screws 8″ o.c. Face layer—laminated with joint compound or attached with 1⅝″ Type S screws 12″ o.c. Face layer—2⅜″ nails 8″ o.c.
Joints:	Taped.
Finish:	3/32″ DIAMOND or IMPERIAL plaster finish both sides.
Perimeter:	May be caulked with SHEETROCK Acoustical Sealant.

2-hr. Rating
UL Design U484
Lath & Plaster System

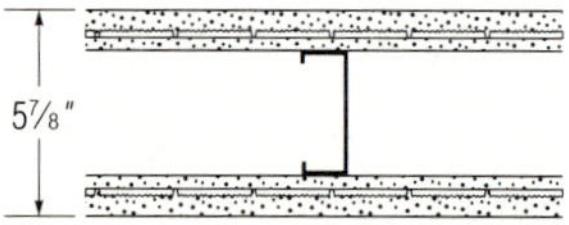

Studs:	Steel 2½″ x 25-ga.
Stud spacing:	16″ o.c.
Gypsum lath:	⅜″ ROCKLATH Plaster Base, each side.
Metal lath:	3.4 lb. UNIMAST Self-Furring Diamond Mesh Lath, each side.
Application:	Gypsum lath applied horizontally.
Attachment:	Gypsum and metal lath attached with 1″ Type S screws 8″ o.c.
Finish:	¾″ scratch and brown coat 100:2 gypsum sand plaster.

Wood Floor/Ceilings

1-hr. Rating
UL Design L501 or L512
Drywall System

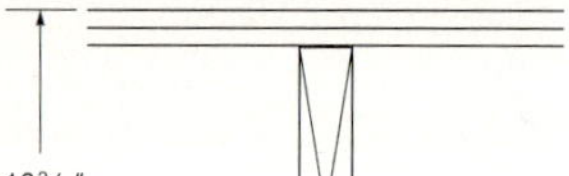

Floor:	1″ nom. wood sub and finished floor.
Joists:	Wood 2x10 (nom.) cross bridged with 1x3 (nom.) lumber.
Joist spacing:	16″ o.c.
Gypsum panel:	⅝″ SHEETROCK brand Gypsum Panel, FIRECODE Core, or ½″ SHEETROCK brand Gypsum Panel, FIRECODE C Core.
Application:	Perpendicular to joists.
Attachment:	1⅞″ cement-coated nails spaced 6″ o.c.
Joints:	Taped and treated.

Veneer Plaster System

Floor: 1″ nom. wood sub and finished floor.

Joists: Wood 2x10 (nom.) cross bridged with 1x3 (nom.) lumber.

Joist spacing: 16″ o.c.

Gypsum panel: 5/8″ IMPERIAL Gypsum Base, FIRECODE Core, or 1/2″ IMPERIAL Gypsum Base, FIRECODE C Core.

Application: Perpendicular to joists.

Attachment: 1 7/8″ cement-coated nails spaced 6″ o.c.

Joints: Taped.

Finish: 3/32″ DIAMOND or IMPERIAL plaster finish both sides.

2-hr. Rating
UL Design L541
Drywall System

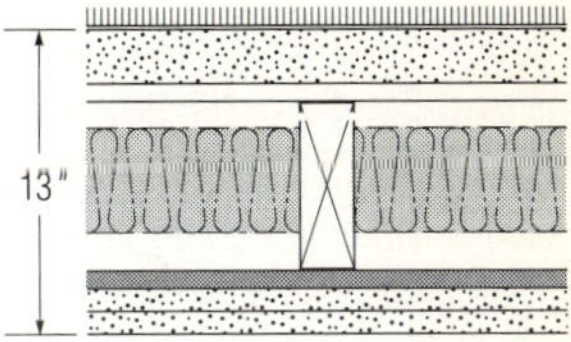

Floor: Carpet/pad, 1 1/2″ gypsum concrete (Type F) flooring, 1/2″ plywood.

Joists: Wood 2x10 (nom.) cross bridged with 1x3 (nom.) lumber.

Joist spacing: 16″ o.c.

Insulation: 3" THERMAFIBER SAFB.

Gypsum panels: 2 layers 5/8″ SHEETROCK brand Gypsum Panels, FIRECODE C Core.

Application: Perpendicular to RC-1 channel.

Attachment: 1″ Type S screws base layer, 1 5/8″ Type S screws face layer.

Joints: Taped and treated.

Steel Floor/Ceilings

3-hr. Rating
G512
Drywall System

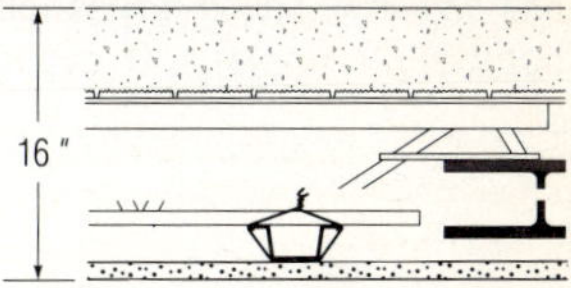

Floor: 2 1/2″ concrete on steel deck or riblath over bar joist—includes 3-hr. unrestrained beam.

Joists: Type 12J2 min. size, spaced 24″ o.c.

Furring Channel: 25-ga. spaced 24″ o.c. perpendicular to joists; 3″ on each side of wallboard end joints—double-strand saddle tied.

Gypsum panel: 5/8″ SHEETROCK brand Gypsum Panel, FIRECODE C Core.

Application: Perpendicular to furring.

Attachment: 1″ Type S screws 12″ o.c.

Joints: Backed with wallboard strips, taped and treated.

Veneer Plaster System

Floor:	2½″ concrete on steel deck or riblath over bar joist—includes 3-hr. unrestrained beam.
Joists:	Type 12J2 min. size, spaced 24″ o.c.
Furring Channel:	25-ga. spaced 24″ o.c. perpendicular to joists; 3″ on each side of wallboard end joints—double-strand saddle tied.
Gypsum panel:	⅝″ IMPERIAL Gypsum Base, FIRECODE C Core.
Application:	Perpendicular to furring.
Attachment:	1″ Type S screws 12 " o.c.
Joints:	Backed with wallboard strips, taped.
Finish:	3/32″ DIAMOND or IMPERIAL plaster finish both sides.

Wood Stud Partitions

Suitable for residential and light-commercial construction where combustible framing is permitted, these designs include single- and double-layer gypsum board facings, single- and double-row studs, those with insulating blankets, and those with resilient attachment. Performance values of up to 2-hr. fire resistance and 59 STC can be obtained.

Steel Stud Partitions

Suitable for all types of construction, these designs include single- and multi-layer gypsum board facings, with and without THERMAFIBER Sound Attenuation Fire Blankets. Performance values of up to 4-hr. fire resistance and 62 STC can be obtained.

Creased Thermafiber Sound Insulation System

Creased THERMAFIBER assemblies are steel-framed, 1-hr. fire-rated systems that offer high sound ratings (50-55 STC) plus the lower in-place cost of lightweight single-layer gypsum board. The systems consist of ⅝″

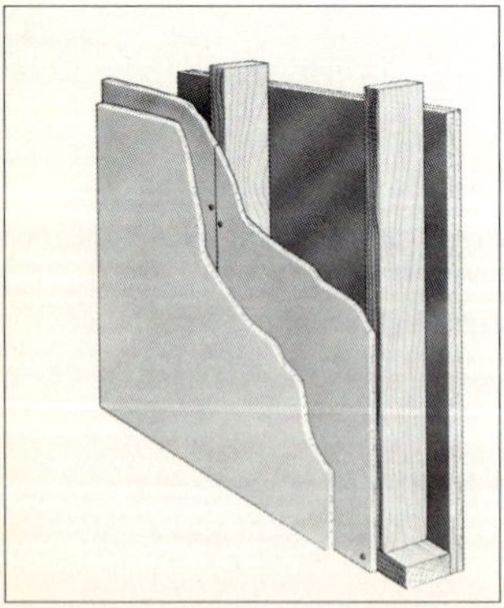

Wood Stud Partitions

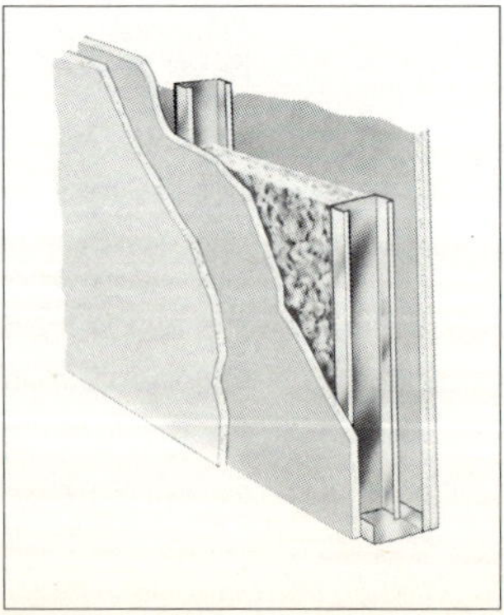

Steel Stud Partitions

SHEETROCK brand Gypsum Panels, FIRECODE C Core; 3⅝″ steel studs spaced 24″ o.c. and set in runners; and THERMAFIBER Sound Attenuation Fire Blankets (SAFB), 25″ wide.

Since the blanket is 1″ wider than the cavity, it is installed with a slit field-cut down the center and partially through the blanket. This allows the blanket to flex or bow in the center, easing the pressure against the studs and transferring it to the face panel, thereby damping sound vibrations more effectively. Panels screw-attach directly or resiliently to the steel framing.

Sound Control Systems

U.S. Gypsum Company fire-rated partition systems offer a range of assemblies that are highly effective in isolating all types of sound. In both wood-framed and steel-framed construction, resilient channel systems offer improved sound attenuation to direct attachment systems.

In steel-framed construction, USG High Sound-Attenuation Steel Framed Systems provide economical sound isolating systems without the excessive weight or space required of masonry construction. Systems are designed to control not only the mid and high frequencies, but also the low frequencies prevalent in music and mechanical equipment environments. Partition systems include both load-bearing and non-load bearing designs.

To properly measure sound attenuation in low frequencies, U.S. Gypsum Company developed the MTC system (Music, Mechanical Equipment Transmission Class). This complementary measurement allows a designer or acoustician to evaluate the differences between systems and specify the desired STC and MTC performance required.

For assistance with specific project requirements, contact your local U.S. Gypsum Company sales representative or the Company's Technical Services Department.

Single-layer resilient system with 3″ THERMAFIBER Insulation delivers 50-STC and a fire rating of 1 hr.

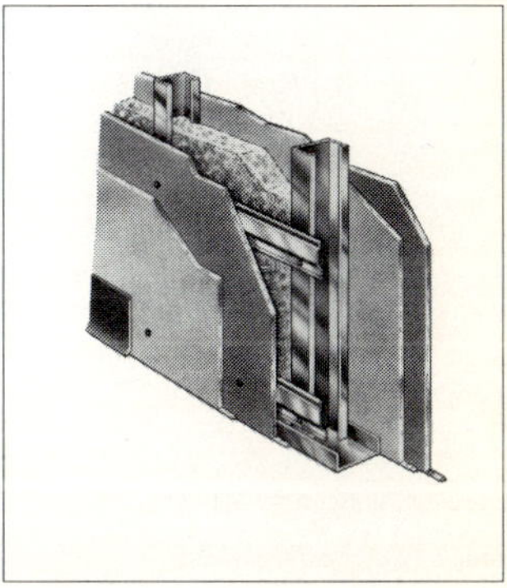

USG High Sound-Attenuation Steel Framed System with sound ratings of 61-STC/57-MTC and a fire rating of 2 hrs.

USG Area Separation Walls

USG Area Separation Walls provide fire and sound barriers between units in multi-family wood-frame buildings. Some code bodies identify this type of wall by other names such as fire wall, lot line wall, party wall or town-house separation wall.

These systems can be used in buildings up to four stories high; offer performance values up to 3-hr. fire resistance and STC 60.

There is a choice of two cost-reducing systems—*Cavity Type* and *Solid Type.* Both systems are continuous, vertical, non-load bearing wall assemblies of gypsum panels and steel framing that extend from foundation through the roof and act as barriers to fire and sound transmission.

Both types of area separation wall systems are engineered for performance with low labor and material costs. Erected by carpenters, the walls offer the same protection as common masonry construction but are significantly thinner and easier to construct.

Solid-Type Separation Wall

The solid-type separation wall is designed for use between independent load-bearing walls. The system consists of two 1″ thick SHEETROCK brand Gypsum Liner Panels installed vertically between 2″ USG Steel C-Runners. Panel edges are inserted in 2″ USG Steel H-Studs spaced 24″ o.c. C-runners are installed at top and bottom of wall and back-to-back between vertical panels cut to a convenient length above each intermediate floor. Studs are attached to wood framing at intermediate floors with 0.063″ USG Aluminum Breakaway Clips that yield and break away when exposed to fire, thus permitting a fire-damaged structure to fall free while the fire barrier remains intact.

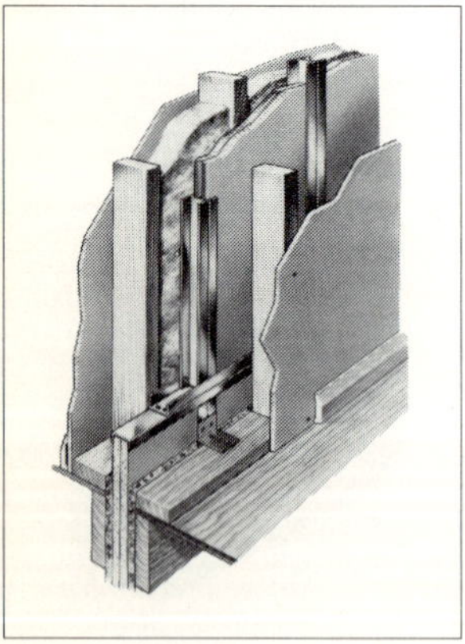

Solid-Type Separation Wall

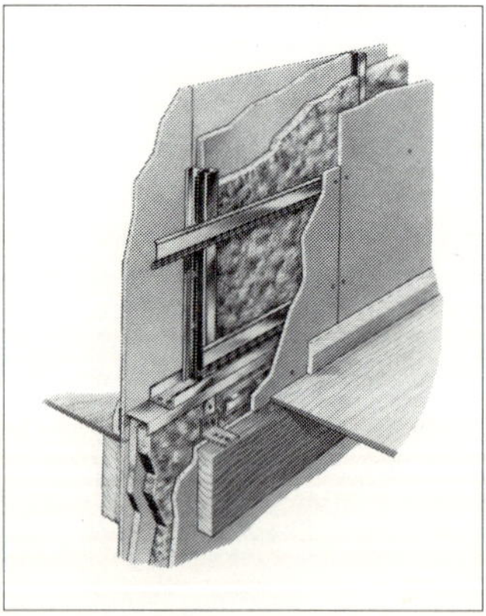

Cavity-Type Separation Wall

An additional clip-to-stud attachment is required 5′ o.c. below the top 23′ of the building. A minimum ¾″ clearance must be maintained between area separation wall and wood framing.

With 25-ga. USG Steel H-Studs, the assemblies are suitable for floor-to-ceiling heights up to 10′ under 5-psf lateral load and up to 8′ as an exterior wall under 15-psf wind load without exceeding L/240 allowable deflection.

With 2″ THERMAFIBER Sound Attenuation Fire Blankets (SAFB) stapled each side of liner panels, the assembly has obtained a 3-hr. fire resistance rating allowing separate selection and construction of tenant walls.

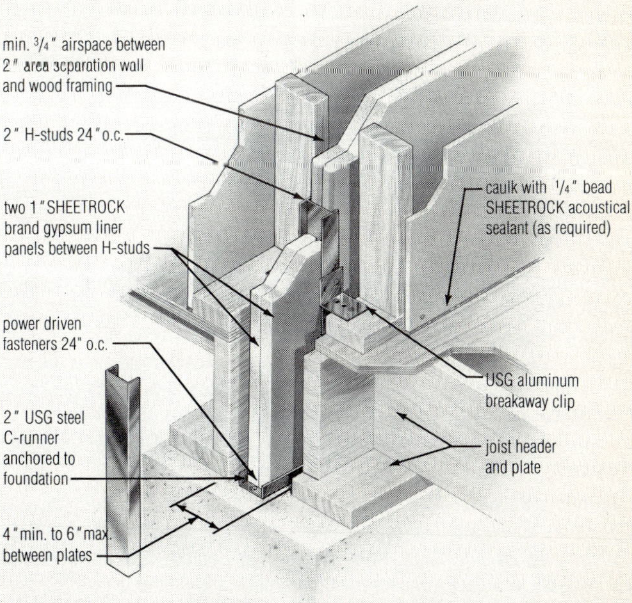

Foundation–solid separation wall

Installation

Layout—A minimum ¾″ clearance must be maintained between area separation wall and wood framing. Three inch space required to accommodate insulation thickness (for 3-hr. wall). THERMAFIBER Insulation fire blocking at intermediate platforms is required in all cases.

Foundation—Position 2″ wide steel USG C-Runner at floor and securely attach runner to foundation with power-driven fasteners at both ends and spaced 24″ o.c. Space adjacent runner sections ¼″ apart. When specified, caulk runner at foundation with min. ¼″ bead of SHEETROCK Acoustical Sealant.

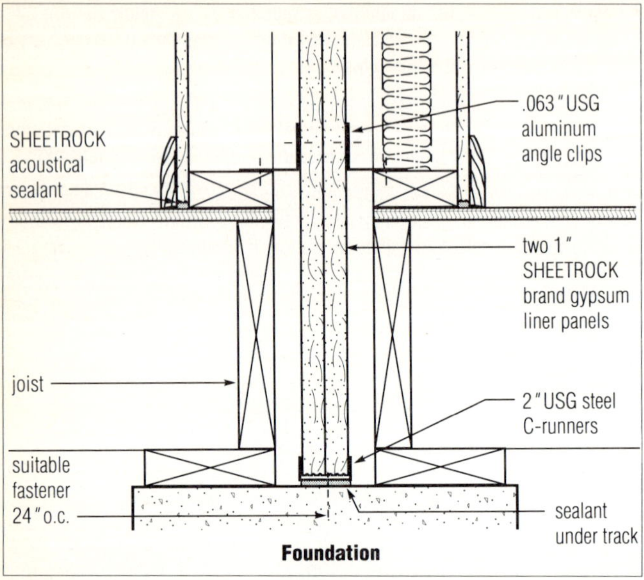

First Floor—Install 1″ liner panels and USG H-Studs cut to convenient length more than floor-to-floor height. Install two thicknesses of 1″ liner panels vertically in USG C-Runner with long edges in USG H-Stud. As an option, USG H-Stud and USG C-Runner may be screw-attached at the end that is fully engaged to runner. Erect USG H-Studs and double-thickness liner panels alternately until wall is completed. Cap ends of run with vertical USG C-Runner and fasten to horizontal USG C-Runner flange with ⅜″ Type S screws.

Intermediate Floors—Cap top of panels and studs with back-to-back USG C-Runners screw-attached together with double ⅜″ Type S screws at ends and spaced 24 o.c. SHEETROCK Acoustical Sealant is recommended between back-to-back runners for tight smoke and sound control. Secure USG H-Studs to framing with .063″ USG Aluminum Breakaway Clips screw-attached to studs with ⅜″ screws and to framing with 1¼″ Type W screws. Attachment may vary for convenience but is required within 6″ of each end of stud and spaced not more than 10′ o.c. Additional clips are required midheight between floors (5′ o.c. max.) for area below the top 23′ of the building. Overall wall height must not exceed 44′. Except at foundation, install fire blocking between joists and fire barrier.

Roof—Continue erecting studs and panels for succeeding stories as previously described. At roof, cap panels with USG C-Runner and fasten to framing within 6″ of roof line with aluminum clips.

Sound Attenuation Fire Blankets—Blankets may be installed in the stud cavity on either side of the firewall to improve the STC rating of the partition. Attach blankets with seven 9⁄16″ staples randomly driven through each blanket. Butt blankets closely and fill all voids.

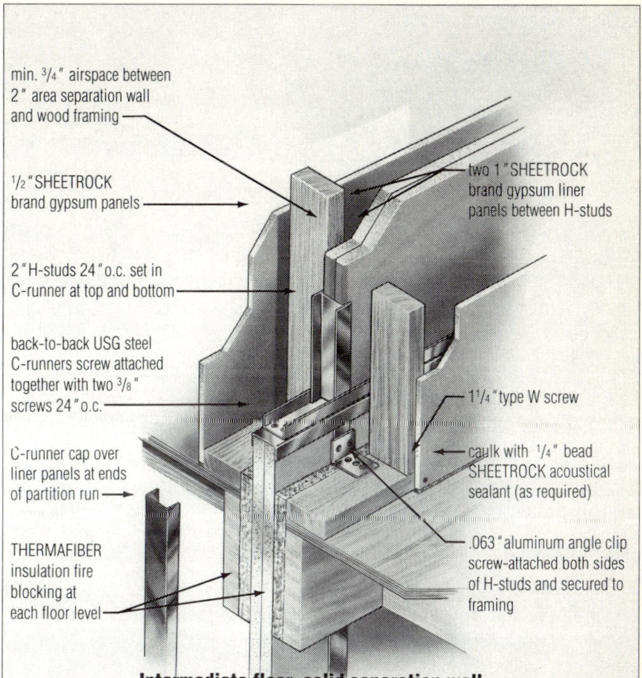

Intermediate floor–solid separation wall

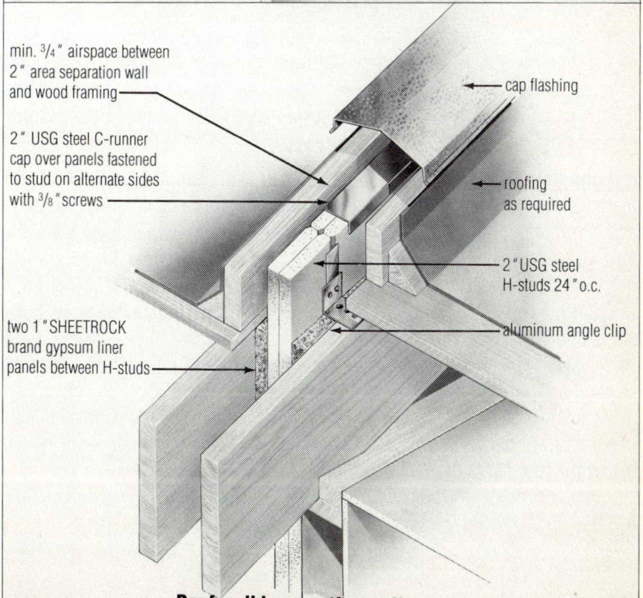

Roof–solid separation wall

Cavity-Type Separation Wall

Cavity Walls are intended for use where a common fire/party wall is accepted by local code at the line of separation between the units. The system consists of 1″ SHEETROCK brand Gypsum Liner Panels set in steel C-runners and formed USG C-H Studs and faced both sides with SHEETROCK brand Gypsum Panels, Water-Resistant, FIRECODE C Core. In this way, the separation wall also serves as the finished interior walls.

This 2-hr. area separation wall is only 3½″ thick for stacked heights up to 23′ using 2½″ USG C-H Studs. For building heights up to 44′, use 4″ 20-ga. USG C-H Studs below the top 23′ of the building. To effect construction of this system and maintain integrity of the area separation wall, the building must be designed with joists running parallel to the area separation wall.

Erection of these assemblies is handled by carpenters during completion of rough framing and setting of roof trusses. This factor allows tight job schedules and faster completion than is possible with masonry fire walls.

STC ratings up to 57 are achievable using RC-1 Resilient Channel and THERMAFIBER SAFB in the construction (see U.S. Gypsum Company Technical Folder SA-925 for details).

Installation

Foundation—Position 2½″ wide steel USG C-Runner at floor and securely attach to foundation with power-driven fasteners at both ends and spaced 24″ o.c. Caulk runner at foundation with min. ¼″ bead of SHEETROCK Acoustical Sealant.

First Floor—Install 1″ SHEETROCK brand Gypsum Liner Panels and steel studs cut to convenient length more than floor-to-floor height. Erect liner panels vertically in USG C-Runner with long edges in groove of USG C-H Stud. Install USG C-H Studs between panels and cap ends of run with USG E-Stud or USG C-Runner. Fasten all corner USG C-Runner (or USG E-Stud) flanges both sides with ⅜″ Type S screws.

Intermediate Floors—Cap top of panels and studs with USG C-Runner and fasten studs to USG C-Runner flanges on alternate sides with ⅜″ Type S screws. Install bottom USG C-Runner for next row of panels over top runner with end joints staggered at least 12″. Fasten runners together with double ⅜″ screws at ends and spaced 24″ o.c. Secure each USG C-H Stud to framing with .063″ USG Aluminum Breakaway Clip, fastened to both sides of each stud with ⅜″ screws and to framing or subfloor with 1¼″ Type W screws. Attachment may vary for convenience but is required within 6″ of each end of stud and spaced not more than 10′ o.c. Overall wall height must not exceed 44′. Except at foundation, install fire blocking between joists and fire barrier. (*Note*: Use 4″ 20-ga. USG C-H Studs on lower floors below the top 23′ of the building.)

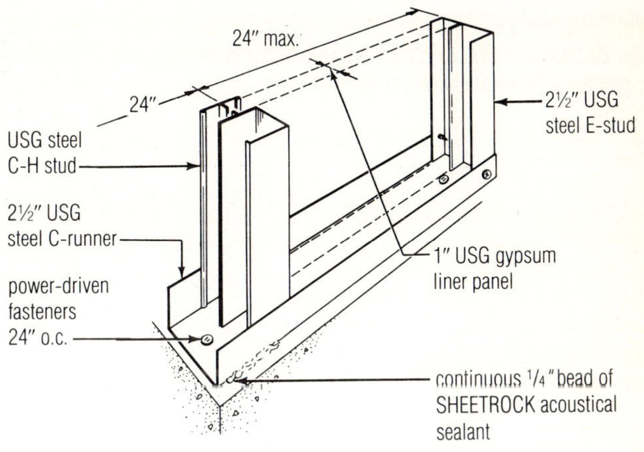

Foundation–cavity wall

Roof—Continue erecting studs and panels for succeeding stories as previously described. At roof, cap panels with USG C-Runner. Fasten studs to framing with aluminum clips within 6″ of roof line.

Sound Attenuation Fire Blankets—Where required by design, install 1½″ blankets between C-H studs and attach to liner panel with five 9⁄16″ staples driven through each blanket, one in center and others spaced 3″ from each corner. Butt ends of blankets closely and fill all voids.

Resilient Channels—Where required by design, install RC-1 Resilient Channels horizontally to face side of studs, 6″ above floor, 6″ below ceiling joists and max. 24″ o.c.

Attach channels to studs with ⅜″ Type S screws driven through holes in mounting flange. Extend channels to ends of runs and attach to USG E-Studs or USG C-Runners. Splice channels by nesting directly over stud; screw attach through both flanges. Reinforce with screws at both ends of splice.

Gypsum Panels—Apply ½″ thick SHEETROCK brand Gypsum Panels, Water-Resistant, FIRECODE C Core, vertically to both sides of studs. Stagger joints on opposite partition sides. Fasten panels with 1″ Type S screws spaced 12″ o.c. in field and along edges and runner flanges.

Resilient Single-Layer—Apply ½″ gypsum boards vertically to resilient channels and fasten with 1¼″ Type S screws placed 6″ away from stud and 12″ o.c. Do not place screws over studs.

Resilient Double-Layer—Apply ⅝″ gypsum board base layer perpendicular to resilient channels with end joints staggered. Fasten with 1¼″ Type S screws placed 6″ away from stud and 12″ o.c. Apply ⅝″ gypsum board face layer vertically over base layer with edge joints staggered and attach with 1⅝″ Type S screws spaced 12″ o.c. and staggered from those in base layer.

Interior Finishing—After construction is closed in, complete the assembly with a U.S. Gypsum Company Joint System or Veneer Plaster System applied according to manufacturer's current directions.

For additional information on USG Area Separation Walls, see U.S. Gypsum Company Technical Folder SA-925.

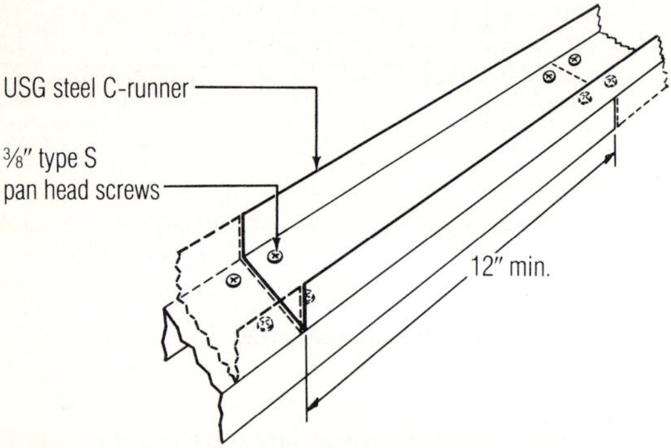

C-runner splice

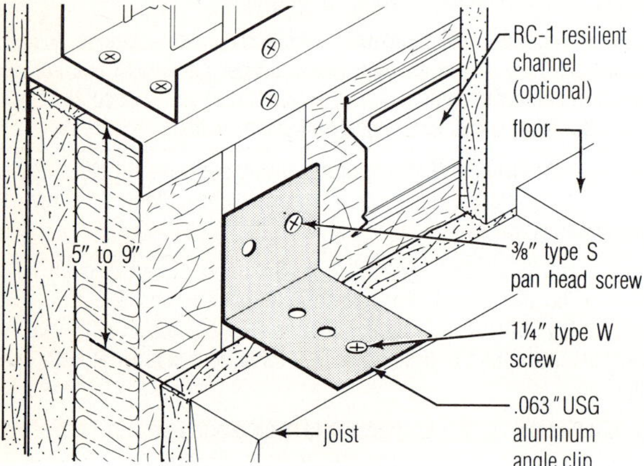

Intermediate floor-cavity wall (50 STC design)

USG Shaft Walls

Cavity Shaft Walls

USG Cavity Shaft Walls are non-load bearing, fire-resistant gypsum board partition systems for enclosing shafts, air ducts and stairwells. Designed for erection from one side, USG Cavity Shaft Walls offer superior performance and greater economy than other designs.

The engineered design of the strong, rigid USG C-H Stud system provides a simpler, thinner, lighter-weight assembly. It offers faster installation and lower material costs which reduce total in-place costs. It also saves on structural framing costs. For example, masonry shaft enclosures in high-rise buildings can weigh up to 45 psf, whereas lightweight USG Shaft Walls range from 9 psf (2-hr. assembly) to 16 psf (3-hr. assembly).

USG Cavity Shaft Walls provide up to 4-hr. fire resistance and sound ratings to 51 STC. Designs are available for intermittent design loads up to 15 psf. For sustained pressure in air returns, design uniform pressure loads should not exceed 10 psf.

Maximum partition heights depend on expected pressures. For elevator shafts, the applied pressure load is selected by the designer based on elevator cab speed and the number of elevators per shaft. Instead of using only deflection criteria, U.S. Gypsum Company design data considers several additional factors in determining limiting partition heights. These include:

Bending stress—the unit force that exceeds the stud strength.

End reaction shear—determined by the amount of force applied to the stud that will bend or shear the J-Runner or cripple the stud.

Deflection—the actual deflection under a load. Allowable deflection is based on the amount of bowing under load that a particular wall can accommodate without adversely affecting the wall finish.

A wide range of product and installation combinations is available to meet performance requirements: intermittent air pressure loading of 5, 7½, 10,

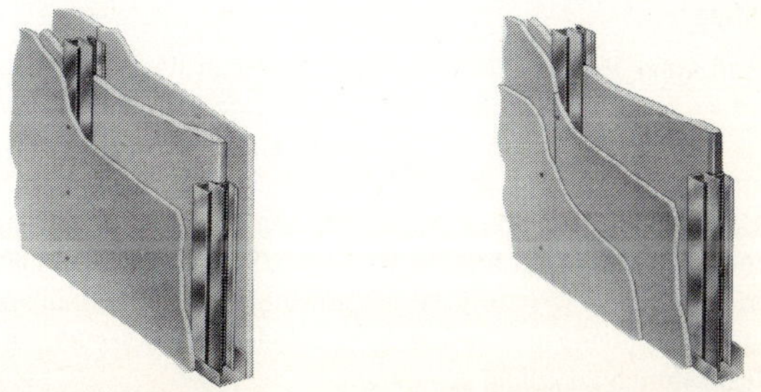

Single layer both sides **Double layer one side**

15 psf; vertical heights in three stud sizes and four steel thicknesses to accommodate lobbies and mechanical rooms. Assemblies can be constructed with fire resistance ratings from 2-hr. to 4-hr. For more information, consult U.S. Gypsum Company Technical Folder SA-926.

Horizontal Shaft Walls

USG Cavity Shaft Walls installed horizontally provide economical construction for fire-resistive duct protection, corridor and other ceilings and stairway soffits. Also ideal for ceilings over office areas in pitched-roof buildings and in modular buildings where ceiling framing is independent of the floor above. With 1″ liner panels inserted into USG C-H Studs 24″ o.c. and triple-layer ½″ SHEETROCK brand Gypsum Panels, FIRECODE C Core, screw attached to studs, the system provides greater spans and 2-hr. protection from fire either inside or outside the duct.

With double-layer ½″ SHEETROCK brand Gypsum Panels, FIRECODE C Core, screw attached to studs, the assembly provides suitable 2-hr. fire-resistive ceiling construction for corridors and stairs. One-hour fire-rated construction is offered with single-layer ⅝″ SHEETROCK brand Gypsum Panels, FIRECODE C Core.

No other drywall shaft assembly provides such an economical horizontal application.

Installation

Studs and Liner Panels—Position steel runners at floor and ceiling with the short leg toward finish side of wall. Securely attach runners to structural supports with power driven fasteners at both ends and max. 24″ o.c. With steel frame construction, install floor and ceiling runners and USG J-Runners or USG E-Studs on columns and beams before it is fireproofed. Remove spray fireproofing from runners and USG E-Studs before installing gypsum liner panels (2-hour steel fireproofing). For other structural steel fireproofing requirements, use Z-shaped stand-off clips secured to structural steel before fireproofing application.

Cut liner panels 1″ less than floor-to-ceiling height and erect vertically between USG J-Runners. Where shaft walls exceed max. available panel height, position liner panel end joints within upper and lower third points of wall. Stagger joints top and bottom in adjacent panels. Screw studs to runners on walls over 16′.

Use steel USG C-H Studs ⅜″ to not more than ½″ less than floor-to-ceiling height, and install between liner panels with liner inserted in the groove. Install full-length steel USG E-Studs or J-Runners vertically at T-intersections, corners, door jambs and columns. Install full-length USG E-Studs over gypsum liner panels both sides of closure panels. For openings, frame with vertical USG E-Stud or J-Runner at edges, horizontal USG J-Runner at head and sill, and reinforcing as specified. Suitably frame all openings to maintain structural support for wall.

Install floor-to-ceiling steel USG J-Runners or USG E-Studs each side of steel-hinged door frames and jamb struts each side of elevator door frames to act as strut studs. Attach strut stud to floor and ceiling runners with two ⅜″ Type S-12 pan head screws. Attach strut studs to jamb anchors with ½″ Type S-12 screws. Over steel doors, install a cut-to-length section of USG J-Runner and attach to strut stud with ⅜″ Type S-12 screws.

Gypsum Panel Attachment—*Single layer one side, one hour:* apply ⅝″ SHEETROCK brand Gypsum Panels, FIRECODE C Core, to C-side of stud. Position gypsum panel vertically and fasten to studs and runners with 1″ Type S screws 12″ o.c. (UL Design U499).

Double layer one side, two hour: apply base layer of ½″ SHEETROCK brand Gypsum Panels, FIRECODE C Core, vertically to studs with 1″ Type S screws 24″ o.c. along edges and in the field of panels. For vertical application, apply face layer of ½″ SHEETROCK brand Gypsum Panels, FIRECODE C Core, vertically and fasten to studs and J-Runners with 1⅝″ Type S screws 12″ along the edges and in the field, staggered from screws in base layer. Joints between base and face layers staggered. For horizontal application, apply face layer horizontally and attach over base layer with 1⅝″ Type S screws 12″ o.c. in the field, along the vertical edges and to the floor and ceiling runners. Face layer attached to base layer with 1½″ long Type G screws midway between studs and 1″ from the horizontal joint (UL Design U438).

Single layer both sides, two hour: apply ½″ SHEETROCK brand Gypsum Panels, FIRECODE C Core, vertically to both sides of studs. Fasten gypsum panels with 1″ Type S screws 12″ o.c. along the vertical edges and in the field (UL Design U467).

Single ¾″ thick layer one side, two hour: apply 1″ SHEETROCK brand Gypsum Liner Panels one side between 4″ USG Steel C-H studs 24″ o.c., install 3″ THERMAFIBER SAFB in cavity, ¾″ SHEETROCK brand Gypsum Panels, ULTRACODE Core, on other side. Position panels vertically and fasten to studs and runners with 1¼″ Type S screws 8″ o.c. (UL Design U492).

Double layer, two hour with DUROCK Cement Board: install 1½″ THERMAFIBER SAFB in stud cavity. Apply base layer of ⅝″ SHEETROCK brand Gypsum

Panels, FIRECODE Core vertically and attach with 1″ Type S screws 24″ o.c. along vertical edges and in the field. Install face layer of ½″ DUROCK Cement Board by lamination to gypsum panels with 4″ wide strip of mastic applied with ¾″ notched trowel midway between studs and fasten to studs with 1⅝″ DUROCK Screws 6″ o.c. (UL Design U459).

Double layer, two hour resilient: apply base layer of ½″ SHEETROCK brand Gypsum Panels, FIRECODE C Core, to resilient channels with end joints staggered; fasten with 1″ Type S screws 12″ o.c. Apply face layer of ½″ SHEETROCK brand Gypsum Panels, FIRECODE C Core, vertically with joints staggered; fasten to channels with 1⅝″ Type S screws 12″ o.c.

Triple layer, three hour: install three layers of ⅝″ SHEETROCK brand Gypsum Panels, FIRECODE C Core, vertically on corridor side of studs. Fasten base layer with 1″ Type S screws 24″ o.c. along vertical edges and in the field; middle layer horizontally with joints staggered with 1⅝″ Type S screws 24″ o.c. along vertical edges and in the field; face layer vertically with joints staggered and fastened with 2¼″ Type S screws 16″ o.c. Attach face layer to J-Runners with 2¼″ Type S screws 12″ o.c.

Horizontal installation, two hour: install three layers of ½″ SHEETROCK brand Gypsum Panels, FIRECODE C Core to horizontally installed C-H and/or E-Studs. Install the base layer with the edges parallel to the studs and attached with 1″ Type S screws 24″ o.c.; the middle layer in the same manner with joints offset 2′ and attached with 1⅝″ Type S screws 24″ o.c.; and the face layer perpendicular to the studs and attached with 2¼″ Type S screws spaced 12″ o.c. Place face layer end joints between studs and secure with 1½″ Type G screws 8″ o.c.

Vent Shaft

USG Vent Shaft System provides a 2-hr. fire-rated enclosure (UL Design U505) for vertical shafts in apartments and other types of multi-story buildings. The assembly is particularly suited for relatively small and widely separated mechanical, service and ventilator shafts. USG Shaft Walls are preferred where service and mechanical lines and equipment are consolidated within the building core.

Installation

Support Member Attachment—Install 1⅜″ x ⅞″ x 22-ga. galvanized steel angles as runners on floor and sidewalls by fastening through their short legs. Steel angles may be used as ceiling runners. Install side angle runners 30″ long and centered for attachment of horizontal bracing angles.

Bracing Angle Attachment—Install 1⅜″ x ⅞″ x 22-ga. galvanized steel bracing angles horizontally at quarter-points between floor and ceiling and spaced max. 5′ o.c. Position long leg vertically for board attachment and fasten to sidewall angles with 1″ Type S screws.

Gypsum Panel and Liner Application—Install ⅝″ gypsum board (FIRECODE C Core) vertically on shaft side and fasten to angles and runners with 1″ Type S screws 16 o.c. Apply SHEETROCK Setting-Type (DURABOND) or SHEETROCK Lightweight Setting-Type (EASY SAND) Joint Compound or SHEETROCK Ready-Mixed Joint Compound—Taping or All Purpose on back

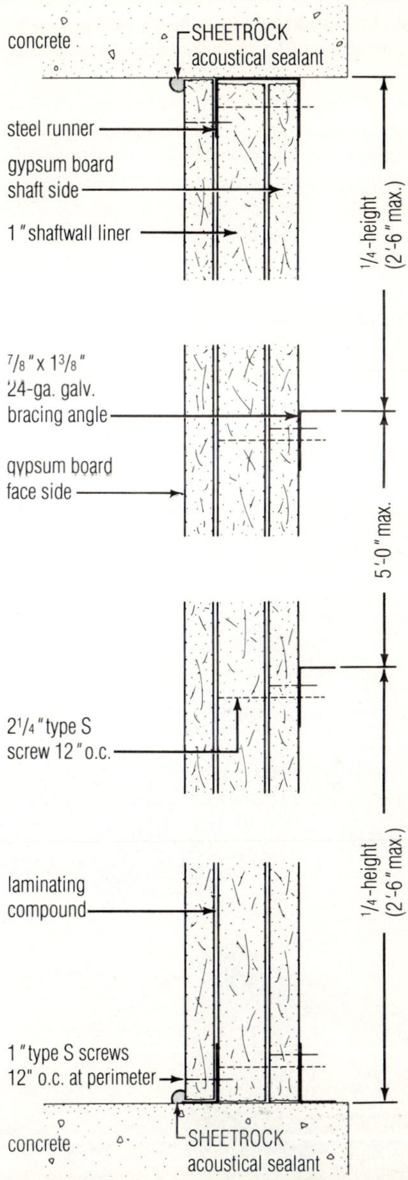

Vent shaft vertical sections

side of liner and strip or sheet-laminate to shaft-side board. Install second set of floor and sidewall angle runners (and ceiling angles, if required) with long legs against liner. Attach liner to runners and angles with 2¼″ Type S screws 12″ o.c. and at least 6″ away from liner edges. Laminate floor-side face board to liners with joint compound and install vertically. Joints should be offset 12″ from one layer to the next and moderate pressure should be applied to ensure good adhesive bond. Fasten to liner with 1½″ Type G screws. Drive screws approx. 24″ from ends of board and 36″ o.c. along lines from vertical edges. Temporary bracing may be used instead of screws to maintain bond until adhesive is hard and dry. Caulk perimeter with SHEETROCK Acoustical Sealant to prevent air infiltration. Complete the assembly with the appropriate drywall or veneer plaster finish application.

Floor/Ceiling Assemblies

Wood Frame Floor/Ceilings

These designs, which are suitable for all types of wood-framed residential and commercial buildings, include those with single and double-layer gypsum board facings, and other assemblies with THERMAFIBER Sound Attenuation Fire Blankets and resilient attachment.

Performance values of up to 2-hr. fire resistance, STC 60 and IIC 69 can be obtained as well as a non-rated fire assembly with STC 57 and IIC 53.

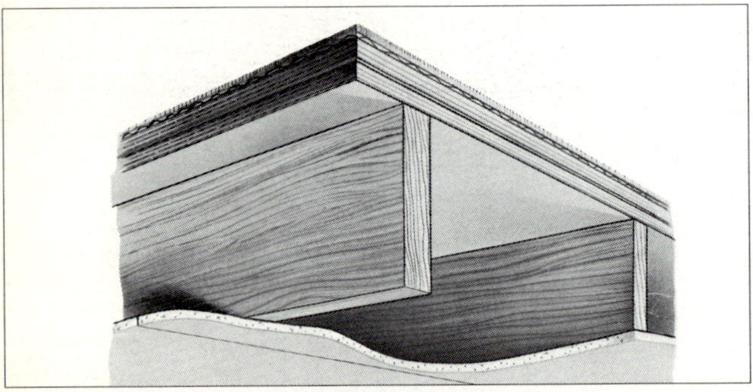

U.S. Gypsum Company publishes data for more than 20 tests conducted on resilient wood-frame ceiling assemblies including the only 1-hr. residential gypsum board system for 48″ joist spacing. For complete listings, refer to Technical Folder SA-924 and USG Construction Selector SA-100.

Sound Control Floor/Ceilings

Several floor/ceiling systems have been developed to provide exceptional sound control as well as fire resistance in wood framed assemblies. The systems require two layers of ⅝″ SHEETROCK brand Gypsum Panels, FIRECODE C Core, applied over RC-1 Resilient Channels and 3″ batts of THERMAFIBER Sound Attenuation Fire Blankets (SAFB) installed within the cavity. More detailed information is provided in Technical Folder SA-924 or Technical Data Sheets WB-1868 and WB-1869.

Noncombustible Floor/Ceilings

Noncombustible ceilings with UNIMAST Metal Furring Channels conceal and protect structural and mechanical elements above a lightweight fire-resistant layer of gypsum board. The furring channels, to which gypsum board is screw-attached, are wire-tied to bar joists or wire-tied to suspended 1½″ main runner channel grillage. Panels are also screw-attached below a direct suspension system. Plaster systems consisting of ROCKLATH Plaster Base or UNIMAST Metal Lath may also be specified.

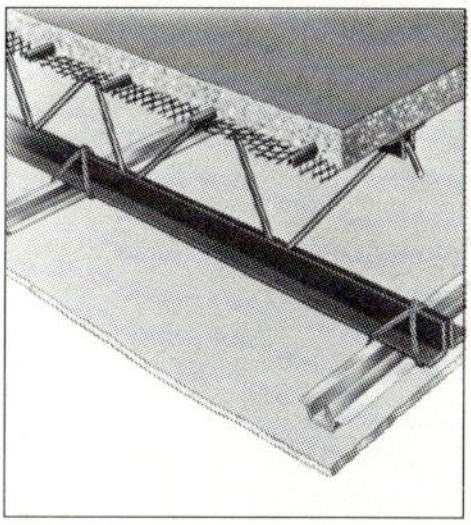

Furred Ceiling

Suspended Ceiling

For long-span suspension beneath large ducts or pipes, steel studs are substituted for furring channels. With foil-back gypsum board the ceiling is effective as a vapor retarder. Also, the board provides a firm base for adhesively applied acoustical tile.

Performance values of up to 3-hr. fire resistance (3-hr. beam) and STC 43 and IIC 60 have been obtained on certain specified systems.

Beam and Column Fire Protection

Beam Fire Protection

Beam fire protection consists of double or triple layers of ⅝″ gypsum board (FIRECODE C Core) screw-attached to framework of steel runners and metal angles. These are lightweight, easily and economically installed assemblies that provide 2-hr. and 3-hr. beam protection.

Installation

Suspension System—Install ceiling runners parallel to and at least ½″ away from beam. Position metal angles with 1⅜″ leg vertical. Fasten ceiling runners to steel floor units with ½″ Type S-12 pan head screws spaced 12″ o.c.

Fabricate channel brackets from 158CR steel runners to allow clearance at bottom of beam for 2-hr. construction and 1″ clearance for 3-hr. assembly. When steel runners are used for corner runners, cope or cut away legs of runner used for brackets to allow insertion of corner runner. When metal angles are used for corner runners, slit channel bracket runner legs and bend runner to right angle. Install channel brackets 24″ o.c. along the length of the beam and fasten to ceiling runner with ½″ Type S-12 pan head screws.

Install lower corner runners parallel to beam. Set steel runner corner runners in coped channel brackets. Apply metal angles to outside of channel brackets with the ⅞″ leg vertical, and fasten with ½″ Type S-12 pan head screws.

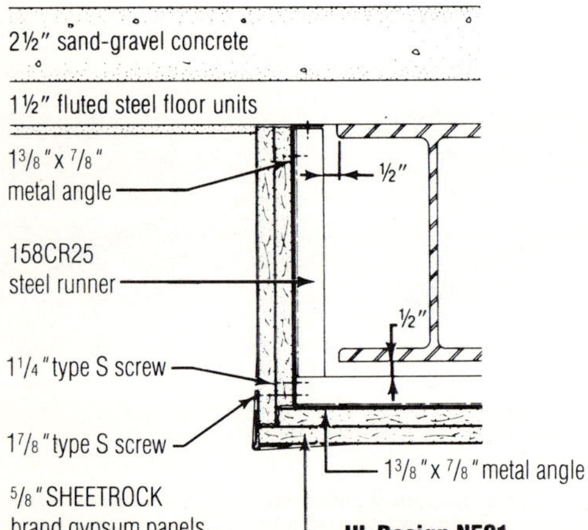

UL Design N501 (beam only)

Gypsum Board—For 2-hr. assemblies, apply vertical base-layer board and attach to ceiling and corner runners with 1¼″ Type S screws spaced 16″ o.c. Install base layer to beam soffit overlapping vertical side panels and fasten with 1¼″ Type S screws 16″ o.c. Apply face-layer boards so soffit board supports vertical side boards. Fasten face layer to runners with 1⅞″ Type S screws spaced 8″ o.c.

For 3-hr. assembly, apply base-layer boards and attach to ceiling and corner runners with 1″ Type S screws spaced 16″ o.c. Apply middle layer over base layer and attach to brackets and runners with 1⅝″ Type S screws spaced 16″ o.c. Install hexagonal mesh over middle layer at beam soffit.

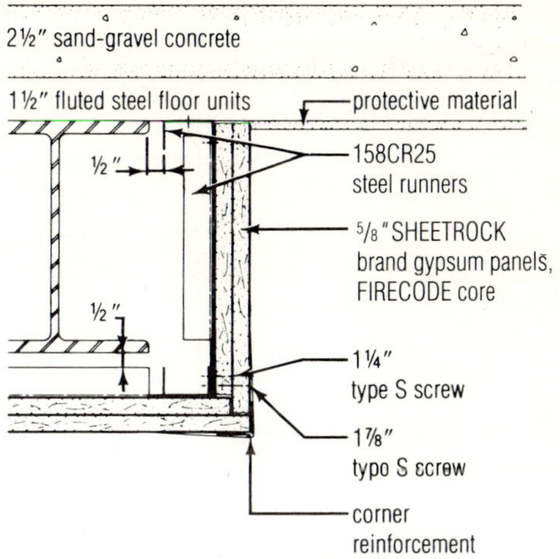

UL Design N505 (beam only)

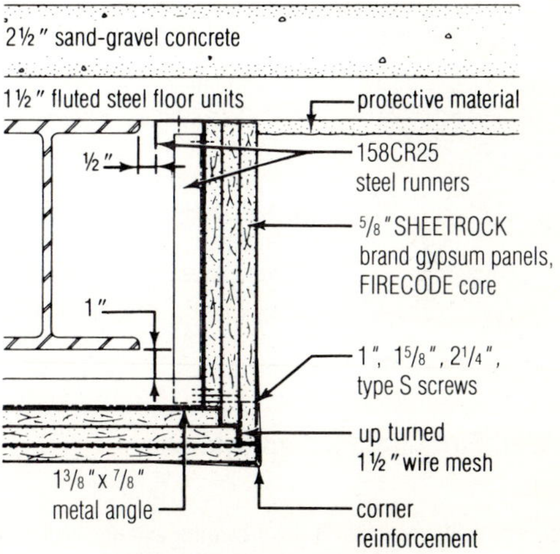

UL Design N502 (beam only)

Extend mesh 1½″ up sides of beam and hold in place with 1⅝″ screws used to attach middle layer. Apply face layer over middle layer and wire mesh, and fasten to brackets and runners with 2¼″ Type S screws spaced 8″ o.c. Apply all layers so soffit panels support vertical side boards.

Finishing Construction—Apply corner bead to bottom outside corners of face layers and finish with joint treatment as directed in Chapter 4.

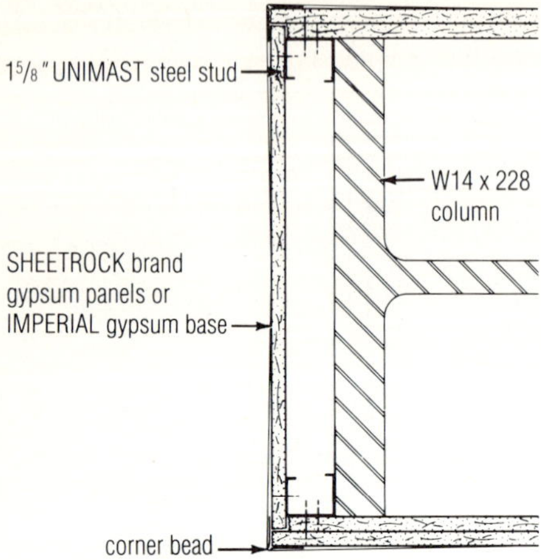

Column fireproofing - UL Des X514

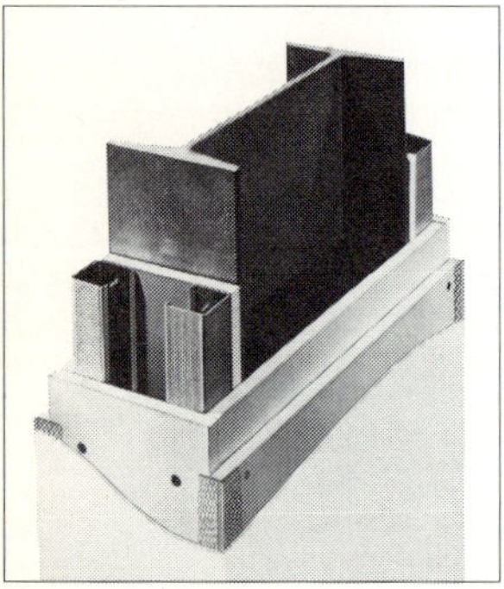

Column Fire Protection

Steel column fire protection with lightweight and compact gypsum board enclosures offers fire ratings of 2, 3 or 4 hours depending upon construction. The board is held in place by a combination of wire, screws, and steel studs. All attachments are mechanical; there's no waiting for adhesives to dry. See U.S. Gypsum Company Technical Folders SA-920 and SA-923 for more detailed information.

Curtain Wall/Fire Containment Systems

Curtain Walls

Exterior curtain walls are nonaxial load-bearing exterior walls. Steel studs used for curtain wall applications are modified channel types roll-formed from five steel thicknesses. The same studs can provide wall framing for both drywall and veneer plaster systems. Exterior surfaces can be brick veneer, portland cement-lime stucco, decorative panels or siding materials.

A wide selection of stud sizes and spacings has been identified to accommodate wind loads to 40 psf, wall heights to 32′ at 15 psf, and a variety of building modules. For load and installation data on curtain walls, design considerations, and exterior or surface finishes, refer to U.S. Gypsum Company Technical Folder SA-923. For curtain wall applications with DUROCK Exterior Cement Board, see Chapter 5 and Technical Folder SA-700.

Curtain Wall Insulation

THERMAFIBER Curtain Wall Insulation provides fire resistance, sound isolation and thermal performance for exterior curtain walls. The insulation batts are installed behind spandrel panels of glass, concrete, marble or granite for thermal control and to ensure spandrel integrity. Proper spandrel panel protection helps eliminate the "leapfrog" effect, keeping fire from lapping around the spandrel panels to re-enter the building at higher floors.

THERMAFIBER Curtain Wall Insulation (CW series) is made of mineral fiber and, when used in conjunction with THERMAFIBER Safing Insulation, offers fire protection to 2,000°F—superior to glass-fiber insulations used for curtain wall applications. For more information on insulation products and their installation, see Technical Folder SA-707.

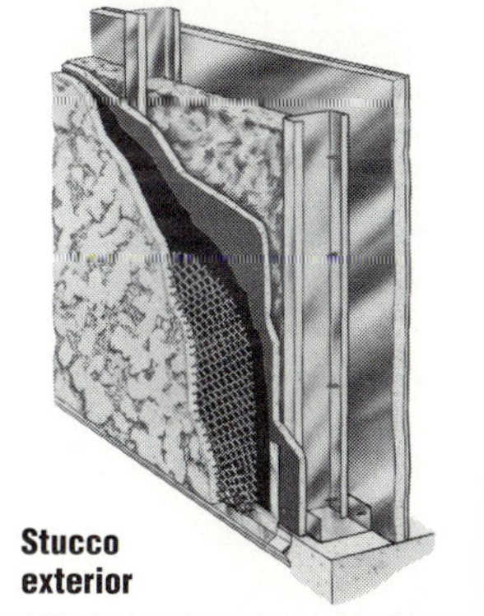

Stucco exterior

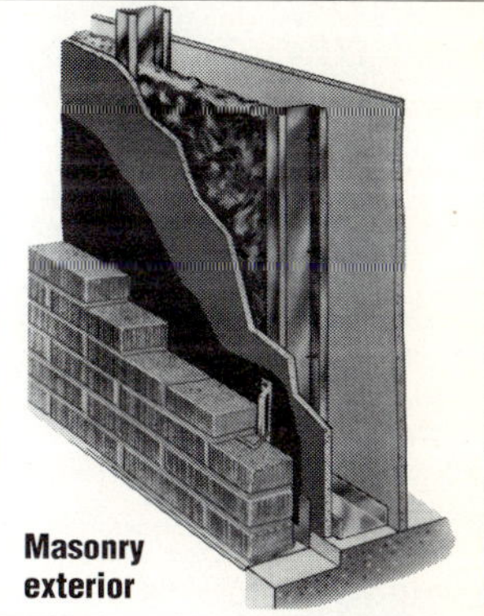

Masonry exterior

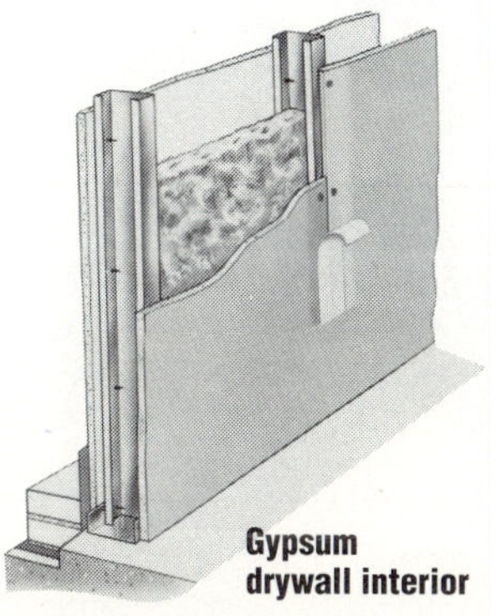

Gypsum drywall interior

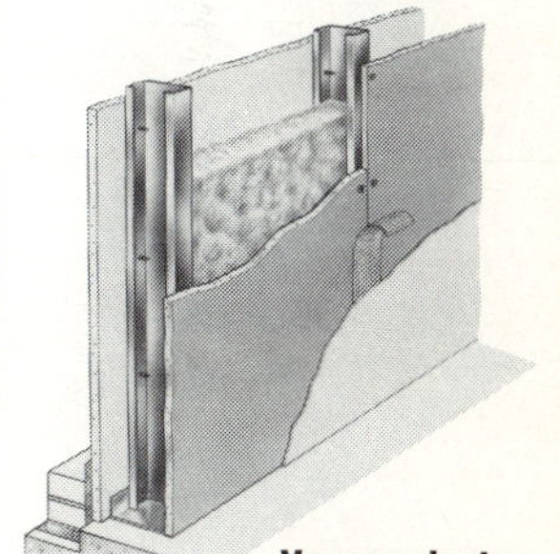

Veneer plaster interior

Safing Insulation

THERMAFIBER Safing Insulation is installed between the floor slab perimeter and the curtain wall assembly, filling all voids. Proper installation, especially with support brackets or impaling clips, eliminates the "chimney" effect and stops passage of fire upward around the floor slab perimeter. See Technical Folder SA-707 for installation information.

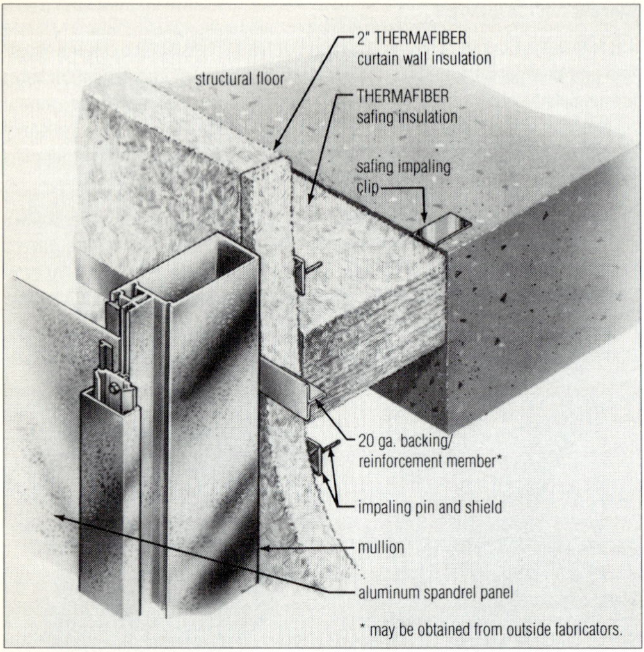

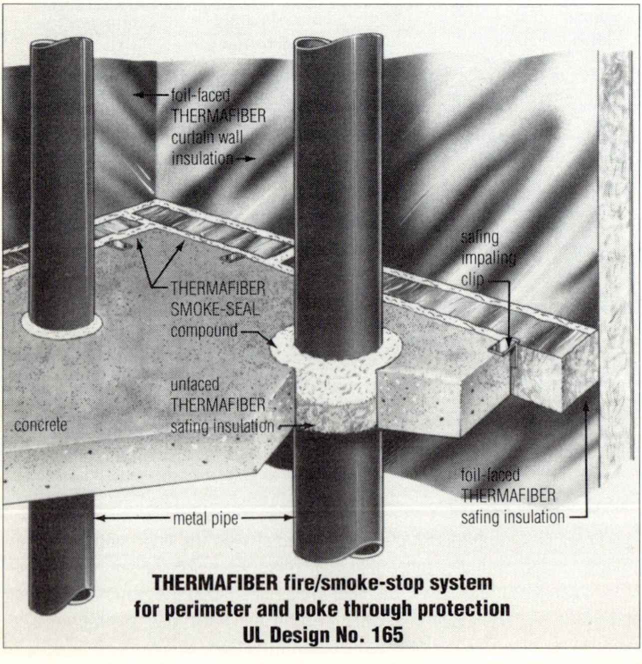

THERMAFIBER fire/smoke-stop system for perimeter and poke through protection UL Design No. 165

THERMAFIBER Fire/Smoke-Stop System for Curtain Wall Perimeters

Curtain wall and safing insulations have effectively stopped fire for many years. However, experience has shown smoke to be often more life-threatening than fire. The THERMAFIBER FIRE/Smoke-Stop System for perimeter protection combines foil-faced THERMAFIBER Curtain Wall and Safing Insulation with a specially designed fire- and smoke-resistant sealant to form an effective barrier to smoke as well as fire.

Perimeter Protection—Install THERMAFIBER Curtain Wall Insulation with foil facing to the inside of the building. Install THERMAFIBER Safing Insulation foil face up. Seal the foil seams between insulation components and at perimeter slab with THERMAFIBER SMOKE SEAL Compound.

Penetration Fire Stops

USG Fire Stop System for Floor and Wall Penetrations

Fire can pass from floor to floor or into adjacent spaces through over-sized floor or wall penetrations, including poke-through openings required for plumbing, telecommunication lines or other utility service.

USG Fire Stop System for Floor and Wall Penetrations employs FIRECODE Compound applied over THERMAFIBER Safing Insulation for a fire stop system to block smoke and flame from passing through openings in concrete floors and gypsum panel walls. Three different UL-classified through-penetration systems are available: No. 449, a 3-hr. fire-rated floor/wall system; No. 450, a 2-hr. fire-rated wall system; and No. 510, a 2-hr. fire-rated wall system. Systems meet ASTM E814 and UL 1479.

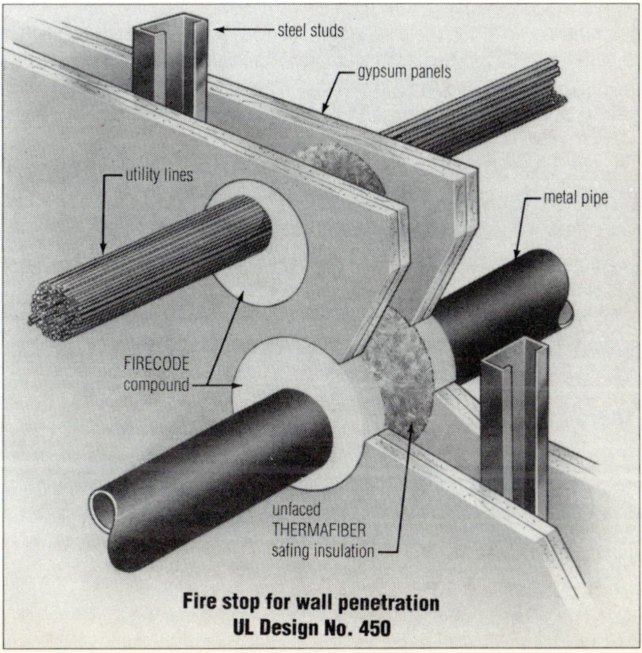

Fire stop for wall penetration
UL Design No. 450

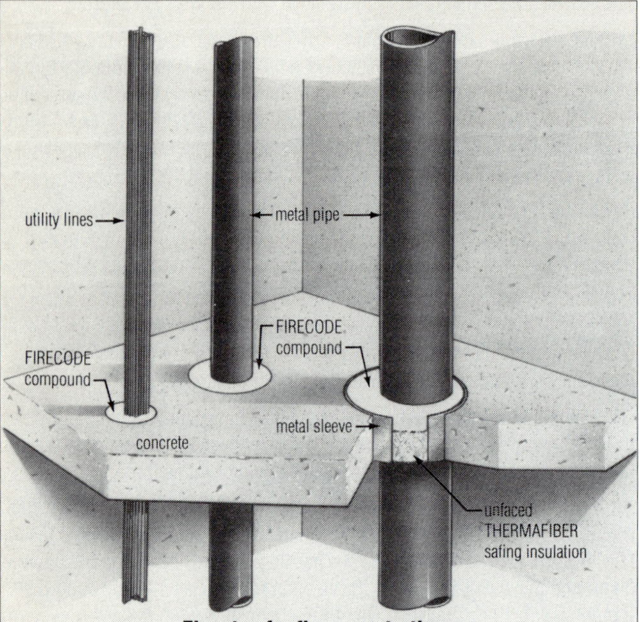

Fire stop for floor penetration
UL Design No. 449

Installation—With a serrated knife, cut THERMAFIBER Safing insulation slightly wider than the opening. Compress and tightly fit min. 3″ thick insulation with min. density of 3.5 pcf completely around penetrant. Mix FIRECODE Compound according to directions on bag. Using a trowel, putty knife or spatula, scoop the compound from its container and work it into the penetration opening. Apply compound to a min. 1″ thickness on top of safing insulation.

THERMAFIBER Fire/Smoke Stop-System for Penetrations

Friction fit 2½″ THERMAFIBER Safing Insulation to fill voids. Apply a 2″ layer of THERMAFIBER SMOKE SEAL Compound on top for floor penetration. Sandwich insulation between 2″ of compound on both sides for wall penetrations. UL-classified system: No. 165. Meets ASTM E814 and UL 1479.

Thermal Insulation

THERMAFIBER FS-15 and FS-25 Commercial Blankets are designed for installation in exterior or load-bearing walls to reduce heat transmission and provide fire protection. Both are ideal for floor/ceilings, walls or crawl spaces in wood or steel-framed assemblies. FS-15 rates flame spread 15, smoke developed 0; FS-25 rates flame spread 25, developed 5.

FIRECODE Compound mixes easily with water at jobsite. There's less waste than with caulking tube products.

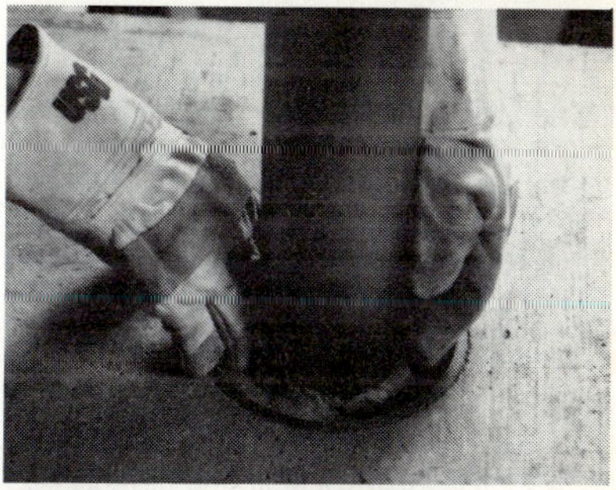

THERMAFIBER Safing Insulation, the forming material, is fit snugly into penetration.

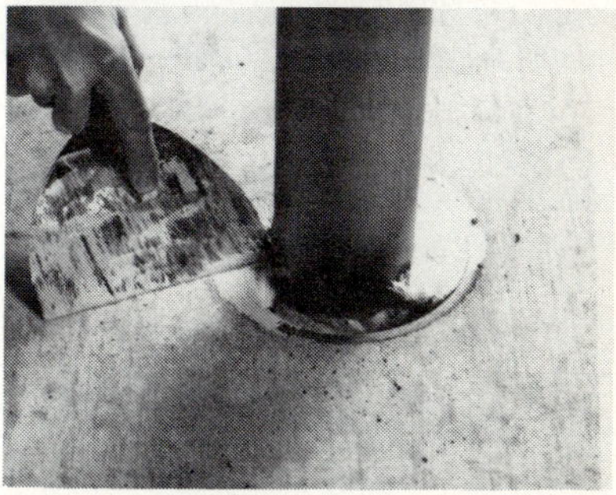

FIRECODE Compound is troweled into penetration to block particulate, fire, sound, smoke and air movement.

Air, Water and Vapor Control

Air and Water Infiltration

Flashing and sealants as shown in construction documents and as selected by the architect and/or structural engineer should be provided to resist air and water infiltration. The flashing and sealants selected shall be installed in a workmanlike manner in appropriate locations to maintain continuity of air/water barriers, particularly at windows, doors and other penetrations of exterior wall.

All gypsum sheathing must be covered with No. 15 asphalt felt or approved water and infiltration barrier to ensure water-tight construction. Asphalt felt should be applied horizontally with 2″ overlap and attached to sheathing.

Sheet should be stapled to sheathing according to manufacturer's directions. Accessories for stucco finishes should be made of zinc alloy with weep holes 12″ o.c.

Vapor Retarders and Air Barriers

Proper use and placement of vapor retarders are important factors in modern, energy-efficient construction. Improper placement of a vapor retarder could produce condensation in exterior wall stud cavities and cause deterioration of the structure.

In cold climates, a vapor retarder is required on the warm interior side of the wall to restrict moisture from the warmer, humid air inside the building from penetrating through wall surfaces and causing condensation on colder surfaces within the cavity.

In climates where high temperature and humidity conditions are sustained, placement of a vapor retarder may be recommended on the exterior side. In any case, location and placement of vapor retarders should be determined by a qualified mechanical engineer.

Two vapor retarders on opposite sides of a single wall can trap water vapor between them and create moisture-related problems in core materials.

When a polyethylene vapor retarder film is installed on ceilings behind gypsum panels under cold conditions, it is recommended that ceiling insulation (batts or blankets) be installed before the board. If loose fill insulation is to be used above the ceiling, it must be installed immediately after the ceiling board is installed during periods of cold weather. Also the plenum or attic space should be properly vented. Failure to follow this procedure can result in moisture condensation behind the gypsum panels, causing board sag.

Note: Although nearly all vapor retarders are also air barriers, not all air barriers are vapor retarders. Standard SHEETROCK brand Gypsum Panels, DUROCK Cement Board, SHEETROCK brand Gypsum Sheathing, No. 15 felt, industry building wrap and other common construction materials serve as air barriers but not as vapor retarders.

Ceiling Sag Precautions

Water-based textures, interior finishing materials and high ambient humidity conditions can produce sag in gypsum ceiling panels if adequate vapor and moisture control is not provided. The following precautions must be observed to minimize sagging of ceiling panels:

1. Where vapor retarder is required in cold weather conditions, the temperature of the gypsum ceiling panels and vapor retarder must remain above the interior air dew point temperature during and after the installation of panels and finishing materials.

2. The interior space must be adequately ventilated and air circulation must be provided to remove water vapor from the structure.

Most sag problems are caused by condensation of water within the gypsum panel. The placement of vapor retarders, climate, insulation levels and ventilation requirements will vary by location and climate, and should be reviewed by a qualified engineer if in question.

Good Design Practices

A common error in buildings with suspended ceilings is to neglect treatment of drywall surfaces within the ceiling plenum on exterior walls. Since the plenum is not visible, care should be taken to make sure that this area is not overlooked. The drywall application and joint treatment should be carried all the way to the spandrel beam or floor structure above. Exterior ceilings and soffits are other areas that may be forgotten. Ceilings, soffits and cutouts for pipe, conduit, knee braces and vent penetrations should be carefully treated to avoid compromising the effectiveness of the vapor retarder and/or air barrier.

Penetrations in the exterior wall for windows, doors, outlets, HVAC and other fixtures or devices must be closed tight with sealant or tape.

Exterior walls insulated with foil-faced THERMAFIBER Curtain Wall Insulation should also use foil tape to close joints, penetrations and damaged areas. If applied in a furred exterior wall without sheathing or backing, insulation must be adequately affixed to the framing so that it won't be dislodged by air movement in the wall cavity. Do not rely on a friction fit.

Control joints should be carefully flashed and/or sealed to prevent water infiltration. Also, particular care should be taken to ensure the integrity of the envelope for airtightness, vapor diffusion and thermal resistance, particularly at intersections and hidden penetrations. Details for floor/wall and roof/wall connections are the most difficult and important design challenges.

General Construction

CHAPTER 9 Planning, Execution & Inspection

Factors Affecting Results

Today's proven-quality products and high-performance systems permit installation of attractive, durable, trouble-free interiors that meet designers' specifications and owners' needs. By using correct installation procedures and equipment, contractors can combine these products into systems with high quality results, thus reducing customer dissatisfaction, poor productivity, callbacks and decreased profitability.

This Chapter identifies product, system, environmental, handling and storage, equipment, installation, workmanship and inspection factors that can affect the end results of a project.

Selection of Materials

In recent years, technological advances in building construction have resulted in new products and systems. Each requires systematic evaluation of performance and appearance characteristics in relation to cost, before selection and use. Evaluation may be done through benefit-cost analysis or life-cycle cost analysis, which considers the total cost of an assembly throughout its useful life. Building materials selection should always be based on total performance, including maintenance, not simply on initial construction cost or a budgeted cost figure. The following items merit consideration in systematically selecting products and systems for gypsum construction.

Satisfy User Needs

To satisfy the owner's functional requirements, it is basic to match products and systems to the end performance desired. For instance, such high-traffic areas as corridors may require hard-wearing, abuse-resistant surfaces available with high-density products. Where quiet surroundings or isolation from noise is needed, systems with high resistance to sound transmission and surfaces that provide sound absorption are essential. Predecorated, low-maintenance surfaces may be justified in the form of vinyl-faced gypsum panels. In common walls between apartments, where greater core widths are needed to enclose plumbing lines, a system with adequate space in the cavity is called for. The objective is to always select products and systems that will improve the total performance of the building components.

Meet Regulatory Requirements

The performance of gypsum construction products and systems must comply with regulatory requirements established by local, state and federal agencies. Local and state building codes and insurance and lending agency requirements should be considered in material selection.

Identify Critical Performance

Any selection of appropriate materials should reflect product or system limitations. Such structural factors as limiting height and span, required number of screws, metal thickness, bracing spacing or maximum frame and fastener spacing should not be exceeded since they affect the flexural properties and strength of an assembly. Yield strength of all steel is not the same. Substitution by size alone is not recommended. System performance following any substitution of material or compromise in assembly design cannot be certified and may result in failure under critical conditions. It is important to note that extreme and continuous high humidity or temperatures may result in sag, joint deformation, poor appearance and possible deterioration of gypsum surface materials. Sealing and painting recommendations are important for proper performance of paints and other finishes.

Establish Performance Requirements

Fire Resistance—Select fire-tested assemblies to comply with regulatory requirements, and construct the assembly according to specifications. If an assembly does not comply, work may be halted by the building inspector or installation may be rejected after completion.

Sound Control—Owner's needs and regulatory requirements dictate the sound control needed. Many assemblies are available to meet requirements. Sound test data is obtained under ideal laboratory conditions per ASTM procedures, except as noted. For assemblies to approach tested performance, strict attention must be given to construction details and installation. The isolation expected from an assembly can be negated by penetrations, perimeter leaks, accidental coupling of decoupled elements, incompatible surrounding structures and other faulty installation practices. The isolation may also be compromised by flanking sound, i.e., structure-borne sound carried via continuous concrete floors and other building elements, bypassing the sound rated assembly.

Structural Strength and Stability—Select systems that provide adequate strength and acceptable deflection under live and dead loads as described in published U.S. Gypsum Company performance tables. Shear or torque loads caused by shelving, sanitary basins, light fixtures and other accessories should also be considered. Shear forces from wind or earthquake may also require consideration. Cracking probably will occur in assemblies of insufficient strength or stiffness if adequate reinforcing is not provided.

Water and Moisture—Choose products and systems that offer adequate resistance to water and high-moisture conditions. Gypsum products are not suitable under conditions of extreme and sustained moisture. DUROCK Cement Board is recommended as a substrate for ceramic tile under these conditions. Products manufactured from steel or other materials subject to corrosion must have a protective coating equal to the service conditions envisioned.

Humidity and Temperature—Determine the environmental conditions to be expected during construction and use. Select products that offer high performance under these conditions or control the job environment. Plaster products should be installed at uniform temperatures above 55°F (13°C). These products may gradually deteriorate under sustained temperatures over 125°F (52°C). High humidity and temperatures may cause problems with veneer plaster finishes, gypsum plasters and gypsum board products.

Durability—High-strength gypsum plaster and veneer plaster products offer high compressive strength and surface hardness to resist damage from impact and abrasion. For long-lasting, problem-free interiors, select products to meet functional needs.

Appearance—Color, texture and surface gloss affect the final appearance of interior surfaces. Texture finishes offer a wide variety of effects for distinctive appearance. High-gloss finishes highlight surface defects; textures hide minor imperfections.

Cleanability and Maintenance—Select products according to functional requirements for washability and resistance to fading, staining and scuffing. Predecorated TEXTONE Vinyl-Faced Gypsum Panels offer a tough, stain-resistant vinyl surface easily cleaned with soap and water. Certain aggregated ceiling texture finishes cannot be washed but can be painted when redecoration is needed.

Light Reflection—Select colors and finishes to meet appearance standards, illumination levels and other functional requirements. Strong side-lighting from windows or surface-mounted light fixtures may reveal even minor surface imperfections. The light strikes the surface obliquely, at a very slight angle, and greatly exaggerates surface irregularities. These conditions, which demand precise installation, increase chances for callbacks and should be avoided. If critical lighting cannot be changed, the effects can be minimized by skim coating the gypsum ceiling panels, finishing the surface with rough-surfaced texture finish or installing draperies and blinds, which soften shadows. As a preventive, use strong parallel to the surface job lights to ensure a flat acceptable joint compound finish prior to priming, texturing and/or painting.

Interface and Compatibility—Materials that come into contact with each other must be compatible. Differences in thermal or hygrometric expansion, strength of substrates or basecoats in relation to finish coats, thermal conductivity and galvanic action are common problem-causing situations. The subject is too complex to be covered in detail here. Contact specific manufacturers for recommendations should questions arise. Following are some precautions of this kind associated with gypsum construction:

1. Gypsum surfaces should be isolated with control joints or other means where necessary to abut other materials, isolate structural movements, changes in shape and gross area limits.

2. Plaster may be applied directly to concrete block, however, where plaster and concrete come into contact, a bonding agent must be used.

3. Due to expansion differences, the application of high-pressure plastic laminates to gypsum panels or plaster generally is not satisfactory.

4. IMPERIAL Gypsum Base and regular SHEETROCK brand Gypsum Panels do not provide sufficient moisture resistance as a base for adhesive application of ceramic tile. Use SHEETROCK brand Gypsum Panels, Water-Resistant, or DUROCK Cement Board.

5. Install resilient thermal gaskets around metal window frames to keep condensation from damaging wall surface materials. The gasket may also reduce galvanic action and resultant corrosion, which occurs when two dissimilar metals contact in the presence of moisture.

Vapor Control—The use and proper placement of vapor retarders is extremely important in modern construction, with its increased use of thermal insulation brought about by the need for energy conservation.

Inattention to proper placement or omission of a vapor retarder with thermal insulation may result in condensation in the exterior wall stud cavities. Cold climates require a vapor retarder on the warm interior side of the wall. A vapor retarder may be required on the outside of the exterior wall for air-conditioned buildings in climates having sustained high outside temperatures and humidity. A qualified mechanical engineer should determine location of the vapor retarder.

Two vapor retarders on opposite sides of a single wall can trap water vapor between them and create moisture-related problems in the core materials.

When a polyethylene vapor retarder film is installed on ceilings behind gypsum panels under cold conditions, it is recommended that ceiling insulation (batts or blankets) be installed before the board. Also the plenum or attic space should be properly vented. Failure to follow this procedure can result in moisture condensation in the back side of the gypsum panels, causing board sag.

Handling and Storage

Even quality products can contribute to problems during application and job failures if not protected from damage and improper handling. Generally, gypsum products should be stored inside at temperatures above freezing, protected from moisture and external damage and used promptly after delivery.

Inspect on Delivery

Products should be inspected for proper quantity and possible damage when delivered on the job. Incorrect quantities may result in job delays due to shortages or extra cost for overages that are wasted. Check products for such physical damage as broken corners or scuffed edges on gypsum board, wet board, bent or corroded steel studs and runners. Inspect containers for evidence of damage that may affect the contents. Look for damaged or torn bags, which could result in waste, lumpy joint compound, preset conventional plaster or veneer plaster finishes. Report any damaged material or shortages immediately.

Store in Enclosed Shelter

Enclosed protection from the weather is required for the storage of all gypsum products. Though not recommended, outdoor storage for up to one month is permissible if products are stored above ground and completely covered. Do not store gypsum products on gypsum risers. Use wood risers to prevent moisture from wicking up and wetting material. Various problems can result when these products get wet or are exposed to direct sunlight for extended periods.

Store gypsum boards flat on a clean, dry floor to prevent permanent sag, damaged or wavy edges or deformed board. Do not store board vertically.

If board is stored on risers, the risers should be at least 2⅞″ wide and placed directly under each other vertically, within 2″ of board ends and no greater than 28″ apart for 14′ board, 23″ apart for 12′ board, 24″ apart for 10′ board, 21″ apart for 9′ board and 25″ apart for 8′ board.

Stack bagged goods and metal components on planks or platforms away from damp floors and walls. Corrosion on corner bead, trim and fasteners may bleed through finishing materials. Ready-mixed joint compounds that have been frozen and thawed repeatedly lose strength, which may weaken the bond.

Protect from Damage

Locate stored stocks of gypsum products away from heavy-traffic areas to prevent damage from other trades. Keep materials in their packages or containers until ready for use, to protect them from dirt, corrosion and distortion. Damaged board edges are more susceptible to ridging after joint treatment. Boards with rough ends will require remedial action before installation, otherwise, deformation or blistering may occur at end joints.

Use Fresh Material

If possible, gypsum construction products should be ordered for delivery to the job just before application. Materials may become damaged by abuse if stored for long periods. To minimize performance problems caused by variable moisture conditions and aging, fresh plaster and veneer plaster finishes should be received on the job frequently.

Job Conditions

Many problems can be directly traced to unfavorable job conditions. These problems may occur during product application or they may not appear until long after job completion.

Recommendations for proper job conditions, given in the appropriate product application chapters here, should be closely followed. If job conditions are unfavorable, correct them before product installation. The following environmental factors can present problems in gypsum construction.

Temperature

Install gypsum products, joint compounds and textures at comfortable working temperatures above 55°F (13°C). In cold weather, provide controlled, well-distributed heat to keep the temperature above minimum levels. For example, if gypsum board is installed at a temperature of 28°F (-2°C), it expands at the rate of ½″ for every 100 lin. ft. when the temperature is raised to 72°F (22°C). At lower temperatures, the working properties and performance of plasters, veneer plaster finishes, joint compounds and textures are seriously affected. They suffer loss of strength and bond if frozen after application and may have to be replaced. Ready-mixed compounds deteriorate from repeated freeze-thaw cycles, lose their workability and may not be usable. Avoid sudden changes in temperature, which may cause cracking from thermal shock.

Humidity

High humidity resulting from atmospheric conditions or from on the job use of such wet materials as concrete, stucco, plaster and spray fireproofing often creates situations for possible problems. In gypsum board, water vapor is absorbed, which softens the gypsum core and expands the paper. As a result, the board may sag between ceiling supports. Sustained high humidity increases chances for galvanized steel components to rust, especially in marine areas where salt air is present. High humidity can cause insufficient drying between coats of joint compounds, which can lead to delayed shrinkage and/or bond failure. Jobs may be delayed because extra time for drying is required between coats of joint compound.

Low humidity speeds drying, especially when combined with high temperatures and air circulation. These conditions may cause dryouts in veneer plaster finishes and conventional plasters. They also reduce working time and may result in edge cracking of the joint treatment. Crusting and possible contamination of fresh compound, check and edge cracking are also caused by hot and dry conditions. Under hot, dry conditions, handle gypsum board carefully to prevent cracking or core damage during erection.

Moisture

Wind-blown rain and standing water on floors increase the humidity in a structure and may cause the problems previously described. Water-soaked gypsum board and plasters have less structural strength and may sag and deform easily. Their surfaces, when damp, are extremely vulnerable to scuffing, damage and mildew.

Ventilation

Ventilation should be provided to remove excess moisture, permit proper drying of conventional gypsum plasters and joint compounds and prevent problems associated with high-humidity conditions. For veneer plaster finishes, to prevent rapid drying and possible shrinkage, poor bond, chalky surfaces and cracking, air circulation should be kept at a minimum level until the finish is set. Rapid drying also creates problems with joint compounds, gypsum plasters and finishes when they dry out before setting fully and, as a result, don't develop full strength.

Sunlight

Strong sunlight for extended periods will discolor gypsum panel face paper and make decoration difficult. The blue face paper on veneer gypsum base will fade to gray or tan from excessive exposure to sunlight or ultraviolet radiation. Applying finishes containing alkali (lime) to this degraded base may result in bond failure unless the base is treated with an alum solution or bonding agent.

Movement in Structures

Today, building frames are much lighter than former heavy masonry or massive concrete structures. Modern structural design uses lighter but stronger materials capable of spanning greater distances and extending buildings higher than ever before. While meeting current standards of building design, these frames are more flexible and offer less resistance to structural movement. This flexibility and resulting structural movement can produce stresses within the usually non-load bearing gypsum assemblies. Unless relief joints are provided to isolate these building movements, when accumulated stresses exceed the strength of the materials in the assembly, they will seek relief by cracking, buckling or crushing the finished surface.

Structural movement and most cracking problems are caused by deflection under load, physical change in materials due to temperature and humidity changes, seismic forces or a combination of these factors.

Concrete Floor Slab Deflection

Dead and live loads cause deflection in the floor slab. If this deflection is excessive, cracks can occur in partitions at the mid-point between supports. If partition installation is delayed for about two months after slabs are completed, perhaps two-thirds of the ultimate creep deflection will have taken place, reducing chances of partition cracking. This is usually a one-time non-cyclical movement.

Wind and Seismic Forces

Wind and seismic forces cause a cyclical shearing action on the building framework, which distorts the rectangular shape to an angled parallelogram. This distortion, called racking, can result in cracking and crushing of partitions adjacent to columns, floors and structural ceilings.

Structural movement

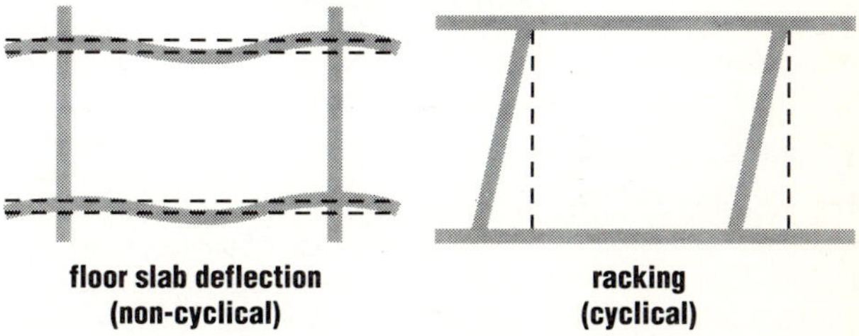

To resist this racking, building frames must be stiffened with shear walls and/or cross-bracing. Light steel-frame buildings are diagonally braced with steel strapping. Wood-frame structures are strengthened with let-in cross-bracing and/or shear diaphragms of structural sheathing. On larger buildings, racking is resisted by shear walls and wind-bracing without considering the strength added by finishing materials. Moreover, the partitions must be isolated from the structure to prevent cracking caused by racking movement and distortion.

Thermal Expansion

All materials expand with an increase in temperature and contract with a decrease. In tall concrete or steel-frame buildings, thermal expansion and contraction may cause cracking problems resulting from racking when exterior columns and beams are exposed or partially exposed to exterior temperatures. Since interior columns remain at a uniform temperature, they do not change in length.

Exposed exterior columns can be subjected to temperatures ranging from over 100° to 0°F (38° to -18°C), and therefore will elongate or contract in length. The amount of expansion or contraction of the exposed columns depends on the temperature difference and several other factors. (Structural movement caused by thermal differentials accumulates to the upper floors.) However, the stiffness of the structure resists the movement and usually full unrestrained expansion is not reached.

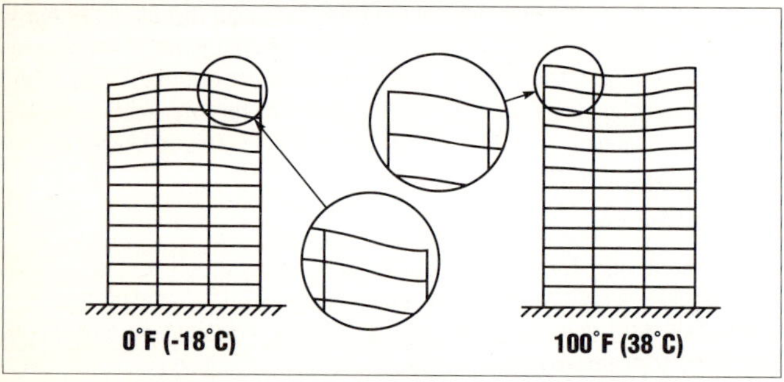

Racking, resulting from thermal movement, is greatest in the outside bays of upper floors in winter when temperature differentials are largest. To prevent major changes as described above, apply proper insulation to exterior structural members. The design should call for control joints to relieve stress and minimize cracking of surfaces.

Hygrometric Expansion

Many building materials absorb moisture from the surrounding air during periods of high humidity and expand; they contract during periods of low humidity. Gypsum, wood and paper products are more readily affected by hygrometric changes than are steel and reinforced concrete. Gypsum boards will expand about ½″ per 100′ with a relative humidity change from 13% RH to 90% RH (see Appendix for coefficients). Unless control joints are provided, hygrometric changes create stresses within the assembly, which result in bowed or wavy walls, sag between supports in ceilings, cracking and other problems.

Relief Joints

Select gypsum assemblies to provide the best structural characteristics to resist stresses imposed on them. As described previously, these systems must resist internal stresses created by expansion and contraction of the components and external stresses caused by movement of the structure. The alternative solution is to provide control and relief joints to eliminate stress buildup and still maintain structural integrity of the assembly.

To control external stresses, partitions and other gypsum construction must be relieved from the structural framework, particularly at columns, ceilings and intersections with dissimilar materials. In long partition runs and large ceiling areas, control joints are recommended to relieve internal stress buildup. Methods for providing relief and control joints are shown in Chapters 3, 5 and 7. These recommendations, for normal situations, provide for ¼″ relief. Relief joints for individual structures should be checked for adequacy by the design engineer to prevent cracking and other deformations.

Cracking in High-Rise Structures

Contractors who install commercial partitions and ceilings should be aware of cracking problems caused by structural movement, deflection, expansion and contraction. These problems, described previously, usually are not due to faulty materials. Anticipated structural movement in the frame and floor system should be taken into account in the design of the building. It is better to solve potential problems with preventive measures before installation rather than attempting repairs afterward.

Some types of construction can be expected to cause cracking in gypsum assemblies if not handled properly. Following are clues to potential problems:

Flat Plate Design—Particularly with column bay sizes exceeding 20′.

Exposed Exterior Columns and Shear Walls—On buildings over 12 floors high and located in a cold climate.

Reinforced Concrete Structures—Erected in cold weather, with partitions installed too soon thereafter. Creep deflection in the floor slab, a cause of partition cracking, is retarded in cold weather and accelerated in warm weather.

Structures Without Shear Walls or Proper Bracing—Particularly if the plan is long and narrow, presenting a large wall area to withstand wind load.

Gypsum Systems Without Expansion Joints—Long partition runs and large ceiling areas must have control joints to compensate for hygrometric and thermal expansion and contraction.

When one or more of these conditions exists, it is wise to notify the owner, architect and general contractor, by letter, of the indicated possible problems and recommend corrective measures. In this way, if corrective action is not taken, fewer questions are likely to arise when problems occur. If corrective measures are effective, all involved will be rewarded with a satisfactory performance, and costly complaints will be avoided.

Structurally Generated Noise

Loads of varying intensity can cause structural movement, which generates noise when two materials rub or work against each other. In high-rise buildings, variable wind pressure can cause a whole structure to drift or sway, causing structural deformation. Such deformation imparts racking stresses to the non-load bearing partition and can create noise.

As another annoyance, lumber shrinkage often results in subfloors and stair treads squeaking under foot traffic. This squeaking can be avoided by using adhesive to provide a tight bond between components and prevent adjacent surfaces from rubbing together.

Acoustical performance values (STC and MTC) are based on laboratory conditions. Such field conditions as lack of sealants, outlet boxes, back-to-back boxes, medicine cabinets, flanking paths, doors, windows and structure borne sound can diminish acoustical performance values. These individual conditions usually require the assessment of an acoustical engineer.

United States Gypsum Company assumes no responsibility for the prevention, cause or repair of these job-related noises.

Lumber Shrinkage

In wood-frame construction, one of the most expensive problems encountered is fastener pops, often caused by lumber shrinkage, in drywall surfaces. Shrinkage occurs as lumber dries. Even "kiln-dried" lumber can shrink, warp, bow and twist, causing board to loosen and fasteners to fail. Gypsum surfaces can also crack, buckle or develop joint deformations when attached across the wide dimension of large wood framing members such as joists. Typically, this installation occurs in stairwells and high wall surfaces where the gypsum finish passes over mid-height floor framing, as in split-level houses.

Framing lumber, as commonly used, has a moisture content of 15% to 19%. After installation, the lumber loses about 10% moisture content and consequently shrinks, particularly during the first heating season.

Wood shrinks most in the direction of the growth rings (flat grain), somewhat less across the growth rings (edge grain) and very little along the grain (longitudinally). Shrinkage tends to be most pronounced away from outside edges and toward the center of the member. When nails are driven toward the central axis, shrinkage leaves a space between the board and the nailing surface, as shown in the drawing below.

Based on experiments conducted by the Forest Products Laboratory and Purdue University, the use of shorter nails results in less space left between the board and nailing surface after shrinkage (shown on next page) than with longer nails having more penetration. Using the shortest nail possible with adequate holding power will result in less popping due to shrinkage. Longer nails, however, usually are required for fire rated construction, as specified by the experiments. Choose the proper nail length from the Selector Guide for Gypsum Board Nails on page 75.

The annular drywall nail, with an overall length of 1¼″, has equivalent holding power to a 1⅝″ coated cooler-type nail, but the shorter length of the nail lessens the chances for nail popping due to lumber shrinkage.

Contractors can take several preventive measures to minimize fastener failures and structural cracking resulting from lumber shrinkage. Type W screws are even better than the nail because they develop greater holding power and thus reduce possibilities for fastener pops. The floating interior angle system effectively reduces angle cracking and nail pops resulting from stresses at intersections of walls and ceilings. Gypsum boards should be floated over the side face of joists and headers and not attached. To minimize buckling and cracking in wall expanses exceeding one floor in height, either float the board over second-floor joists using resilient channels or install a horizontal control joint at this point.

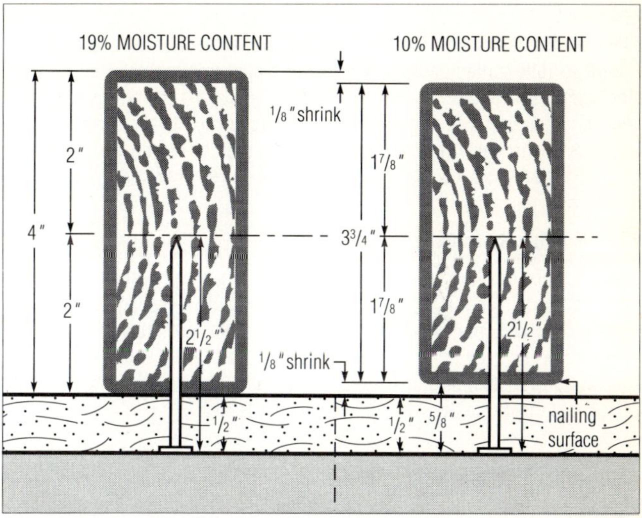

Using 2½" nails.

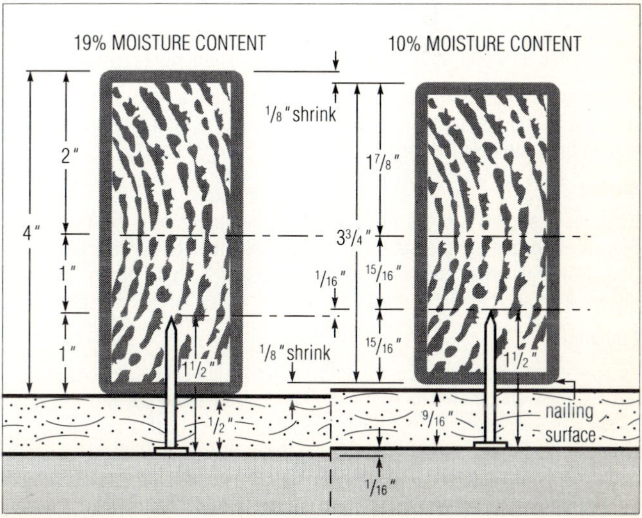

Using 1½" nails.

Workmanship

U.S. Gypsum Company products are quality-tested and job-proven for fast, economical installation and problem-free results. Unfortunately, sometimes these products fail to achieve optimum performance after installation due to improper or unspecified application.

Follow Current Directions

The major cause of job problems and poor performance after application is failure to follow manufacturer's directions and architect's specifications. Application procedures should be checked regularly to conform with current manufacturer's recommendations. Product modifications to upgrade in-place performance may require slight changes in mixing or application methods. New products may require the adoption of entire new procedures and techniques.

Meet Specifications Fully

Building specifications are designed to provide a given result, but unless specified construction materials and methods are used and the proper details followed, the actual job performance will probably fall short of requirements. Excessive water usage, oversanding, improper surface preparation, substitution of materials, skimping and shortcuts should not be tolerated because they lead, inevitably, to problems.

Equipment Selection

A large selection of equipment is available for gypsum construction and particularly for mechanical application of veneer plaster and texture finishes, conventional plasters and joint treatment. The mechanical spray equipment chosen should be based on the type of material and the situations presented on each job. The size of the job, delivery volume required, portability and access through doorways also deserve consideration. Low maintenance and accessibility of parts for cleanup are important factors.

Using the wrong equipment for the job can cause serious problems. Improper equipment affects production as well as strength, workability, setting time and finished appearance.

Mixing

Equipment should provide the correct mixing action and mixing speed. Equally important are proportioning and mixing procedures required for the particular material as shown in Chapters 4, 5 and 7. Poor mixing practices adversely affect material performance and can cause various problems.

Pumping

Equipment should have capacity sufficient for the job, hose size and pumping distance, and should be kept in good repair. To minimize abrasive wear in the pump mechanism, the pump type should be suitable for the aggregate and mixes being used. High plaster/sand ratios, small-diameter hoses and leakage increase the possibility of aggregate packing in the pump and hose. Use large-diameter hoses and no more hose length than needed. Small-diameter, long hoses cause pumps to wear faster and may lead to quick-set and low-strength problems in fluid materials.

Spraying

Nozzle or orifice size of the spray gun and air pressure used must be suitable for the material being applied. Improper nozzles and incorrect air pressures affect the spray pattern and may cause stoppage and aggregate fallout. With most veneer plaster finishing, a catalyst tank with metering device is required to adjust setting time.

Product Quality

Gypsum construction products from U.S. Gypsum Company provide the essential requirements of economy, problem-free installation and high performance in fire and sound-rated systems. During manufacture, these products are carefully controlled to meet specific performance standards when applied according to directions and under proper job conditions.

Complaint Procedure

Should a probable product deficiency appear, stop using the suspected defective material immediately and ask your supplier to notify U.S. Gypsum Company at once so that a representative can investigate the complaint and take remedial action. Do not continue to use improperly performing materials because the labor cost of replacement or reworking far exceeds the material cost.

Sampling

For analyzing suspected materials, obtain samples of the material that fully represent the complaint condition. Save bags, wrappers and packages that will identify place and time of manufacture. For some complaints, samples of related materials such as aggregates, are also necessary. Weather conditions, mixing times and proportions of ingredients should be fully reported.

Substitution and Certification

U.S. Gypsum Company will provide test certification for published fire, sound and structural data covering systems designed and constructed according to its specifications. Tests on U.S. Gypsum Company products are conducted to meet the exact performance requirements of established test procedures specified by various building code agencies. Any substitution of materials or compromise in assembly design cannot be certified and may result in failure of a system in service, especially under critical conditions of load or fire exposure.

How to Inspect a Job

Proper job inspection during installation many times reveals potential problem areas or procedures that produce unsatisfactory results. Corrective action taken immediately is usually less costly than callbacks to repair and perhaps rebuild walls and ceilings after the job is completed.

A complete understanding of job details, schedules and specifications is necessary to conduct proper inspection. If the assembly is to meet fire and sound-rating requirements, then construction details must also be known.

All walls and ceilings must be judged by these criteria and the contract conditions. Thus, it is important that drawings and specifications be complete, accurate and easily understood.

The job inspection phase of supervision is most important and, in many cases, will determine the success of the job. An accurate check should be made of the following major categories so that best results can be obtained.

Schedule of Inspection

Make job inspections at the following stages:

1. When job is almost ready for materials delivery, in order to check environmental conditions and plan for delivery.

2. When materials are delivered to the job.

3. When framing is erected but before board or lath application.

4. When gypsum board base layer and/or face layer are applied.

5. When joints are treated; when veneer plaster finish or conventional plaster is applied.

6. When job is completed.

Delivery and Storage

When materials are delivered, check the following:

1. See that materials meet specifications and are in good condition.

2. Store gypsum boards flat, on the floor; store plasters and bagged goods flat, on a raised platform. Protect from moisture and damage by abuse.

3. Protect framing materials from damage and moisture.

Framing Inspection

Framing members, either wood or metal, must meet architect's specifications and be free of defects. During and after framing construction, make the following inspections:

1. See that wood and steel framing materials meet specifications as required by local building codes, regulations and standards.

2. Check accuracy of alignment and position of framing, including bracing if required, according to plans and details. Make sure load-bearing studs are directly underneath the members they support.

3. See that partitions are straight and true; ceilings level.

4. Measure spacing of studs and joists. Spacing should not exceed maximum allowable for the system.

5. Look for protrusions of blocking, bridging or piping, and twisted studs and joists that would create an uneven surface. Correct situation before board attachment.

6. Make sure there is appropriate blocking and support for fixtures and board.

7. See that window and door frames, electrical and plumbing fixtures are set for the board thickness used.

8. Check for proper position and attachment of resilient and furring channels.

9. Review all wood and steel framing for compliance with minimum framing requirements outlined in Chapter 2.

10. Examine steel studs at corners, intersections, terminals, shelf-walls, door and borrowed light frames for positive attachment to floor and ceiling runners. All load-bearing and curtain wall studs must be attached to runner each side, top and bottom. All load-bearing studs should sit tight against web of runner.

11. Inspect spliced steel components for proper assembly. (Shaft wall and curtain wall studs and load-bearing framing should not be spliced.)

12. See that steel stud flanges in field all face the same direction.

13. See that preset door frames are independently fastened to floor slab and that borrowed light frames are securely attached to stud and runner rough framing at all jamb anchors.

14. Make sure that door and borrowed light frames will be spot-grouted, as required.

Suspended Grillage

1. Measure spacing of hangers, channels and studs to see that they are within allowable limits.

2. Check ends of main runner and furring channels. They should not be let into or supported by abutting walls, and should extend to within 6″ of the wall to support a furring channel.

3. Make sure furring channel clips are alternated and that furring channel splices are properly made.

4. See that mechanical equipment is independently supported and does not depend upon the grillage for support.

5. Inspect construction around light fixtures and openings to see that recommended reinforced channel support is provided.

Inspecting Drywall and Veneer Plaster Installations

Base Layer

1. Verify that material being used complies with specifications and requirements of fire or sound rating.

2. Make sure that proper perpendicular or parallel application of board is being used and that end joints are staggered.

3. Check for cracked and damaged-edge panels; see that they are not used.

4. See that the recommended fasteners are being used, spaced and set properly.

5. Check for proper use of acoustical sealant.

6. Inspect installation to make sure thermal insulating or sound attenuation fire blankets are properly attached and fitted.

7. Be certain vapor retarder is installed and sealed as required.

8. Review appropriate system construction and application, and inspect for compliance with laminating recommendations and other construction procedures. On fire-rated assemblies, be sure a SHEETROCK Setting-Type (DURABOND) or Lightweight Setting-Type (EASY SAND) Joint Compound is used for lamination.

9. See that required control joints are properly located and installed.

Face Layer

1. Verify material compliance.

2. Look for high-quality workmanship. Cracked or damaged-edge boards should not be used. Board surfaces should be free of defects; joints correctly butted and staggered.

3. Check for proper application method—perpendicular or parallel.

4. Examine fasteners for compliance with specifications, proper spacing and application.

5. Review adhesive application method and see that recommendations and specifications are being followed. Under adverse drying conditions resulting from high humidity, at either high or low temperature, drying of the laminating compound could be prolonged. Consult the drying time table on page 200 for guidance.

6. Inspect trim, corner beads and related components for alignment, grounds, secure attachment and proper installation.

7. Make sure that acoustical sealant is applied around electrical outlets and other penetrations and that it completely seals the void.

Fasteners

1. Make sure recommended or specified fasteners are used.

2. See that fasteners are applied in such a manner that the board hangs flat against the framing without binding.

3. Observe whether board is held tightly against framing during application. Test for loose board by pushing adjacent to the fastener. See that face paper is not broken when fastener is driven. If necessary, a second fastener should be driven within 1½″ of the faulty one.

4. Examine fastener positions. Fasteners should be at least ⅜″ in from edges and ends.

5. Make sure that fastener heads in veneer plaster assemblies are flush with the gypsum base surface, not dimpled.

Adhesives

1. See that adhesive is applied to clean, dry surfaces only.

2. So proper bond can be obtained, make sure that board is erected within allowable time limit after adhesive is applied.

3. Measure size of bead and spacing, and see that a sufficient quantity is applied.

4. Observe impacting blows for proper spacing and positioning.

5. Make sure temporary fastening and shoring holds panel tightly in place.

6. Review appropriate adhesive application methods (see Chapter 3) and inspect for compliance.

Joint Treatment—Drywall

1. Make sure panel surface is ready for joint treatment. Fastener heads should be properly seated below panel surface. Protrusions should be sanded below level of surface. Joints between panels should be filled with joint compound before taping.

2. See that recommended mixing directions are followed (see Chapter 4). Only clean water and mixing equipment should be used. SHEETROCK Setting-Type (DURABOND) and Lightweight Setting-Type (EASY SAND) Joint Compounds cannot be held over or retempered.

3. Inspect joints and corners to see that tape is properly embedded and covered promptly with a thin coat of joint compound. Only compounds suitable for embedding should be used. Avoid heavy fills.

4. Make sure compound is used at its heaviest workable consistency and not overthinned with water.

5. Make sure joint compound is allowed to dry thoroughly between coats (see drying time guide on page 200). Exception: SHEETROCK Setting-Type (DURABOND) and Lightweight Setting-Type (EASY SAND) Joint Compounds need only be set prior to a subsequent application.

6. Inspect second and third coats over joints for smoothness and proper edge feathering.

7. See that fastener heads and metal trim are completely covered.

8. See that the paper surface of the gypsum board has not been damaged by sanding.

9. Make sure that all finished joints are smooth, dry, dust free and sealed before decoration.

Inspecting Joint Treatment and Finish—Veneer Plaster

1. See that corner bead is properly attached and aligned at all outside corners.

2. See that control joints are properly installed where required.

3. See that proper joint reinforcement is used—IMPERIAL Tape for normal conditions over wood framing. For abnormal job or weather conditions and jobs with steel stud framing, joint treatment must be SHEETROCK Joint Tape covered with a thin coat of SHEETROCK Setting Type (DURABOND) Joint Compound.

4. See that IMPERIAL Tape is not overlapped at intersections.

5. Be sure that all taped, preset IMPERIAL Base joints are set before finish application begins.

6. Be sure that no gypsum base with faded face paper is installed.

Inspecting Conventional Plaster Installations

Plaster Base

1. See that material being used complies with specifications and fire or sound-tested construction.

2. Review appropriate system construction and application, and inspect for proper installation practices.

3. Check for proper application of base perpendicular to framing members, and see that end joints are staggered.

4. Check for cracked and damaged edges of plaster base. These should not be used.

5. Be sure recommended fasteners or clips are used and spaced properly.

6. Check for proper use of acoustical sealant.

7. Inspect installation to make sure that insulating blankets are properly attached and fitted.

8. Be sure adequate supports are in place for fixture and cabinet application.

Grounds for Plastering

The thickness of basecoat plaster is one of the most important elements of a good plaster job. To ensure proper thickness of plaster, grounds should be properly set and followed. Check the following points:

1. All openings should have specified plaster grounds applied as directed.

2. If plaster screeds are used, the dots and continuous strips of plaster forming the screed must be applied to the ground thickness to permit proper plumbing and leveling.

3. Grounds should be set for recommended minimum thickness for particular plaster base being used (see Chapter 7).

4. Control joints should be installed as required for materials and construction with lath separated behind joint.

Job Conditions for Plastering

This phase of inspection is also important. Periodically make an accurate check of the following points:

1. At no time should plastering be permitted without proper heating and ventilation. Circulation of air is necessary to carry off excess moisture in the plaster, and a uniform temperature in a comfortable working range helps to avoid structural movement due to temperature differential.

2. To prevent "dryouts," precautions must be taken against rapid drying before plaster set has occurred. Check temperature during damp, cold weather where artificial heat is provided. During hot, dry summer weather, cover window and door openings to prevent rapid drying due to uneven air circulation.

Plaster Application

After determining what materials are to be used on the job, refer to correct mixing and application procedures described in Chapter 7.

The visible success of the job is at stake with the finish plaster coat, and required measures should be taken to finish correctly:

1. Check plaster type and mixing operation.

2. See that proper plaster thickness is maintained.

3. Inspect plaster surfaces during drying. Setting of basecoat plaster is indicated by hardening of plaster and darkening of surface as set takes place. Plaster that has set but not yet thoroughly dried will be darker in color than the unset portion. This accounts for the mottled effect as the plaster sets.

4. Consult architect's specifications to see that proper surface finish method is being used.

5. Check temperature of building for proper finish plaster drying conditions.

Cleanup

For a complete job, cleanup is the final stage. All scaffolding, empty containers and excess materials should be removed from the job site. Floors should be swept and the building and site left in good condition for decoration and finishing.

General Construction

CHAPTER 10 Problems, Remedies & Preventive Measures

Chapter 9 discussed problems associated with gypsum construction, many of which are beyond the control of contractors working from construction documents.

Other problems, resulting from improper job conditions and application practices, are the direct responsibility of the contractor and are controllable. In this Chapter these problems as well as corrective remedies and preventive measures are discussed.

Drywall Construction

Almost invariably, unsatisfactory results show up first in the areas over joints or fastener heads. Improper application of either the board or joint treatment may be at fault, but other conditions existing on the job can be equally responsible for reducing the quality of the finished gypsum board surface.

To help determine the cause of a problem, what follows is physical description of each defect along with a discussion of the factors causing unsatisfactory results. Also provided is a checklist that identifies possible causes for the irregularity as well as an index to the numerically listed problems, causes, remedies and preventions. By checking each numerical item listed for the defect, the exact problem cause can be determined and corrected.

Description of Defect

Fastener Imperfections—A common defect, which takes on many forms. May appear as darkening, localized cracking; a depression over fastener heads; pop or protrusion of the fastener or the surface area immediately surrounding the fastener. Usually caused by improper framing or fastener application.

Joint Defects—Generally occur in a straight-line pattern and appear as ridges, depressions or blisters at the joints, or darkening over the joints or in adjacent panel areas. Imperfections may result from incorrect framing or joint treatment, or climatic conditions if remedial action has not been taken.

Loose Panels—Board does not have tight contact with framing, rattles when impacted or moves when pressure is applied to the surface. Caused by improper application of panels, framing out of alignment or improper fastening.

Joint Cracking—Appears either directly over the long edge or butt ends of boards, or may appear along the edge of taped joints. Often caused by structural movement and/or hygrometric and thermal expansion and contraction, or by excessively fast drying of joint compounds.

Field Cracking—Usually appears as diagonal crack originating from a corner of a partition or intersection with structural elements. Also seen directly over a structural element in center of a partition. May originate from corners of doors, light fixtures and other weak areas in the surface created by penetration. Caused by structural movement described previously. Also, see "Door and Window Openings" on page 119 for use of control joints to minimize cracking.

Angle Cracking—Appears directly in the apex of wall-ceiling or interior angles where partitions intersect. Also can appear as cracking at edge of paper reinforcing tape near surface intersections. Can be caused by structural movement or improper application of joint compound in corner angle.

Bead Cracking—Shows up along edge of flange. Caused by improper bead attachment, faulty bead or joint compound application.

Wavy Surfaces—Boards are not flat but have a bowed or undulating surface. Caused by improper board fit, misaligned framing, hygrometric or thermal expansion. Also see "Handling and Storage" on page 128 for proper procedure to keep boards flat before installation.

Board Sag—Occurs in ceilings, usually under high-humidity conditions. Caused by insufficient framing support for board; board too thin for span; poor job conditions; improperly installed or mislocated vapor retarder; use of unsupported insulation directly on ceiling panels; or improperly fitted panels. Refer to appropriate chapters for proper job ventilation, storage and frame spacing, particularly with water-based texture finishes.

Surface Defects—Fractured, damaged or crushed boards after application may be caused by abuse or lumber shrinkage. Also, see "Discoloration" below.

Discoloration—Board surface has slight difference in color over joints, supports or fasteners. Caused by improper paint finishing, uneven soiling and darkening from aging or ultraviolet light.

Water Damage—Stains, paper bond failure, softness in board core or mildew growth are caused by sustained high humidity, standing water and improper protection from water leakage during transit and storage. See pages 128-129 for proper handling, storage and environmental conditions.

Checklist for Drywall Problems

To find the specific cause for a problem described above, check, on the following pages, all numerical references listed in the particular category.

Fastener imperfections	5, 6, 7, 10, 11, 12, 13, 14, 24, 25, 26, 29
Joint defects	1, 6, 8, 9, 16, 20, 21, 22, 23, 26, 27, 28, 29
Loose panels	5, 6, 7, 9, 10, 11, 12, 13, 14, 24, 25
Joint cracking	6, 9, 17, 18, 19, 23
Field cracking	15
Angle cracking	17, 19
Bead cracking	17
Wavy surfaces	5, 9, 19
Board sag	9, 12, 31
Surface defects	2, 15, 27, 28, 29, 30
Discoloration	26, 27, 28, 29
Water damage	2, 4

1. Panels—Damaged Edges

Cause: Paper-bound edges have been damaged or abused; may result in ply separation along edge or in loosening of paper from gypsum core, or may fracture or powder the core itself. Damaged edges are more susceptible to ridging after joint treatment (Fig. 1).

Remedy: Cut back any severely damaged edges to sound board before application.

Prevention: Avoid using board with damaged edges that may easily be compressed or can swell upon contact with moisture. Handle gypsum panels with reasonable care.

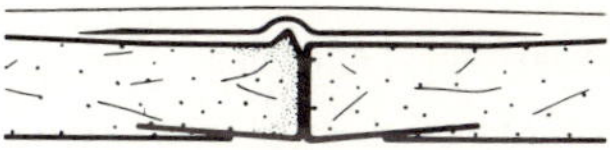

Fig. 1

2. Panels—Water-Damaged

Cause: During transit or storage, water has damaged panels from heavy rain, floods, broken pipes, etc. Water-damaged panels may be subject to scuffing and may develop paper bond failure or paper delamination from the gypsum core after application. Dissolved glue from bundling tapes may damage board faces and cause them to stick together. If stored wet, may be subject to mildew. Prolonged soaking or exposure to water can soften gypsum core and destroy bond of the paper to the core.

Remedy: The amount of water exposure and the length of time exposed are both critical factors in preventing excessive losses. As soon as possible, dry wet board completely before using. Moisture damage delamination should not be present after thorough drying. If it is, remove loose paper and skim coat area with SHEETROCK Setting-Type (DURABOND) Joint Compound. Replace board if there is extensive loose paper. Handle board cautiously and re-pile with bundles separated by spacer strips of gypsum board. Check incoming board for water stains or dampness. Protect carefully during shipment and storage. Do not erect damp panels; this may result in paper bond failure. Replace boards that have soft cores.

Prevention: Protect from high moisture conditions of any kind.

3. Panels—Paper Delamination

Cause: Manufacturing conditions, water damage.

Remedy: Manufacturing conditions or water damage causing delamination often can be treated as above. If board is received on job with paper delaminating, inspect delivery to determine extent of damage. Do not install or finish prior to contacting U.S. Gypsum Company representative. If delamination is minor, peel back paper to where it soundly bonds to board and treat with joint compound.

Prevention: None.

4. Panels—Mildew

Cause: Mildew can occur on almost any surface depending on heat and humidity conditions. Gypsum panels that have become wet for any reason are susceptible to mildew growth.

Remedy: A liquid bleach solution of 1 cup bleach to 3 cups of water may be used to clean moderately affected surfaces. Proper ventilation and/or heat should be used to thoroughly dry the affected area. Mildew growth may occur again if proper conditions are not maintained.

Prevention: Keep gypsum panels and the job site area as dry as possible to prevent mildew spores from blooming.

---

5. Framing—Members Out of Alignment

Cause: Due to misaligned top plate and stud, hammering at points "X" (Fig. 2) as panels are applied on both sides of partition will probably result in nailheads puncturing paper or cracking board. Framing members more than ¼" out of alignment with adjacent members make it difficult to bring panels into firm contact with all nailing surfaces.

Remedy: Remove or drive in problem fasteners and only drive new fasteners into members in solid contact with board.

Prevention: Check alignment of studs, joists, headers, blocking and plates before applying panels, and correct before proceeding. Straighten badly bowed or crowned members. Shim out flush with adjoining surfaces. Use adhesive attachment.

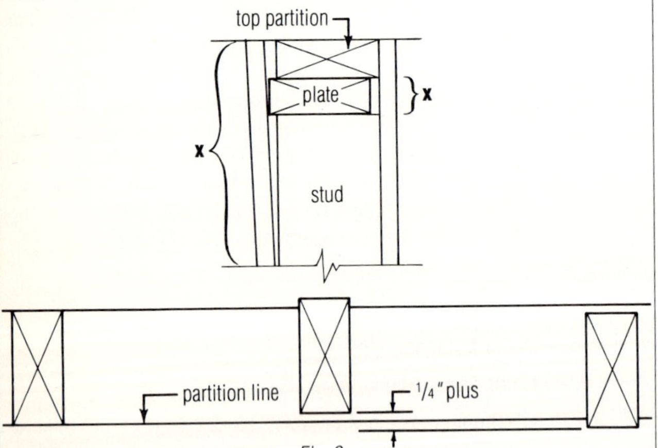

Fig. 2

6. Framing—Members Twisted

Cause: Framing members have not been properly squared with plates, presenting angular nailing surface (Fig. 3). When panels are applied, there is danger of puncturing paper with fastener heads or of reverse twisting of member as it dries out, with consequent loosening of board and probable fastener pops. Warped or wet dimension lumber may contribute to deformity.

Remedy: After moisture content in framing has stabilized, remove problem fasteners and re-fasten with carefully driven Type W screws.

Prevention: Align all twisted framing members before board application. Also, see wood framing requirements on page 100.

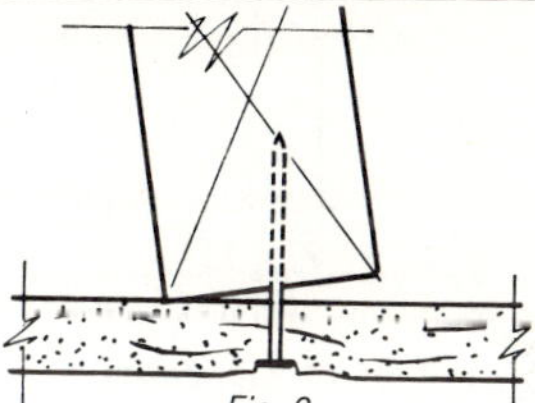

Fig. 3

7. Framing—Protrusions

Cause: Bridging, headers, fire stops or mechanical lines have been installed improperly so as to project beyond face of framing, preventing panels from contacting nailing surface (Fig. 4). Result will be loose board, and fasteners driven in area of protrusion will probably puncture face paper.

Remedy and Prevention: Same as for "Framing—Members Twisted," above.

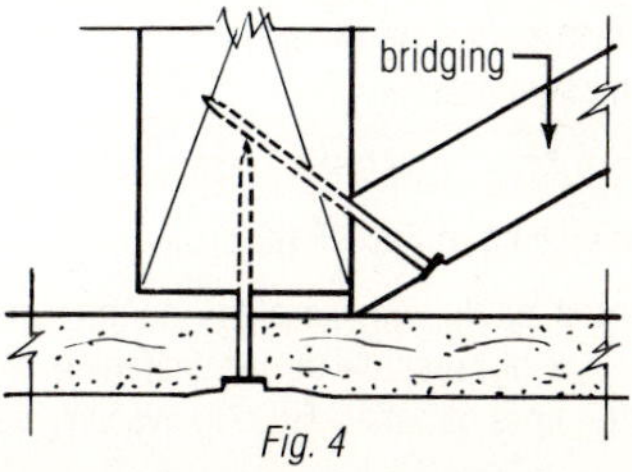

Fig. 4

8. Framing (Steel)—Panel Edges Out of Alignment

Cause: Improper placement of steel studs or advancing in the wrong direction when installing panels can cause misalignment of panel edges and give the appearance of ridging when finished.

Remedy: Fill and feather out joint with joint treatment.

Prevention: Install steel studs with all flanges pointed in the same direction. Then install panels by advancing in the direction opposite the flange direction (Fig. 5).

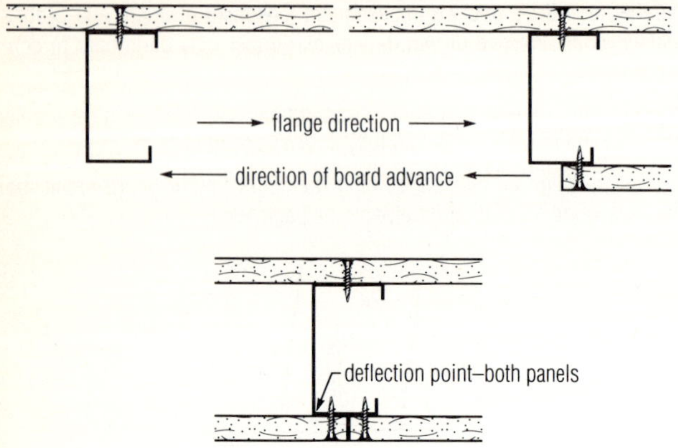

Correct application

Fig. 5

9. Panels—Improperly Fitted

Cause: Forcibly wedging an oversize panel into place bows the panel and builds in stresses preventing it from contacting the framing (Fig. 6). The result: following fastening, a high percentage of fasteners on the central studs probably will puncture the paper. May also cause joint deformation.

Remedy: Remove panel, cut to fit properly and replace. Fasten panels so that the board hangs flat against framing without binding against previously installed panels or framing. Apply pressure to hold panel tightly against framing while driving fasteners.

10. Fasteners—Puncturing of Face Paper

Cause: Poorly formed nailheads, careless nailing, excessively dry face paper or soft core, lack of pressure during fastening. Nailheads that puncture paper and shatter core of panel (Fig. 7) have very little grip on board.

Remedy: Remove improperly driven fastener, hold panel tightly and properly drive new fastener.

Prevention: Correction of faulty framing (see previous Framing Problems) and properly driven nails produce tight attachment with slight uniform dimple (Fig. 8). Nailhead bears on paper and holds panel securely against framing member. Use proper fastener or adhesive application. Screws with specially contoured head are best fastener known to eliminate cutting and fracturing. If face paper becomes dry and brittle, its low moisture content may aggravate nail cutting. Raise moisture content of board and humidity in work area.

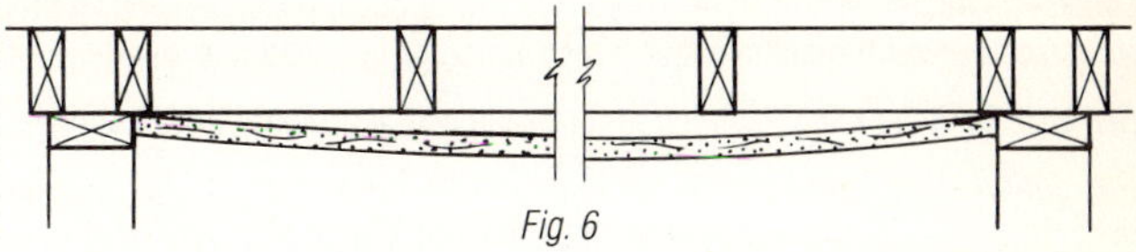

Fig. 6

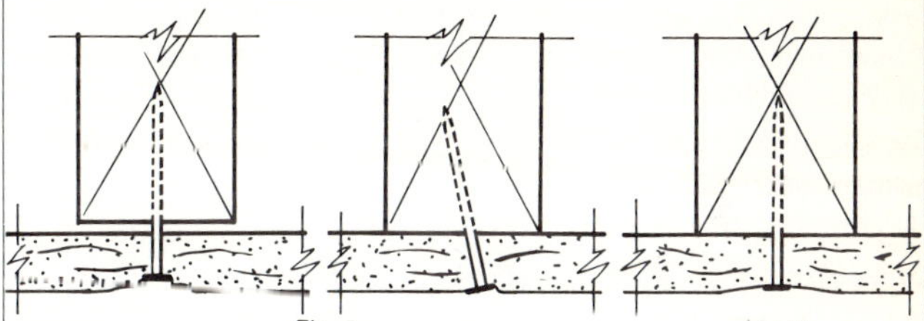

Fig. 7 *Fig. 8*

11. Fasteners—Nails Loosened by Pounding

Cause: Applying panels to the second side of a partition can loosen nails on opposite side (lack of hand pressure during fastening). Particularly true when lightweight, soft lumber, undersized studs or furring are used.

Remedy: Check panels for tightness on the partition side where panels were first applied. If looseness is detected, strike each nailhead an additional hammer blow, being careful to not overdrive the nail.

Prevention: Use proper framing, Type W screws or adhesive application.

---

12. Fasteners—Unseated Nails

Cause: Flexible or extremely hard framing or furring does not permit nails to be properly driven. May result from undersized framing members, type of wood used, supports that exceed maximum allowable frame spacing or lack of hand pressure during fastening.

Remedy: Replace nails with 1¼″ Type W screws.

Prevention: Use proper framing (see Chapter 2), Type W screws or adhesive application. Apply pressure to hold panel tight against framing while driving fasteners.

13. Fasteners—Loose Screws

Cause: Using the wrong type screw for the application or an improperly adjusted screwgun results in a screw stripping or not seating properly.

Remedy: Remove faulty fastener and replace with a properly driven screw.

Prevention: Use screws with combination high/low threads for greater resistance to stripping and pullout; set screwgun clutch to proper depth.

---

14. Panels—Loosely Fastened

Cause: Framing members are uneven because of misalignment or warping; lack of hand pressure on panel during fastening. Head of fastener alone cannot pull panel into firm contact with uneven members. Also, see "Panels—Improperly Fitted."

Remedy: With nail attachment, during final blows of hammer, apply additional pressure with hand to panel adjacent to nail (Fig. 9) to bring panel into contact with framing.

Prevention: Correct framing imperfections before applying panels; for a more solid attachment, use 1¼″ Type W screws or use adhesive method (see Chapter 3). Apply pressure to hold panel tightly against framing while driving fasteners.

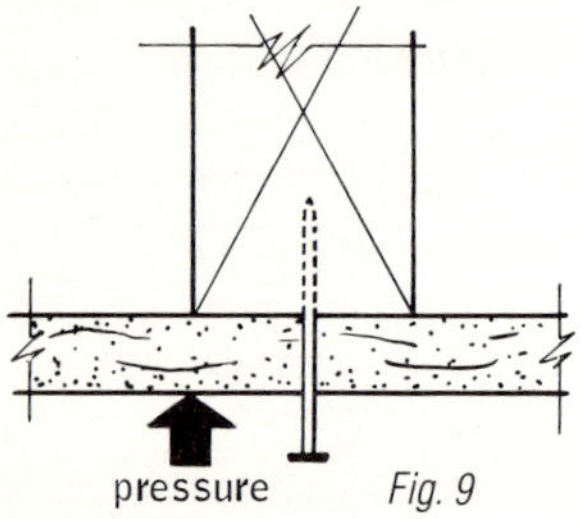

Fig. 9

15. Panels—Surface Fractured After Application

a. *Cause:* Heavy blows or other abuse has fractured finished wall surface; too large a break for repair with joint compound.

Remedy: Cut a square-shaped or triangular section around damaged area, with a utility or keyhole saw (Fig. 10); use a rasp or sanding block to slope edges inward at 45°. Cut corresponding plug from sound gypsum panel, sand edges to exact fit (Fig. 11). If necessary, cement extra slat of gypsum panel to back of face layer to serve as brace. Butter edges (Fig. 12) and finish as a butt joint with joint compound (Fig. 13).

b. *Cause:* Attaching panel directly to flat grain of wide-dimensional wood framing members such as floor joists and headers. Shrinkage of wood causes fracture of board.

Remedy: As above, where appropriate, or repair as for joint ridging.

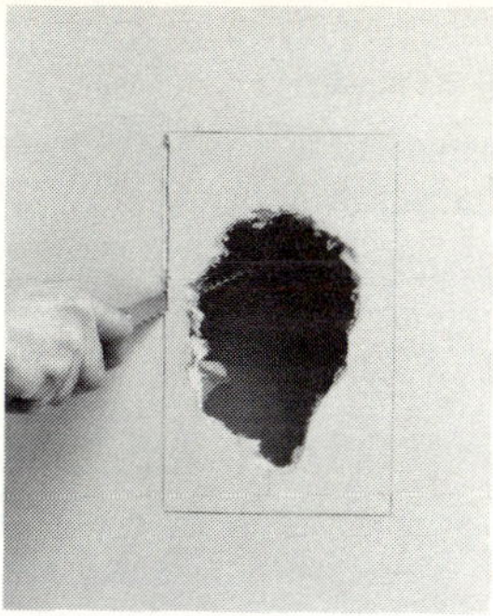

Fig. 10

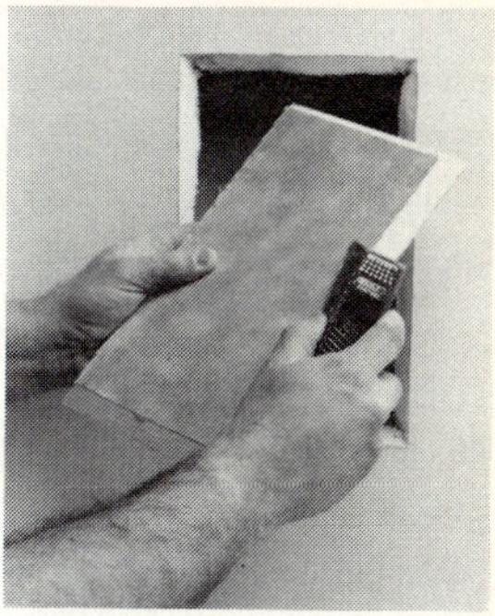

Fig. 11

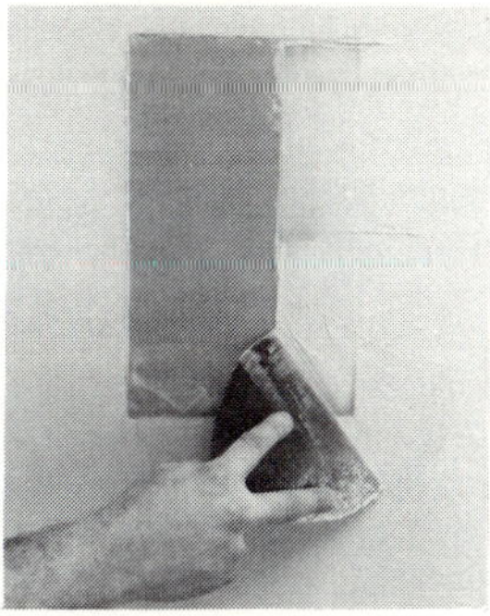

Fig. 12

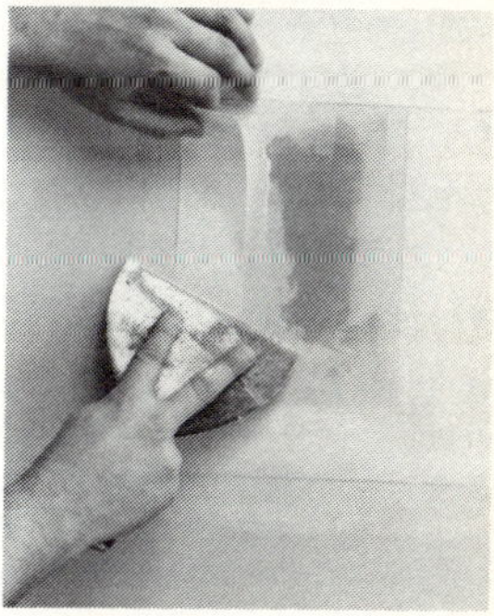

Fig. 13

Prevention: To provide a flexible base to allow for movement of framing, attach RC-1 Resilient Channel to framing members and apply panels. Allow ½" space at bottom edges of board for movement. Or attach board directly to studs but allow ¼" separation between panels, and install SHEETROCK Zinc Control Joint No. 093 (see single-layer application on page 142).

c. *Cause:* Knife scoring beyond corner of cutout for electrical boxes, light fixtures and door and window openings produces cracks in panel surface.

Remedy: Repair cuts with joint compound and tape before finishing.

Prevention: Stop score marks at corners, cut openings accurately.

d. *Cause:* Abnormal stress buildup resulting from structural deflection or racking discussed previously.

Remedy: Relieve stress, provide adequate isolation and retape, feathering joint compound over board area to disguise buildup.

Prevention: Provide proper isolation from structure to prevent stress buildup.

e. *Cause:* Excessive stresses resulting from hygrometric and/or thermal expansion and contraction discussed previously.

Remedy: Correct unsatisfactory environmental conditions, provide sufficient relief. Retape, feathering joint compound over broad area.

Prevention: Correct improper job conditions and install control joints for relief in long partition runs and large ceiling areas (see page 158).

---

16. Joints—Blisters in Tape

Cause: Insufficient or overly thin compound was used under the tape; tape was not initially pressed into good contact with the compound; or too much compound was forced from under tape by application of excessive tool pressure when embedding.

Remedy: Open up blistered area by slitting tape. Fill cut with joint compound and press tape back in place with knife blade. When dry, smooth to level finish.

Prevention: Provide sufficient compound under entire tape.

---

17. Joints—Edge Cracking

a. *Cause:* After joint treatment, straight narrow cracks along edges of tape result from: too rapid drying because of high temperature accompanied by low humidity or excessive drafts; improper application, such as overdilution of joint compound; use of wrong compound (topping instead of taping); excessive joint compound under tape; failure to follow embedding with a thin coat over tape; or cold, wet application conditions, which also may cause poor bond.

This problem, difficult to see when it first occurs, may not be discovered until decoration begins. However, the cause can be attributed to some aspect of the taping operation.

Remedy: Especially when poor atmospheric conditions exist, carefully examine all joints after taping and skimming applications have dried; repairs are more economical at this stage. Cut away any weakly bonded tape edges. Fill hairline cracks with cut shellac (2 to 3 lb.); groove out larger cracks with sharp tool; coat with shellac and allow to dry, then refill with joint compound; or cover cracks with complete joint treatment, including reinforcing tape; feather to surface level with plane of board.

Prevention: Use either SHEETROCK Setting-Type (DURABOND) or SHEETROCK Lightweight Setting-Type (EASY SAND) Joint Compound, which have the best built-in resistance to cracks. Place shielding devices over room openings to prevent drafts. Do not apply joint treatment over hot surfaces. Wet down floors if room humidity is too low. During cold weather, control heat at min. 55°F (13°C) and supply good ventilation. Avoid practices listed under "Cause," above.

b. *Cause:* After joint treatment, cracks along edges of corner bead or trim can result from the same unsatisfactory conditions listed above for tape. Also can be caused by impact on the bead.

Remedy: Remove applied joint compound. Securely fasten corner bead or trim to framing beneath panels. Refinish bead with joint compound.

Prevention: Use SHEETROCK Corner Bead No. 800 and SHEETROCK Metal Trim with expanded flanges, which minimize cracking.

---

18. Joints—Center Cracking

a. *Cause:* Abnormal stress buildup resulting from structural deflection or racking discussed previously.

Remedy: Relieve stress. Provide adequate isolation and retape, feathering joint compound over broad area to disguise buildup.

Prevention: Provide proper isolation from structure to prevent stress buildup.

b. *Cause:* Excessive stresses resulting from hygrometric and/or thermal expansion and contraction discussed previously.

Remedy: Correct unsatisfactory environmental conditions. Provide sufficient relief; retape, feathering joint compound over broad area.

Prevention: Correct improper job conditions and install control joints for relief in long partition runs and large ceiling areas (see page 158).

---

19. Joints—Angle Cracking

a. *Cause:* Too much compound applied over tape at apex of angle.

Remedy: After compound is completely dry, smooth out excess compound at apex. Fill only hairline cracks with compound. Do not apply additional compound, which will build up.

Prevention: Keep excess compound from corner, leaving only a small amount or no compound in apex.

b. *Cause:* Slitting or scoring reinforcing tape during application. May result from use of improper tool.

Remedy: If crack extends through the tape, retape and finish.

Prevention: Use proper tool for corner treatment.

c. *Cause:* Structural movement from two separate supports or framing members, which react independently to applied loads. Often occurs in wall-ceiling angles where wall is attached to top plate and ceiling is attached to floor or ceiling joists running parallel to top plate.

Remedy: Remove fasteners closer than 6" from angle, retape and finish.

Prevention: Use "Floating Interior Angle" application described on page 177.

d. *Cause:* Structural or thermal movement resulting from two dissimilar materials or constructions.

Remedy: Remove tape, provide relief, finish with angle edge trim and caulk.

Prevention: Use channel-type or angle edge trim over gypsum board where two dissimilar surfaces interface.

e. *Cause:* Excessive paint thickness application of paint under poor conditions.

Remedy: Correct unsatisfactory job conditions. Scrape away cracked paint. Fill and feather with joint compound. Prime and paint.

Prevention: Provide proper job conditions. Apply recommended thickness of prime and finish coats of paint.

---

20. Joints—High Crowns

Cause: Excessive piling of compound over joint; compound not feathered out beyond shoulders, improper bedding of tape; framing out of alignment or panel edges not tight against framing; improper adjustment of tools; misuse of or worn tools.

Remedy: Sand joints to flush surface. Take care to avoid scuffing paper by oversanding.

Prevention: Embed tape properly, using only enough compound to cover tape and fill taper depression or tape itself at butt joints; feather compound far enough to conceal.

---

21. Joints—Excessive and/or Delayed Shrinkage

Cause: (1) Atmospheric conditions—slow drying and high humidity; (2) Insufficient drying time between coats of compound; (3) Excessive water added in mixing compound; (4) Heavy fills.

Remedy: See "Starved Joints," below.

Prevention: Allow each coat of joint compound to dry thoroughly before applying succeeding coat, or use a low-shrinkage SHEETROCK Setting-Type (DURABOND) or SHEETROCK Lightweight Setting-Type (EASY SAND) Joint Compound.

---

22. Joints—Starved Joints

Cause: This is a form of delayed shrinkage caused chiefly by insufficient drying time between coats of compound. May also be caused by insufficient compound applied over tape to fill taper, overthinning or oversmoothing of compound. Shrinkage usually progresses until drying is complete.

Remedy: Use fast-setting SHEETROCK Setting-Type (DURABOND) or SHEETROCK Lightweight Setting-Type (EASY SAND) Joint Compound or reapply a full cover coat of heavy-mixed compound over tape. Since this is heaviest application, most shrinkage will take place in this coat, making it easier to fill taper properly. Finish by standard procedure.

Prevention: Allow each coat of joint compound to dry thoroughly before applying succeeding coat, or use a low-shrinkage SHEETROCK Setting-Type (DURABOND) or SHEETROCK Lightweight Setting-Type (EASY SAND) Joint Compound.

---

23. Joints—Ridging

Cause: All building materials grow or shrink in response to changes in temperature and humidity. When they are confined to a specific space, such as gypsum panels in a partition or ceiling, they are put under stress, either compression or tension, depending on the temperature or humidity conditions. These stresses are relieved when the panel bends outward in the region of the joint. Once this bending takes place, the system takes a set and never returns to normal. It becomes progressively worse with each change of temperature or humidity. This progressive deformation appears as a continuous ridge along the length of joint, with a uniform fine, ridgelike pattern at center.

Remedy: (1) Let ridge develop fully before undertaking repairs; usually six months is sufficient. Make repairs under hot and dry conditions; (2) Smooth ridge down to reinforcing tape without cutting through tape. Fill concave areas on either side of ridge with light fill of thick-mix compound. After this is dry, float very thin film of compound over entire area; (3) Examine area with strong sidelighting to make certain that ridge has been concealed. If not, use additional feathering coats of compound. Redecorate. Ridging can recur, but is usually less severe. Continuous wetting will aggravate condition.

Prevention: Where available, use SHEETROCK brand Gypsum Panels, SW Edge, with the exclusive rounded edge designed to prevent ridging. Follow general recommendations for joint treatment (see Chapter 1) and approved application procedure, which includes back-blocking and laminated double-layer application to minimize potential ridging problems (see Chapter 3). Pay particular attention to temperature, ventilation, consistency of compound, prompt covering coat over tape, minimum width of fill, finish coats and required drying time between coats.

---

24. Fasteners—Nail Pops from Lumber Shrinkage

Cause: Improper application, lumber shrinkage or a combination of both. With panels held reasonably tight against framing member and with proper-length nails, only severe shrinkage of the lumber normally will cause nail pops. But if nailed loosely, any inward pressure on panel will push nailhead through its thin covering pad of compound. Pops resulting

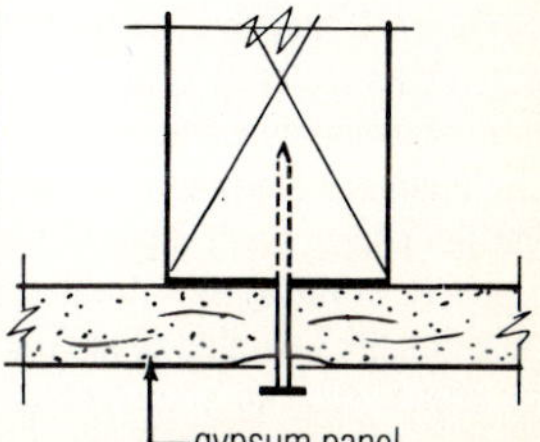

Fig. 14

from "nail creep" occur when shrinkage of the wood framing exposes nail shank and consequently loosens panel (see "Lumber Shrinkage" on page 374).

Remedy: Repairs usually are necessary only for pops which protrude .005″ or more from face of board (Fig. 14). Smaller protrusions may require repair if they occur in a smooth gloss surface or flat-painted surface under extreme lighting conditions. Those that appear before or during decoration should be repaired immediately. Pops that occur after one month's heating or more are usually caused wholly or partly by wood shrinkage and should not be repaired until near end of heating season. An often effective procedure for resetting a popped nail is to place a 4″ broad knife over the nail and hit with hammer to seat flush with surface. A more permanent method is to drive proper nail or Type W screw about 1½″ from popped nail while applying sufficient pressure adjacent to nailhead to bring panel in firm contact with framing. Strike popped nail lightly to seat it below surface of board. Remove loose compound, apply finish coats of compound and paint.

Prevention: Proper nail application; use of lumber meeting framing requirements (see page 100); attachment with Type W screws or by adhesive application (see Chapter 3).

---

25. Fasteners—Bulge Around Fastener

Cause: Overdriving fasteners, driving them with the wrong tool or failing to hold board firmly against framing while driving fasteners can puncture and bulge face paper and damage core of board. Following application of joint compound or texture finish that wets the board paper can result in board bulging or swelling around fastener.

Remedy: Drive screw fastener close to damaged area, clean out damaged paper core, repair with SHEETROCK Setting-Type (DURABOND) or SHEETROCK Lightweight Setting-Type (EASY SAND) Joint Compound, and refinish.

Prevention: Use correct tool and drive fasteners properly. Also, see "Panels—Loosely Fastened."

---

26. Finish—Discoloration

a. *Cause:* Differences in suction of panel paper and joint compound may lighten paint color or change gloss or sheen in higher-suction areas; most common when conventional oil-based paints are used. Also caused by texture differences between the face paper and finished joint compound or by overthinning of paint. May also occur over fasteners in ceilings subjected to severe artificial or natural side lighting. Suction differences may also cause greater amounts of texturing material to be deposited over high-suction areas, causing color differences when viewed from an angle. Before painting, face panel paper may be darkened from exposure to sunlight.

Remedy: Redecorate.

Prevention: Before painting or texturing, apply a prime coat of SHEETROCK First Coat or undiluted interior flat latex paint with high solids content. Avoid roughening surface paper when sanding joint compound. Use strong parallel to the surface job lights to ensure a flat acceptable joint compound finish prior to priming, texturing and/or painting.

b. *Cause:* The use of preservatives in paint formulation. As the scientific community learns more about health hazards, additives to many products are changed or eliminated. Mercury, for example, was banned for use in paints in August, 1990. Some of these additives could cause a reaction resulting in an off-color appearance.

Remedy: A high-quality finish coat or coats will, generally, produce an acceptable finish.

Prevention: A good quality primer or SHEETROCK First Coat and finish coat properly applied according to the paint manufacturer's recommendations will prevent most discoloration problems.

---

27. Finish—Gloss Variation with High Gloss Paints

Cause: Differences in suction of panel paper and joint compound or texture differences between the face paper and finished joint compound (as stated in #26-a). Problem is accentuated by strong side lighting with slight angle of incidence to ceiling or wall surface.

Remedy: Redecorate.

Prevention: Before painting with a high gloss paint, apply a skim coat of joint compound over the entire wall surface or use a veneer plaster system. If skim coating is not done, apply a prime coat of SHEETROCK First Coat or an undiluted interior flat latex paint with high solids content.

---

28. Finish—Joint Darkening

Cause: This condition occurs most commonly with color-tinted paint rather than white. Most severe when applied in humid weather when joints have not fully dried.

Remedy: Apply a prime coat of SHEETROCK First Coat or an undiluted interior flat latex paint with high solids content. Repaint only after joints are thoroughly dry.

Prevention: Be sure joints are thoroughly dry before painting (see "Drying Time" on page 200).

---

29. Finish—Shadowing

Cause: Temperature differentials in outside walls or top-floor ceilings causes collection of airborne dust on colder spots of interior surface. Results in photographing or shadowing over fasteners, furring or framing. Most severe with great indoor-outdoor temperature variation.

Remedy: Wash painted surfaces, remove spots with wallpaper cleaner, or redecorate surfaces. Change air filters regularly.

Prevention: Use double-layer application with adhesively applied face layer. Use separately framed free-standing interior wall surface and insulate in void to reduce temperature difference between steel or wood components and panels.

---

30. Finish—Decorating Wallboard Damaged by Over-Sanding

Cause: Paper fibers of the exposed gypsum panel surface are scuffed and raised by over-sanding from the joints or use of sandpaper that is too coarse.

Remedy: Correct severe paper fiber raise with a light skim coat of a setting-type or drying-type joint compound. Minor fiber raise may be treated by light sanding with very fine sandpaper or wiped down with a damp sponge or cloth.

Prevention: Good finish work with compound, feathered edges, etc. reduces the need for sanding, in general, limiting exposure to this kind of problem.

---

31. Panels—Ceiling Sag after Installation

a. *Cause:* Too much weight from overlaid insulation; exposure to sustained high humidity; vapor retarder improperly installed or wetting causes ceiling panels to sag after installation. Also caused by installing board too thin for framing spacing.

Remedy: Remove sagged board or fur ceiling using RC-1 Resilient Channels; apply another layer of board. If the sag condition is judged to be the result of high moisture conditions and not improper framing or unusual weight loads, the affected areas may possibly be filled with SHEETROCK Lightweight All Purpose Joint Compound Ready-Mixed (PLUS 3) to achieve a smooth surface. Care must be taken to achieve proper ventilation and drying conditions.

Prevention: Follow recommended frame spacing and attachment procedures and use recommended products only. Use SHEETROCK brand Interior Gypsum Ceiling Board, where available. See "Ceiling Sag Precautions" on page 362.

b. *Cause:* Water-based textures wet face paper and weaken gypsum core, causing ceiling panels to sag after installation.

Remedy: Same as above.

Prevention: See Chapter 3 for proper frame spacing and application procedures. See "Ceiling Sag Precautions" on page 362.

Veneer Plaster Construction

Many problems associated with veneer plaster construction have the same cause, remedy and prevention as with drywall systems. Similarity of problems appears in application of the base, framing irregularities, cracking due to structural movement, hygrometric and thermal expansion and fastener imperfections. The additional problems that follow are those specifically relating to veneer plaster finish construction. If solutions to your problems in veneer plaster construction are not described, check similar problems for drywall construction found previously in this Chapter.

Application Problems

1. Mixing—Foaming Action in Mixer

Cause: Use of alum as an accelerator when limestone aggregate is used.

Remedy: None. Dispose of batch.

Prevention: Use moulding plaster or quick-set gauging plaster as an accelerator when limestone aggregate is used. Or use sand aggregate.

---

2. Setting Time—Variable Set Time Within Batch

Cause: Insufficient or excessive mixing.

Remedy: None. Dispose of batch.

Prevention: Use proper drill speed; follow recommended mixing times (see pages 207-208).

---

3. Slow Set—IMPERIAL Finish Plaster/Basecoat; DIAMOND Interior Finish Plaster/Basecoat

a. Cause: High air and mixing water temperature, in excess of 100°F (38°C).

Remedy: Proper use of accelerator.

Prevention: Avoid extremes of air and water temperatures.

b. Cause: Contaminated mixing water.

Remedy: None.

Prevention: Use drinkable water only.

c. Cause: Excessive use of retarder.

Remedy: None – dispose of batch.

Prevention: Follow recommendations for mix proportions and use of additives.

---

4. Quick Set—IMPERIAL Finish Plaster/Basecoat; DIAMOND Interior Finish Plaster/Basecoat

a. *Cause:* Low air and water temperature below 40°F (4°C).

Remedy: None. Dispose of batch.

Prevention: Avoid extremes of air and water temperatures.

b. *Cause:* Contaminated mixing water; dirty mixing equipment.

Remedy: None. Dispose of batch.

Prevention: Use drinkable water only. Clean set plaster residue from equipment after each batch. Always use clean mix equipment.

c. Cause: Excessive use of accelerator.

Remedy: None—dispose of batch.

Prevention: Follow recommendations for mix proportions and use of additives.

---

5. Quick Set—IMPERIAL Finish Plaster and DIAMOND Interior Finish Plaster Only

Cause: Contamination; excessive use of aggregate and/or accelerator.

Remedy: None. Dispose of batch.

Prevention: Use drinkable water. Clean set plaster residue from equipment after each batch. Always use clean mix equipment. Follow recommendations for mix proportions and use of additives.

---

6. Workability—Stiff Working

Cause: Mixing action has improper or insufficient shear.

Remedy: None. Use remainder of batch, if at all workable.

Prevention: Follow recommendations for mixing time, drill speed and type of mixing paddle (see pages 207-208).

In-Place Problems

7. Bond Failure—Delamination of Finish Coat

Cause: Basecoat not left rough and open; finish coat not properly scratched into basecoat to have necessary keying.

Remedy: Remove loose material, brush basecoat thoroughly, apply bonding agent and refinish.

Prevention: Follow application recommendations (see Chapter 4).

---

8. Bond Failure—DIAMOND Interior Finish Plaster

Cause: Application over faded (not normal blue color) gypsum base.

Remedy: Remove loose material, brush base clean, apply bonding agent and refinish.

Prevention: Do not store or apply base where it will be exposed to sunlight for an extended time. Where exposed, spray faded base with alum solution or bonding agent before finish is applied.

---

9. Cracks—Joint Cracking

a. *Cause:* IMPERIAL Tape overlapped at joint intersections.

Remedy: Large cracks: apply SHEETROCK Joint Tape and SHEETROCK Ready-Mixed Joint Compound (All Purpose or Taping) over the cracks. Minor cracks: flush out area with SHEETROCK Ready-Mixed Joint Compound (All Purpose or Taping).

Prevention: Avoid overlapping tape at all joint intersections, including those at angles.

b. *Cause:* Improper steel stud placement. Gypsum base application advanced in wrong direction relative to flange direction.

Remedy: Repair with SHEETROCK Joint Tape and SHEETROCK Ready-Mixed Joint Compound (All Purpose or Taping).

Prevention: Install steel studs with all flanges pointing in same direction. Arrange gypsum base application so lead edge of base is attached to open edge of flange first (See "Framing (Steel)—Panel Edges Out of Alignment" on page 389).

c. *Cause:* Overly rapid drying conditions.

Remedy: Repair with SHEETROCK Joint Tape and SHEETROCK Ready-Mixed Compound (All Purpose or Taping).

Prevention: Liberally sprinkle floor with water to raise humidity. Use SHEETROCK Joint Tape and SHEETROCK Setting-Type (DURABOND) Joint Compound on all joints. Allow compound to dry thoroughly before applying finish. Where steel stud framing is used, apply SHEETROCK Joint Tape and SHEETROCK Setting-Type (DURABOND) Joint Compound.

d. *Cause:* IMPERIAL Tape with steel framing.

Remedy: Repair with SHEETROCK Joint Tape and SHEETROCK Ready-Mixed Compound (All Purpose or Taping).

Prevention: Use only SHEETROCK Joint Tape and SHEETROCK Setting-Type Joint Compound (DURABOND).

---

10. Cracks—Field Cracking

Cause: Gypsum base installed with vertical joints extending from corners of door and window openings.

Remedy: Apply SHEETROCK Joint Tape and SHEETROCK Setting-Type (DURABOND) Joint Compound finished with SHEETROCK Ready-Mixed Joint Compound (All Purpose or Taping). Prime and seal. This is a cosmetic treatment; there is no guarantee that cracks will not reopen.

Prevention: Install SHEETROCK Zinc Control Joint No. 093 or cut base to fit around openings with joints centered above openings, not at corners.

---

11. Cracks—Craze and Map Cracking

Cause: Veneer plaster application too thin. Also can be caused by too rapid drying.

Remedy: Apply spackling putty. Prime and seal.

Prevention: Apply recommended thicknesses for both one- and two-coat work. Avoid excessive ventilation, which may cause rapid drying. When weather is hot and dry, sprinkle floor with water to raise humidity.

---

12. Blemishes—Blistering

a. *Cause:* Loose paper on gypsum base as a result of improper cutting or from "peelers" caused by careless handling.

Remedy: Cut and remove unbonded paper, apply bonding agent if gypsum core is exposed, and refinish.

Prevention: Follow proper handling and cutting procedures.

b. *Cause:* Troweling too early and lack of absorption; excessive material buildup.

Remedy: Minimize troweling and allow finish to become firm. Finish-trowel over freshly set surface to eliminate blisters.

Prevention: Apply material in uniform thickness with minimum amount of troweling to produce smooth surface.

---

13. Blemishes—Joint Ridging and Beading

Cause: Joints not preset; excessive ventilation and poor heat control. Most likely to occur with one-coat applications.

Remedy: Repair with SHEETROCK Ready-Mixed Joint Compound (All Purpose or Taping).

Prevention: Preset all joints before veneer plaster finish application; keep ventilation to minimum and control heat. In extremely hot, dry weather use SHEETROCK Joint Tape and SHEETROCK Setting-Type (DURABOND) Joint Compound as alternative joint reinforcement.

---

14. Blemishes—Spalling at Exterior Corners

Cause: Use of solid-flange drywall corner bead.

Remedy: Remove all loose material and corner bead. Install expanded-flange corner bead and refinish.

Prevention: Use expanded-flange corner bead.

---

15. Stains—Staining and Rusting

Cause: Use of improper fasteners, or exposed, improperly prepared metal trim.

Remedy: Apply rust-locking primer over stains.

Prevention: Use recommended coated fasteners (see Chapter 1). Apply rust-locking primer to all exposed metal.

16. Soft, Weak Surface—Dryouts

Cause: Too rapid drying conditions.

Remedy: Fog-spray surface with water or alum solution to provide setting action. When set, apply SHEETROCK Ready-Mixed Joint Compound (All Purpose or Taping) for acceptably smooth surface.

Prevention: Avoid extending set and/or temperature and humidity conditions, which cause rapid drying.

---

17. Soft, Weak Surface—Crumbly Areas

Cause: Use of excessive amount of sand aggregate and/or retarder.

Remedy: Treat soft areas with penetrating sealer.

Prevention: Allow minimum ventilation; use recommended amounts only of aggregate or retarder. Avoid prolonged set.

Cement Board Construction

DUROCK Cement Board and Exterior Cement Board Systems require careful adherence to published installation procedures and high standards of workmanship

1. Exterior System—Panel Joint Cracking

a. *Cause:* Abnormal stress build-up resulting from structural deflection or racking.

Remedy: Relieve stress. Provide control joints and proper bracing.

Prevention: Provide relief at perimeter of panels where panels abut to foundations, floors, roofs, columns and beams. Provide adequate lateral and diagonal bracing to limit deflection of framing.

b. *Cause:* Improper joint reinforcement, rapid drying, freezing or wetting of joint.

Remedy: Remove finish and loose mortar or adhesive with a belt sander, grinder, router or saw. Do not cut board surface. Reapply ceramic tile setting material or DUROCK Basecoat, embed DUROCK Tape, and finish to match adjacent areas.

Prevention: Use DUROCK Tape 2″ wide for interior ceramic tile and 4″ wide for DUROCK Exterior Ceramic Tile or Finish System. Carefully center tape over joint. Apply mortar or basecoat and embed tape. Protect treated joint from excessive changes in temperature, maintain above 45° for a minimum of 24 hours, moist cure for 24 hours by misting with water before and after application.

2. Exterior System—Cold Joining

Cause: Partially finishing a section of basecoated cement board and allowing the DUROCK Exterior Finish to dry before reaching a control joint or other such termination point.

Remedy: Refinish surface and do not allow the material to dry before reaching a termination point.

Prevention: Schedule your work to avoid finishing in direct sunlight. Use tarps, if necessary, to avoid working in direct sunlight. Have a large enough crew size to properly handle the wall section. Pay close attention to potential cold joining at different scaffold heights.

---

3. Exterior Finish System—Color and Texture Differences

Cause: Using different quantities of water in thinning the finish. Plasterers using dissimilar floating techniques. Irregularities in the basecoated surface.

Remedy: Refinish the entire panel from control joint to control joint, or other such termination point.

Prevention: Always use the same amount of water when thinning the DUROCK Exterior Finish. Make sure that all plasterers are using the same floating technique. Check the basecoated surface to ensure that it is flat. Remember: Texture is color and color is texture.

---

4. Exterior Finish System—Surface Fractured

Cause: Heavy impact punctures from moving equipment or vandalism. Holes resulting from previous fixture attachment.

Remedy: Cut out a section around the damaged area with edges centered on the framing members. Use a carbide-tipped blade in an electric circular saw set to a depth slightly less than the finished panel thickness. Make final cut with utility knife being careful not to cut the water barrier behind the DUROCK Panel. Remove finish and a portion of the basecoat for approximately 4″ around all four sides of the hole using an electric grinder or belt sander. Cut a patch of DUROCK Panel and attach to framing with appropriate fasteners. Fill joints around patch with basecoat, embed DUROCK Tape, and level joints. Let cure for at least 4 hours. Apply basecoat to a level slightly below adjacent finish and let cure for 24 hours. Apply DUROCK Exterior Finish to the patched area and float level with existing surface. After 24 hours, apply finish to entire wall area, working within corners, control joints, or other area dividers.

5. Interior Tile System—Surface Fractured

Cause: Heavy impact punctures from moving equipment or vandalism. Holes resulting from previous fixture attachment.

Remedy: Where tile can be removed without damaging the DUROCK Cement Board—Remove damaged tile by cutting and chipping tile. Scrape or grind bond coat down to skim coat. Tape any cracks with DUROCK Tape. Reset tile and grout.

Where tile cannot be removed without causing excessive damage to the DUROCK Cement Board—Remove the damaged section by cutting through the tile and DUROCK Cement Board. Install additional framing so that the perimeter of existing and new panels will be supported. Cut a patch of DUROCK Cement Board that closely fits the opening. Apply a generous amount of organic adhesive to the edges of existing and new panel. Install panel and attach to framing with appropriate fasteners. Smooth and level adhesive at panel joints. Let adhesive cure for 24 hours before setting and grouting new tile.

---

6. Delamination—Basecoat or Mortar

Cause: Improper mixing procedures or improper basecoating techniques.

Remedy: Remove all material that is not properly bonded to the DUROCK Panel surface. Apply a bonding agent such as Larsen's WELD-CRETE; then reapply basecoat to the area.

Prevention: Always use the proper amount of clean potable water when mixing the material. Too much water will significantly reduce the bond strength of the material. Apply the material using a scratch and double back method. The tight scratch coat keys the material into the DUROCK Panel surface.

---

7. Delamination—Tile and Thin Brick

Cause: Allowing the mortar to skin over before setting the tile or brick. Not back-buttering the tile.

Remedy: Scrape the mortar from the DUROCK Panel surface. Apply a bonding agent to the DUROCK Panel surface and allow to dry. Apply fresh mortar to the DUROCK Panel and back-butter the tile and push into place.

Prevention: Do not allow the mortar to skin over. Back-butter the tile and beat-in to ensure 100% coverage.

Texture Finishes

U.S. Gypsum Company texturing materials offer a wide range of decorative yet practical finishes. Properly used, they can provide interest and variety in decoration while covering minor defects in the base surface. However, certain working conditions, application techniques or equipment problems can cause unsatisfactory results. The following list describes the problem, probable cause, remedy and prevention for particular situations.

1. Mixing—Lumping of Wet Mix

Cause: Too much water added to initial mix, making lumps difficult to break up.

Remedy and Prevention: Initial amount of water added to mix should be slightly less than recommended. After lumps are broken up, add remaining water.

---

2. Mixing—Slow Solution Time

Cause: Insufficient soaking and/or use of very cold water.

Remedy and Prevention: Allow materials to soak for up to two hours, as necessary, if using cold water.

---

3. Mixing—Wet Mix Too Thin

Cause: Addition of excessive water during initial mix. Also, insufficient soaking time in cold water.

Remedy and Prevention: Use recommended amount of water to ensure proper consistency. Allow materials to soak up to two hours, if necessary, when using cold water.

---

4. Application—Excessive Aggregate Fallout in Spraying

Cause: Excessive air pressure at nozzle; holding gun too close to surface being sprayed.

Remedy and Prevention: Use proper air pressure for type of material to be sprayed: low for IMPERIAL QT Spray Texture Finish; high for USG Spray Texture Finish. (Consult appropriate U.S. Gypsum Company Data Sheet for recommended air pressure.) Hold spray gun at proper distance from surface to prevent excessive bounce or fallout of aggregate.

---

5. Application—Flotation of Aggregate

Cause: Overdilution of job mix and/or lack of adequate mixing after water is added to control consistency.

Remedy and Prevention: Add correct amount of water as directed on bag to ensure proper suspension of materials in mix. Make certain that water is blended into mix.

6. Application—Poor Coverage with Spray Finishes

Cause: Not enough water to bring texture material to proper spray viscosity and/or improper application, such as moving spray gun too slowly, overloading spray surface and using incorrect spray pressures.

Remedy and Prevention: Add proper amount of water as directed on bag. Use correct spray gun pressures and application technique to ensure uniform texture with maximum coverage.

---

7. Application—Poor Hide

Cause: Overdilution of mix causing reduction in hiding power. Insufficient water in spray finishes causes poor material atomization, resulting in surface show-through. Also caused by overextending material or choosing incorrect spray pressures.

Remedy and Prevention: See above.

---

8. Application—Poor Bond or Hardness

Cause: Overdilution of job-mix results in thinning out of binder in texture. Contamination or intermixing with other than recommended materials can destroy bonding power.

Remedy and Prevention: Add proper amount of water as stated in bag directions. Always use clean mixing vessel and water. Never intermix with other products (except materials as recommended).

---

9. Application—Stoppage of Spray Equipment

Cause: Contamination of material or oversize particles can cause clogging of spray nozzle orifices. Also caused by using incorrect nozzle size for aggregate being sprayed.

Remedy and Prevention: Prevent contamination during mixing. Use correct nozzle for aggregate size of texture material.

---

10. Application—Unsatisfactory Texture Pattern

Cause: Improper spray pressures and/or worn spray equipment, either fluid or air nozzle. Also improper spraying consistency of mix and/or spraying technique.

Remedy and Prevention: Use recommended amount of water to ensure proper spraying consistency. Handle spray equipment correctly to achieve best results. Make certain that spray accessories are in good working condition; replace when necessary.

11. Application—Unsatisfactory Pumping Properties

Cause: Mix too heavy. Pump equipment worn or of insufficient size and power to handle particular type of texture.

Remedy and Prevention: Use recommended amount of water to ensure proper spraying consistency. Make sure that equipment is in good repair and capable of pumping heavy materials.

---

12. Application—Texture Buildup

Cause: Texturing over a high-suction drywall joint (surface not properly sealed) and/or allowing too much time between roller or brush application and texturing operation. Overdilution of texture material will produce texture buildup over joint.

Remedy and Prevention: Before texturing, apply a prime coat of SHEETROCK First Coat or an undiluted, interior flat latex paint with high solids content. Use correct amount of water when mixing texture material. Allow safe time interval between application and final texturing.

---

13. Finish Surface—Poor Touchup

Cause: Touchup of a textured surface to completely blend with the surrounding texture is extremely difficult. A conspicuous touchup is caused either by texture or color variance.

Remedy: Perform touchup operation with extreme care; otherwise, re-texture entire wall or ceiling area.

---

14. Finish Surface—Joint Show-Through

Cause: Overthinned or overextended texturing material does not adequately hide the normal contrast between joint and gypsum panel paper. Also caused by improperly primed surface.

Remedy: Use correct amount of water when mixing texture material and apply at recommended rate of coverage until joint is concealed.

Prevention: Before texturing, apply a prime coat of SHEETROCK First Coat or undiluted, interior flat latex paint with high solids content.

---

15. Finish Surface—White Band or Flashing Over Gypsum Panels

Cause: High-suction gypsum panel joint causes a texture variation, which often appears as a color contrast.

Remedy: Allow texture to dry, and paint entire surface.

Prevention: Before texturing, apply a prime coat of SHEETROCK First Coat or undiluted, interior flat latex paint with high solids content.

16. Finish Surface—White Band or Flashing Over Concrete

Cause: Damp concrete surface on which leveling compound has dried completely can produce results similar to those of high-suction joint.

Remedy: Allow texture to dry, and paint entire surface.

Prevention: Allow concrete to age for at least 60 days for complete dryout. Before texturing, apply a prime coat of SHEETROCK First Coat or undiluted, interior flat latex paint with high solids content.

---

17. Finish Surface—Joint Darkening

Cause: Application over damp joint compound, especially in cold, humid conditions.

Remedy: Allow texture to dry completely, and paint entire surface.

Prevention: Before texturing, apply a prime coat of SHEETROCK First Coat or an undiluted, interior flat latex paint with high solids content.

Conventional Plaster Construction

All U.S. Gypsum Company basecoat and finish plasters are carefully manufactured and thoroughly tested before shipment. Along with the functional characteristics offered, U.S. Gypsum Company plasters are carefully formulated for use under normal, prevailing weather conditions and with aggregates commonly used in the market.

Plasters are adversely affected by aging and abnormal storage conditions, use of the wrong aggregate and improper proportioning, all of which may affect the set, hardness and working properties of the material. Most plaster problems result from the following situations:

1. Adverse atmospheric and job conditions.
2. Set conditions too fast or slow.
3. Poor quality and incorrect proportioning of aggregate.
4. Improper mixing, application or thickness of basecoat or finish.
5. Incorrect lathing practices.
6. Dirty or worn mixing or pumping equipment.

Basecoat and finish coat plasters are so closely interrelated that problems pertaining to their use are treated together. No attempt is made here to discuss problems that might occur due to structural deficiencies. These were covered previously in this Chapter. Plaster problems are classified under the specific existing condition. These are discussed in order, in the following groups:

1. Plaster cracks.
2. Blemishes, color variation, surface stains.
3. Weak, soft walls.

4. Bond failure.
5. Set and working qualities.
6. Gauged lime putty problems.

---

1. Cracks—Connecting vertical, horizontal cracks at somewhat regular intervals, often in "stepped" pattern; also diagonal in appearance.

Material: Plaster over metal or gypsum lath.

a. *Cause:* Plaster too thin, insufficient plaster grounds.

Remedy: Patch.

Prevention: Apply plaster to proper thickness.

b. *Cause:* Weak plaster (through dryout or slow set).

Remedy: Spray with alum solution to accelerate set.

Prevention: Add accelerator to plaster mix to bring setting time to normal range.

c. *Cause:* Excessive use of aggregate.

Remedy: Patch.

Prevention: Use proper proportions of aggregate and plaster.

d. *Cause:* Failure to use Striplath reinforcement at potential weak points.

Remedy: Cut out, reinforce and repair.

Prevention: Install proper reinforcing.

e. *Cause:* Expansion of rough wood frames.

Remedy: Remove plaster and lath as necessary. Seal frames and patch.

Prevention: Seal frames. Cut basecoat along grounds prior to set. Install control joints over frames.

Material: Plaster over unit masonry.

Cause: Structural movement of masonry units.

Remedy and Prevention: Correct masonry construction, install control joint, patch.

Material: Plaster over brick, clay tile or concrete block at door openings.

Cause: Poor lintel construction, improper frame construction.

Remedy: Patch.

Prevention: Use proper frame and lintel construction with self-furring metal lath reinforcement.

---

2. Cracks—Fine cracks, random pattern, generally 1″ to 3″ apart. Includes shrinkage cracks, crazing, alligatoring, chip cracks.

Material: Gauged lime-putty finish over gypsum basecoat, used with any plaster base.

Cause: Insufficient gauging plaster-shrinkage of lime. Insufficient troweling during setting. Applied finish too thick. Basecoat too wet or too dry and too little or too much suction.

Remedy: Apply spackling putty and primer-sealer.

Prevention: Use sufficient gauging plaster, trowel sufficiently or properly condition basecoat before applying finish.

---

3. Cracks—Fine cracks, irregular pattern, generally 6″ to 14″ apart; "map cracking."

Material: Trowel finishes over gypsum basecoat—unit masonry plaster base.

a. *Cause:* Finish coat applied too thick.

Remedy: Patch.

Prevention: Apply finish coat to 1/16″ thickness but not more than 1/8″.

b. *Cause:* Improper timing of final troweling.

Remedy: Patch.

Prevention: Water-trowel as final set takes place (not before) to provide dense, smooth surface.

c. *Cause:* Retempering gauged lime putty.

Remedy: Discard batch; make up new gauge.

Prevention: Gauged lime putty should not be retempered once it has started final set.

---

4. Cracks—Random pattern, usually less than 12″ apart, called map, shrinkage or fissure cracking.

Material: Basecoat over masonry.

a. *Cause:* High suction of masonry base.

Remedy: If bond to base is sound and cracks are open 1/16″ or more, fill by troweling across cracks with properly aggregated plasters. If bond is sound, finish over fine cracks with highly gauged trowel finish or float finish. If curled at edges and bond is unsound, remove and reapply using proper plaster method.

Prevention: Wet masonry with water to reduce suction before basecoat application.

b. *Cause:* Under-aggregating of basecoat; slow set.

Remedy: Same as above.

Prevention: Use 3 cu. ft. of aggregate per 100 lb. gypsum plaster (See Chapter 4 for proper proportion of aggregates). Discontinue use of job-added retarder and accelerator, if necessary, to obtain proper set.

c. *Cause:* Dryout condition.

Remedy: Spray basecoat either with water or alum solution to thoroughly wet plaster. Proceed same as above.

Prevention: In hot, dry weather, protect plaster from drying too rapidly before set. Spray plaster during set time if necessary.

---

5. Cracks—At wall or ceiling angles.

Material: Plaster over gypsum lath.

a. *Cause:* Thin plaster.

Remedy: Cut out and patch.

Prevention: Follow correct application procedure.

b. *Cause:* Failure to use Cornerite reinforcement.

Remedy and Prevention: Same as above.

---

6. Blemish—Water-soluble, powdery crystals on surface, generally white but may be colored. Can be brushed off.

Material: Basecoat and finish plaster over concrete block or clay tile.

Cause: Efflorescence. As masonry units dry, water-soluble salts from units or mortar joints leach out and are deposited on the plaster surface.

Remedy: After plaster surfaces are thoroughly dry, brush off efflorescence, apply oil-based sealer and paint.

Prevention: On interior walls, eliminate source of moisture, remove efflorescence before plastering; decorate with oil-based sealer and paint. On exterior walls, eliminate moisture source, fur out, lath and plaster.

---

7. Blemish—"Pops" or peak-like projections which fall out and create little craters or pits; often with fine radial cracks.

Material: Gauged lime-putty finish.

a. *Cause:* Unslaked lime in mortar which slakes and swells after it is applied.

Remedy: Remove core of "pops" and patch after popping has ceased.

Prevention: Allow sufficient soaking time for normal hydrated lime or use double-hydrated lime or a prepared finish such as RED TOP Finish Plaster.

b. *Cause:* Contamination from foreign matter.

Remedy: Same as above.

Prevention: Eliminate source of impurity.

Material: Gypsum basecoat and finish.

c. *Cause:* Lumpy or undissolved retarder added on job. Retarder lumps swell or "pop" when wet.

Remedy: Cut out spots and patch.

Prevention: Completely disperse retarder before adding to mix water; mix well to distribute retarder throughout plaster.

---

8. Blemish—Blisters in finish coat occur during or immediately after application.

Material: Gauged lime putty finish.

a. *Cause:* Base too green (wet); insufficient suction; too much water used in troweling.

Remedy: After finish has set, trowel with very little water.

Prevention: Do not apply finish coat over green basecoat.

b. *Cause:* Finish too plastic.

Remedy: Same as above.

Prevention: Add small amount of very fine white sand to putty or increase amount of gauging plaster.

---

9. Blemish—Excess material ("slobbers") on finish surface.

Material: Gauged lime-putty finish.

Cause: Improper joining technique, excessive or improper troweling leaves excess material on finished surface.

Remedy: Scrape off excess material before decoration. Seal surface when plaster is dry.

Prevention: Previously applied finish should be cut square for completion of finish. Avoid excessive troweling at joining.

---

10. Blemish—Peeling paint.

Material: Gauged lime-putty finish.

a. *Cause:* Paint applied over wet plaster.

Remedy: Scrape off peeled paint, allow plaster to dry, and redecorate.

Prevention: Be sure plaster is dry before decorating, and use a breather-type paint.

b. *Cause:* Weak finish. Plaster worked through set.

Remedy: Scrape off peeled paint, patch and decorate.

Prevention: Do not retemper or trowel finish after set.

---

11. Color Variations—Streaks and discoloration.

Material: Lime finishes, gauged with gauging plaster or Keenes Cement.

a. *Cause:* Lime and gauging plaster not thoroughly mixed.

Remedy: Seal and decorate.

Prevention: Follow recommended mixing procedures.

b. *Cause:* Too much water used in troweling.

Remedy: Same as above.

Prevention: Apply as little water as possible in troweling.

c. *Cause:* Dirty tools or water.

Remedy: Same as above.

Prevention: Wash tools and use clean water.

---

12. Color Variations—Light and dark spots.

Material: Float finish.

a. *Cause:* Improper technique or too much water used in floating.

Remedy: Seal and paint to get uniform color.

Prevention: Follow recommended application procedures.

b. *Cause:* Spotty suction on basecoat which was dampened unevenly by throwing water on with a brush rather than by spraying with a fine nozzle.

Remedy: Same as above.

Prevention: Dampen basecoat uniformly using a fine spray.

---

13. Color Variations—Light or flat spots in light-color paint.

Material: Oil-based paint over gauged lime-putty finish.

a. *Cause:* Surface painted too soon after plastering (alkali in lime saponifies paint); paint pigments not limeproof.

Remedy and Prevention: Apply primer-sealer and repaint.

Material: Any colored paint over any plaster finish.

b. *Cause:* Non-uniform absorption results in uneven surface gloss and coloration.

Remedy and Prevention: Apply primer-sealer and repaint.

14. Surface Stains—Yellow, brown or pink staining—"yellowing."

Material: Any lime-putty finish over any basecoat and plaster base; generally occurs while surface is damp.

a. *Cause:* Contaminated aggregate.

Remedy: Apply primer-sealer and repaint.

Prevention: Use clean aggregate.

b. *Cause:* By-product of combustion from unvented fossil fuel space heaters.

Remedy: Same as above.

Prevention: Vent heaters to outside.

c. *Cause:* Tarpaper behind plaster base; creosote-treated framing lumber; tar or tar derivatives used around job; sulphur or chemical fumes.

Remedy: Same as above.

Prevention: Use asphalted paper. Remove source of air contamination.

---

15. Surface Stains—Rust.

Material: Plaster over any plaster base.

Cause: Rusty accessories; or any protruding metal.

Remedy: Apply rust-locking primer-sealer and decorate.

Prevention: Use accessories made of zinc alloy or with hot-dip galvanizing. Do not use accessories that show rust prior to installation.

---

16. Soft Walls—Soft, white, chalky surfaces, occurring during hot, dry weather, usually near an opening.

Material: Gypsum basecoat over any plaster base.

Cause: Dryout. Too much water has been removed before plaster can set.

Remedy: Spray with alum solution or plain water to set up dryout areas.

Prevention: Screen openings in hot, dry weather; spray plaster during set; raise humidity by sprinkling floor with water.

---

17. Soft Walls—Soft, dark, damp surfaces occurring during damp weather.

Material: Gypsum basecoat over any plaster base.

Cause: Sweat-out. Too little ventilation allows water to remain in wall for an extended period after plaster set. Some plaster has re-dissolved.

Remedy: Dry walls with heat and ventilation. If sweat-out condition continues, there is no remedy; remove and replaster.

Prevention: Properly heat and ventilate during plastering.

18. Soft Walls—Soft, dark, damp surfaces, occurring in freezing weather.

Material: Gypsum basecoat over any plaster base.

Cause: Frozen plaster.

Remedy: If plaster freezes before set, no remedy except to remove and replaster.

Prevention: Close building, supply heat.

---

19. Soft Weak Walls—General condition, not spotty or due to slow set.

Material: Gypsum basecoat over any plaster base.

Cause: Too much aggregate or fine, poorly graded aggregate.

Remedy: No remedy; remove and replaster.

Prevention: Use properly graded aggregate and correct proportioning.

---

20. Weak Walls—Weak plaster.

Material: Gypsum basecoat.

Cause: Extremely slow set.

Remedy: Spray with alum solution to accelerate set.

Prevention: Add accelerator to plaster mix to bring setting time within normal range.

Material: Gauged lime-putty finish over any basecoat.

Cause: Too little gauging with insufficient troweling; retempering; basecoat too wet.

Remedy: No remedy; remove and replaster.

Prevention: Use proper ratio of gauging to lime putty. Do not retemper plaster. Trowel adequately to ensure desired hardness.

---

21. Bond Failure—Basecoat separation.

Material: Gypsum basecoat over gypsum or metal lath.

Cause: Too much aggregate; plaster application over frost on lath; freezing of plaster before set; addition of lime or portland cement; excessive delay in plaster application after mixing; extremely slow set; retempering.

Remedy: No remedy except to replaster.

Prevention: Provide proper job conditions during plastering. Follow correct mixing and application procedures.

22. Bond Failure—Brown coat separation from scratch coat.

Material: Gypsum basecoat plasters.

a. *Cause:* Weak scratch coat.

Remedy: None; remove and replaster.

Prevention: Use proper aggregate amount. Avoid retempering.

b. *Cause:* Failure to provide mechanical key in scratch coat.

Remedy: Roughen scratch coat and replaster.

Prevention: Cross-rake scratch coat to provide rough surface for brown coat.

c. *Cause:* Dryout of scratch coat.

Remedy: Water-spray scratch coat for thorough set before brown coat application.

Prevention: Provide proper job conditions during plastering; screen openings in hot, dry weather. Water-spray plaster during set. Raise humidity by sprinkling floor with water.

---

23. Bond Failure—Finish coat separation.

Material: Gauged lime putty finish applied over gypsum brown coat.

a. *Cause:* Brown coat too smooth, too dry, wet or weak; finish improperly applied.

Remedy and Prevention: Strip off finish, correct condition of brown coat and replaster.

b. *Cause:* Frozen finish coat.

Remedy and Prevention: Remove finish, provide sufficient heat during plastering, reapply finish.

c. *Cause:* Incomplete hydration of finish lime.

Remedy and Prevention: Remove finish; using properly proportioned double-hydrated lime or a prepared finish, reapply finish.

---

24. Slow Set—See Soft, Weak Walls.

---

25. Quick Set—Plaster sets before it can be properly applied and worked.

Material: Gypsum basecoat over any plaster base.

a. *Cause:* Dirty water, tools or mixing equipment; excessive use of accelerator.

Remedy: Discard material as soon as it begins to stiffen; do not retemper.

Prevention: Use clean water, tools and equipment.

b. *Cause:* Mixing too long.

Remedy: See above.

Prevention: Reduce mixing time.

c. *Cause:* Poor aggregate.

Remedy: See above.

Prevention: Use clean, properly graded aggregate or add retarder.

d. *Cause:* Error in manufacture.

Remedy: See above. Send samples to manufacturer's representative.

Prevention: Add retarder.

e. *Cause:* Machine-pumping and application that exceed limits of time and distance pumped for plaster being used.

Remedy: See above.

Prevention: Add retarder. Use plaster designed for machine application.

---

26. Erratic Set—Lack of uniformity in set.

Material: Gauged lime putty over gypsum basecoat.

Cause: Variable temperature.

Remedy and Prevention: Maintain uniform job temperature. In cold weather, heat building to min. 55°F (13°C).

---

27. Poor Working—Works hard or "short," loses plasticity and spreadability. Does not carry proper amount of aggregate.

Material: Gypsum basecoat over any plaster base.

a. *Cause:* Aged or badly stored plaster.

Remedy: Obtain fresh plaster and mix equal parts with aged plaster or use less aggregate.

Prevention: Use fresh plaster.

b. *Cause:* Over-aggregating.

Remedy: None.

Prevention: Use proper proportioning.

Material: Gauged lime putty over gypsum basecoat.

a. *Cause:* Aged lime, partially carbonated; warehoused too long or improperly.

Remedy: None.

Prevention: Use fresh material.

b. *Cause:* Improper soaking, slaking. Low temperature during putty preparation.

Remedy: None.

Prevention: Use proper lime-putty preparation procedure. Do not soak at temperatures below 40°F (4°C).

---

28. Soupy Lime—Material too fluid for proper gauging and application.

Material: Lime putty.

a. *Cause:* Incorrect soaking.

Remedy: None.

Prevention: Follow directions for type of lime being used.

b. *Cause:* Cold weather, cold mixing water.

Remedy: None.

Prevention: The gelling action of lime is retarded when material is soaked in temperatures less than 40°F (4°C) with cold water. Use warm water to quicken gelling.

---

29. Lumpy Lime—Material too lumpy for proper blending with gauging plaster.

Material: Lime putty.

a. *Cause:* Old lime.

Remedy: None.

Prevention: Use fresh lime.

b. *Cause:* Damp lime.

Remedy: None.

Prevention: Protect lime from moisture on job and in storage.

c. *Cause:* Incorrect soaking.

Remedy: None.

Prevention: Follow soaking directions for type of lime used.

d. *Cause:* Excessive evaporation.

Remedy: Add proper quantity of water and allow to soak.

Prevention: Cover lime box with tarpaulin to reduce evaporation.

General Construction

CHAPTER 11 Tools & Equipment

The Tools You Need

United States Gypsum Company does not manufacture or distribute tools or equipment. Suitably designed tools are essential for high-quality workmanship. Using the right tools for specific jobs can improve efficiency and reduce labor costs. This Chapter contains an extensive sampling of tools designed to meet the needs of drywall, veneer plaster and plastering contractors. Some of the more commonly used hand tools can be found at building material dealers, hardware stores and home centers.

Framing Tools

Laser Alignment Tool—An extremely precise "sighting" device that utilizes a laser beam for all construction alignment jobs. Provides maximum accuracy and speed for partition layouts and leveling of suspended ceiling grillage (right).

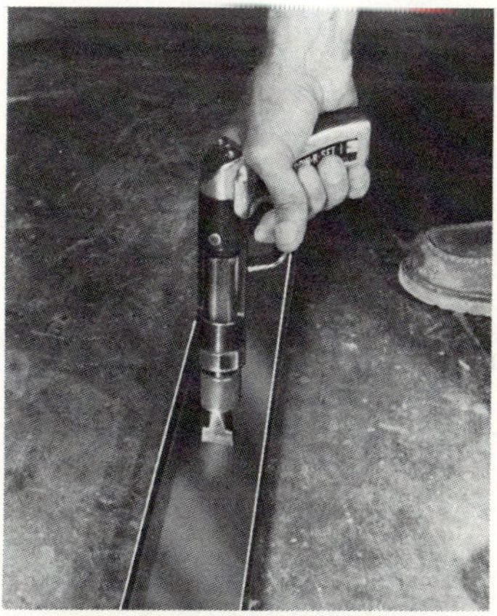

Stud Driver—Used to drive fasteners in concrete for attachment of runners. Power-driven model shown. Also available in hammer-driven models (left).

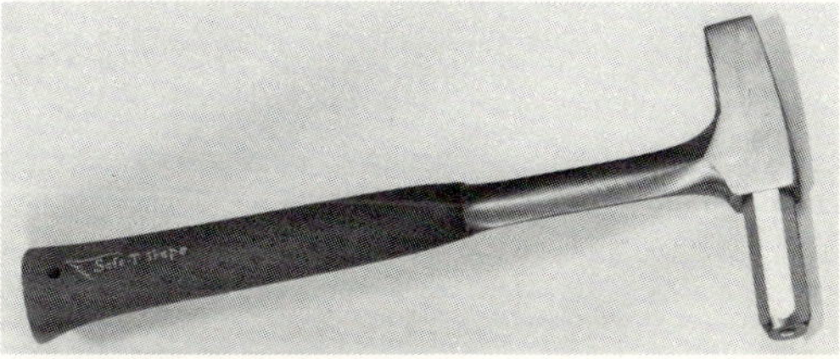

Magnetic Hammer—Magnetic head holds nails to speed attachment of floor and ceiling runners. Replaceable stainless steel magnet.

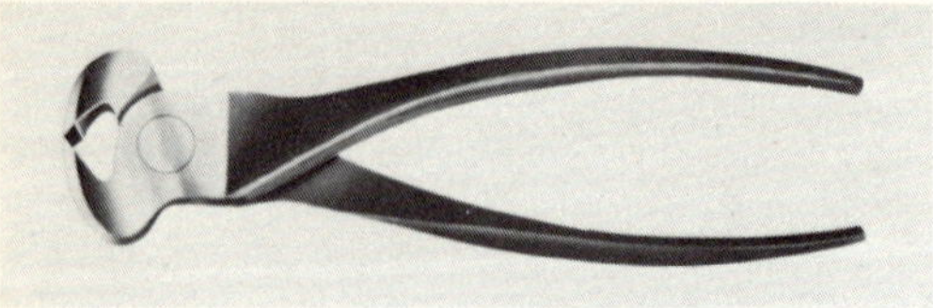

End Cut Nippers—Lather's nippers for wire-tied attachments of metal lath, grillage and framing components.

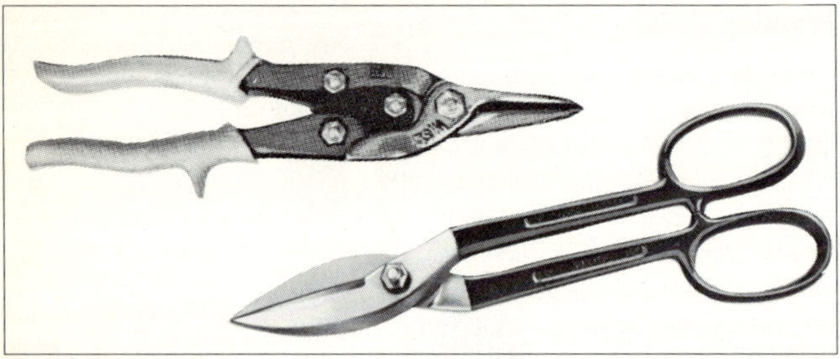

Metal Snips—Make straight and curved cuts in steel framing components. Several sizes and styles available.

Channel Stud Shear—Cuts steel studs and runners quickly, cleanly without deforming. Has fixed guides for 1⅝″, 2½″ and 3⅝″ sizes. For use with a maximum steel thickness of 20 ga.

Circular Saw—Cuts steel studs, runners and joists of various gauges with appropriate blade. Hand-held and portable, it ensures easy on-site cutting and trimming. Use a carbide-tipped blade for cutting DUROCK Cement Board.

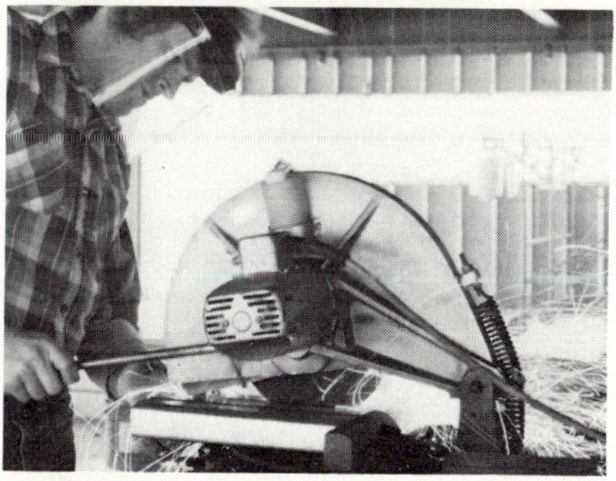

Chop Saw—The chop saw's abrasive wheel cuts all framing members. Its steel base can be placed on a bench, saw horse or floor for fast and efficient gang-cutting of members.

Electric Router—Used for irregular cuts and providing holes in gypsum board for electrical boxes, heating ducts and grilles, and other small openings.

Band Saw—A variety of models are available for use in cutting steel framing members—both bench and floor models with wet or dry systems. Variable blade speeds and vertical cutting options provide on-site flexibility.

Cut-off Saw—Hand-held, this saw uses an abrasive wheel and provides more power than a circular saw. For on-site work with heavier-gauge members.

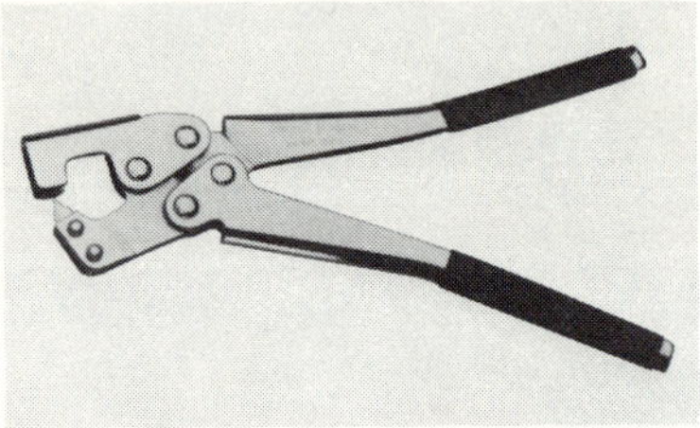

Stud Crimper—For setting and splicing metal studs, roughing-in door holders and window headers, setting electrical boxes and punching hanger-wire holes in ceiling grids.

Combination Chalk Line Box and Plumb Bob—A plumb-bob shaped device that holds retractable 100′ chalk line and chalk. Single tool plumbs floor-ceiling alignments, snaps chalk line. Not shown.

Magnetic Punch—Holds and drives nails for floor and ceiling channels in hard-to-get-at areas. Available in several sizes. Not shown.

Board and Lath Application Tools

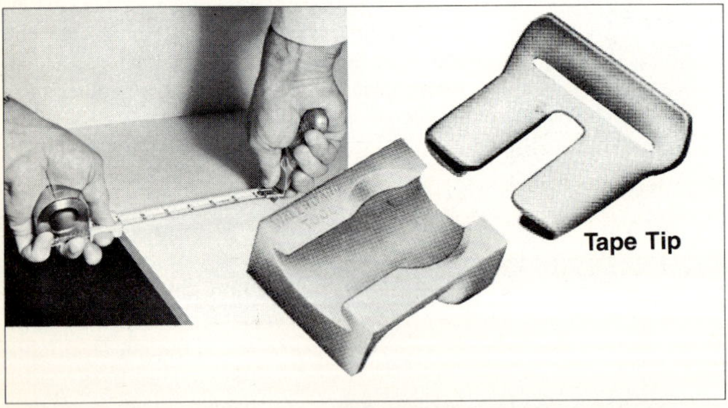

Steel Rule with Cutting Guide—The adjustable tape guide protects fingers and provides sure grip for holding accurate measurements.

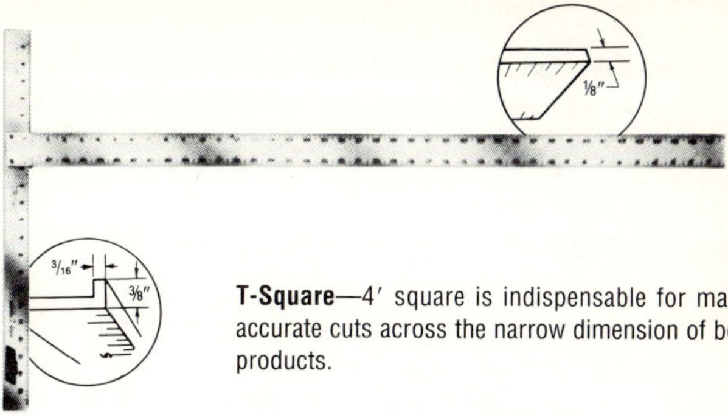

T-Square—4′ square is indispensable for making accurate cuts across the narrow dimension of board products.

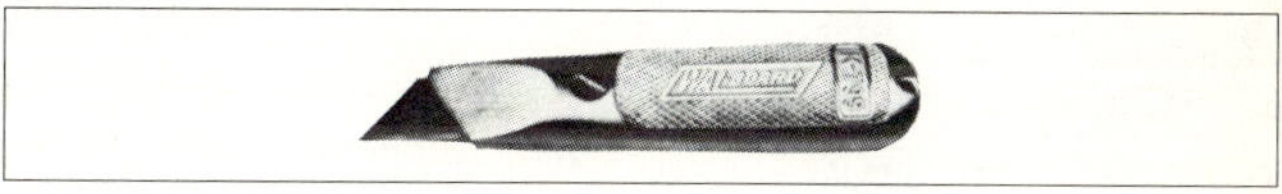

Utility Knife—The standard knife for cutting board products. Has replaceable blade; extra blades stored in handle.

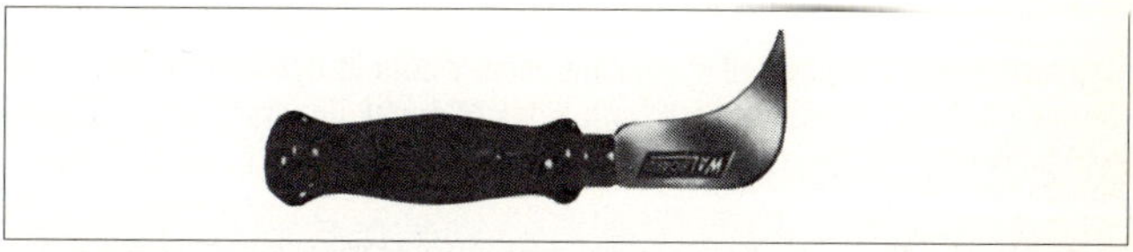

Hook-Bill Knife—Useful for trimming gypsum boards and for odd-shaped cuts. (Also known as linoleum knife.) Use a carbide-tipped scoring tool for DUROCK Cement Board.

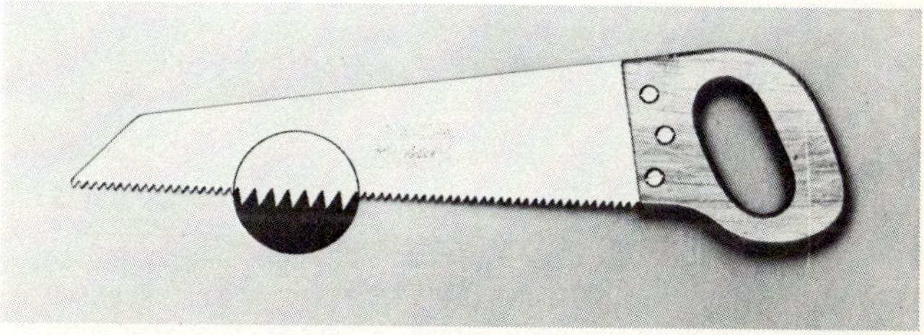

Drywall Saw—Short blade and coarse teeth (inset above) cut gypsum boards quickly and easily.

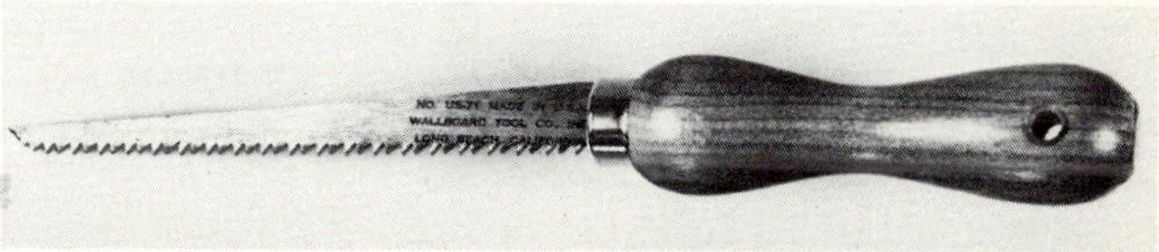

Utility Saw-Keyhole—type saw for cutting small openings and making odd-shaped cuts. Sharp point and stiff blade can be punched through board for starting cut.

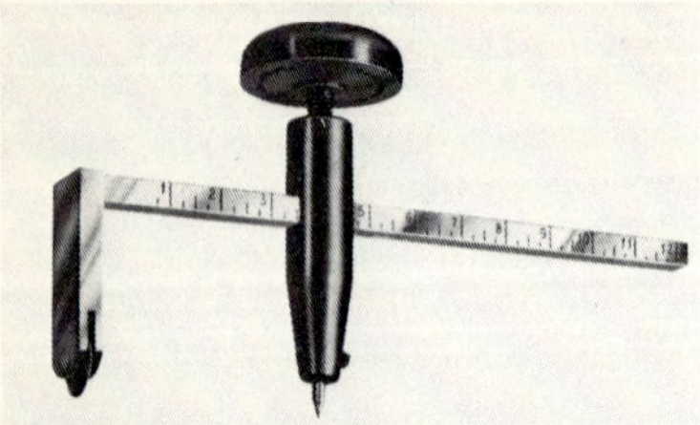

Circle Cutter—Calibrated steel shaft allows accurate cuts up to 16" diameter. Cutter wheel and center pin are heat-treated.

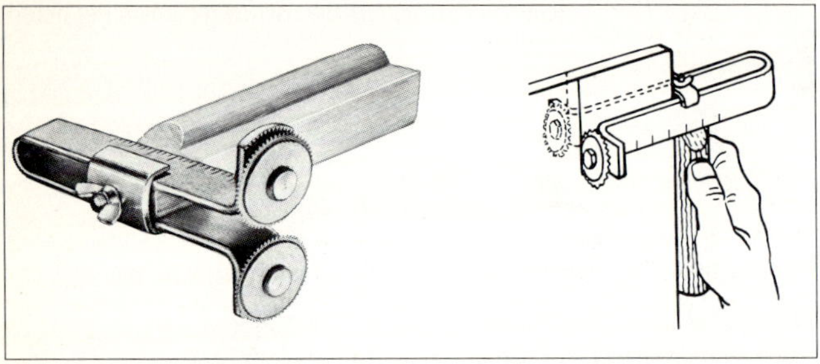

Gypsum Board Stripper—For making narrow cuts in gypsum boards. Tool has two cutting wheels that cut both sides of panel simultaneously. Handle serves as edge guide and its position adjusts to make cuts up to 4½" wide.

Rasp—Quickly and efficiently smooths rough-cut edges of gypsum boards. Manufactured model at left features replaceable blade and clean-cut slot to prevent clogging. Job-made model at right consists of metal lath stapled to a 2" x 4" wood block.

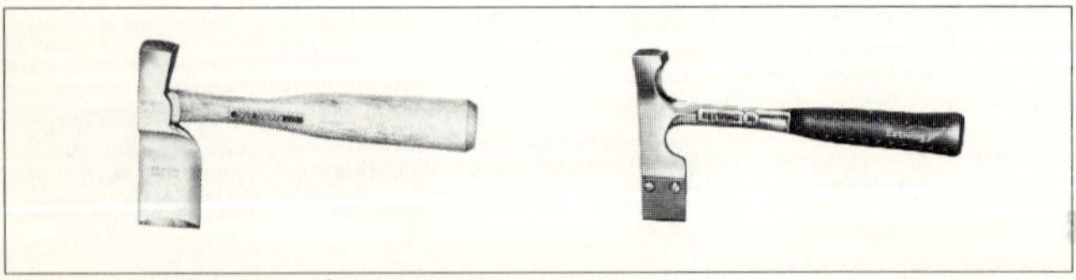

Lather's Hatchet—The standard nailing and cutting tool for gypsum lath. Available with fixed or replaceable knife edge.

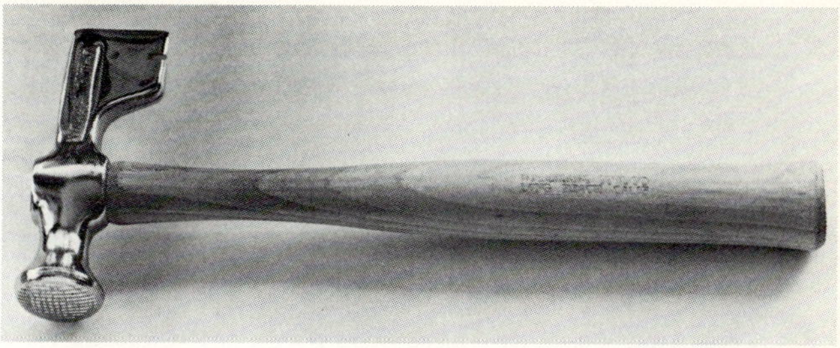

Drywall Hammer—Has symmetrical convex face designed to compress gypsum panel face and leave desired "dimple." Blade end is not for cutting but for wedging and prying panel. Not for veneer plaster bases, which require a tool with a flatter head.

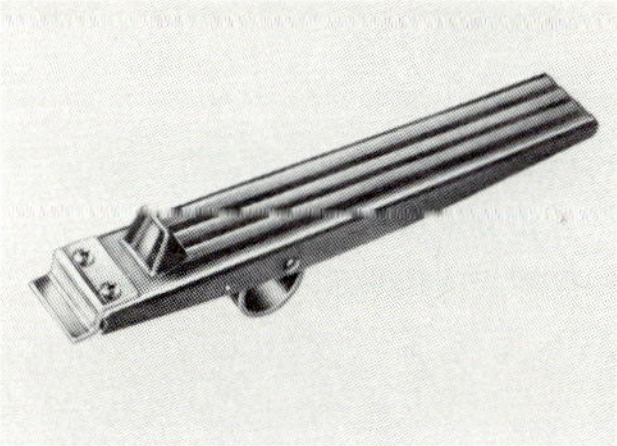

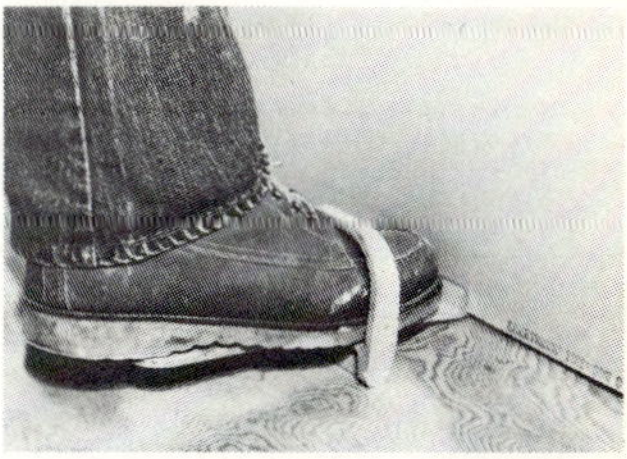

Gypsum Panel Lifters—Device is designed to move the panel forward as it lifts. Can be used for panels applied either perpendicular or parallel. Two types are available, shown above and below.

Cradle-type lifter allows one person application of drywall to sidewalls and sloped ceilings as well as level ceilings. Tripod base with rollers for easy movement.

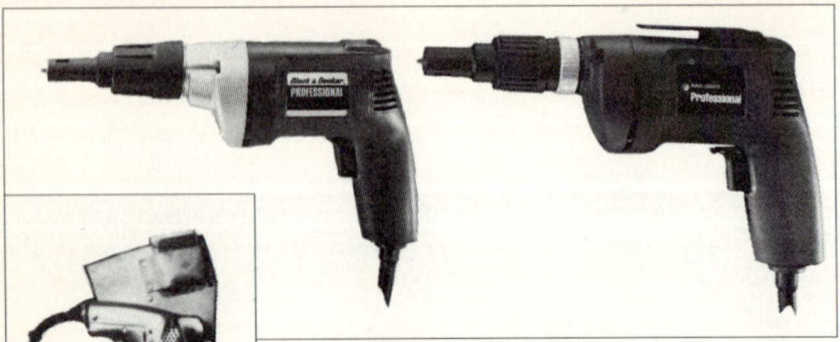

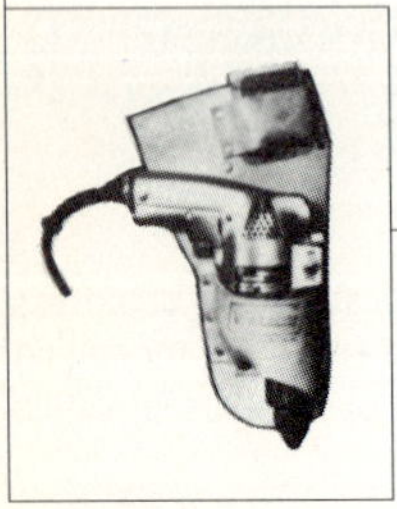

Two types of power screwguns (above) drive drywall screws in gypsum board attachment or lamination. Screwgun holster (left) is worn on worker's belt.

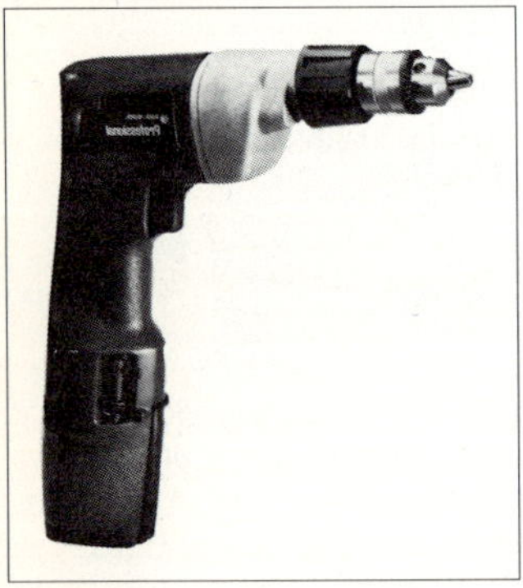

Cordless Screwgun—Operates with power from battery pack which recharges in one hour.

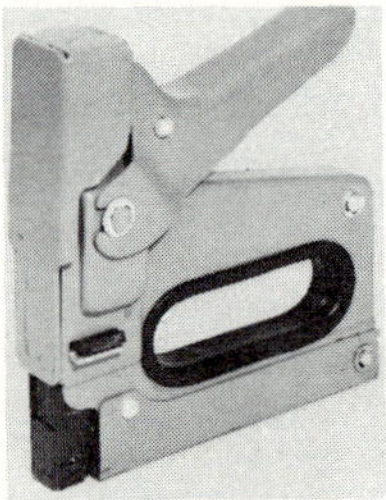

Pistol-Type Stapler—For attachment of THERMAFIBER Insulating Blankets to wood studs and to the inner face of gypsum boards in steel-framed assemblies. Also for attachment of corner beads, Striplath, Cornerite and IMPERIAL Tape.

Caulking Equipment—Both manual and power equipment are available for application of adhesives and acoustical sealants:

Cartridge-Type Hand Gun: Used with 29-oz. cartridges. Bead size determined by cut of nozzle. Aids uniform application of adhesive. Applicator capacity: ⅒ gal., 1 qt.

Bulk-Type Hand Gun: Has trigger mechanism, withstands rough usage and offers minimum resistance to large bulk load of adhesive. Applicator has 1-qt. capacity; may be filled using Alemite Loader Pump.

Note: Bulk guns and pumping equipment are often used to minimize theft of cartridges.

Alemite Loader Pump, Model 7537: Clamps on 5-gal. container to mechanically load bulk-type adhesive hand guns. Eliminates waste of hand and paddle loading.

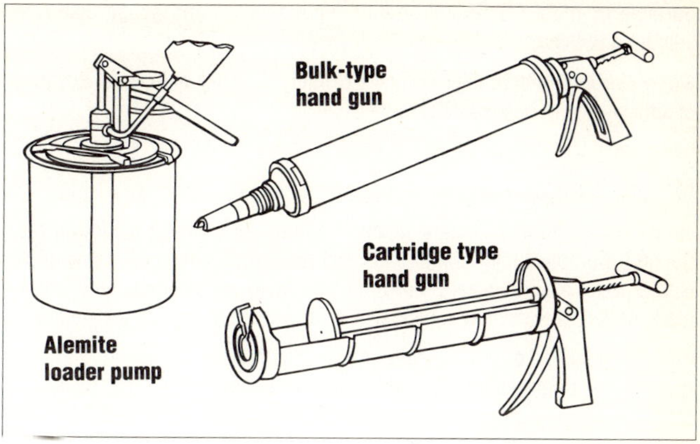

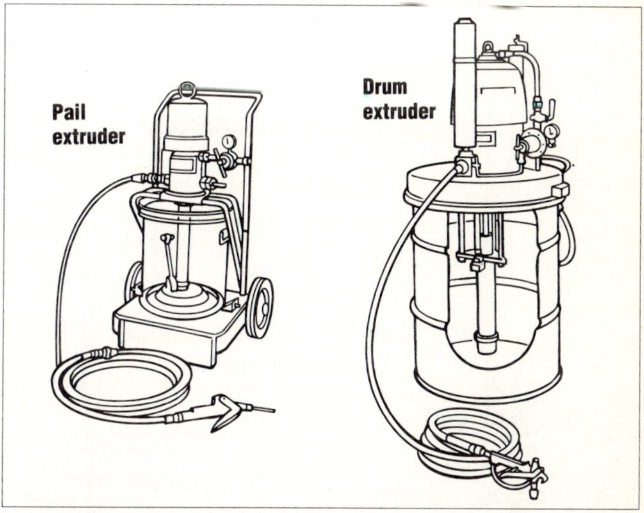

Pumping Equipment: Because of their high-volume output, pumping machines provide greater efficiency and production in the transfer, flow and spray of adhesives. Nail and screw attachment are supplemented or replaced by this type of adhesive application equipment in many building operations such as flooring, partitions and ceilings.

Large pumping equipment permits purchasing in bulk and high-volume application; contributes to job economy by reducing waste and application time.

Most machine dispensing systems are available with a selection of pumps, flow valves, nozzles and accessories. Equipment manufacturers offer a wide choice of components to provide the exact system for the job.

Pail Extruder: For high-volume extrusion of adhesives from pails. Air power depends on viscosity (low, medium or high) of the material. Offered in portable or mobile units with pump, air regulators and gauge, pail ram, adapter and hose.

Drum Extruder: Comparable to pail extruder; used for high-ratio extrusion of adhesive from bulk containers.

Mixing Equipment

Joint Compound and Texture Mixing Paddle—While hand mixing of U.S. Gypsum Company joint compounds and textures is adequate, many applicators prefer electric mixers. Power mixing saves considerable time, particularly on large jobs where mixing in a central location is most convenient.

Power is supplied by a ½" heavy-duty electric drill with a speed of max. 400 rpm. for joint compounds, 300-600 rpm for textures. Drills that operate at high speeds will whip air bubbles into the mix, rendering it unfit for finish coat purposes.

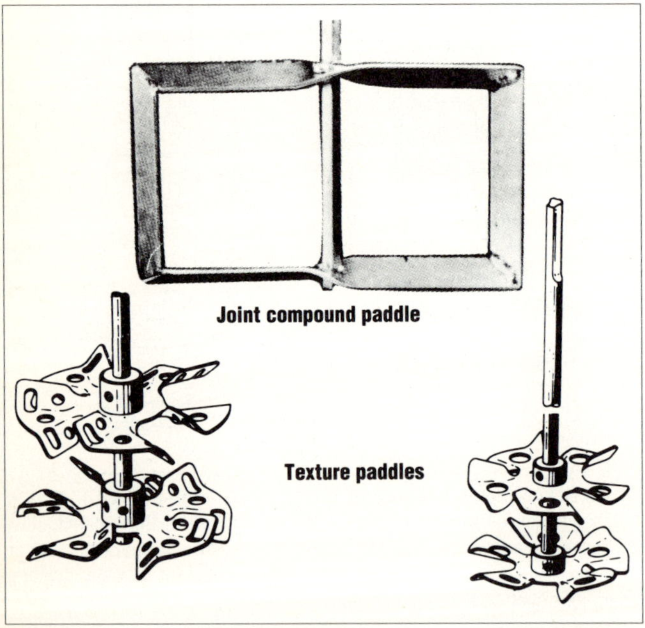

Mixing paddles are available in various styles, such as the typical examples shown above. Paddles designed for joint compounds and textures, however, should *not* be used for mixing veneer plaster finishes. The latter require a special cage-type paddle, shown on opposite page.

Potato Masher—For hand-mixing joint compounds, available in several styles. Models shown on top of next page are especially effective for removing dry material from sides of mixing bucket.

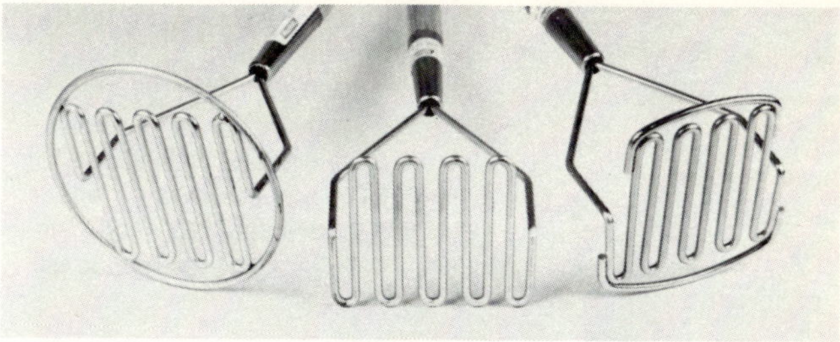

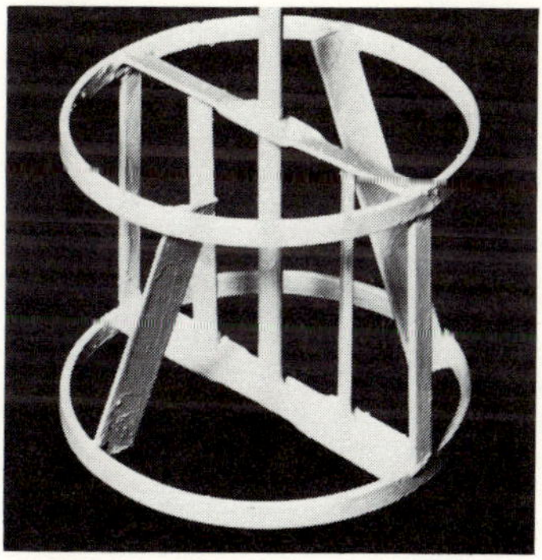

Veneer Plaster Mixer—The recommended mixer for U.S. Gypsum Company veneer plaster finishes. This cage-type paddle provides high shear action, necessary for proper dispersion of plaster ingredients in mixing water, and to develop high plasticity in the mix. Operated at proper speed, the paddle mixes thoroughly, producing a virtually air-free plaster.

A heavy-duty ½″ electric drill with a no-load rating of 900 to 1,000 rpm is delivers sufficient power and speed for mixing U.S. Gypsum Company veneer plasters.

Plaster and Stucco Mixer—Standard paddle-type mixer for stucco and all conventional plasters (not suitable for veneer plaster finishes). Available with capacities from 5 to 7 cu. ft. in either electric or gasoline-powered models.

Vertical Drum Mixers—Consist of an electric motor (which drives shaft-mounted paddles) mounted atop an open-end drum. Models are available for mixing double-hydrated lime and for joint compound. Lime mixers made in one- and three-bag sizes; joint compound mixers in made 16- and 30-gal. sizes.

Finishing Tools

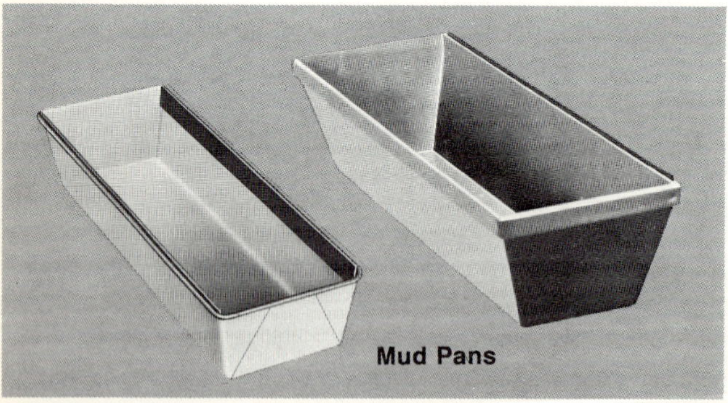

Mud Pan—Joint compound carrier for the hand finisher. Wide range of sizes, with or without a knife-cleaning blade. Available in plastic, stainless steel, galvanized steel and tinplate.

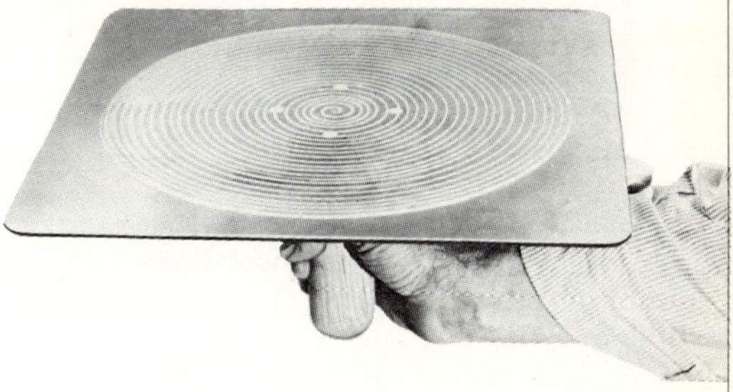

Hawk—Suitable for carrying any cementitious material by a hand applicator—joint compound, plaster, veneer finishes and stucco. Available in sizes from 8″ x 8″ to 14″ x 14″ and in aluminum and magnesium. Smooth surface model is preferred for joint compound.

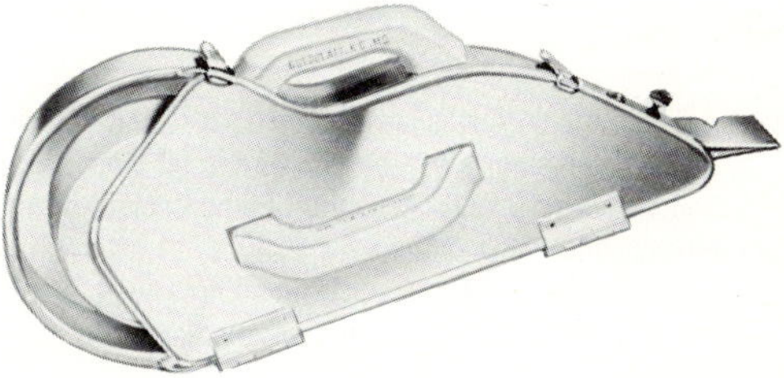

Banjo—Applies paper tape and joint compound simultaneously to flat joints. In two models: one stores tape and compound in separate compartments; the other stores tape and compound together.

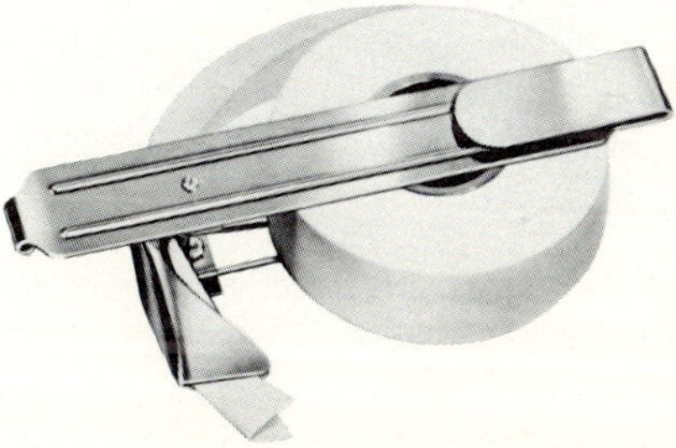

Tape Dispenser and Creaser—Holds tape rolls up to 500′. Has built-in creaser to fold tape for corner application.

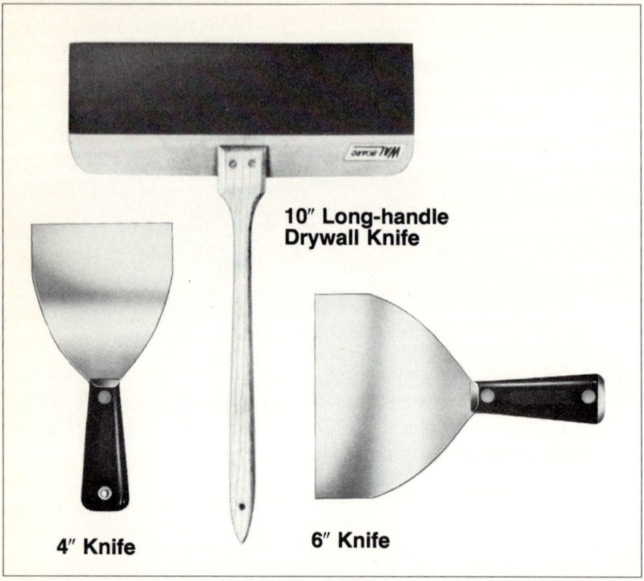

Taping Knives—4″, 5″ and 6″ knives are designed for taping, fastener spotting, angle taping and finishing; a 10″ or wider knife for finish coating. All have square corners needed for corner work. The two narrower knives are available with either plain handle (shown on 4″ knife) or with hammerhead handle (shown on 6″ knife). Other drywall finishing knives are available with blade widths up to 24″. Long-handle models also available.

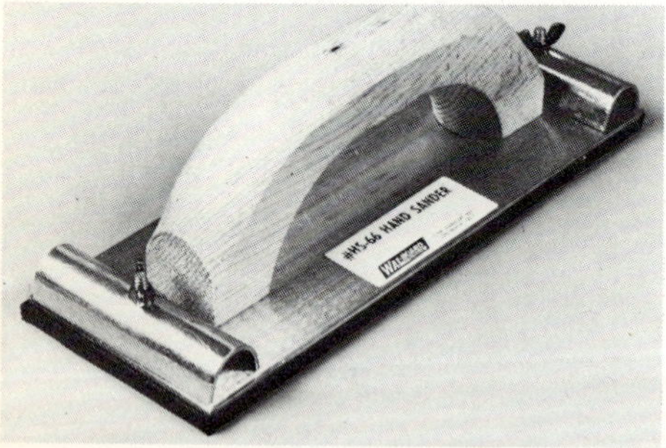

Hand Sander—3¼″ x 9¼″ base plate speeds joint sanding. Models include those with wood or aluminum handles.

Pole Sander—Long handle permits sanding of ceiling joints without necessity of worker using stilts.

Universal Angle Sander—Spring-loaded center hinge adjusts this tool automatically to fit corner angles from 82° to 100° to sand both sides at same time.

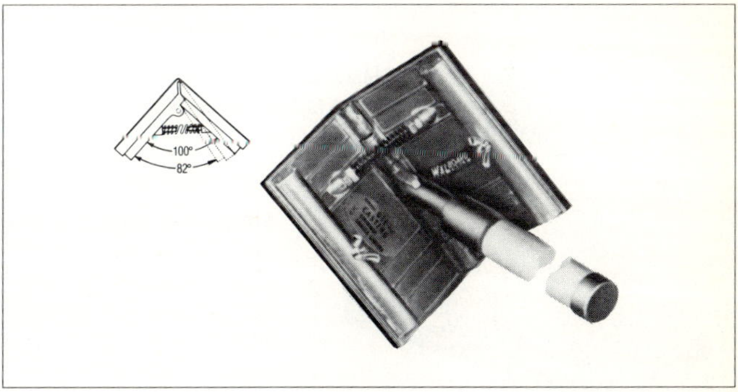

Angle Plow—For interior corner finishing of veneer plaster jobs. Similar tool with narrower blades available for conventional plaster.

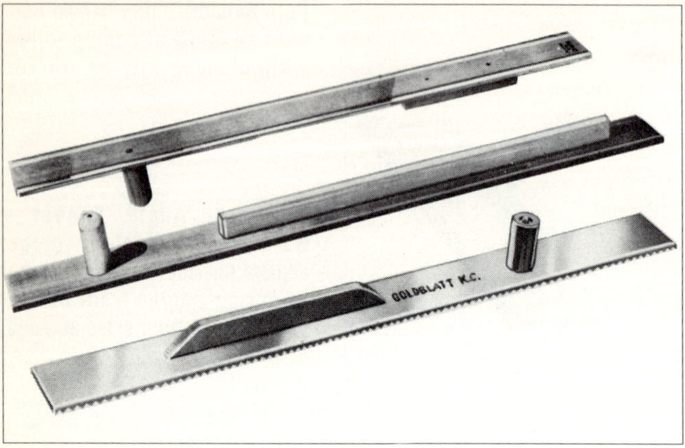

Darby—For smoothing or floating plaster brown coat. Also for finish coat where an especially true and even surface is desired. Made in wood, metal-edged wood or all metal. Notched darby at bottom is for undercoats.

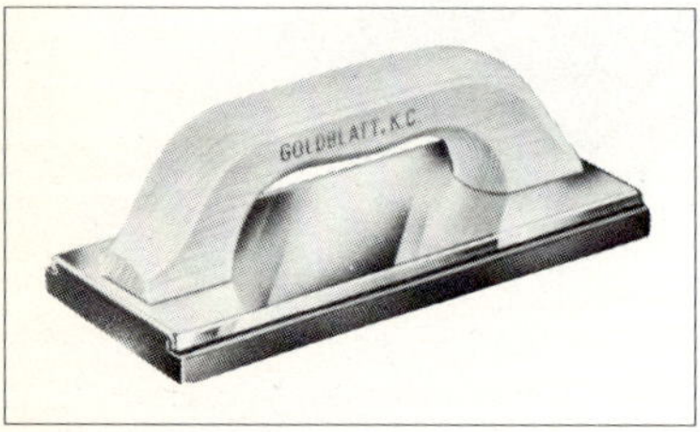

Float—For producing a sand-finish effect on plaster and veneer plaster surfaces. Floats may be faced with sponge rubber (as shown), cork, felt or carpet.

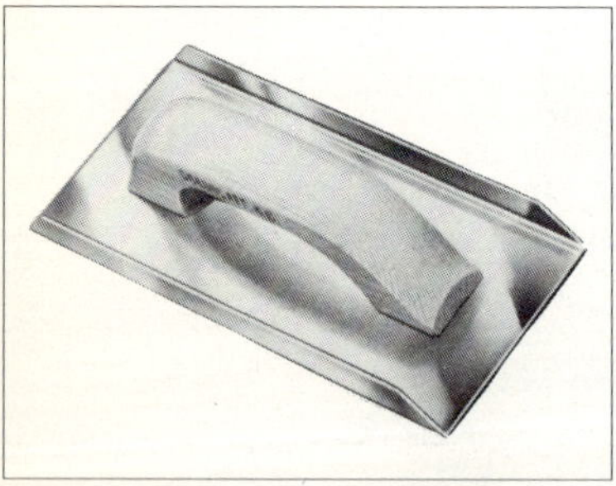

Angle Float—Should not be confused with floats used to raise sand-float textures. Angle floats for corner work with conventional plasters can be used for either brown or finish coat.

Trowels—Available in several styles and in lengths from 10″ to 16″. *Curved* trowels are preferred by many drywall tapers because the blade has a shallow curve (approx. 5⁄32″ deep) that helps feather and finish joints to a low, inconspicuous crown.

Flat trowels are the standard tools for veneer plaster and conventional plaster work.

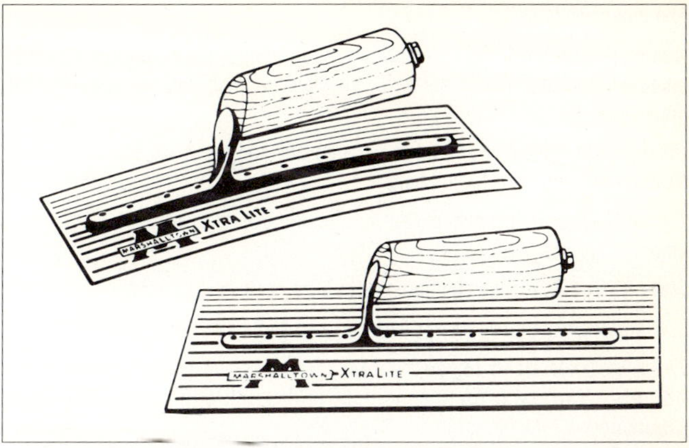

Taping and Finishing Tools—Sometimes called "mechanical" tools, this line of equipment includes specialized tools for every phase of drywall joint treatment work.

The types available are:

Automatic Taper: Applies tape and proper amount of joint compound simultaneously to flat joints or corners. Designed for high-volume machine tool application.

Corner Roller: Used to embed tape in corner and force excess compound from under tape prior to finishing.

Corner Finisher: Distributes excess compound evenly over tape and feathers edges.

Hand Pump: Fills mechanical tools from 5-gal. pail.

Flat Applicator: Applies compound for hand application of tape.

Finishing Tools: Used for application of successive coats of joint compounds. Also used for mechanical tool application of COVER COAT Compound.

Hand Texture Equipment

a. **Glitter Gun**—For embedding glitter in wet texture ceilings. Hand-crank model shown is most economical but is not as efficient as air-powered type (not shown).

b. **Mixing Paddles**—Available in various styles, paddles are used with heavy-duty ½″ electric drill for thorough, time-saving mixing of joint compounds and texture products.

c. **Stucco Brush**—For creating a variety of textures from stipple to swirl. Other variations can be achieved with thicker application and deeper texturing.

d. **Texture Brush**—Available in many sizes and styles, tandem-mounted brushes cover large area to speed texturing job.

e. **Wipedown Blade**—Tool has hardened steel blade and long handle to speed cleaning of walls and floors after application of joint compound or texture materials. Blade has rounded corners to avoid gouging.

f. **Roller**—Standard paint roller is adapted to particular type of finish required. Roller sleeves available include short-nap, long-nap and carpet type in professional widths.

g. **Roller Pan**—For use with roller. Some models can hold up to 25-lb. supply of mixed texture.

h. **Masking Tape Dispenser**—For holding large rolls of masking tape when applying polyvinyl film. Especially useful for respray work when extensive masking is required (not shown).

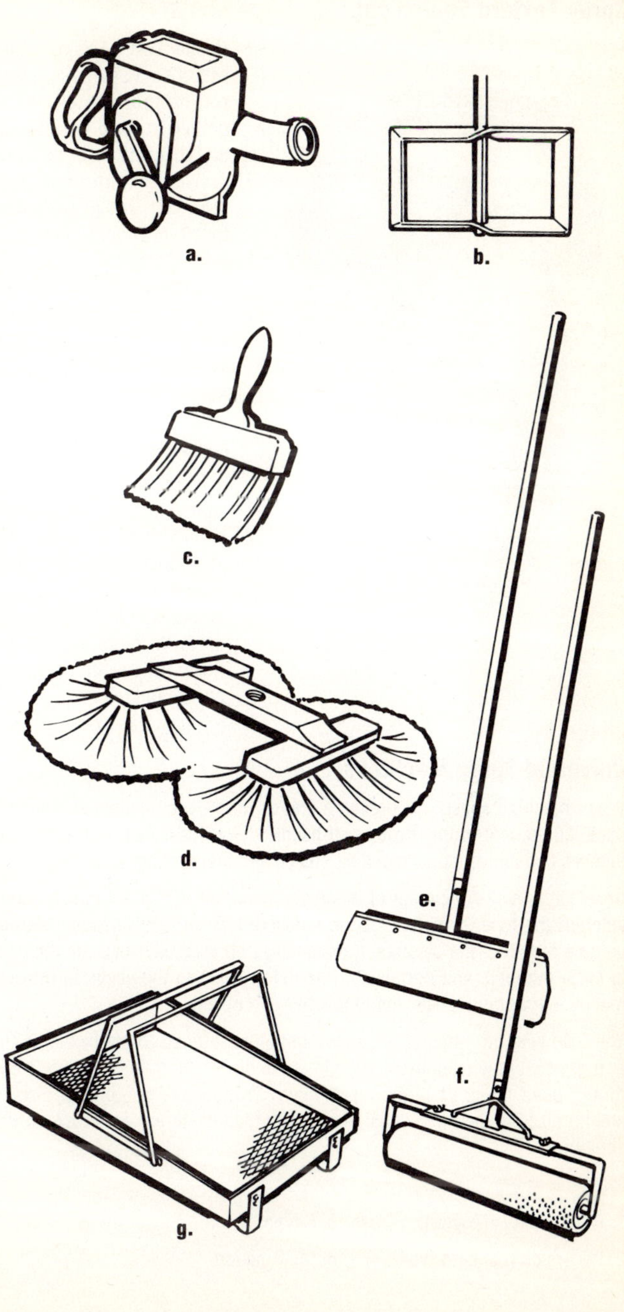
a.
b.
c.
d.
e.
f.
g.

Spray Texture Equipment

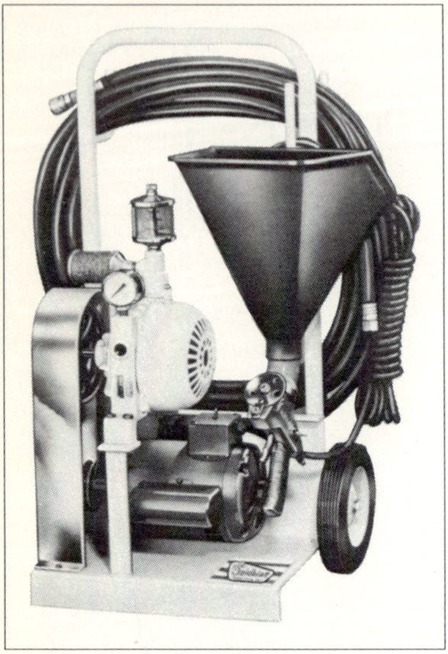

Hopper Gun—This machine, with a spray gun and material hopper mounted together to form an integral unit, handles most types of drywall texture materials. Material is gravity-fed through a hand-held hopper. Compressed air is introduced at the spray-nozzle orifice where texture material is atomized and applied to substrate. The same type of gun also available with larger motor and compressor.

A third and heavier-duty type of texture machine features a material pump. The gun furnished with this model is hopperless and accepts two hoses—a material hose and an air hose. A separate compressor is required.

Universal Spray Machines

When machine speed, air pressure and/or nozzle are adapted to material used, equipment in this group can handle drywall textures, veneer plaster finishes and conventional plasters, stucco and fireproofing materials.

In selecting new equipment of this type, a number of factors must be considered: the type of material to be sprayed, type of finish desired, output volume required, the distance (horizontally and vertically) that the material is to be pumped, and portability. Portability refers to the ability to move a machine through the halls and doorways in a building.

The following information is general in nature, offered to aid in the selection of new spray equipment. Equipment is discussed in terms of the commonly used types of pumping devices. Prospective equipment buyers should discuss their individual needs with manufacturers and users of the equipment.

Pump types are:

1. Rotor-stator (Moyno).

2. Peristaltic (squeeze-type).

3. Single-piston.

4. Multi-piston.

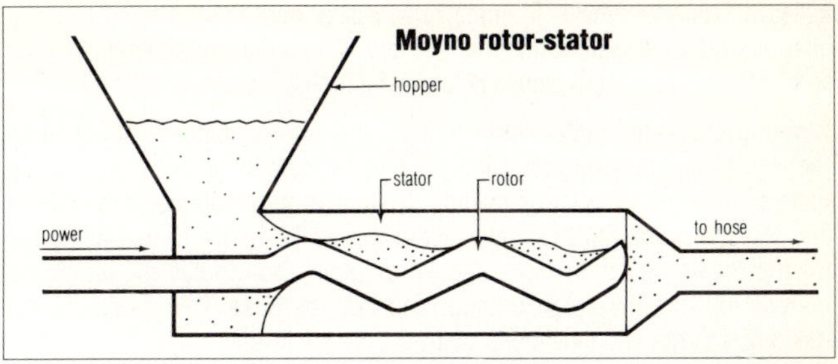

Rotor-stator (Moyno)—This pump resembles a screw turning inside a cylinder. A threaded metal rotor turns inside a stationary metal sleeve (the stator) that is lined with a pliable material such as rubber or neoprene. The threaded rotor is connected to the power source and when in motion draws material from the hopper and drives it through the hose.

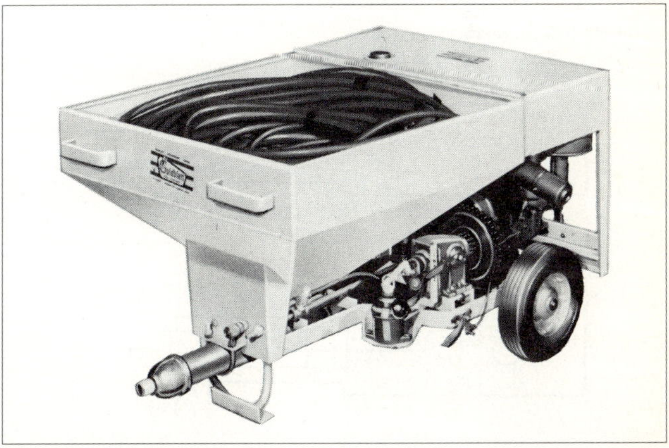

Acoustic spray machine—7-hp rotor-stator model, 100-psi compressor, 35-gal. hopper, 75 ft. hose for pole gun.

Rotor-stator pumps have a relatively high wear incidence with abrasive aggregates such as sand or perlite. However, they are particularly suited for pumping textures with polystyrene aggregates since these aggregates introduce "slip" into the mix and reduce pumping resistance. In addition, the smooth, constant delivery action makes rotor-stator pumps a good choice for very fine textures.

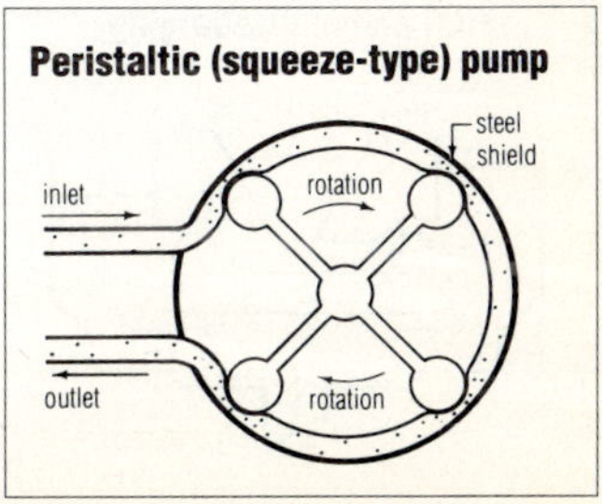

Peristaltic—The action of this type of pump is like a wheel running lengthwise over a hose, squeezing material in the hose forward (the pump is sometimes called a "squeeze-type"). Multiple rollers pass over the pumping line and ensure smooth, constant material delivery. Offers the same benefits as the rotor-stator pump. Also relatively low-volume, high-wear.

Single-piston—In single-piston pumps, the main cylinder is about twice the size of its companion surge cylinder. As material is forced out of the main cylinder into the surge cylinder, only half of its volume is forced out and into the hose. As the main piston makes its return stroke to fill the main cylinder with material, a second or ram piston forces the remaining material out the the surge cylinder and into the hose. This double action minimizes surge and pulsations to an acceptable level.

Although single-piston pumps do deliver materials with some amount of surge, many operators who specialize in perlite texture work prefer them because of their low-wear, low-maintenance performance.

These are high-volume pumps that can be metered for moderately fine textures.

Single-piston pump

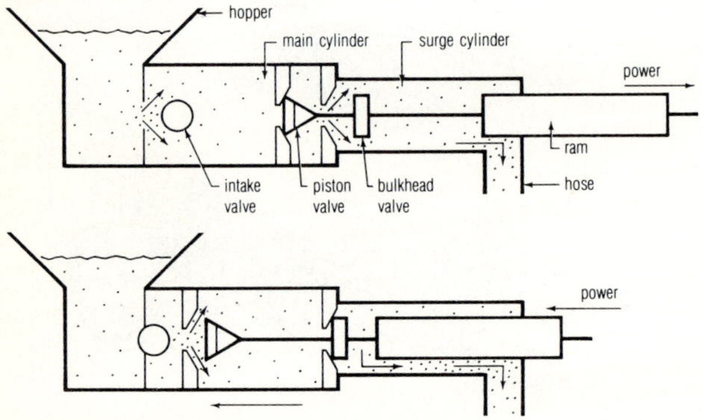

Multi-piston—Pumps having two or more pistons share a common feature. All are designed to reduce surge to the lowest possible level. While one piston is discharging material into a manifold (which in turn connects

Double-piston pump

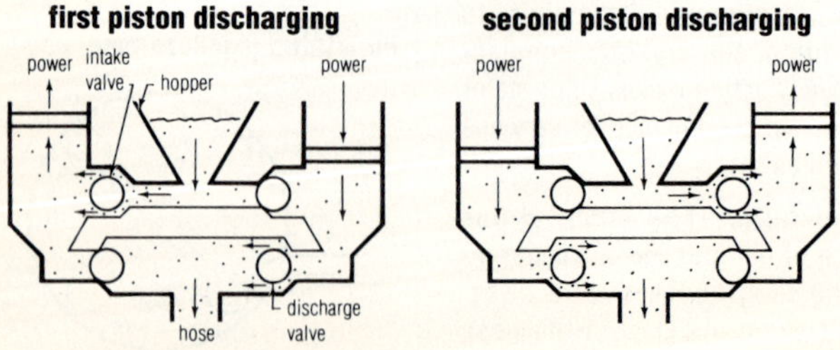

to the material hose), the other cylinder (or cylinders) is drawing material from the hopper in preparation for the pump stroke.

In comparative terms, multi-piston pumps deliver the highest volume of material of all pump types. Like single-piston machines, these pumps can be metered down for a moderately fine texture.

Large-volume acoustic and texture machine—18-hp air piston model, 53-cfm compressor, 440-gal. capacity.

Custom-mounted acoustic and texture machine—air piston model powered by truck engine, 85-cfm compressor, 755-gal. capacity.

Hoses, Guns, Nozzles

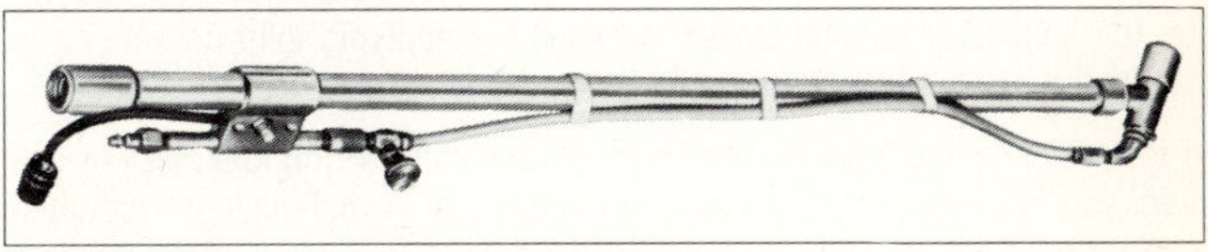

Hoses—Used to carry material from pump to nozzle. They vary in type and generally have a diameter of ¾ to 1 in.

Pole Guns—Used with any universal spray machines as well as largest of drywall texture machines described earlier in this section. Their length allows any operator to spray moderately high ceilings without scaffolding or stilts. Model shown has electric start-stop control. Also available with air start-stop control.

Texture Guns—Professional-type equipment for specific texture applications, the Binks Model 7E2 Type Texturing Gun is used for high volume or heavy texture designs. Binks Model 18D Type for lighter textures. Comparable equipment is available from other spray-equipment manufacturers.

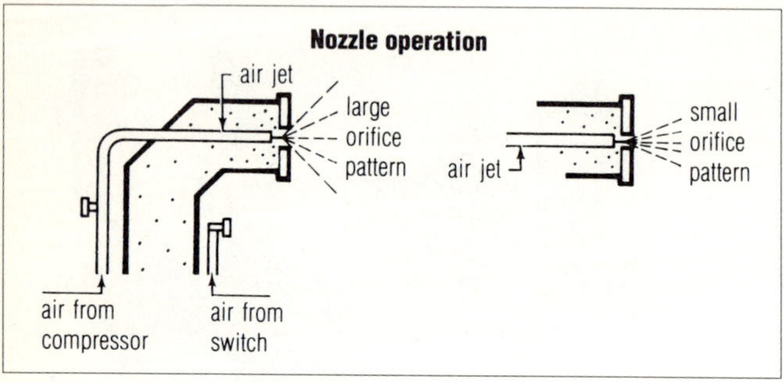

Nozzles—Provide for a variety of spray textures, and vary in orifice openings from ¼″ to ⅝″ Those used for conventional texturing are never larger than ½″.

Miscellaneous Equipment

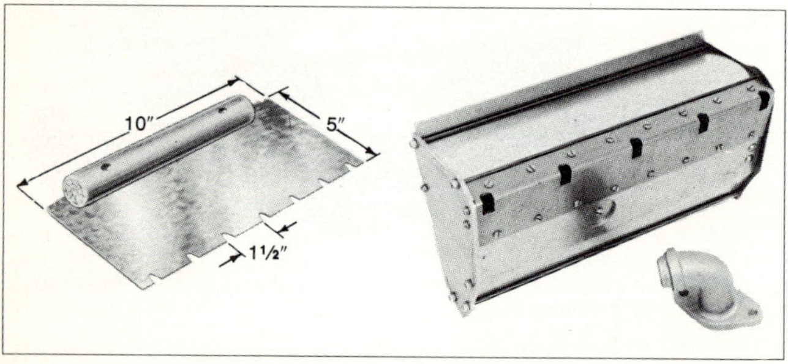

Joint Compound and Adhesive Spreaders—Made either commercially or by the applicator himself, these are used for applying joint compound in laminated gypsum panel assemblies.

The spreader shown (above left) is easily made on the job. Stainless or galvanized sheet steel make the best spreaders. Other materials are *not* satisfactory because compound tends to accumulate and dry in the notches. A good spreader blade has about the same stiffness as a plasterer's trowel.

Notches should be an inverted "V" shape, ½″ deep, ⅜″ wide at the base and spaced 1½″ to 2″ o.c. A piece of wood dowel or window stop attached near top edge of blade provides a grip.

The tool shown (above right) is a laminating spreader that applies properly sized beads of adhesive at correct spacings.

Gypsum Board Dolly—For efficient transport of gypsum boards around the floors of a building. The load, centered over large side wheels, is easily steered and moved by one worker.

Folding Trestle—Top surface, 9½″ x 48″, provides work surface or stand-on work platform. Legs adjust from 18″ to 32″ in 2″ increments.

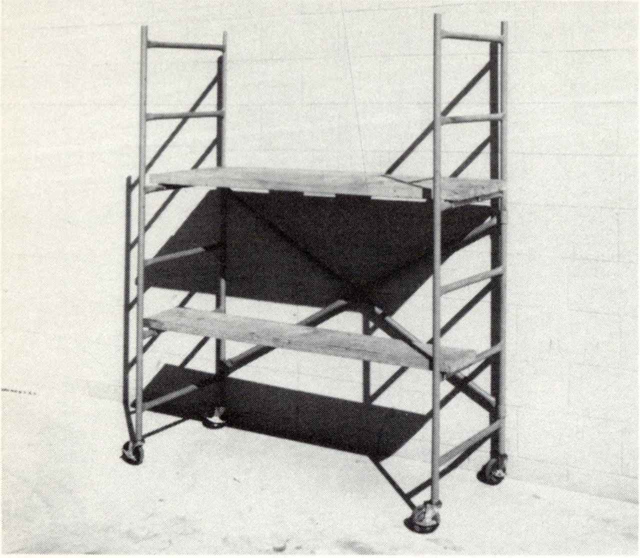

Midget Scaffold—Portable and easy to set up. Ideal for jobs that do not require full scaffolding.

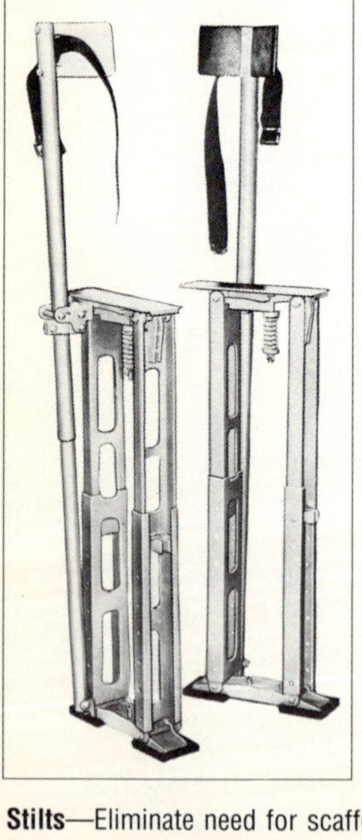

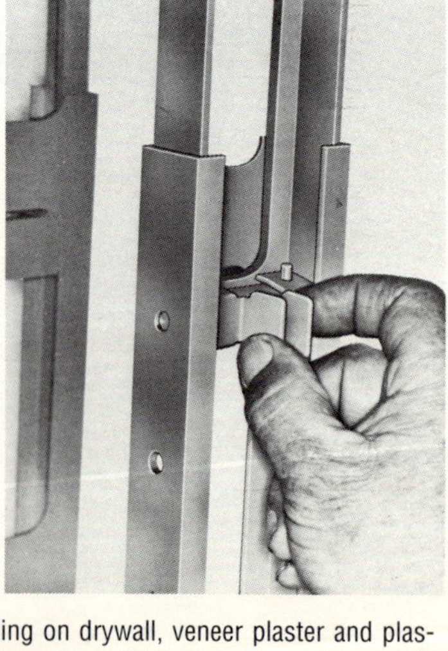

Stilts—Eliminate need for scaffolding on drywall, veneer plaster and plaster jobs. Gives applicator full mobility plus height needed for ceiling work.

Some stilts have articulated joints to flex with ankle movement. Available in fixed-height and adjustable-height types (adjustable, articulated model shown).

Spray Shield—36″ wide aluminum shield protects abutting wall or ceiling against overspray during spraying operation.

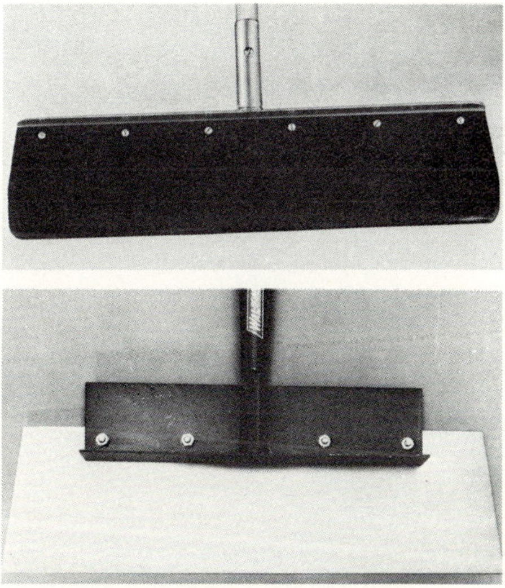

Wall and Floor Scrapers—Both tools have hardened steel blades and long handles to speed cleaning of walls and floors after application of joint compound, plaster or texture materials. Wall scraper (top photo) has rounded corners to avoid gouging.

Appendix

Agencies and Associations

ACI	American Concrete Institute P.O. Box 19150 Detroit, Mich. 48219 313-532-2600
AGC	Associated General Contractors of America 1957 E St., N.W. Washington, D.C. 20006 202-393-2040
AIA	The American Institute of Architects 1735 New York Ave., N.W. Washington, D.C. 20006 202-626-7300
A Ins. A	American Insurance Association 1130 Connecticut Ave., N.W. Washington, D.C. 20036 202-828-7100
AISI	American Iron & Steel Institute 1101 17th St., N.W. Washington, D.C. 20036 202-452-7100
ANSI	American National Standards Institute 11 W. 42nd St. New York, N.Y. 10036 212-642-4900
APA	American Plywood Association P.O. Box 11700 Tacoma, Wash. 98411 206-565-6600
ASA	Acoustical Society of America 500 Sunnyside Blvd. Woodbury, N.Y. 11797 516-576-2360
ASC	Adhesive and Sealant Council 1627 K St., N.W., Suite 1000 Washington, D.C. 20006 202-452-1500
ASHRAE	American Society of Heating, Refrigerating & Air Conditioning Engineers, Inc. 1791 Tullie Circle, N.E. Atlanta, Ga. 30329 404-636-8400

ASTM	ASTM 1916 Race St. Philadelphia, Pa. 19103 215-299-5400
AWCI	Association of the Wall & Ceiling Industries International 1600 Cameron St., Suite 200 Alexandria, Va. 22314 703-684-2924
BIA	Brick Institute of America 11490 Commerce Park Drive Reston, Va. 22091 703-620-0010
BOCA	Building Officials & Code Administrators International, Inc. 4051 W. Flossmoor Rd. Country Club Hills, Ill. 60478 708-799-2300
CABO	Council of American Building Officials 5203 Leesburg Pike, Suite 708 Falls Church, Va. 22041 703-931-4533
CISCA	Ceilings & Interior Systems Construction Association 579 W. North Ave. Elmhurst, Ill. 60126 708-833-1919
DFCI	Drywall Finishing Council, Inc. 345 West Meats Avenue Orange, Calif. 92665 714-637-2770
EIMA	Exterior Insulation Manufacturers Association 2759 State Rd. 580 Clearwater, Fla. 34621 813-726-6477
GA	Gypsum Association 810 First St., N.E., Suite 510 Washington, D.C. 20002 202-289-5440
GSA	General Services Administration 18th & F Streets, N.W. Washington, D.C. 20405 202-708-5082

HUD	Dept. of Housing & Urban Development HUD Building 451 Seventh St., S.W. Washington, D.C. 20410 202-708-1422
ICBO	International Conference of Building Officials 5360 South Workman Mill Rd. Whittier, Calif. 90601 310-699-0541
ML/SFA	Metal Lath/Steel Framing Association 600 S. Federal St., Suite 400 Chicago, IL 60605 312-922-6222
NAHB	National Association of Home Builders 15th & M Streets, N.W. Washington, D.C 20005 202-822-0200
NCMA	National Concrete Masonry Association 2302 Horse Pen Rd. Herndon, Va. 22071 703-713-1900
NCSBCS	National Conference of States on Building Codes and Standards 505 Huntmar Park Drive, Suite 210 Herndon, Va. 22070 703-436-0100
NEMA	National Electrical Manufacturers Association 2101 L St., N.W. Washington, D.C. 20037 202-457-8400
NFiPA	National Fire Protection Association 1 Batterymarch Park Quincy, Mass. 02269 1-800-344-3555
NFoPA	National Forest Products Association 1250 Connecticut Ave., N.W., Suite 200 Washington, D.C. 20036 202-463-2700
NIBS	National Institute of Building Sciences 1201 L Street, N.W., Suite 400 Washington, D.C. 20005 202-289-7800
NLA	National Lime Association 3601 N. Fairfax Drive Arlington, Va. 22201 703-243-5463

NTIS	National Technical Information Service U.S. Dept. of Commerce 5285 Port Royal Road Springfield, Va. 22161 703-487-4650
PCA	Portland Cement Association 5420 Old Orchard Road Skokie, Ill. 60077 708-966-6200
RAL	Riverbank Acoustical Laboratories 1512 Batavia Ave. Geneva, Ill. 60134 708-232-0104
SBCCI	Southern Building Code Congress, International 900 Montclair Rd. Birmingham, Ala. 35213 205-591-1853
TCA	Tile Council of America P.O. Box 326 Princeton, N.J. 08542 609-921-7050
TPI	Truss Plate Institute 583 D'Onofrio Drive, Suite 200 Madison, Wis. 53719 608-833-5900
UL	Underwriters Laboratories Inc. 333 Pfingsten Road Northbrook, Ill. 60062 708-272-8800
WHI	Warnock Hersey International Inc. 530 Garica Ave. Pittsburgh, Calif. 94565 510-432-7344

Rating Fire Endurance

(ASTM E119, UL 263 and NFPA 251)

This is the standard test for rating the fire resistance of columns, girders, beams, and wall-partition, floor-ceiling and roof-ceiling assemblies. It is published by three organizations, designated above, and is essentially the same for all three.

The test procedure consists of the fire endurance test for all assemblies (not individual products) and, in addition, a hose stream test for partition and wall assemblies. The test specimen assembly must meet the following requirements:

1. Structural elements subjected to the test must support the maximum design loads applied throughout the test period. Columns, beams, girders and structural decks must carry the load without failure.

This test does not imply that the test specimen will be suitable for use after the exposure. Some specimens are so damaged after one hour of exposure that they would require replacement, even though they meet all of the requirements for a 4-hr. rating.

2. No openings may develop in an assembly that will permit flames or hot gases to penetrate and ignite combusitbles on the other side.

3. An assembly must resist heat transmission so that temperatures on the side opposite the fire are maintained below designated values. The temperature of the unexposed surface is measured by thermocouples covered with dry refractory filter pads attached directly to the surface. In the case of walls and partitions, one thermocouple is located at the center of the assembly, one in center of each quarter-section, and the other four at the discretion of the testing authority.

The integrity of walls and partitions is evaluated in the hose stream test that examines the construction's ability to resist disintegration under adverse conditions. The hose stream test subjects a duplicate sample to one-half of the indicated fire exposure (but not more than one hour), then immediately to a stream of water from a fire nozzle at a prescribed pressure and distance. This test simulates the effect water would have on the exposed surface under real fire conditions. If there is a breakthrough on the unexposed side, sufficient to pass a stream of water, the result is test failure.

The time-temperature curve used for the fire endurance test is shown on next page. The temperature of the furnace is obtained from the average readings of nine thermocouples, symmetrically located, and placed 6″ from the exposed surface of walls and partitions, or 12″ from the exposed surface of floors, ceilings and columns.

Conditions for Hose Stream Test

	Water Pressure At Base of Nozzle		Duration of Application, Min. per 100 ft² (9.29²) Exposed Area
Resistance Period	**lbf/in²**	**kPa**	
8 hr. and over	45	310	6
4 hr. and over if less than 8 hr.	45	310	5
2 hr. and over if less than 4 hr.	30	207	2½
1½ hr. and over if less than 2 hr.	30	207	1½
1 hr. and over if less than 1½ hr.	30	207	1
Less than 1 hr., if desired	30	207	1

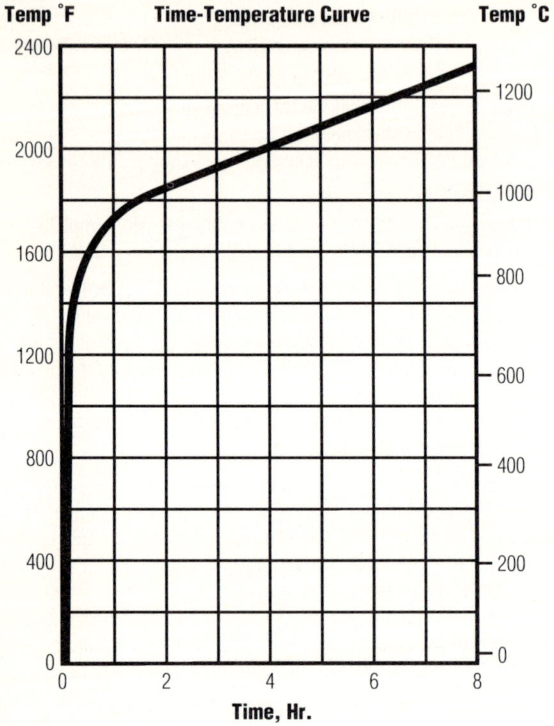

Through-Penetration Fire Stops

(ASTM E814)

ASTM E119 is the guide for assessing fire performance of most construction products and assemblies. However, ASTM E814 Fire Tests of Through-Penetration Fire Stops has been developed in recognition of the special role these constructions play in fire protection.

This standard test is applicable to through-penetration fire stops of various materials and construction. Fire stops are intended for use in openings in fire-resistive walls and floors. They consist of materials that fill the opening around penetrating items such as cables, cable trays, conduits, ducts, and pipes and their means of support.

The test method considers the resistance of fire stops to an external force simulated by a hose stream. Two ratings are established for each fire stop. An F Rating is based upon flame occurrence on the unexposed surface, while the T Rating is based upon the temperature rise as well as flame occurrence on the unexposed side of the fire stop.

A fire stop shall be considered as meeting the requirements for an F Rating when it remains in the opening during the fire test and hose stream test within the following limitations:

1. The fire stops shall have withstood the fire test for the rating period without permitting the passage of flame through openings, or the occurrence of flaming on any element of the unexposed side of the fire stops.
2. During the hose stream test, the fire stop shall not develop any opening that would permit a projection of water from the stream beyond the unexposed side.

A fire stop shall be considered as meeting the requirements for a T Rating when it remains in the opening during the fire test and hose stream test within the above limitations for F Rating plus the following:
1. The transmission of heat through the fire stops during the rating period shall not have been such as to raise the temperature of any thermocouple on the unexposed surface of the fire stop or on any penetrating item more than 325°F above its initial temperature.

Surface Burning Characteristics

(ASTM E84, ANSI 2.5, NFPA 225 and UL 723)

The characteristics of interior finish materials that are related to fire protection are:

—ability to spread fire

—quantity of smoke developed when burning

Materials that have high flame spread and produce large quantities of smoke are considered undesirable, especially when used in areas where people assemble or are confined.

The flame spread test (Surface Burning Characteristics of Building Materials) is often referred to as the Steiner Tunnel Test, after its originator.

In the test, a 20″ x 25′ sample, forming the roof of a rectangular furnace, is

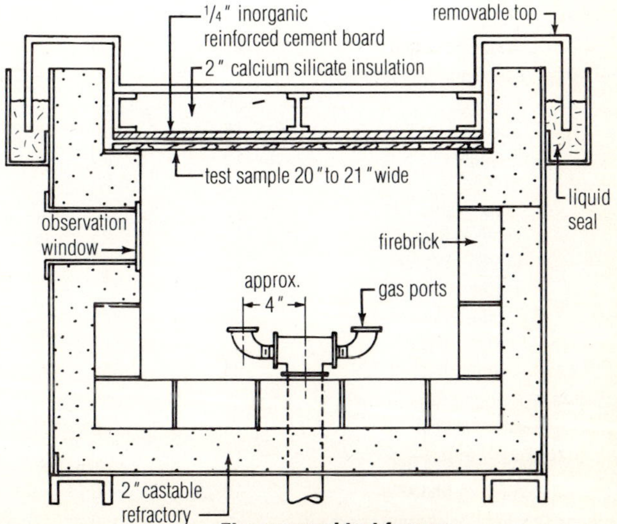

Flame spread test furnace

subjected to a fire of controlled severity, placed 12″ from one end of the sample. Where the flame contacts the sample is considered to be 4½′ from the fire, so the test is actually conducted over 19½′ of the sample.

The time required for the flame to travel the 19′ to the end of the sample, along with the smoke and heat produced, is compared with similar figures for red oak which is arbitrarily given the value of 100 for these two characteristics, and inorganic reinforced board which is given the value of 0.

Smoke developed is measured by means of a photoelectric cell connected to an ammeter which indicates changes in smoke density.

Obviously, the indices developed in the tunnel test are relative, but enough is known about the burning characteristics of materials to make these indices reliable for building code specifications.

Most building codes divide materials into four classes, based on the Flame Spread Indices. The numbering and range of each class varies with the different codes, but they generally follow this pattern:

Class I (Class A)—0-25 | Class III (Class C)—76-200
Class II (Class B)—26-75 | Class IV (Class D)—over 200

Surface Burning Characteristics (per ASTM E84)

Product	Flame Spread	Smoke Developed
SHEETROCK brand Gypsum Panels	15	0
SHEETROCK brand Interior Gypsum Ceiling Board	15	0
SHEETROCK brand Exterior Gypsum Ceiling Board	20	0
SHEETROCK brand Gypsum Panels, Water-Resistant	20	0
TEXTONE Vinyl-Faced Gypsum Panels		
Pumice	20	25
Suede	15	25
Presidio	15	25
Granite	15	25
Woodgrain	20	15
Linen Pattern	15	25
Country Weave	20	35
Textile (Type I, Fabric-Backed)*	25	70
Brittany (Type I, Fabric-Backed)*	25	55
THERMAFIBER Sound Attenuation Fire Blankets	15	0
THERMAFIBER FS-15 Blankets	15	0
THERMAFIBER FS-25 Blankets	10-25	0
DUROCK Cement Board, Underlayment and Exterior Cement Board	5	0

*Comply with Federal Specification CCC-2-408C, Type I.

Determination of Sound Transmission Class (STC)

Testing for airborne sound transmission is performed under rigidly established procedures set up by the American Society for Testing and Materials (ASTM procedure E90-90). Several independent acoustical laboratories across the nation are qualified to perform the tests. Although all are presumably reliable and follow the ASTM procedure, the results tend to vary slightly. For this reason, test results from more than one laboratory should never be compared on an exact basis.

Tests are conducted on a sample assembly, at least 2.4m x 2.4m in size. The assembly is installed between two rooms constructed in such a way that sound transmitted between the rooms by paths other than through the assembly is insignificant. Background noise in the rooms is monitored to ensure it does not affect test results.

The sound source consists of an electronic device and loudspeaker which produce a continuous random noise covering a minimum frequency range of 113 to 4,450 Hz (Hertz—cycles per second). Panel diffusers and/or rotating vanes are set up so noise is diffused and the sound level is measured at several microphone positions in each room. Readings are taken at sixteen ⅓-octave frequency-band intervals. Average sound levels in the receiving room are subtracted from the corresponding sound levels in the source room. The differences (sound levels of the actual transmission) are recorded as transmission-loss values (adjustments are made for test room absorption and test assembly size).

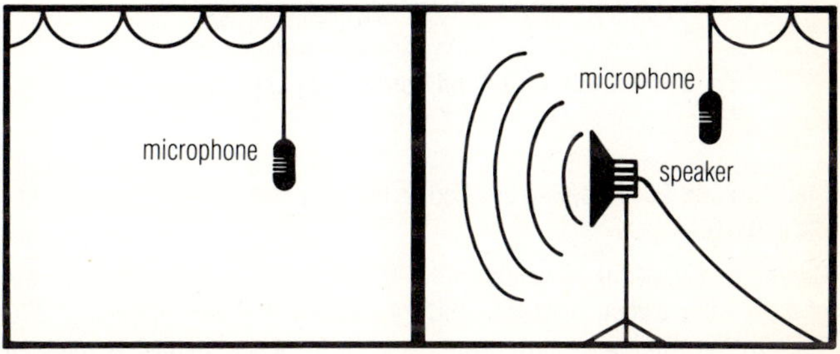

Sound Test Sample Assembly

These transmission-loss values are then plotted on a frequency band-sound pressure level graph and the resulting curve is compared to a standard reference contour. The Sound Transmission Class (STC), as defined by the rating procedure set forth in ASTM E413-87, is determined by adjusting the reference contour vertically until the decibel (dB) total of all frequency bands on the test curve that are below the reference contour does not exceed 32, and no point on the test curve is more than 8 dB below the reference contour. Then, with the reference contour adjusted to meet these standards, its transmission loss at 500 Hz (500 cycles per second) is taken as the STC (dropping the dB unit).

An alternative procedure, frequently used for the measurement of sound transmission loss under field conditions, is given in ASTM Standard Test

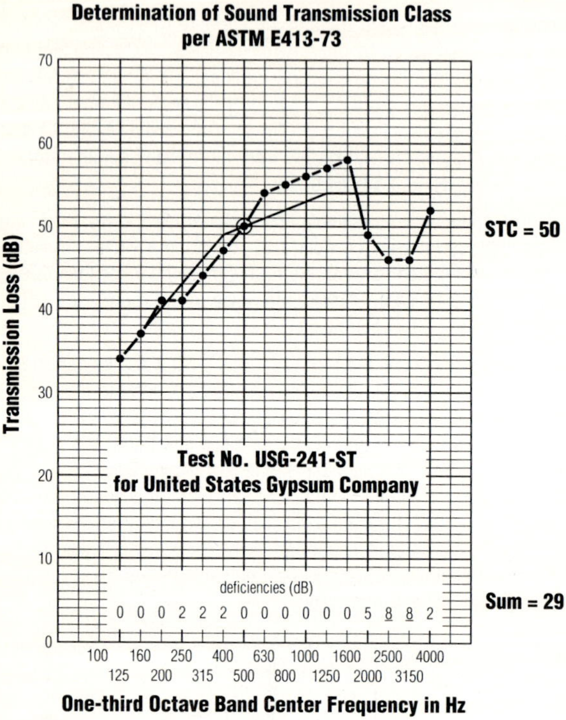

Method E336-90. This may be used to obtain a Field Sound Transmission Class (FSTC).

Reproduced above is the graph of an actual sound transmission-loss test of a drywall partition, Test No. USG-241-ST. The partition is rated at STC 50 with the reference contour adjusted to meet the standards outlined above. The deficiencies at 2,500 Hz and 3,150 Hz are 8, the allowable maximum.

The total of all points below the criterion curve is 29, three points less than the 32 allowed.

The reference contour itself is plotted to allow for subjective human response to sound pressure at the 16 frequency bands measured. Because the human ear is less sensitive to low-frequency sound pressure than to high frequencies, the reference contour has been adjusted to allow some additional noise at low frequencies. This avoids down-rating test results because of noise levels that are least objectionable to people. The ASTM test procedure explains the use of STC in the following excerpt from E413:

"These single-number ratings correlate in a general way with subjective impressions of sound transmission for speech, radio, television, and similar sources of noise in offices and buildings. This classification method is not appropriate for sound sources with spectra significantly different from those sources listed above. Such sources include machinery, industrial processes, bowling alleys, power transformers, musical instruments, many music systems and transportation noises such as motor vehicles, aircraft and trains. For these sources, accurate assessment of sound transmission requires a detailed analysis in frequency bands."

Music/Machinery Transmission Class (MTC)

While STC is a widely used rating of sound attenuation performance, it does not accurately measure music, machinery/mechanical equipment noise or any sound with a substantial portion of low-frequency energy.

To provide a more accurate rating of attenuation of these sound sources, U.S. Gypsum Company developed MTC (Music/Machinery Transmission Class). The MTC rating does not replace STC but complements it. Use STC to isolate speech sound sources; use MTC to isolate music and machinery/mechanical equipment sound sources.

The MTC of an assembly can be less than, the same as, or greater than its STC. Ratings make acoustical design more manageable. And using both STC and MTC ratings improves the precision for selecting the best acoustical system.

Determination of Impact Insulation Class (IIC)

Impact sound originates when one body strikes another, such as in the case of footsteps, hammering and objects falling. Even though some of the sound energy is eventually conducted to the air, the sound is still classified as impact.

Impact sound travels through the structure with little loss of energy if the structure is continuous and rigid. Thus, tenants without enough heat can pound on a radiator and notify the superintendent (and all other tenants as well) of the situation.

Transmission of impact sound can be controlled by isolation, absorption and elimination of flanking paths, and offset by the introduction of masking sound. Limpness in the construction affects transmission of impact sound, but is difficult to introduce because of the structural requirements of the assembly.

Mass plays a secondary role in the isolation of impact sound. The benefit of mass in a sound-control construction is its resistance to being set into vibration. In retarding airborne sound, this is very effective because the sound energy is small. With impact sound, the energy is greater and is applied directly to the construction by the sound source with little energy loss. Thus, the mass of that surface is immediately set into motion. For this reason, concrete slab construction at 100 $lb/ft.^2$ is only slightly more effective in retarding impact sound than simple wood frame construction at 10 $lb/ft.^2$.

Although leaks in a floor-ceiling assembly must be sealed to stop transmission of the airborne sound associated with impact, they play little part in retarding the transmission of structure-borne sound.

Absorbing Impact Sound

The use of sound attenuation blankets is as effective in controlling impact sound as for airborne sound. Of course, unless the opposite surfaces of the assembly (floor and ceiling) are isolated or decoupled, sound travels through the connecting structure.

Structural Flanking Paths

One of the most frequent causes of sound performance failure in a floor-ceiling assembly is flanking paths. Impact sound produces high energy at the source. This energy follows any rigid connection between construction elements with little loss. For example, in a child's tin-can telephone, sound travels better through the tight string stretched between the cans than through the surrounding air.

Some of the most common flanking paths are supplied by plumbing pipes, air ducts and electrical conduit rigidly connected between floor and ceiling. Continuous walls between floors, columns or any other continuous structural elements will act as flanking paths for impact sound. In fact, any rigid connection between the two diaphragms transmits impact sound.

Methods of Impact Rating

Assemblies designed to retard transmission of impact sound are tested for performance as prescribed by ASTM Standard Method E492-90. The floor-ceiling assembly is constructed between two isolated rooms, and microphones are positioned in the receiving room to record the pressure of transmitted sound.

The impact sound source is a standard tapping machine. It rests on the floor of the test assembly and drops hammers at a uniform rate and impact

USG Acoustical Research Facility at Round Lake, Ill.

Determination of Impact Insulation Class

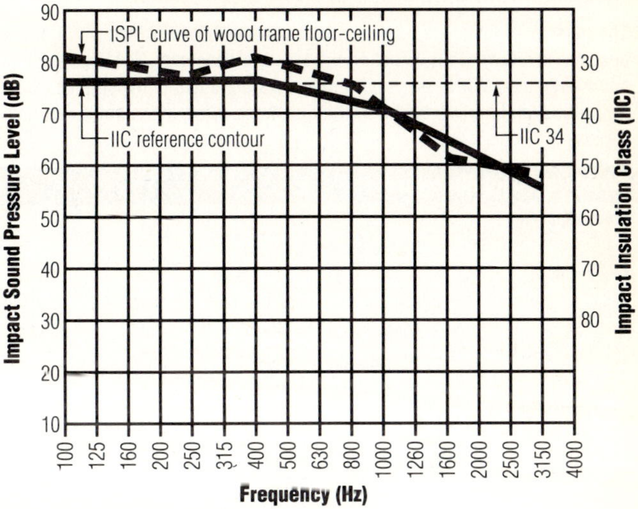

energy. The sound produced depends to a large extent on the floor surface material. Carpet and pad, for example, greatly improve IIC ratings. The transmitted sound is measured and recorded at several microphone locations and four locations of the tapping machine. Results are corrected to a standard absorption so that results from different laboratories may be compared.

These results, recorded at sixteen ⅓-octave bands, are plotted and compared with a standard reference contour in much the same manner as Sound Transmission Class determinations, except that deficiencies lie above the contour.

Impact sound rating methods were established by the Federal Housing Administration (now HUD). The earliest was a single-number rating system called Impact Noise Rating (INR) and published in FHA 750.

The current rating system is described in E989-89. To determine this Impact Insulation Class (IIC), the ISPL curve is plotted on a graph as shown above. The reference contour is then shifted to the lowest point where no point on the ISPL curve is more than 8 dB above it, and the sum of all ISPL deviations above it is no more than 32 dB. The location of the reference contour at 500 Hz is projected to the IIC scale, right of graph, to read IIC Rating.

The IIC relates to STC ratings with respect to acceptability, and is a positive number. IIC values will usually be 51 points above the corresponding former INR values, but some deviations can occur. Tests must be analyzed individually against IIC criteria.

Gypsum Board Screw Usage

The number of fasteners used to install gypsum board varies with framing spacing, screw spacing, panel orientation and panel size. The charts below show estimated screw usage per thousand square feet of gypsum board for both horizontal and vertical board attachment. Allowance should be made for loss.

Horizontal Board Attachment

Frame Spacing	Screw Spacing (Inches)			
4′ x 8′ Board	**8**	**12**	**16**	**24**
8″	2844	2031	1625	1219
12″	1969	1406	1125	844
16″	1531	1094	875	656
24″	1094	781	625	469
4′ x 10′ Board				
8″	2800	2000	1600	1200
12″	1925	1375	1100	825
16″	1488	1063	850	638
24″	1050	750	600	450
4′ x 12′ Board				
8″	2780	1980	1590	1190
12″	1900	1360	1090	820
16″	1460	1050	840	630
24″	1030	730	590	440

Vertical Board Attachment

Frame Spacing	Screw Spacing (Inches)			
4′ x 8′ Board	**8**	**12**	**16**	**24**
8″	2844	1969	1531	1094
12″	2031	1406	1094	781
16″	1625	1125	875	625
24″	1219	844	656	469
4′ x 10′ Board				
8″	2800	1925	1488	1050
12″	2000	1375	1063	750
16″	1600	1100	850	600
24″	1200	825	638	450
4′ x 12′ Board				
8″	2771	1896	1458	1021
12″	1979	1354	1042	729
16″	1583	1083	833	583
24″	1188	813	625	438

Packaging and Coverage Data

The following tables show packaging and approximate coverage data for drywall and veneer plaster construction, cement board construction, and conventional lath and plaster construction. Always consult directions on product package for most up-to-date coverage information.

Drywall and Veneer Plaster Construction

Joint Treatments

Product	Approx. Coverage	Packaging
SHEETROCK Setting-Type Joint Compounds (DURABOND)		
20	83 lb./1000 ft.^{2}	25 lb. bag
45	83 lb./1000 ft.^{2}	25 lb. bag
90	83 lb./1000 ft.^{2}	25 lb. bag
210	83 lb./1000 ft.^{2}	25 lb. bag
300	83 lb./1000 ft.^{2}	25 lb. bag
SHEETROCK Lightweight Setting-Type Joint Compounds (EASY SAND)		
20	52 lb./1000 ft.^{2}	18 lb. bag
45	52 lb./1000 ft.^{2}	18 lb. bag
90	52 lb./1000 ft.^{2}	18 lb. bag
210	52 lb./1000 ft.^{2}	18 lb. bag
300	52 lb./1000 ft.^{2}	18 lb. bag
SHEETROCK Ready-Mixed Joint Compounds		
Taping, Topping, All Purpose	138 lb./1000 ft.^{2}	12 lb., 42 lb. or 61.7 lb. pail; 48 lb., 50 lb., 61.7 lb. cartons
PLUS 3 (Lightweight All Purpose)	9.4 gal./1000 ft.^{2}	1 gal., 4.5 gal. pail; 3.5 and 4.5 gal. cartons
SHEETROCK Drying-Type Powder Joint Compounds		
Taping, Topping, All Purpose	83 lb./1000 ft.^{2}	25 lb. bag
AP LITE (Lightweight All Purpose)	67 lb./1000 ft.^{2}	20 lb. bag
SHEETROCK Joint Tape		
75′ roll	370 ft./1000^2	24 per carton
250′ roll	370 ft./1000^2	20 per carton
500′ roll	370 ft./1000^2	10 per carton
IMPERIAL Joint Tape		
300′ roll	370 ft./1000^2	12 per carton

Drywall and Veneer Plaster Construction (cont.)

Fire Stops

Product	Approx. Coverage	Packaging
FIRECODE Compound		
	517.4 in.²/1″ deep (walls)	15 lb. bag
	537 in.²/1″ deep (floors)	
THERMAFIBER Smoke Seal Compound		
	41′/⅜″ bead	30 oz. cartridge
	609′/⅜″ bead	3½ gal.
	870′/⅜″ bead	5 gal.

Sealant

Product	Approx. Coverage	Packaging
SHEETROCK Acoustical Sealant		
	89′ (¼″ bead)	29 oz. cartridge
	40′ (⅜″ bead)	
	22′ (½″ bead)	
	392′ (¼″ bead/gal.)	5 gal. pail
	174′ (⅜″ bead/gal.)	
	98′ (½″ bead/gal.)	

Veneer Plaster

Product	Approx. Coverage	Packaging
IMPERIAL Basecoat Plaster		
	3250-4250 ft.²/ton (gypsum base)	80 lb. bag
DIAMOND Veneer Basecoat Plaster		
	4000-5000 ft.²/ton (gypsum base)	79.4 lb. bag
IMPERIAL Finish Plaster		
	3500-4000 ft.²/ton (1-coat: gypsum base)	80 lb. bag
	3200-3600 ft.²/ton (2-coat: gypsum base)	
DIAMOND Interior Finish Plaster		
	Depends on mix ratio. See page 89.	50 lb. bag

Concrete Covering

Product	Approx. Coverage	Packaging
COVER COAT Compound		
	¹⁄₁₆″ thick: 1½-2 ft.²/lb.	61.7 lb. pail, 61.7 lb. carton
	¹⁄₁₆″ thick: 2-2½ ft.²/lb. with sand.	

Decorating Products

Product	Approx. Coverage	Packaging
SHEETROCK First Coat		
	300-500 ft.²/gal.	1 and 5 gal. pails; 25 lb. bags
IMPERIAL QT Spray Texture Finish		
	Up to 8 ft.²/lb.	32 lb. and 40 lb. bags
USG Spray Texture Finish		
	15-25 ft.²/lb.	40 lb. and 50 lb. bags
SHEETROCK Wall And Ceiling Spray Texture (TUF TEX)		
	20-40 ft.²/lb.	40 lb. bag
USG Multi-Purpose Texture Finish		
	10-20 ft.²/lb.	25 lb. and 50 lb. bags
USG Texture XII Drywall Surfacer		
	20-35 ft.²/lb.	25 lb. bag
USG QUIK & EASY Wall and Ceiling Texture		
	Up to 250 ft.²/3.5 gal.	3.5 gal. pail
USG Acoustical Finish		
	¼″ thick.: 3 ft.²/lb	30 lb. bag
	½″ thick: 1.5 ft.²/lb.	

Cement Board Construction

Product	Approx. Coverage	Packaging
DUROCK Interior Tape		
2″ wide	75′/roll	24 rolls/carton
DUROCK Exterior Tape		
4″ wide	150′/roll	4 rolls/carton
DUROCK Wood Screws		
1¼″*	1,550-1,650 pcs./1000 ft.²	5,000 per carton
1⅝″	1,550-1,650 pcs./1000 ft.²	4,000 per carton
2¼″	1,550-1,650 pcs./1000 ft.²	3,000 per carton
DUROCK Steel Screws		
1¼″*	1,550-1,650 pcs./1000 ft.²	5,000 per carton
1⅝″	1,550-1,650 pcs./1000 ft.²	4,000 per carton
2¼″	1,550-1,650 pcs./1000 ft.²	2,000 per carton
DUROCK Insulation Board		
2′x4′x1″ thick	192 ft.²	24 pcs./bag
2′x4′x1½″ thick	128 ft.²	16 pcs./bag
2′x4′x2″ thick	96 ft.²	12 pcs./bag
DUROCK Reinforcing Mesh		
38″x50 yd. roll	475 ft.²/roll	4 rolls/carton

*Also available in twenty 150-piece boxes per carton.

Cement Board Construction (cont.)

Product	Approx. Coverage	Packaging
DUROCK Trim		
J-Trim	8′	50 per carton
L-Trim	8′	50 per carton
DUROCK Corner Bead		
	8′	60 per carton
DUROCK Control Joint		
	8′	25 per carton
DUROCK Exterior Basecoat		
	80-90 ft.^{2} for 1⁄16″ thick.	50 lb. bag
DUROCK Latex Fortified Mortar		
Skim Coat Bond Coat	40-50 ft.^{2} for 1⁄8″ thick. 80-90 ft.^{2}	50 lb. bag
DUROCK Latex Fortified Grout		
	60-80 ft.^{2} (1⁄2″x6″x6″ tile with 3⁄8″ wide joints)	50 lb. bag
	30-45 ft.^{2} (1⁄2″x2 1⁄4″x8″ thin brick with 3⁄8″ wide joints)	
DUROCK Exterior Finish		
Coarse Texture	100-125 ft.^{2}/pail	67.5 lb. pail
Fine Texture	150-175 ft.^{2}/pail	
DUROCK Over-Coat		
Over Coarse Texture	175 ft.^{2}/gal.	46 lb. pail
Over Fine Texture	200 ft.^{2}/gal.	

Conventional Lath and Plaster

Basecoat Plasters

Product	Approx. Coverage	Packaging
RED TOP Gypsum Plaster		
Regular, Lightweight, Machine Application	Depends on mix ratio. See page 285.	100 lb. bag
RED TOP Two-Purpose Plaster		
	Depends on mix ratio. See page 285.	100 lb. bag
STRUCTO-BASE Gypsum Plaster		
Regular, Machine Application	Depends on mix ratio. See page 285.	100 lb. bag
STRUCTO-LITE Gypsum Plaster		
Regular, Masonry, Type S	Depends on mix ratio. See page 285.	80 lb. bag
RED TOP Wood Fiber Plaster		
	Depends on mix ratio. See page 285.	100 lb. bag

Gauging Plasters

Product	Approx. Coverage	Packaging
RED TOP Gauging Plaster		
Regular Quick Set and Slow Set, "Quality," Quick Set and Slow Set	Depends on mix ratio. See page 290.	100 lb. bag
CHAMPION Gauging Plaster		
Regular-Unaggregated, "Quality"	Depends on mix ratio. See page 290.	50 lb. and 100 lb. bags
STAR Gauging Plaster		
Regular, "Quality"	Depends on mix ratio. See page 290.	50 lb. and 100 lb. bags
STRUCTO-GAUGE Gauging Plaster		
Quick Set, Slow Set	Depends on mix ratio. See page 290.	100 lb. bag
RED TOP Keenes Cement		
Quick Trowel, Regular	Depends on mix ratio. See page 290.	100 lb. bag

Finish Limes

Product	Approx. Coverage	Packaging
IVORY Finish Lime		
Double-Hydrate	Blended with gauging plasters. See page 290.	50 lb. bag
SNOWDRIFT Finish Lime		
Double-Hydrate	Blended with gauging plasters. See page 290.	50 lb. bag
RED TOP Finish Lime		
Single-Hydrate	Blended with gauging plasters. See page 290.	50 lb. bag
GRAND PRIZE Finish Lime		
Single-Hydrate	Blended with gauging plasters. See page 290.	50 lb. bag

Prepared Finishes

Product	Approx. Coverage	Packaging
IMPERIAL Finish Plaster		
	3200-4000 ft.^{2}/ton*	80 lb. bag
DIAMOND Interior Finish Plaster		
	5000-6000 ft.^{2}/ton (neat)* 3000-3500 ft.^{2}/ton (sanded 1:1)*	50 lb. bag
RED TOP Finish		
Regular, Quick Set	3300-3900 ft.^{2}/ton*	50 lb. bag

*Depending on base material.

Conventional Lath and Plaster (cont.)

Ornamental Plasters

Product	Approx. Coverage	Packaging
USG Moulding Plaster		
	1.5 cu. ft./100 lbs.	50 lb. and 100 lb. bags
HYDROCAL White Gypsum Cement		
	N/A	50 lb. and 100 lb. bags
HYDROCAL FGR 95 Gypsum Cement		
	N/A	50 lb. and 100 lb. bags

Special Additives

Product	Approx. Coverage	Packaging
RED TOP Retarder		
	N/A	1½ lb. bag
RED TOP Accelerator		
	N/A	2 lb. bag

Prepared Exterior Finish

Product	Approx. Coverage	Packaging
ORIENTAL Exterior Finish Stucco		
	⅛" thick.: 150-200 yd.^{2}/ton	100 lb. bag

Masons and Stucco Lime

Product	Approx. Coverage	Packaging
BONDCRETE Air-Entraining Masons and Stucco Lime		
	N/A	50 lb. bag
MORTASEAL Autoclaved Masons Lime		
	N/A	50 lb. bag

Comparing Plaster Systems

The chart below compares conventional plaster and veneer plaster systems to help in selection for specific job applications.

Characteristics	Comments			
1. Conventional Plaster Best system to attain a uniform, monolithic, blemish-free, smooth surface with excellent wear resistance.				
2. IMPERIAL Basecoat with selected finish shown below, "A" through "E"	**Finish Plaster Rating (No. 1 Best—No. 4 Acceptable)**			
	Productivity	**Hardness**	**Workability**	**Ease to Achieve Smooth Surface**
A. IMPERIAL Finish Ultimate in surface hardness and abrasion resistance. Easily textured. Low productivity and hard to achieve a completely smooth finish.	4	1	4	4
B. DIAMOND Interior Finish Plaster Single-bag, ready-to-use finish. Moderate high strength. Acceptable workability. Extremely adaptable to textured finishes. Satisfactory smooth finish.	2	2	2	3
C. Regular Gauging Lime Putty Highest productivity. Best workability. Joinable, easiest to achieve a monolithic finish. Only moderate surface hardness.	1	4	1	1
D. STRUCTO-GAUGE Gauging Lime Putty (1:1) Hardest dense putty finish. Moderate workability and ease of application. Excellent finish appearance.	2	3	2	2
E. RED TOP Keenes Cement, Lime Putty and Sand Unique, only truly retemperable material. Best choice for coloring or tinting large plaster wall areas. Ultimate choice for texturing. Can be floated for extended time period.	Due to its unique nature, Keenes is not rated with above finishes.			
3. IMPERIAL Finish (one-coat) Monolithic, smooth or textured appearance. Ultimate in surface hardness. Primarily intended for direct application to plaster base. Achieves high productivity due to compatibility with absorbent surface of plaster base. Ready for finishing in 48 hours with favorable drying conditions.	Fast completion shortens construction time, brings in paying tenants faster, thus reducing interest paid on project construction loan.			
4. DIAMOND Interior Finish Plaster Monolithic appearing. Hard, wear-resistant surface. Provides texture desired. Ready for final finish in as little as 48 hours under favorable drying conditions. Greatest coverage for single coat application over special absorbent surface of plaster base. Lowest cost veneer system.	See comment on IMPERIAL Finish			

Metric Terms

Basic Units

Quantity	Metric (SI) Unit	Symbol	U.S.A. equivalent (nom.)[1]
Length	millimeter	mm	0.039 in.
	meter	m	3.281 ft.
			1.094 yd.
Area	meter	m^2	10.763 $ft.^2$
			1.195 $yd.^2$
Volume	meter	m^3	35.314 ft^3
			1.307 $yd.^3$
Volume (Fluid)	liter	L	33.815 oz.
			0.264 gal.
Mass (Weight)	gram	g	0.035 oz.
	kilogram	kg	2.205 lb.
	ton	t	2,204.600 lb.
			1.102 tons
Force	newton	N	0.225 lbf.
Temperature (Interval)	kelvin	K	1.8°F
	degree celsius	°C	1.8°F
Temperature	celsius	°C	(°F-32)5/9
Thermal Resistance		$\frac{K \cdot m^2}{W}$	$\frac{5.679\ ft^2 \cdot hr \cdot °F}{Btu}$
Heat Transfer	watt	W	3.412 Btu/hr.
Pressure	kilopascal	kPa	0.145 lb./$in.^2$ (psi)
	pascal	Pa	20.890 lb./$ft.^2$ (psf)

(1) To convert U.S.A. units to SI units, divide by U.S.A. equivalent.

Prefixes (Order of Magnitude)

Prefix	Symbol	Factor
mega	M	1000000 = 10^{+6}
kilo	k	1000 = 10^{+3}
centi[1]	c	0.01 = 10^{-2}
milli	m	0.001 = 10^{-3}
micro	μ(mu)	0.000001 = 10^{-6}

(1) Limited use only.

Specification Standards

The listings following contain existing standard specifications, classified as Federal, Army, Navy, etc., that apply to U.S. Gypsum Company materials described in this Handbook. Where ASTM, local codes, etc., require product variance, consult your U.S. Gypsum Company representative.

Specification Standards

Product	Federal Specification	ASTM Designation
Plaster		
RED TOP gypsum plaster		C28—gypsum neat plaster
RED TOP two-purpose gypsum plaster		C28—gypsum neat plaster
RED TOP wood fiber plaster		C28—gypsum wood fiber
STRUCTO-LITE plaster perlite aggregate		C28—gypsum ready-mix plaster C35
RED TOP gauging plaster		C28—gypsum gauging for finish coat
RED TOP keenes cement regular quick trowel		 C61 C61
STRUCTO-GAUGE plaster		C28—gypsum gauging for finish coat
STRUCTO-BASE plaster		C28—gypsum neat plaster
IMPERIAL plaster		C587—gypsum veneer plaster
DIAMOND plaster		C587—gypsum veneer plaster
Lime		
RED TOP and GRAND PRIZE finish limes		C6 type N
IVORY finish lime		C206 type S
RED TOP masons hydrate		C207 type N
Gypsum Bases		
ROCKLATH plaster base—⅜″ & ½″		C37
IMPERIAL gypsum base—½″ & ⅝″		C588
UNIMAST Metal Lath & Accessories*		
Bases, metal: (for) plaster, lath and stucco constr.	QQ-L-101C	D847 (metal lath only)
3.4# galv. diamond mesh lath, 2.5# and	class 3 (flat diam. mesh)	
3.4# ptd. ⅛″	class 3 (self-furring diam. mesh)	
4-mesh z-riblath	class 4 (⅛″ flat rib)	
2.75# and 3.4#; ⅜″ riblath	class 1 (⅜″ rib)	
Paper—Paper-back lath	UU-13-790A, Type 1 Grade Style II (paper)	
Hanger wire—tie wire	QQ-W-461H class 1 finish 5 (1006 type steel)	A641

*Studs, runners and other accessories identified as UNIMAST products in this chart are marketed by United States Gypsum Company as integral components of its construction systems.

Specification Standards (cont.)

Product	Federal Specification	ASTM Designation
Gypsum Panels		
SHEETROCK brand gypsum panels		C36
TEXTONE vinyl-covered		C960
SHEETROCK brand, water-resistant		C630
SHEETROCK brand coreboard		C442
SHEETROCK brand liner panels		Exceeds C442
SHEETROCK brand exterior gypsum ceiling board		C931
SHEETROCK brand interior gypsum ceiling board		C36
Sheathing		
SHEETROCK brand gypsum sheathing		C79
GYP-LAP sheathing		C79
USG triple-sealed sheathing		C79
Joint Treatment		
SHEETROCK joint compounds		C475
Firestopping		
FIRECODE compound		E814
Metal Accessories		
UNIMAST SJ studs, CR runners	QQ-S-775E type 1, class e (steel)	C645, C955, A568, A525 (galv. coating), A792 (alum.-zinc coating), A591 (galv. coating)
UNIMAST ST25/22 studs, CR25/22 runners	QQ-S-775E, type 1, class f (steel)	C645, A568 (steel), A525 (galv. coating), A463 (alum. coating), A792 (alum.-zinc coating), A591 (galv. coating)
UNIMAST ST20 studs, CR20 runners	QQ-S-775E, type 1, class e	C645, A568 (steel), A446 (steel), A525 (galv. coating), A792 (alum.-zinc coating), A591 (galv. coating)
RC-1 resilient channels	QQ-S-775E, type 1, class f (steel)	A568 (steel), A525 (galv. coating), A792 (alum.-zinc coating)
USG shaft wall/area separation wall studs		A446 (steel), A525 (galv. coating), A792 (alum.-zinc coating), A591 (galv. coating)
STRUCTOCORE steel accessories		A526 (steel sheet)
Drywall screws		C1002 (type S and SUPER-TITE)
SUPER-TITE screws		C954 (type S-12 and SUPER-TITE DRILLERS)

Specification Standards (cont.)

Product	Federal Specification	ASTM Designation
Sealant		
SHEETROCK acoustical sealant		C919
Ceiling Suspension System		
DONN grid		C635, C636, C645, C841, E119, C1264
Mineral Fiber Insulation		
THERMAFIBER sound atten. fire blanket		C665
THERMAFIBER safing insulation, curtain wall insulation, mineral felt fireproofing	HH-I-558B form A, classes 1 & 2	C665

Products/UL Designations

The United States Gypsum Company products listed below are identified in the UL Fire Resistance Directory by the designations shown.

Product	UL Desig.
SHEETROCK brand Gypsum Panels	R
SHEETROCK brand Gypsum Panels, FIRECODE Core	SCX
SHEETROCK brand Gypsum Panels, FIRECODE C Core	C
SHEETROCK brand Gypsum Panels, FIRECODE Core, Water-Resistant	WRX
SHEETROCK brand Gypsum Panels, FIRECODE C Core, Water-Resistant	WRC
SHEETROCK brand Gypsum Liner Panels	SLX
SHEETROCK brand Gypsum Sheathing, FIRECODE Core	SHX
SHEETROCK brand Gypsum Panels, FIRECODE Core, Vinyl-Covered	FCV
SHEETROCK brand Formboard	FB
IMPERIAL Plaster Base	IPR
IMPERIAL Plaster Base (Type X)	IP-X1
IMPERIAL Plaster Base (Type C)	IP-X2
DUROCK Cement Board	CB
DUROCK Exterior Cement Board	ECB
FIRECODE Compound	FC
THERMAFIBER Safing Insulation	SAF

Permeance—U.S. Gypsum Company Products

Moisture Vapor Permeance

Product[1]	Finish	Perms[2,3]
Gypsum Panels		
⅜″ SHEETROCK brand Regular		35.3
½″ SHEETROCK brand Regular		34.2
½″ SHEETROCK brand Regular	1-coat flat latex paint	28.3
½″ SHEETROCK brand Regular	2-coats flat latex paint	28.4
½″ SHEETROCK brand Regular	2-coats gloss enamel (oil)	1.0
⅝″ SHEETROCK brand Regular		26.6
⅝″ SHEETROCK brand FIRECODE Core		28.6
½″ SHEETROCK brand FIRECODE C Core		31.8
⅝″ SHEETROCK brand FIRECODE C Core		25.9
½″ SHEETROCK brand, Water-Resistant		30.2
⅝″ SHEETROCK brand, Water-Resistant FIRECODE C Core		30.2
⅝″ SHEETROCK brand Water-Resistant FIRECODE Core		26.7
½″ TEXTONE		
Pumice Pattern		0.8
Suede		0.6
Presidio		0.6
Granite		0.6
Woodgrain		0.6
Linen Pattern		0.5
Country Weave Pattern		0.8
Textile (Type I, Fabric-Backed)(4)		1.0
Brittany (Type I, Fabric-Backed)(4)		2.1
1″ SHEETROCK brand Gypsum Liner Panel		24.0
Gypsum Base		
½″ IMPERIAL		28.8
½″ IMPERIAL	DIAMOND Interior Finish Plaster	24.4
½″ IMPERIAL	1 coat IMPERIAL Finish Plaster	5.3
½″ IMPERIAL	IMPERIAL Basecoat/ IMPERIAL Finish Plaster	8.0
⅝″ IMPERIAL		26.9
½″ IMPERIAL FIRECODE C		30.0
⅝″ IMPERIAL FIRECODE C		26.2
⅜″ gypsum base and ½″ gypsum plaster; metal lath and ¾″ gypsum plaster		20.0

Moisture Vapor Permeance *(cont.)*

Product[1]	Finish	Perms[2,3]
Gypsum Sheathing		
½″ SHEETROCK brand Gypsum Sheathing, Regular		23.3

(1) All foil-back products, less than 0.06 perm.
(2) All tests comply with ASTM E96 (desiccant method).
(3) Grain per sq. ft. per hr. per in. of water vapor pressure difference (grain/ft^2-h-in-Hg.).
(4) Comply with Federal Specification CCC-2-408C, Type I.

Thermal and Hygrometric Expansion of Building Materials

Thermal Coefficients of Linear Expansion of Common Building Materials

Unrestrained 40°—100°F. (4°—38 C.)

Material	Coefficient x10^{-6}in/(in.°F)	Coefficient x10^{-6}mm/(mm.°C)
Gypsum Panels and Bases	9.0	16.2
Gypsum Plaster (sanded 100:2, 100:3)	7.0	12.6
Wood Fiber Plaster (sanded 100:1)	8.0	14.4
Aluminum, Wrought	12.8	23.0
Steel, Medium	6.7	12.1
Brick, Masonry	3.1	5.6
Cement, Portland	5.9	10.6
Concrete	7.9	14.2
Fir (parallel to fiber)	2.1	3.8
Fir (perpendicular to fiber)	3.2	5.8

Hygrometric Coefficients of Expansion (unrestrained)

Inches/Inch/% R. H. (5%—90% R. H.)	
Gypsum Panels and Bases	7.2x10^{-6}
Gypsum Plaster (sanded 100:2, 100:3)	1.5x10^{-6}
Wood Fiber Plaster (sanded 100:1)	2.8x10^{-6}
STRUCTO-LITE Plaster	4.8x10^{-6}
Vermiculite Gypsum Plaster (sanded 100:2)	3.8x10^{-6}

Thermal Resistance Coefficients of Building and Insulating Materials[1]

Thickness		Product	Density		Resistance	
in	mm		lb/ft³	kg/m³	in.ft.².F/Btu	K.m²/W
2-2¾	50.8-69.9	THERMAFIBER Mineral Fiber Insulation (SAFB & FS)	3.0	48.1	7.00	1.23
3-3½	76.2-88.9	THERMAFIBER Mineral Fiber Insulation (SAFB & FS)	3.0	48.1	11.00	1.94
5¼-6½	133.4-165.1	THERMAFIBER Mineral Fiber Insulation (SAFB & FS)	3.0	48.1	19.00	3.35
1	25.4	Extruded Polystyrene Insulation	2.2	35.2	5.00	0.88
½	12.7	SHEETROCK brand Gypsum Panels	43	690.2	0.45	0.08
⅝	15.9	SHEETROCK brand Gypsum Panels	43	690.2	0.56	0.10
½	12.7	SHEETROCK brand FIRECODE C Core Panels	50	800.9	0.45	0.08
⅝	15.9	SHEETROCK brand FIRECODE and FIRECODE C Core Panels	50	800.9	0.56	0.10
⅜	9.5	ROCKLATH Plaster Base	50	800.9	0.32	0.06
½	12.7	SHEETROCK brand Gypsum Sheathing	50	800.9	0.45	0.08
½	12.7	Sanded Plaster	105	1681.9	0.09	0.02
½	12.7	Plaster with Lightweight Aggregate	45	720.8	0.32	0.06
4	101.6	Common Brick	120	1922.2	0.80	0.14
½	12.7	DUROCK Cement Board	72	1153.3	0.26	0.05
½	12.7	DUROCK Exterior Cement Board	72	1153.3	0.26	0.05
4	101.6	Face Brick	130	2082.4	0.44	0.08
1	25.4	Portland Cement Stucco with Sand Aggregate	116	1858.1	0.20	0.04
4	101.6	Concrete Block, 3-oval Core, Cinder Aggregate			1.11	0.20
8	203.2	Concrete Block, 3-oval Core, Cinder Aggregate			1.72	0.30
12	304.8	Concrete Block, 3-oval Core, Cinder Aggregate			1.89	0.33
—	—	Vapor-Permeable Felt			0.06	0.01
—	—	Vapor-Retarder Plastic Film			Negl.	—
1	25.4	Stone			0.08	0.01
1x8	25.4-203.2	Wood Drop Siding			0.79	0.14
¾x10	19.1-254.0	Beveled Wood Siding			1.05	0.18
¾-3½	19.1- 88.9	Plain Air Space, non-reflective[2]			0.92	0.17

(1) All factors based on data from 1981 ASHRAE Handbook of Fundamentals. Factors at 75°, mean temperature. (2) Conditions: heat flow horizontal; mean temperature 50°F. Temperature differential 30°F; E (emissivity) 0.82.

U.S. Gypsum Company Literature

Complete technical data on U.S. Gypsum Company products and systems can be found in the USG Corporation Architectural Technical Literature series. Those folders applying to drywall, cement board, insulation and plaster construction are listed below with their appropriate CSI numbers. Copies of all folders are available through U.S. Gypsum Company sales offices.

Folder No. & Description		CSI No.
General		
SA-100	Construction Selector	—
Steel Framing		
UN-30	UNIMAST Steel Framing Systems	05400
Insulation		
SA-707	THERMAFIBER Life-Safety Insulation Systems	07200
SA-727	USG Fire Stop System	07270
Exterior Finish Systems		
SA-700	DUROCK Exterior Cement Board Systems	07240
Lath, Plaster		
SA-920	USG Plaster Products, Accessories & Systems	09200
Gypsum Drywall		
SA-921	USG High-Attenuation Systems	09250
SA-923	Drywall/Steel-Framed Systems	09250
SA-924	Drywall/Wood-Framed Systems	09250
SA-925	USG Area Separation Wall Systems	09250
SA-926	USG Cavity Shaft Wall Systems	09250
SA-927	Gypsum Panels & Accessories	09250
Prefinished Panels		
SA-928	TEXTONE Vinyl-Faced Gypsum Panels	09985
Tile Accessories		
SA-932	DUROCK Cement Board Systems	09390
Finishing Materials		
SA-933	Texture and Finish Products	09200
Security Walls		
SA-1119	STRUCTOCORE Security Wall Systems	11190

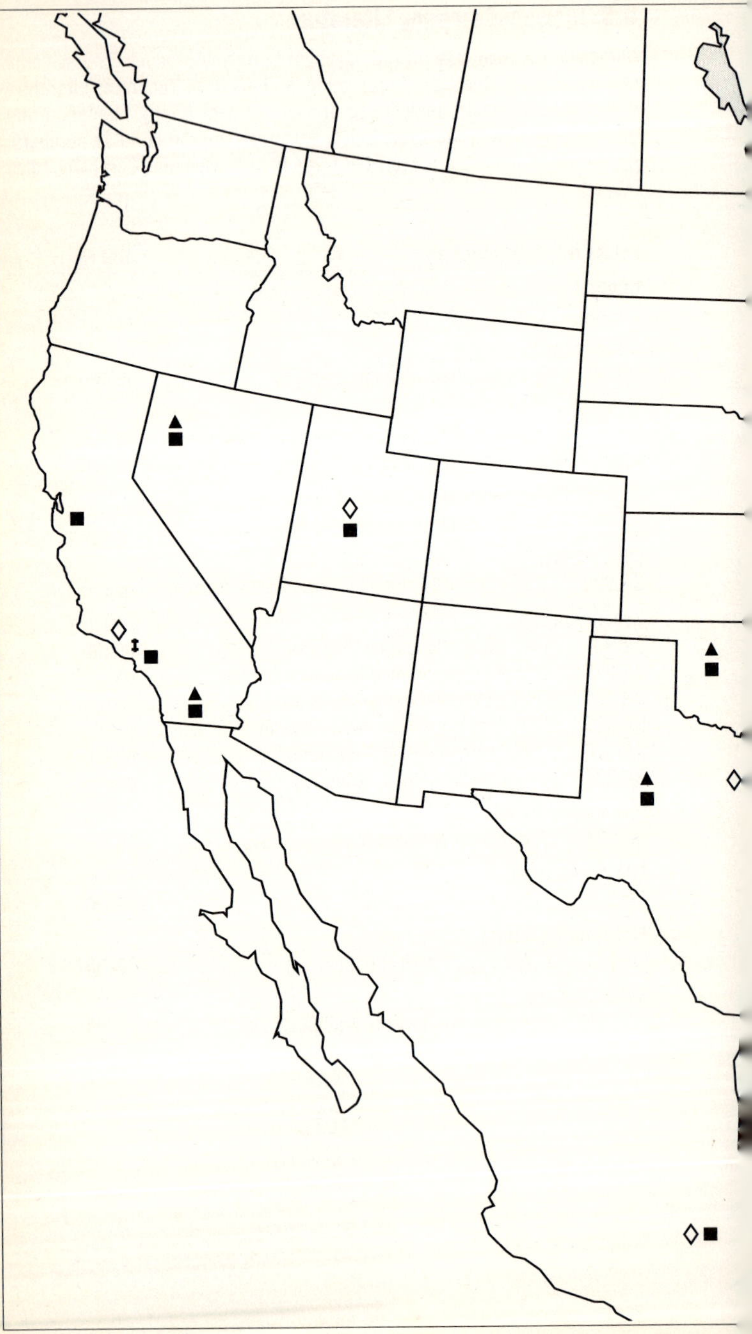

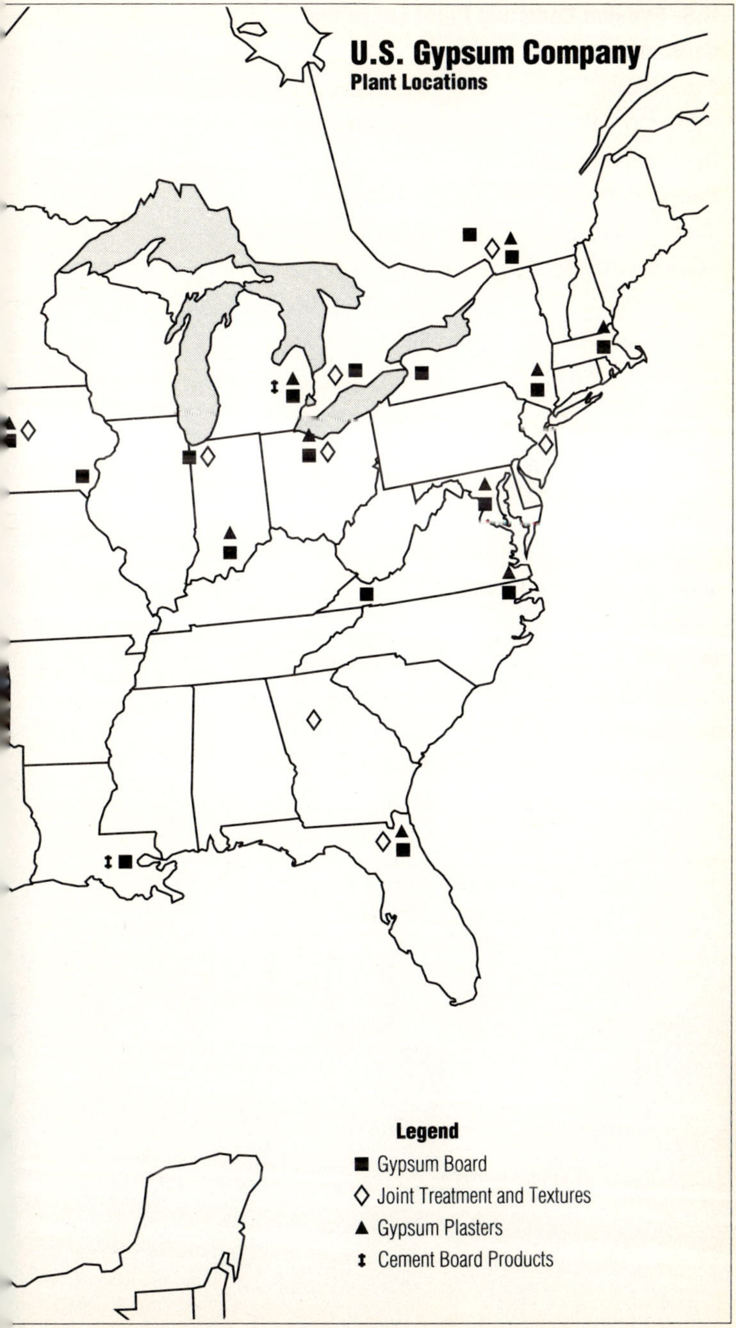
U.S. Gypsum Company
Plant Locations
Legend
Gypsum Board
Joint Treatment and Textures
Gypsum Plasters
Cement Board Products

U.S. Gypsum Company Plant Locations

GYPSUM BOARD

Fremont, Calif.
Plaster City, Calif.
Santa Fe Springs, Calif.
Hagersville, Ontario, Can.
Montreal, Quebec, Can.
St. Jerome, Quebec, Can.
Jacksonville, Fla.
East Chicago, Ind.
Fort Dodge, Iowa
Baltimore, Md.
Boston, Mass.
Puebla, Mexico
Detroit, Mich.
New Orleans, La.
Empire, Nev.
Oakfield, N.Y.
Stony Point, N.Y.
Gypsum, Ohio
Southard, Okla.
Galena Park, Tex.
Sweetwater, Tex.
Sigurd, Utah
Norfolk, Va.
Plasterco, Va.
Shoals, Va.

JOINT TREATMENT & TEXTURES

Hagersville, Ontario, Can.
Monteral, Quebec, Can.
Torrance, Calif.
Jacksonville, Fla.
Chamblee, Ga.
East Chicago, Ind.
Fort Dodge, Iowa
Puebla, Mexico
Port Reading, N.J.
Gypsum, Ohio
Dallas, Tex.
Sigurd, Utah

GYPSUM PLASTERS

Plaster City, Calif.
Montreal, Quebec, Can.
Jacksonville, Fla.
Shoals, Ind.
Fort Dodge, Iowa
Baltimore, Md.
Boston, Mass.
Detroit, Mich.
Empire, Nev.
Stony Point, N.Y.
Gypsum, Ohio
Southard, Okla.
Sweetwater, Tex.
Norfolk, Va.

CEMENT BOARD

Santa Fe Springs, Calif.
New Orleans, La.
Detroit, Mich.

UNIMAST Incorporated Plant Locations

CONSTRUCTION STEEL

Morrow, GA.
Franklin Park, Ill.
Boonton, N.J.
Warren, Ohio

Glossary

Acoustics—Science dealing with the production, control, transmission, reception and effects of sound, and the process of hearing.

Aggregate—Sand, gravel, crushed stone or other material that is a main constituent of portland cement concrete and aggregated gypsum plaster. Also, polystyrene, perlite and vermiculite particles used in texture finishes.

AIA—American Insurance Assn., successor to the National Board of Fire Underwriters and a nonprofit organization of insurance companies. Also, American Institute of Architects.

Airborne Sound—Sound traveling through the medium of air.

Anchor—Metal securing device embedded or driven into masonry, concrete, steel or wood.

Anchor Bolt—Heavy, threaded bolt embedded in the foundation to secure sill to foundation wall or bottom plate of exterior wall to concrete floor slab.

ANSI—American National Standards Institute, a nonprofit, national technical association that publishes standards covering definitions, test methods, recommended practices and specifications of materials. Formerly American Standards Assn. (ASA) and United States of America Standards Institute (USASI).

Annular Ring Nail—A deformed shank nail with improved holding qualities specially designed for use with gypsum board.

Area Separation Wall—Residential fire walls, usually with a 2- to 4-hour rating, designed to prevent spread of fire from an adjoining occupancy; extends from foundation to or through the roof. Identified by codes as either "fire wall," "party wall" or "townhouse separation wall."

ASA—Formerly American Standards Assn., now American National Standards Institute (ANSI).

ASTM—Formerly American Society for Testing and Materials, now ASTM, a nonprofit, national technical society that publishes definitions, standards, test methods, recommended installation practices and specifications for materials.

Attenuation—Reduction in sound level.

Back Blocking—A short piece of gypsum board adhesively laminated behind the joints between each framing member to reinforce the joint.

Backup Strips—Pieces of wood nailed at the ceiling-sidewall corner to provide fastening for ends of plaster base or gypsum panels.

Balloon Frame—Method of framing outside walls in which studs extend the full length or height of the wall.

Bar Joist—Open-web, flat truss structural member used to support floor or roof structure. Web section is made from bar or rod stock, and chords are usually fabricated from "T" or angle sections.

Batten—Narrow strip of wood, plastic, metal or gypsum board used to conceal an open joint.

BCMC—Board for the Coordination of Model Codes; part of the Council of American Building Officials Association (CABO).

Beam—Loadbearing member spanning a distance between supports.

Bearing—Support area upon which something rests, such as the point on bearing walls where the weight of the floor joist or roof rafter bears.

Bed—To set firmly and permanently in place.

Bending—Bowing of a member that results when a load or loads are applied laterally between supports.

Board Foot (Bd. Ft.)—Volume of a piece of wood, nominal 1″ x 12″ x 1′. All lumber is sold by the board-foot measure.

BOCA—Building Officials Conference of America, a nonprofit organization that publishes the National Building Code.

Brick Veneer—Non-loadbearing brick facing applied to a wall to give appearance of solid-brick construction; bricks are fastened to backup structure with metal ties embedded in mortar joints.

Bridging—Members attached between floor joists to distribute concentrated loads over more than one joist and to prevent rotation of the joist. Solid bridging consists of joist-depth lumber installed perpendicular to and between the joists. Cross-bridging consists of pairs of braces set in an "X" form between joists.

CABO—Council of American Building Officials Association, made up of representatives from three model codes. Issues National Research Board (NRB) research reports.

Camber—Curvature built into a beam or truss to compensate for loads that will be encountered when in place and load is applied. The crown is placed upward. Insufficient camber results in unwanted deflection when the member is loaded.

Cant Beam—Beam with edges chamfered or beveled.

Cant Strip—Triangular section laid at the intersection of two surfaces to ease or eliminate effect of a sharp angle or projection.

Carrying Channel—Main supporting member of a suspended ceiling system to which furring members or channels attach.

Casement—Glazed sash or frame hung to open like a door.

Casing—The trim around windows, doors, columns or piers.

Cement Board—A factory-manufactured panel, ¼″ to ¾″ thick, 32″ to 48″ wide, and 3′ to 10′ long, made from aggregated and reinforced portland cement.

Chalk Line—Straight working line made by snapping a chalked cord stretched between two points, transferring chalk to work surface.

Cladding—Gypsum panels, gypsum bases, gypsum sheathing, cement board, etc. applied to framing.

Coefficient of Thermal Conductance (C)—Amount of heat (in Btu) that passes through a specific thickness of a material (either homogeneous or heterogeneous) per hr., per sq. ft., per °F. Measured as temperature difference between surfaces.

The "C" value of a homogeneous material equals the "k" value divided by the material thickness:

C = k/t where t = thickness of material in inches

It is impractical to determine a "k" value for some materials such as building paper or those only used or formed as a thin membrane, so only "C" values are given for them.

Coefficient of Thermal Conductivity (k)—Convenient factor represents the amount of heat (in Btu) that passes by conduction through a one inch thickness of homogeneous material, per hr., per sq. ft., per °F. Measured as temperature difference between the two surfaces of the material.

Coefficient of Heat Transmission (U)—Total amount of heat that passes through an assembly of materials, including air spaces and surface air films. Expressed in Btu per hr., per sq. ft., per °F temperature difference between inside and outside air (beyond the surface air films). "U" values are often used to represent wall and ceiling assemblies, floors and windows.

Note: "k" and "C" values *cannot* simply be added to obtain "U" values. "U" can only be obtained by adding the thermal resistance (reciprocal of "C") of individual items and dividing the total into 1.

Coefficient of Hygrometric Expansion—See Hygrometric Expansion.

Coefficient of Thermal Expansion—See Thermal Expansion.

Column—Vertical loadbearing member.

Compression—Force that presses particles of a body closer together.

Compressive Strength—Measures maximum unit resistance of a material to crushing load. Expressed as force per unit cross-sectional area, e.g., pounds per square inch (psi).

Concrete Footing—Generally, the wide, lower part of a foundation wall that spreads the weight of the building over a larger area. Its width and thickness vary according to weight of building and type of soil on which building is erected.

Conduction, Thermal—Transfer of heat from one part of a body to another part of that body, or to another body in contact, without any movement of bodies involved. The hot handle of a skillet is an example. The heat travels from the bottom of the skillet to the handle by conduction.

Convection—Process of heat carried from one point to another by movement of a liquid or a gas (i.e., air). Natural convection is caused by expansion of the liquid or gas when heated. Expansion reduces the density of the medium, causing it to rise above the cooler, more dense portions of the medium.

Gravity heating systems are examples of the profitable use of natural convection. The air, heated by the furnace, becomes less dense (consequently lighter) and rises, distributing heat to the various areas of the house without any type of blower. When a blower is used, the heat transfer method is called "forced convection."

Corner Brace—Structural framing member used to resist diagonal loads that cause racking of walls and panels due to wind and seismic forces. May consist of a panel or diaphragm, or diagonal flat strap or rod. Bracing must function in both tension and compression. If brace only performs in tension, two diagonal tension members must be employed in opposing directions as "X" bracing.

Corner Post—Timber or other member forming the corner of a frame. May be solid or built-up as a multi-piece member.

Cripple—Short stud such as that used between a door or window header and the top plate.

Curtain Wall—Exterior wall of a building that is supported by the structure and carries no part of the vertical load except its own. Curtain walls must be designed to withstand wind loads and transfer them to the structure.

Cycle (Acoustic)—One full repetition of a motion sequence during periodic vibration. Movement from zero to +1 back to zero to -1 back to zero. Frequency of vibration is expressed in Hertz (cycles per second—see Frequency).

Dead Load—Load on a building element contributed by the weight of the building materials.

Decibel (dB)—Adopted for convenience in representing vastly different sound pressures. The sound pressure level (SPL) in decibels is 10 times the logarithm to the base 10 of the squared ratio of the sound pressure to a reference pressure of 20 micropascals. This reference pressure is considered the lowest value at 100 Hz that the ear can detect. For every 10 dB increase or decrease in SPL, a sound is generally judged to be about twice or half as loud as before the change.

Decoupling—Separation of elements to reduce or eliminate the transfer of sound, heat or physical loads from one element to the other.

Deflection—Displacement that occurs when a load is applied to a member or assembly. The dead load of the member or assembly itself causes some deflection as may occur in roofs or floors at mid-span. Under applied wind loads maximum deflection occurs at mid-height in partitions and walls.

Deflection Limitation—Maximum allowable deflection is dictated by the bending limit of the finish material under the required design load (e.g., usually 5 psf for interior partitions). Often expressed as ratio of span (L) divided by criterion factor (120, 180, 240, 360). For example, in a 10′ or 120″ high wall, allowable deflection under L/240 criterion equals 120″/240 or ½″ maximum.

Selection of limiting heights and spans are frequently based on minimum code requirements and accepted industry practice as follows: (a) L/120 for gypsum panel surfaces and veneer plaster finish surfaces, (b) L/240 for conventional lath and plaster surfaces, (c) L/360 for mechanically attached marble or heavy stone to walls; however, support for its own weight should be from the floor or separate supports. Although some building codes permit these deflections, more conservative criteria are frequently advised so that applied loads are not visible or aesthetically unacceptable.

Deformation—Change in shape of a body brought about by the application of a force internal or external. Internal forces may result from temperature, humidity or chemical changes. External forces from applied loads can also cause deformation.

Design Load—Combination of weight (dead load) and other applied forces (live loads) for which a building or part of a building is designed. Based on the worst possible combination of loads.

Dew Point—The temperature at which air becomes saturated with moisture and below which condensation occurs.

Door Buck—Structural element of a door opening. May be the same element as the frame if frame is structural, as in the case of heavy steel frames.

Double-Hung Window—Window sash that slides vertically and is offset in a double track.

Drip—Interruption or offset in an exterior horizontal surface, such as a soffit, immediately adjacent to the fascia. Designed to prevent the migration of water back along the surface.

Drywall—Generic term for interior surfacing material, such as gypsum panels, applied to framing using dry construction methods, e.g., mechanical fasteners or adhesive. See SHEETROCK brand Gypsum Panels.

Exterior Insulation and Finish System (EIFS)—Exterior cladding assembly consisting of a polymer finish over a reinforcement adhered to foam plastic insulation that is fastened to masonry, concrete, building sheathing or directly to the structural framing. The sheathing may be cement board or gypsum sheathing.

Extrapolate—To project tested values, assuming a continuity of an established pattern, to obtain values beyond the limit of the test results. Not necessarily reliable.

Factor of Safety—Ratio of the ultimate unit stress to the working or allowable stress.

Fascia Board—Board fastened to the ends of the rafters or joists forming part of a cornice.

Fast Track—Method that telescopes or overlaps traditional design-construction process. Overlapping phases as opposed to sequential phases is keynote of the concept.

Fatigue—Condition of material under stress that has lost, to some degree, its power of resistance as a result of repeated application of stress, particularly if stress reversals occur as with positive and negative cyclical loading.

Fire Endurance—Measure of elapsed time during which an assembly continues to exhibit fire resistance under specified conditions of test and performance. As applied to elements of buildings, it shall be measured by the methods and to the criteria defined in ASTM Methods E119, Fire Tests of Building Construction and Materials; ASTM Methods E152, Fire Tests of Door Assemblies; ASTM Methods E814, Fire Test of Through-Penetration Fire Stops; or ASTM Methods E163, Fire Tests of Window Assemblies.

Fireproof—Use of this term in reference to buildings is discouraged because few, if any, building materials can withstand extreme heat for an extended time without some effect. The term "fire-resistive" or "resistant" is more descriptive.

Fire Resistance—Relative term, used with a numerical rating or modifying adjective to indicate the extent to which a material or structure resists the effect of fire.

Fire-Resistive—Refers to properties or designs to resist effects of any fire to which a material or structure may be expected to be subjected.

Fire-Retardant—Denotes substantially lower degree of fire resistance than "fire-resistive." Often used to describe materials that are combustible but have been treated to retard ignition or spread of fire under conditions for which they were designed.

Fire Stop—Obstruction in a cavity designed to resist the passage of flame, sometimes referred to as "fire blocking."

Fire Taping—The taping of gypsum board joints without subsequent finishing coats. A treatment method used in attic, plenum or mechanical areas where aesthetics are not important.

Fire Wall—Fire-resistant partition extending to or through the roof of a building to retard spread of fire. See Area Separation Wall.

Flame Spread—Index of the capacity of a material to spread fire under test conditions, as defined by ASTM Standard E84. Materials are rated by comparison with the flame-spread index of red oak flooring assigned a value of 100 and inorganic reinforced cement board assigned a value of 0.

Flammable—Capability of a combustible material to ignite easily, burn intensely or have rapid rate of flame spread.

Flanking Paths—Paths by which sound travels around an element intended to impede it, usually some structural component that is continuous between rooms and rigid enough to transmit the sound. For example, a partition separating two rooms can be "flanked" by the floor, ceiling or walls surrounding the partition if they run uninterrupted from one room to the other. Ducts, conduits, openings, structural elements, rigid ties, etc., can be sound flanking paths. The acoustic effect of sound flanking paths is dependent on many factors.

Flashing—Strips of metal or waterproof material used to make joints waterproof, as in the joining of curtain wall panels.

Footing—Lower extremity of a foundation or loadbearing member that transmits load to load-bearing substrate.

Force—Amount of applied energy to cause motion, deformation or displacement and stress in a body.

Foundation—Component that transfers weight of building and occupants to the earth.

Frequency (Sound)—Number of complete vibrations or cycles or periodic motion per unit of time.

Furring—Member or means of supporting a finished surfacing material away from the structural wall or framing. Used to level uneven or damaged surfaces or to provide space between substrates. Also an element for mechanical or adhesive attachment of paneling.

Gable—Uppermost portion of the end wall of a building that comes to a triangular point under a sloping roof.

Gauging Plaster—Combine with lime putty to provide setting properties, to increase dimensional stability during drying, and to provide initial surface hardness in lime finish coats.

Girder—Beam, especially a long, heavy one; the main beam supporting floor joists or other smaller beams.

Gusset—Wood or metal plate riveted, bolted, glued or pressed (wood trusses) over joints to transfer stresses between connected members.

Header—Horizontal framing member across the ends of the joists. Also the member over a door or window opening in a wall.

Heat—Form of energy thought to be characterized by the rate of vibration of the molecules of a substance. The hotter the substance, the faster the molecules vibrate. On the other hand, when there is no heat present it is thought the molecules will be at rest, which theoretically occurs at absolute zero, –459.7°F (–273.15°C or 0.0°K).

Heat Quantity (Btu)—Common unit of measure of the quantity of heat is the British Thermal Unit (Btu). One Btu is the amount of heat required to raise one pound of water from 63° to 64°F (1 Btu = 1055.06 J). This is about the amount of heat given off by one wooden match. A pound of coal can produce 13,000 Btu.

Heat Transfer—Heat always flows toward a substance of lower temperature until the temperatures of the two substances equalize. It travels by one or more of three methods: conduction, convection or radiation.

Heel of Rafter—Seat cut in a rafter that rests on the wall plate.

Hertz—The units of measure of sound frequency, named for Heinrich H. Hertz. One Hertz equals one cycle per second.

Honeycomb—Any substance having cells suggesting a mass of cells such as those built by the honeybee. Some hollow-core doors use the honeycomb principle in their construction.

HUD—Housing and Urban Development, federal agency.

HUD Mobile Home Standards—Officially, the National Mobile Home Construction and Safety Standards Act of 1974 for construction of mobile homes. Includes the following agencies: DAPIA—Design Approval Primary Inspection Agency and IPIA—Production Inspection Primary Inspection Agency.

HVAC—Heating, ventilating and air conditioning. (ASHRAE Guide is the technical reference source.)

Hygrometric Expansion—All materials, particularly those of organic origin, expand and contract in relation to their moisture content, which varies with environment. The Hygrometric Coefficient of Expansion is expressed in "Inches Per Inch Per Percent Of Relative Humidity." Example: gypsum board has a coefficient of 7.2×10^{-6} in. per in. per %rh. This means that with an increase in relative humidity of from 10% to 50%, a gypsum board wall 300 ft. long will have an unrestrained linear expansion of 1.0368″ or 1 1/32″.

ICBO—International Conference of Building Officials, a nonprofit organization that publishes the Uniform Building Code.

Impact Insulation Class (IIC)—Single-number rating used to compare and evaluate the performance of floor-ceiling constructions in isolating impact noise. The advantages of this rating system are positive values and the correlation with Sound Transmission Class (STC) values—both providing approximately equal isolation at a particular value. The IIC rating is used by building agencies for specifying minimum sound-control performance of assemblies in residential construction.

Impact Noise Rating (INR)—Obsolete rating system for floor-ceiling construction in isolating impact noise. INR ratings can be converted to approximate IIC ratings by adding 51 points; however, a variation of 1 or 2 points may occur.

Incombustible—See Noncombustible.

Insulation (Thermal)—Any material that measurably retards heat transfer. There is wide variation in the insulating value of different materials. A material having a low density (weight/volume) will usually be a good thermal insulator.

Interpolate—To estimate untested values that fall between tested values.

ISO—International Standards Organization, an organization similar in nature to ASTM.

Jamb—One of the finished upright sides of a door or window frame.

Jamb Stud—Wood or metal stud adjacent to the door jamb.

Joist—Small beam that supports part of the floor, ceiling or roof of a building.

Joist Hanger—Metal shape formed for hanging on the main beam to provide support for the end of a joist.

Kiln-Dried Lumber—Lumber that has been dried and seasoned with carefully controlled heat in a kiln.

Label Service (UL)—Program allowing a manufacturer to place Underwriters Laboratories Inc. labels on his products that have met UL requirements. A UL representative visits the manufacturing location to obtain samples of the products for testing by UL. In some cases, samples are also purchased on the open market for testing. The public is thereby assured that products bearing the UL label continually meet UL specifications.

Leaks (Sound)—Small openings at electrical boxes and plumbing, cracks around doors, loose-fitting trim and closures all create leaks that allow sound to pass through, reducing the acoustical isolation of a wall, floor or ceiling system.

Ledger Strip—Strip fastened to the bottom edge of a flush girder to help support the floor joists.

Life-Cycle Costing—Selection of the most economical material and systems based on initial costs, maintenance costs and operating costs for the life of the building.

Limiting Height—Maximum height for design and construction of a partition or wall without exceeding the structural capacity or allowable deflection under given design loads.

Lintel—Horizontal member spanning an opening such as a window or door. Also referred to as a Header.

Live Load—Part of the total load on structural members that is not a permanent part of the structure. May be variable, as in the case of loads contributed by the occupancy, and wind and snow loads.

Load—Force provided by weight, external or environmental sources such as wind, water and temperature, or other sources of energy.

Loudness—Subjective response to sound pressure, but not linearly related thereto. A sound with twice the pressure is not twice as loud. See Decibel.

Louver—Opening with slanted fins (to keep out rain and snow) used to ventilate attics, crawl spaces and wall openings.

Mass—Property of a body that resists acceleration and produces the effect of inertia. The weight of a body is the result of the pull of gravity on the body's mass.

Metric Terms—Metric units shown as equivalents in this Handbook are from the International System of Units in use throughout the world, as established by the General Conference of Weights and Measures in 1960. Their use here complies with the Metric Conversion Act of 1975, which committed the United States to a coordinated voluntary conversion to the metric system of measurement.

Refer to the page 472 in Appendix for metric units and conversion factors applicable to subjects covered in this Handbook. For additional information, refer to ASTM E380-76, Standard for Metric Practice.

Miter—Joint formed by two pieces of material cut to meet at an angle.

Model Code—Building code, written and published by a building-official association, available to states, counties and municipalities for adoption (for a fee) in lieu of their own, e.g., Uniform Building Code, Standard Building Code, National Building Code.

Module—(1) In architecture, a selected unit of measure used as a basis for building layout; (2) In industrialized housing, a three-dimensional section of a building, factory-built, shipped as a unit and interconnected with other modules to form the complete building. Single-family units factory-built in two halves are usually referred to as "sectionals."

Modulus of Elasticity (E)—Ratio between stress and unit deformation, a measure of the stiffness of a material.

Moment of Inertia (I)—Calculated numerical relationship (expressed in in.^{4}) of the resistance to bending of a member, a function of the cross-sectional shape and size. A measure of the stiffness of a member based on its shape. Larger moments of inertia indicate greater resistance to bending for a given material.

Moulding (also Molding)—Narrow decorative strip applied to a surface.

Mullion—Vertical bar or division in a window frame separating two or more panes.

Muntin—Horizontal bar or division in a window frame separating multiple panes or lights.

Music/Machinery Transmission Class (MTC)—Rating developed by U.S. Gypsum Company to isolate music and machinery/mechanical equipment noise or any sound with a substantial portion of low frequency energy. MTC does not replace Sound Transmission Class (STC) but complements it.

Nail Pop—The protrusion of the nail usually attributed to the shrinkage of or use of improperly cured wood framing.

NBFU—National Board of Fire Underwriters, now merged into the American Insurance Assn.

NBS—National Bureau of Standards, a federal agency.

NCSBCS—National Conference of States on Building Codes and Standards, a nonprofit organization formed to increase interstate cooperation and coordinate intergovernmental reforms of building codes.

Neutral Axis—The plane through a member (at the geometric center of the section in symmetrical members) where the fibers are neither under tensile nor compressive stress.

NFiPA—National Fire Protection Assn., an international technical society that disseminates fire prevention, fighting and protection information. NFiPA technical standards include the National Electrical Code which is widely adopted.

NFoPA—National Forest Products Association.

Noise Reduction Coefficient (NRC)—Arithmetic average of sound absorption coefficients at 250, 500, 1000 and 2000 Hz.

Nominal—Term indicating that the full measurement is not used; usually slightly less than the full net measurement, as with 2″ x 4″ studs that have an actual size when dry of 1½″ x 3½″.

Noncombustible—Definition excerpted from the ICBO Uniform Building Code:

1. Material of which no part will ignite and burn when subjected to fire.

2. Material having a structural base of noncombustible material as defined, with a surface not over ⅛″ thick that has a flame spread rating of 50 or less.

The term does not apply to surface finish materials.

Octave—Interval between two sounds having a basic frequency ratio of two. The formula is 2^n times the frequency, where n is the desired octave interval. The octave band frequency given in sound test results is usually the band center frequency, thus the 1000 Hz octave band encompasses frequencies from 707 Hz to 1414 Hz (n = ± ½). The 1000 Hz one-third-octave band encompasses frequencies from 891 Hz to 1122 Hz (n = ± ⅙).

OSU—Ohio State University, an independent fire-testing laboratory which is currently inactive.

Parapet Wall—Extension of an exterior wall above and/or through the roof surface.

Penny (d)—Suffix designating the size of nails, such as 6d (penny) nail, originally indicating the price, in English pence, per 100 nails. Does not designate a constant length or size, and will vary by type (e.g., common and box nails).

Performance Specification—States how a building element must perform as opposed to describing equipment, products or systems by name.

Perm—A unit of measurement of Water Vapor Permenance (ASTM E96).

Pilaster—Projecting, square column or stiffener forming part of a wall.

Pillar—Column supporting a structure.

Pitch of Roof—Slope of the surface, generally expressed in inches of vertical rise per 12″ horizontal distance, such as "4-in-12 pitch."

Plate—"Top" plate is the horizontal member fastened to the top of the studs or wall on which the rafters, joists or trusses rest; "sole" plate is positioned at bottom of studs or wall.

Platform—Floor surface raised above the ground or floor level.

Platform Framing—Technique of framing where walls can be built and tilted-up on a platform floor, and in multi-story construction are erected sequentially from one platform to another. Also known as "Western" framing.

Plenum—Chamber in which the pressure of the air is higher (as in a forced-air furnace system) than that of the surrounding air. Frequently a description of the space above a suspended ceiling.

Portland Cement—Hydraulic cement produced by pulverizing clinker consisting essentially of hydraulic calcium silicates, usually containing one or more forms of calcium sulfate as an interground addition.

Prescription Specification—Traditional procedure used on building projects to describe by name products, equipment or systems to be used.

Purlin—Horizontal member in a roof supporting common rafters, such as at the break in a gambrel roof. Also, horizontal structural member perpendicular to main beams in a flat roof.

Racking—Forcing out of plumb of structural components, usually by wind, seismic stress or thermal expansion or contraction.

Radiation—Transfer of heat energy through space by wave motion. Although the radiant energy of heat is transmitted through space, no heat is present until this energy strikes and is absorbed by an object. Not all of the radiant heat energy is absorbed; some is reflected to travel in a new direction until it strikes another object. The amount reflected depends on the nature of the surface that the energy strikes. This fact explains the principle of insulating foil and other similar products that depend on reflection of radiant heat for their insulating value.

Radiant heat travels in straight lines in all directions at about the speed of light. In radiant heating systems, heat is often radiated down from the ceiling. As it strikes objects in the room, some is absorbed and some reflected to other objects. The heat that is absorbed warms the object, which, in turn, warms the surrounding air by conduction. This warmed air sets up gentle convection currents that circulate throughout the room.

Rafter—That member forming the slanting frame of a roof or top chord of a truss. Also known as hip, jack or valley rafter depending on its location and use.

Rafter Tail—That part of a rafter that extends beyond the wall plate—the overhang.

Reflected Heat—See Radiation.

Reflected Sound—Sound that has struck a surface and "bounced off." Sound reflects at the same angle as light reflects in a mirror; the angle of incidence equals the angle of reflection.

Large curved surfaces tend to focus (concave) or diffuse (convex) the sound when reflected. However, when the radius of the reflecting surface is less than the wavelength of the sound, this does not hold true. Thus, a rough textured surface has little effect on diffusion of sound.

Reflective Insulation—Material that reflects and thus retards the flow of radiant heat. The most common type of reflective insulation is aluminum foil. The effectiveness of reflective barriers is diminished by the accumulation of dirt and by surface oxidation.

Reverberation—Persistence of sound after the source stops. When one hears the 10th, 20th, 50th, 100th, etc., reflection of a sound, one hears reverberation.

Reverberation Time—Essentially the number of seconds it takes a loud sound to decay to inaudibility after the source stops. Strictly, the time required for a sound to decay 60 dB in level.

Ridge—Peak of a roof where the roof surfaces meet at an angle. Also may refer to the framing member that runs along the ridge and supports the rafters.

Rise—Measurement in height of an object; the amount it rises. The converse is “fall.”

Riser—Vertical face of a step supporting the tread in a staircase.

Rough Framing—Structural elements of a building or the process of assembling elements to form a supporting structure where finish appearance is not critical.

Sabin—Measure of sound absorption of a surface, equivalent to 1 sq. ft. of a perfectly absorptive surface.

Safing—Fire stop material in the space between floor slab and curtain wall in multi-story construction.

Safing Off—Installation of fire safety insulation around floor perimeters, between floor slab and spandrel panels. Insulation helps retain integrity of fire resistance ratings.

SBCCI—Southern Building Code Congress International, nonprofit organization that publishes the Standard Building Code.

Scab—Small piece or block of wood that bridges several members or provides a connection or fastening between them.

Section Modulus (S)—Numerical relationship, expressed in in.^{3}, of the resistance to stress of a member. It is equal to the moment of inertia divided by the perpendicular distance from the neutral axis to the extremity of the member.

Shaft Wall—Fire-resistant wall that isolates the elevator, stairwell and vertical mechanical chase in high-rise construction. This wall must withstand the fluctuating (positive and negative) air-pressure loads created by elevators or air distribution systems.

Shadowing—An undesirable condition where the joint finish shows through the surface decoration.

Shear—Force that tends to slide or rupture one part of a body from another part of the body or from attached objects.

Sheathing—Plywood, gypsum, wood fiber, expanded plastic or composition boards encasing walls, ceilings, floors and roofs of framed buildings. May be structural or non-structural, thermal-insulating or non-insulating, fire-resistant or combustible.

SHEETROCK—Leading brand of gypsum panel for interior wall and ceiling surfaces, developed and improved by United States Gypsum Company. There is only one SHEETROCK brand Gypsum Panel.

Shoring—Temporary member placed to support part of a building during construction, repair or alteration; also may support the walls of an excavation.

Sill—Horizontal member at the bottom of door or window frames to provide support and closure.

Sill Plate—Horizontal member laid directly on a foundation on which the framework of a building is erected.

Slab—Flat (although sometimes ribbed on the underside) reinforced concrete element of a building that provides the base for the floor or roofing materials.

Soffit—Undersurface of a projection or opening; bottom of a cornice between the fascia board and the outside of the building; underside of a stair, floor or lintel.

Sole Plate—See Plate.

Sound Absorption—Conversion of acoustic or sound energy to another form of energy, usually heat.

Sound Insulation, Isolation—Use of building materials or constructions that will reduce or resist the transmission of sound.

Sound Intensity—Amount of sound power per unit area.

Sound Pressure Level (SPL)—Expressed in decibels, the SPL is 20 times the logarithm to the base 10 of the ratio of the sound pressure to a reference pressure of 20 micropascals. See Decibel.

Sound Transmission Class (STC)—Single-number rating for evaluating the effectiveness of a construction in isolating audible airborne sound transmission across 16 frequencies. Higher numbers indicate more effectiveness. Tested per ASTM E90.

Span—Distance between supports, usually a beam or joist.

Spandrel Beam—Horizontal member, spanning between exterior columns, that supports the floor or roof.

Spandrel Wall—Exterior wall panel, usually between columns, that extends from the window opening on one floor to one on the next floor.

Speed of Sound—Speed of sound in air varies with atmospheric pressure and temperature, but is the same at all frequencies. For most architectural work, the speed of sound should be taken as 1,130 ft./second.

Stile—Vertical outside member in a piece of mill work, as a door or sash.

Stirrup—Hanger to support the end of the joist at the beam.

Stop—Strip of wood fastened to the jambs and head of a door or window frame against which the door or window closes.

Strain—Unit deformation in a body that results from stress.

Stress—Unit resistance of a body to an outside force that tends to deform the body by tension, compression or shear.

Stringer—Heavy horizontal timber supporting other members of the frame in a wood or brick structure; a support also for steps.

Structure-borne Sound—Sound energy imparted directly to and transmitted by solid materials, such as building structures.

Strut—Slender structural element that resists compressive forces acting lengthwise.

Stud—Vertical load-bearing or non-load bearing framing member.

Subfloor—Rough or structural floor placed directly on the floor joists or beams to which the finished floor is applied. As with resilient flooring, an underlayment may be required between subfloor and finished floor.

Substrate—Underlying material to which a finish is applied or by which it is supported.

Surface Burning Characteristic—Rating of interior and surface finish material providing indexes for flame spread and smoke developed, based on testing conducted according to ASTM Standard E84.

Temperature—Measurement of the intensity (not quantity) of heat. The Fahrenheit (°F) scale places the freezing point of water at 32° and the boiling point at 212°. The Centigrade or Celsius (°C) scale, used by most countries and in scientific work, places the freezing point of water at 0° and the boiling point at 100°. On the Kelvin (K) scale, the unit of measurement equals the Celsius degree and measurement begins at absolute zero 0° (–273°C).

Tensile Strength—Maximum tensile stress that can be developed in a given material under axial tensile loading. Also the measure of a material's ability to withstand stretching.

Tension—Force that tends to pull the particles of a body apart.

Thermal Expansion—All materials expand and contract to some extent with changes in temperature. The Thermal Coefficient of Linear Expansion is expressed in "Inches Per Inch Per Degree Fahrenheit." Example: gypsum board has a coefficient of 9.0×10^{-6} in. per in. per °F. This means that with an increase in temperature of 50°, a gypsum board wall 100 ft. in length will have a linear expansion of .54″ or an excess of ½″. The expansion characteristics of some other building materials are more pronounced; a 50° temperature increase would produce expansion in a 100′ length of approx. ¾″ in aluminum, ⅜″ in steel and ½″ in concrete.

Thermal Resistance (R)—Resistance of a material or assembly to the flow of heat. It is the reciprocal of the heat transfer coefficient:

(1/C, or 1/U)

For insulating purposes, low "C" and "U" values and high "R" values are the most desirable.

Threshold—Raised member at the floor within the door jamb. Its purpose is to provide a divider between dissimilar flooring materials or serve as a thermal, sound or water barrier.

Through-penetration Fire Stop—A system for sealing through-penetrations in fire-resistant floors, walls and ceilings.

Through-penetrations—Through-penetration, or "poke-through" openings as they are sometimes called, are holes that penetrate an entire floor or wall assembly to allow the passage of piping, ducts, conduit, cable trays, electrical cables, communications wiring, etc.

Time-Temperature Curve—Rate of rise of temperature in a fire-testing furnace.

Toenail—Method of fastening two boards together as in a "T" by driving nails into the board that forms the stem of the "T" at an angle so they enter the other board and cross each other.

Tongue-and-Groove Joint—Joint where the projection or "tongue" of one member engages the mating groove of the adjacent member to minimize relative deflection and air infiltration; widely used in sheathing, flooring and paneling. Tongues may be in "V," round or square shapes.

Transmission Loss (TL)—Essentially the amount, in decibels, by which sound power is attenuated by passing from one side of a structure to the other. TL is independent of the rooms on each side of the structure and theoretically independent of the area and edge conditions of the structure.

Tread—Horizontal plane or surface of a stair step.

Trimmer—Double joists or rafters framing the opening of a stairway well, dormer opening, etc.

Truss—Open, lightweight framework of members, usually designed to replace a large beam where spans are great.

UBC—Uniform Building Code—document promulgated by the International Conference of Building Officials.

U of C—University of California, an independent fire-testing laboratory.

"U" Factor—Coefficient of heat transfer, "U" equals 1 divided by (hence, the reciprocal of) the total of the resistances of the various materials, air spaces and surface air films in an assembly. See Thermal Resistance.

UL—Underwriters Laboratories Inc., founded by NBFU, and now operated in affiliation with American Insurance Assn. UL is a not for profit laboratory operated for the purpose of testing devices, systems and materials as to their relation to life, fire and casualty hazard in the interest of public safety.

USASI—United States of America Standards Institute, now American National Standards Institute.

Vapor Retarder—Material used to retard the flow of water vapor through walls and other spaces where this vapor may condense at a lower temperature.

Veneer Plaster—Calcined gypsum plaster specially formulated to provide specific workability, strength, hardness and abrasion resistance characteristics when applied in thin coats (1/16" to 3/32" nom.) over veneer gypsum base or other approved base. The term thin coat plaster is sometimes used in reference to veneer plaster.

Wavelength (Sound)—Wave is one complete cycle of sound vibration passing through a medium (such as air) from compression through rarefaction and back to compression again. The physical length of this cycle is termed the wavelength. Wavelengths in air vary from about 11⁄16″ for a 20,000-cycle per sec. (see Frequency) sound, to approximately 56½′ for a 20-cycle per sec. sound—the two approximate extremes of human hearing sensitivity. There are waves outside of this range, but generally, they cannot be heard by humans.

Weep Hole—Small aperture at the base of an exterior wall cavity intended to drain out trapped moisture.

Wet Sand—To smooth a finished joint with a small-celled wet sponge. A preferred method to reduce dust created in the dry sanding method.

WHI—Warnock Hershey International, an independent fire-testing laboratory.

Index

A

B

D

E

F

G

H

J

K

L

M

N

O

P

Q

R

S

T

U

V